Ganzha · Mayr · Vorozhtsov (Eds.)

Computer Algebra in Scientific Computing

Springer-Verlag Berlin Heidelberg GmbH

Victor G. Ganzha   Ernst W. Mayr
Evgenii V. Vorozhtsov (Eds.)

Computer Algebra in Scientific Computing

CASC 2000

Proceedings of the Third Workshop on
Computer Algebra in Scientific Computing,
Samarkand, October 5-9, 2000

Springer

Editors

Victor G. Ganzha
Ernst W. Mayr

Institut für Informatik
Technische Universität München
80290 München, Germany

e-mail: ganzha@informatik.tu-muenchen.de
mayr@in.tum.de

Evgenii V. Vorozhtsov

Institute of Theoretical and Applied Mathematics
Russian Academy of Sciences
Novosibirsk 630090, Russia

e-mail: vorozh@itam.nsc.ru

Mathematics Subject Classification (2000): 68Q40, 65M06, 13P10, 12Y05, 34A25, 20C40, 34D20, 68T35

Cataloging-in-Publication Data applied for

Die Deutsche Bibliothek - CIP-Einheitsaufnahme

Computer algebra in scientific computing : proceedings of the Third Workshop on Computer Algebra in Scientific Computing, Samarkand, October 5-9, 2000 / CASC 2000. Victor G. Ganzha... (ed.). - Berlin; Heidelberg; New York; Barcelona; Hong Kong; London; Milan; Paris; Singapore; Tokio: Springer, 2000
ISBN 3-540-41040-6

ISBN 978-3-642-62490-2 ISBN 978-3-642-57201-2 (eBook)
DOI 10.1007/978-3-642-57201-2

Originally published by Springer-Verlag Berlin Heidelberg New York in 2000

Production: PRO EDIT GmbH, 69126 Heidelberg
Typesetting by the authors using a Springer $\TeX$ Macro Package
Cover Design: design & production GmbH, Heidelberg

SPIN: 10778663 46/3142 5 4 3 2 1 0

Preface

The vast area of Scientific Computing, which is concerned with the computer-aided simulation of various processes in engineering, natural, economical, or social sciences, now enjoys rapid progress owing to the development of new efficient symbolic, numeric, and symbolic/numeric *algorithms*.

There has already been for a long time a worldwide recognition of the fact that the mathematical term *algorithm* takes its origin from the Latin word *algoritmi*, which is in turn a Latin transliteration of the Arab name "Al Khoresmi" of the Khoresmian mathematician Moukhammad Khoresmi, who lived in the Khoresm khanate during the years 780 – 850. The Khoresm khanate took significant parts of the territories of present-day Turkmenistan and Uzbekistan. Such towns of the Khoresm khanate as Bukhara and Marakanda (the present-day Samarkand) were the centers of mathematical science and astronomy. The great Khoresmian mathematician M. Khoresmi introduced the Indian decimal positional system into everyday's life; this system is based on using the familiar digits 1,2,3,4,5,6,7,8,9,0. M. Khoresmi had presented the arithmetic in the decimal positional calculus (prior to him, the Indian positional system was the subject only for jokes and witty disputes). Khoresmi's *Book of Addition and Subtraction by Indian Method (Arithmetic)* differs little from present-day arithmetic. This book was translated into Latin in 1150; the last reprint was produced in Rome in 1957.

Moukhammad Khoresmi also introduced the term "al jabr", which transformed to *algebra* in Europe. In the year 827, M. Khoresmi measured the meridian arc length in the Syrian desert, and he thus calculated the size of our planet. In addition, M. Khoresmi discovered two techniques for the solution of algebraic equations, which now belong to the basics of algebra. Abu Ali Houssain ibn Abdoullah ibn Ali ibn Hassan ibn Sina (Avitsenna) was one of his famous disciples.

Therefore, our wish to organize CASC'2000, the third conference devoted to applications of Computer Algebra in Scientific Computing, in Samarkand was not incidental. The two earlier conferences in this sequence, CASC'98 and CASC'99, had been held in St. Petersburg, Russia, and in Munich, Germany, and proved to be very successful.

This volume contains revised versions of the papers submitted to the conference by the participants and accepted by the program committee after a thorough reviewing process. The collection of papers included in the proceedings covers various topics of computer algebra methods, algorithms and software applied to scientific computing: symbolic-numeric analysis, solving differential equations, fast computations with polynomials, matrices, and other related objects, complexity analysis of algorithms and programs, special purpose programming environments, application to physics, mechanics, and to other areas.

In particular, a significant group of papers deals with applications of Computer Algebra methods for the solution of current problems in group theory, which mostly arise in mathematical physics (including the Navier-Stokes equations).

Another important trend which may be seen from the present collection of papers, is the application of Computer Algebra methods to the development of new efficient analytic and numerical solvers, both for ordinary and partial differential equations.

A number of papers are devoted to algorithmic and software aspects associated with the efficient use of Computer Algebra methods when solving problems in technology (e.g., robotics, gyroscopic systems).

Some papers deal with interfacing several computer algebra packages using OpenMath as well as computer algebra software with automatic theorem provers and study the problem singularities in plotting with help of the Component-based Web Architecture.

The CASC'2000 workshop was supported financially by a generous grant from the Deutsche Forschungsgemeinschaft (DFG). We are grateful to M. Mnuk for his technical help in the preparation of the camera ready manuscript for this volume. We also thank the publisher, Springer Verlag, for their support in preparing these proceedings, and the University of Samarkand for hosting the workshop. Here, our particular thanks are due to the members of the local organizing committee in Samarkand: Akhmadjon Soleev (chair), Ulugbek Narzullaev (secretary), and Shavgat Ayupov.

Munich, July 2000

V.G. Ganzha
E.W. Mayr
E.V. Vorozhtsov

Workshop Organization

CASC'00 was organized jointly by the Technische Universität München, Germany, and the Samarkand State University, Uzbekistan.

Workshop Chairs

Vladimir Gerdt (JINR, Dubna)
Ernst W. Mayr (TU München)

Program Commitee

Laurent Bernardin (Zurich)
Victor Edneral (Moscow)
Victor Ganzha (Munich, co-chair)
Jaime Gutierrez (Santander)
Simon Gray (Ashland)
Ilias Kotsireas (London Ontario)
Robert Kragler (Weingarten)
Michal Mnuk (Munich)
Bernard Mourrain (Sophia-Antipolis)
Hirokazu Murao (Tokyo)
Eugenio Roanes-Lozano (Madrid)
Werner Seiler (Mannheim)
Akhmadjon Soleev (Samarkand)
Stanly Steinberg (Albuquerque)
Nikolay Vassiliev (St. Petersburg)
Evgenii Vorozhtsov (Novosibirsk, co-chair)
Volker Weispfenning (Passau)
Andreas Weber (Tubingen)
Franz Winkler (Linz)
Christoph Zenger (Munich)

Organizing Commitee

Michal Mnuk (Munich, chair)
Annelies Schmidt (Munich)
Klaus Wimmer (Munich)
Shavgat Ayupov (Samarkand)
Ulugbek Narzullaev (Samarkand)
Akhmadjon Soleev (Samarkand)

Electronic Address

WWW site: `http://www14.in.tum.de/CASC2000`

Table of Contents

Fast Matrix Computation of Subresultant Polynomial Remainder Sequences

Alkiviadis G. Akritas[1] and Gennadi I. Malaschonok[2]

[1] University of Thessaly, Volos, Greece ***
[2] Tambov University, Tambov, Russia

Abstract. We present an improved (faster) variant of the matrix-triangularization subresultant prs method for the computation of a greatest common divisor of two polynomials A and B (of degrees d_A and d_B, respectively) along with their polynomial remainder sequence [1]. The computing time of our fast method is $O(n^{2+\beta} \log \|C\|^2)$, for standard arithmetic and $O(((n^{1+\beta} + n^3 \log \|C\|)(\log n + \log \|C\|)^2)$ for the Chinese remainder method, where $n = d_A + d_B$, $\|C\|$ is the maximal coefficient of the two polynomials and the best known $\beta < 2.356$. By comparison, the computing time of the old version is $O(n^5 \log \|C\|^2)$.
Our improvement is based on the work of Malaschonok [12] who proposed a new, recursive method for the solution of systems of linear equations in integral domains with complexity $O(n^\beta)$ over the integers (same as the complexity of matrix multiplication).
In this paper we present an overview of the two methods mentioned above and show how they are combined into a fast matrix method for computing polynomial remainder sequences. An example is also presented to clarify the concepts.

1 Introduction

Let Z be an integral domain, and let

$$A_i = \sum_{j=1}^{m} c_{ij} x^{m-j},$$

where $c_{ij} \in Z$, $i = 1, 2, \ldots, n$; then

$$mat(A_1, A_2, \ldots, A_n)$$

denotes the matrix (c_{ij}) of order $n \times m$. Moreover, let $A, B \in Z[x], \deg A = d_A, \deg B = d_B$ and let

$$M_k = mat(x^{d_B-k-1}A, x^{d_B-k-2}A, \ldots, A, x^{d_A-k-1}B, x^{d_A-k-2}B, \ldots, B),$$
$$0 \le k < \min(d_A, d_B)$$

*** Akritas is on leave from the University of Kansas, USA

be the matrix of order $(d_A+d_B-2k)\times(d_A+d_B-k)$, where M_0 is the well-known Sylvester's matrix. Then, kth subresultant polynomial of A and B is called the polynomial

$$S_k = \sum_{i=0}^{k} M_k^i x^i,$$

of degree $\leq k$, where M_k^i is a minor of the matrix M_k of order $d_A + d_B - 2k$, formed by the elements of columns $1, 2, \ldots, d_A + d_B - 2k - 1$ and column $d_A + d_B - k - i$. Habicht's known theorem [8] establishes a relation between the subresultant polynomials $S_0, S_1, \ldots, S_{\min(d_A,d_B)-1}$ and the polynomial remainder sequence(prs) of A and B, and also demonstrates the so-called *gap* structure.

According to the matrix-triangularization subresultant prs method (see for example Akritas' book [2] or papers [3], [4]) all the subresultant polynomials of A and B can be computed *within sign* by transforming the matrix (suggested by Sylvester [14])

$$mat(x^{\max(d_A,d_B)-1}A, x^{\max(d_A,d_B)-1}B, x^{\max(d_A,d_B)-2}A, x^{\max(d_A,d_B)-2}B, \ldots, A, B),$$

of order $2\cdot\max(d_A, d_B)$, into its upper triangular form with the help of Dodgson's integer preserving transformations [7]; they are then located using an extension of a theorem by Van Vleck [3], [15]. (We depart from established practice and we give credit to Dodgson, and not to Bareiss [5], for the integer preserving transformations; see also the work of Waugh and Dwyer [16] where they use the same method as Bareiss, but 23 years earlier, and they name Dodgson as their source–differing from him only in the choice of the pivot element ([16], page 266). Charles Lutwidge Dodgson (1832–1898) is the same person widely known for his writing *Alice in Wonderland* under the pseudonym Lewis Carroll.)

In Section 2 we present an overview of the matrix-triangularization subresultant prs method allowing us to *exactly* compute and locate the members of the prs (*without* using Van Vleck's theorem [15]) by applying Dodgson's integer preserving transformations to a matrix of order $d_A + d_B$.

In Section 3 we present an overview of the recursive method for the solution of systems of linear equations in integral domains. Its complexity is the same as that of matrix multiplication.

In Section 4 we prove that the recursive method allows us to compute and locate the members of the prs and gcd.

Finally, in Sections 5 and 6 we present the computing time analysis for our method and an example to clarify the concepts.

2 The "Slow" Matrix Computation of Subresultant Polynomial Remainder Sequences

We assume that $\deg A = d_A \geq \deg B = d_B$ and we denote by M the following matrix

$$M = mat(x^{d_A-1}B, x^{d_A-2}B, \ldots, x^{d_B-1}B, xc^{d_B-1}A, x^{d_B-2}B, x^{d_B-2}A, \ldots, B, A)$$

of order d_A+d_B with elements $a_{ij}(j,i=1,2,\ldots,d_A+d_B)$. (This matrix can be obtained from Sylvester's matrix M_0 after a rearrangement of its rows.)

Dodgson's integer preserving transformations

$$a_{ij}^{k+1}=\frac{(a_{ij}^k a_{kk}^k - a_{ik}^k a_{kj}^k)}{a_{k-1,k-1}^{k-1}} \qquad \text{(D)}$$

(see [5], [7], [10] or [16]) where we set $a_{00}^0=1$, $a_{ij}^1=a_{ij}$ and it is assumed that $a_{kk}^k\neq 0, k=1,2,\ldots,d_A+d_B$, are applied to the matrix $M=(a_{ij})$ and transform it to the upper-triangular matrix $M_D=(b_{ij}),(i,j=1,2,\ldots,d_A+d_B)$, where

$$b_{ij}=\begin{cases} 0 & \text{for } i>j \\ a_{ij}^i & \text{for } i\leq j \end{cases}$$

and, in general,

$$a_{ij}^k=\begin{vmatrix} a_{11} & \ldots & a_{1,k-1} & a_{1j} \\ \vdots & \ddots & \vdots & \vdots \\ a_{k-1,1} & \ldots & a_{k-1,k-1} & a_{k-1,j} \\ a_{i1} & \ldots & a_{i,k-1} & a_{ij} \end{vmatrix}$$

with $1\leq k\leq d_A+d_B$, and $k\leq i,\ j\leq d_A+d_B$.

The upper-triangular matrix $M_D=(b_{ij})$ will be called Dodgson matrix.

The following two theorems can be used to locate the members of the prs in the rows of M_D. The *correct* sign is computed.

Case 1: If none of the diagonal minors of the matrix M is equal to zero, then we have:

Theorem 1. Dodgson's integer preserving transformation will transform matrix M to the upper triangular matrix M_D, which contains all n subresultants (located in rows $d_A+d_B-2k, k=0,1,2,\ldots,d_B-1$) $S_k=\sum_{i=0}^k M_k^i x^i$,

$$M_k^i=(-1)^{\sigma(k)}a_{d_A+d_B-2k,d_A+d_B-k-i}^{d_A+d_B-2k}$$

and $\sigma(k)=(d_A-d_B+1)+\cdots+(d_A-k)=(1/2)(d_B-k)(2d_A-d_B-k+1)$, $k=0,1,\ldots,d_B-1$.

Case 2: If *not* all diagonal minors of the matrix M are nonzero, then we have the following theorem (the term *bubble pivot*, used below, means that, after pivoting, row i_p is *immediately* below row j_p):

Theorem 2. Dodgson's integer preserving transformations with *bubble pivot* and choice of the pivot element by column, will transform matrix M to the upper triangular matrix M_D, and at the same time will compute all subresultants S_k; if, in the process, s row replacements take place, namely row j_1 replaces row i_1, j_2 replaces i_2, …, j_s replaces i_s, (and after each replacement row i_p is immediately below row j_p, $p=1,2,\ldots,s$), then **(a)** $S_k=0$, for all k such that $\frac{(d_A+d_B-i_p)}{2}>k>\frac{(d_A+d_B-j_p)}{2}$ and for all $p=1,2,\ldots,s$. **(b)** for all $p=$

$1, 2, \ldots, s$, if $k = \frac{(d_A + d_B - i_p)}{2}$ is an integer number not in (a), S_k is located in row i_p *before* it is replaced by row j_p. **(c)** for the remaining $k, (k = 0, 1, \ldots, d_B - 1$ and those not in (a) or (b)) S_k is located in row $j = d_A + d_B - 2k$.

Moreover, in (b) and (c) the subresultant $S_k = \sum_{i=0}^{k} M_k^i x^i$, is located in row j in such a way that

$$M_k^i = (-1)^{\sigma(k)+\sigma(j)} a^j_{j,j+k-i}$$

$\sigma(k) = (1/2)(d_B - k)(2d_A - d_B - k + 1)$, $\sigma(j) = \sum_{p=1}^{s} j_p - \sum_{p=1}^{s} i_p$, $j_p \leq j$, $i_p \leq j$.

Note that in cases (b) and (c) Theorem 2 reduces to Theorem 1 in the case of a complete prs, and due to the fact that rows above row j change places, the sign changes by a factor $(-1)^{\sigma(j)}$.

For the given polynomials over the integers in this discussion, the complexity of this method is

$$O(n^5 \log \|C\|^2))$$

where $n = d_A + d_B$ and $\|C\|$ is the maximal coefficient of the two polynomials.

3 The Recursive Method

Let $\sum_{j=1}^{m-1} a_{ij} x_j = a_{im}$, $i = 1, 2, \ldots, n$ be the system of linear equations with extended coefficients matrix

$$A = (a_{ij}),\ i = 1, \ldots, n,\ j = 1, \ldots, m.$$

whose coefficients are in the integral domain $\mathbf{Z}$: $A \in \mathbf{Z}^{n \times m}$.

The solution of such a system may be written according to Cramer's rule

$$x_j = \frac{\delta^n_{jm} - \sum_{p=n+1}^{m-1} x_p \delta^n_{jp}}{\delta^n},\ j = 1, \ldots, n,$$

where x_p, $p = n+1, \ldots, m$, are free variables and $\delta^n \neq 0$. $\delta^n = |a_{ij}|$, $i = 1, \ldots, n$, $j = 1, \ldots, n$, - denote the corner minors of the matrix A of order n, and δ^n_{ij} - denote the minors obtained by substituting the column i of the matrix A by the column j in the minors δ^n, $i = 1, \ldots, n$, $j = n+1, \ldots, m$. So we need to construct an algorithm for computing the minor δ^n and the matrix $G = (\delta^n_{ij})$, $i = 1, \ldots, n$, $j = n+1, n+2, \ldots, m$.

This means that we must make the reduction of the matrix A to the diagonal form

$$A \to (\delta^n I_n, G).$$

(I_n denotes the unit matrix of order n.)

For the extended coefficients matrix $\mathbf{A}$ we use the following notation:

$$\mathbf{A}^k_{ij} = \begin{pmatrix} a_{11} & a_{12} & \cdots & a_{1,k-1} & a_{1j} \\ a_{21} & a_{22} & \cdots & a_{2,k-1} & a_{2j} \\ \vdots & \vdots & \ddots & \vdots & \vdots \\ a_{k-1,1} & a_{k-1,2} & \cdots & a_{k-1,k-1} & a_{k-1,j} \\ a_{i1} & a_{i2} & \cdots & a_{i,k-1} & a_{ij} \end{pmatrix}$$

which is the matrix, formed by surrounding the sub-matrix of order $k-1$ in the upper left corner by row i and column j,

$$a_{ij}^k = \det \mathbf{A}_{ij}^k;$$

$a_{ij}^1 = a_{ij}$, $\delta^0 = 1$, $\delta^k = a_{kk}^k$, δ_{ij}^k is the determinant of the matrix, that is obtained from the matrix $\mathbf{A}_{kk}^k$ after the substitution of column i by column j.

We shall use the minors δ_{ij}^k and a_{ij}^k for the construction of the matrices

$$A_{k,c}^{r,l,(p)} = \begin{pmatrix} a_{r+1,k+1}^p & a_{r+1,k+2}^p & \cdots & a_{r+1,c}^p \\ a_{r+2,k+1}^p & a_{r+2,k+2}^p & \cdots & a_{r+2,c}^p \\ \vdots & \vdots & \ddots & \vdots \\ a_{l,k+1}^p & a_{l,k+2}^p & \cdots & a_{l,c}^p \end{pmatrix}$$

and

$$G_{k,c}^{r,l,(p)} = \begin{pmatrix} \delta_{r+1,k+1}^p & \delta_{r+1,k+2}^p & \cdots & \delta_{r+1,c}^p \\ \delta_{r+2,k+1}^p & \delta_{r+2,k+2}^p & \cdots & \delta_{r+2,c}^p \\ \vdots & \vdots & \ddots & \vdots \\ \delta_{l,k+1}^p & \delta_{l,k+2}^p & \cdots & \delta_{l,c}^p \end{pmatrix}$$

$G_{k,c}^{r,l,(p)}, A_{k,c}^{r,l,(p)} \in \mathbf{Z}^{(l-r)\times(c-k)}$, $0 \le k < n$, $k < c \le n$, $0 \le r < m$, $r < l \le m$, $1 \le p \le n$.

We describe one recursive step, that makes the following reduction of the matrix $\tilde{A}$ to the diagonal form

$$\tilde{A} \to (\delta^l I_{l-k}, G_{l,c}^{k,l,(l)}), \ 0 \le k < c \le m, \ k < l \le n, \ l < c.$$

Note that if $k = 0$, $l = n$ and $c = m$, then we obtain the solution of the system.

We choose an arbitrary integer s: $k < s < l$ and write the matrix $\tilde{A} = A_{k,c}^{k,l,(k+1)}$ as follows:

$$\tilde{A} = \begin{pmatrix} A_{k,c}^{k,s(k+1)} \\ A_{k,c}^{s,l,(k+1)} \end{pmatrix}$$

where $A_{k,c}^{k,s,(k+1)}$ is the upper part of the matrix $\tilde{A}$ consisting of $s-k$ rows and $A_{k,c}^{s,l,(k+1)}$ is the lower part of the matrix $\tilde{A}$.

Then, one recursive step is briefly described by the following reduction of the matrix $A_{kc}^{kl(k+1)}$:

$$A_{kc}^{kl(k+1)} = \begin{pmatrix} A_{k,c}^{k,s(k+1)} \\ A_{k,c}^{s,l,(k+1)} \end{pmatrix} \to_1 \begin{pmatrix} \delta^s I_{s-k} & G_{sc}^{ks(s)} \\ A_{k,s}^{s,l,(k+1)} & A_{s,c}^{s,l,(k+1)} \end{pmatrix} \to_2 \begin{pmatrix} \delta^s I_{s-k} & G_{sc}^{ks(s)} \\ 0 & A_{sc}^{sl(s+1)} \end{pmatrix} \to_3$$

$$\to_3 \begin{pmatrix} \delta^s I_{s-k} & G_{sl}^{ks(s)} & G_{lc}^{ks(s)} \\ 0 & \delta^l I_{l-s} & G_{lc}^{sl(l)} \end{pmatrix} \to_4 \begin{pmatrix} \delta^s I_{s-k} & 0 & G_{lc}^{ks(l)} \\ 0 & \delta^l I_{l-s} & G_{lc}^{sl(l)} \end{pmatrix} \to \begin{pmatrix} \delta^l I_{s-k} & G_{lc}^{kl(l)} \end{pmatrix}$$

Expanding we have the following four steps:

The first step

Recursively we make the following reduction of the matrix $A_{k,c}^{k,s(k+1)} \in \mathbf{Z}^{(s-k)\times(c-k)}$ to the diagonal form

$$A_{k,c}^{k,s(k+1)} \to (\delta^s I_{s-k}, G_{s,c}^{k,s,(s)}).$$

The second step

We write the matrix $A_{k,c}^{s,l,(k+1)}$ in the following way:

$$A_{k,c}^{s,l,(k+1)} = (A_{k,s}^{s,l,(k+1)}, A_{s,c}^{s,l,(k+1)})$$

where $A_{k,s}^{s,l,(k+1)}$ consists of the first $s-k$ columns and $A_{s,c}^{s,l,(k+1)}$ consists of the last $c-s$ columns of the matrix $A_{k,c}^{s,l,(k+1)}$.

The matrix $A_{s,c}^{s,l,(s+1)}$ is obtained from the matrix identity [12]:

$$\delta^k \cdot A_{s,c}^{s,l,(s+1)} = \delta^s \cdot A_{s,c}^{s,l,(k+1)} - A_{k,s}^{s,l,(k+1)} \cdot G_{s,c}^{k,s,(s)}.$$

The minors δ^k must not equal zero.

The third step

Recursively we make the following reduction of the matrix $A_{s,c}^{s,l,(s+1)} \in \mathbf{Z}^{(l-s)\times(c-s)}$ to the diagonal form

$$A_{s,c}^{s,l,(s+1)} \to (\delta^l I_{l-s}, G_{l,c}^{s,l,(l)}).$$

The fourth step

We write the matrix $G_{s,c}^{k,s,(s)}$ in the following way:

$$G_{s,c}^{k,s,(s)} = (G_{s,l}^{k,s,(s)}, G_{l,c}^{k,s,(s)})$$

where $G_{s,l}^{k,s,(s)}$ consists of the first $l-s$ columns and $G_{l,c}^{k,s,(s)}$ consists of the last $c-l$ columns of the matrix $G_{s,c}^{k,s,(s)}$.

The matrix $G_{l,c}^{k,s,(l)}$ is obtained from the matrix identity [12]:

$$\delta^s \cdot G_{l,c}^{k,s,(l)} = \delta^l \cdot G_{l,c}^{k,s,(s)} - G_{s,l}^{k,s,(s)} \cdot G_{l,c}^{s,l,(l)}.$$

The minors δ^s must not equal zero.

So we get

$$G_{l,c}^{k,l,(l)} = \begin{pmatrix} G_{l,c}^{k,s,(l)} \\ G_{l,c}^{s,l,(l)} \end{pmatrix}$$

and δ^l.

4 Dichotomous Recursion

If during each step of the partition process the number of rows in the upper and lower submatrixes is the same and equal to a power of 2, then we will call such a process - the dichotomous recursion process.

Let us denote by

$$A_k^{2^p} = A_{kc}^{k,k+2^p(k+1)}, \quad G_k^{2^p} = G_{kc}^{k-2^p,k(k)}$$

the matrices with 2^p rows and $c-k$ columns. For any natural number s we denote by $b[s]$ the maximal power of 2, that divides the nubmer s, i.e. for $s = \sum_{i=0}^{j} c_i 2^i, c_i \in \{0,1\}$ $b[s] = 2^w$, where $w = min\{i|c_i \neq 0\}$.

Then one dichotomous recursive step is as follows:

$$A_{t2^p}^{2^p} \rightarrow_1 \begin{pmatrix} \delta^{(2t+1)2^{p-1}} I_{2^{p-1}} & G_{(2t+1)2^{p-1}}^{2^{p-1}} \\ * & * \end{pmatrix} \rightarrow_2 \begin{pmatrix} \delta^{(2t+1)2^{p-1}} I_{2^{p-1}} & G_{(2t+1)2^{p-1}}^{2^{p-1}} \\ 0 & A_{(2t+1)2^{p-1}}^{2^{p-1}} \end{pmatrix}$$

$$\rightarrow_3 \begin{pmatrix} * & * & * \\ 0 & \delta^{(t+1)2^p} I_{2^{p-1}} & G_{(t+1)2^p}^{2^{p-1}} \end{pmatrix} \rightarrow_4 \begin{pmatrix} \delta^{(t+1)2^p} I_{2^p} & G_{(t+1)2^p}^{2^p} \end{pmatrix}$$

$$p = 1,2,\ldots,N, \quad t = 0,1,\ldots,2^{N-p}-1, \quad n = 2^N.$$

We obtain the following sequence of matrices:

$A_0^2; G_2^2, A_2^2; G_4^2, G_4^4, A_4^4; G_6^2, A_6^2, G_8^2, G_8^4, G_8^8, A_8^8; G_{10}^2, A_{10}^2, G_{12}^2, G_{12}^4, A_{12}^4, G_{14}^2, A_{14}^2,$ $G_{16}^2, G_{16}^4, G_{16}^8, G_{16}^{16}, \ldots A_{4s}^{b[4s]}, G_{4s+2}^2, A_{4s+2}^2, G_{4s+4}^2, G_{4s+4}^4, \ldots, G_{4s+4}^{b[4s+4]}, \ldots$ which we call the *dichotomous sequence of matrices.*

Theorem 3. The dichotomous sequence of matrices contains all the rows of the Dodgson matrix M_D.

Proof: The dichotomous sequence of matrices includes the matrices $A_t^{b[t]}$, $t = 2,4,\ldots,n-2$, $\;G_t^2$, $t = 2,4,\ldots,n$.

The matrix $A_t^{b[t]}$, $t = 2,4,\ldots,n-2$, has the next first row

$$(a_{t+1,t+1}^{t+1}, a_{t+1,t+2}^{t+1}, \ldots, a_{t+1,m}^{t+1}),$$

the matrix G_t^2, $t = 2,4,\ldots,n-2$, has the next last row

$$(\delta_{t,t}^t, \delta_{t,t+1}^t, \ldots, \delta_{t,m}^t) = (a_{t,t}^{t+1}, a_{t,t+1}^{t+1}, \ldots, a_{t,m}^{t+1})$$

There are rows of the Dodgson matrix M_D with the numbers $t+1$ ($t = 2,4,\ldots,n-2$) and t ($t = 2,4,\ldots,n$) respectively. So the dichotomous sequence of matrices contains all the rows of the Dodgson matrix M_D . □

Therefore, we can use the dichotomous sequence of matrices for computing the Dodgson matrix with the prs's and gcd.

Comparing with Dodgson's method, care must be taken not to make bubble-pivot in each row with zero diagonal element but just in the even rows, because just the diagonal elements in the even rows will be the pivots. Details can be seen in the example that follows taken from our previous paper [1].

5 Complexity Issue

Let $\|C\|$ denote the maximal coefficient of the two polynomials, n^β denote the complexity of matrix multiplication of order n, $n = d_A + d_B$, where d_A and d_B are the degrees of the two polynomials. The best known $\beta < 2.356$. Since the fast method of matrix transformation required n^β operations with integer numbers so it is easy to obtain the complexity of our method.

Our method requires

$$O(n^{2+\beta}\log^2\|C\|)$$

bit operations for standard arithmetic and

$$O(((n^{1+\beta} + n^3\log\|C\|)(\log n + \log\|C\|)^2)$$

for the Chinese remainder method. (The cost of reducing modulo a prime is included.)

Depending on the exact value of β (which is an open problem) the complexity of our algorithm is clearly better/worse than $O(n^4\log^2\|C\|)$, the complexity of the classical subresultant prs algorithm [9].

6 Example of Fast Matrix Transformation

We present an example to clarify the concepts discussed above. It is the same example taken from our previous paper [1].

For the polynomials

$$\begin{aligned} A &= 2x^4 + 5x^3 + 5x^2 - 2x + 1 \\ B &= 3x^3 + 3x^2 + 3x - 4 \end{aligned}$$

we obtain the following matrices:

$$\begin{pmatrix} 3 & 3 & 3 & 4 & 0 & 0 & 0 \\ 2 & 5 & 5 & -2 & 1 & 0 & 0 \\ 0 & 3 & 3 & 3 & -4 & 0 & 0 \\ 0 & 2 & 5 & 5 & -2 & 1 & 0 \\ 0 & 0 & 3 & 3 & 3 & -4 & 0 \\ 0 & 0 & 2 & 5 & 5 & -2 & 1 \\ 0 & 0 & 0 & 3 & 3 & 3 & -4 \end{pmatrix} \to_1 \begin{pmatrix} 9 & 0 & 0 & -14 & -3 & 0 & 0 \\ 0 & 9 & 9 & 2 & 3 & 0 & 0 \\ 0 & 3 & 3 & 3 & -4 & 0 & 0 \\ 0 & 2 & 5 & 5 & -2 & 1 & 0 \\ 0 & 0 & 3 & 3 & 3 & -4 & 0 \\ 0 & 0 & 2 & 5 & 5 & -2 & 1 \\ 0 & 0 & 0 & 3 & 3 & 3 & -4 \end{pmatrix} \to_2$$

$$\begin{pmatrix} 9 & 0 & 0 & -14 & -3 & 0 & 0 \\ 0 & 9 & 9 & 2 & 3 & 0 & 0 \\ 0 & 0 & 0 & 21 & 45 & 0 & 0 \\ 0 & 0 & 27 & 41 & -24 & 9 & 0 \\ 0 & 0 & 3 & 3 & 3 & -4 & 0 \\ 0 & 0 & 2 & 5 & 5 & -2 & 1 \\ 0 & 0 & 0 & 3 & 3 & 3 & -4 \end{pmatrix} \to_3 \begin{pmatrix} 9 & 0 & 0 & -14 & -3 & 0 & 0 \\ 0 & 9 & 9 & 2 & 3 & 0 & 0 \\ 0 & 0 & -63 & 0 & -149 & -21 & 0 \\ 0 & 0 & 0 & -63 & 135 & 0 & 0 \\ 0 & 0 & 3 & 3 & 3 & -4 & 0 \\ 0 & 0 & 2 & 5 & 5 & -2 & 1 \\ 0 & 0 & 0 & 3 & 3 & 3 & -4 \end{pmatrix}$$

$$\rightarrow_4 \begin{pmatrix} -63 & 0 & 0 & 0 & 231 & 0 & 0 \\ 0 & -63 & 0 & 0 & 98 & 21 & 0 \\ 0 & 0 & -63 & 0 & -149 & -21 & 0 \\ 0 & 0 & 0 & -63 & 135 & 0 & 0 \\ 0 & 0 & 3 & 3 & 3 & -4 & 0 \\ 0 & 0 & 2 & 5 & 5 & -2 & 1 \\ 0 & 0 & 0 & 3 & 3 & 3 & -4 \end{pmatrix} \rightarrow_5$$

$$\begin{pmatrix} -63 & 0 & 0 & 0 & 231 & 0 & 0 \\ 0 & -63 & 0 & 0 & 98 & 21 & 0 \\ 0 & 0 & -63 & 0 & -149 & -21 & 0 \\ 0 & 0 & 0 & -63 & 135 & 0 & 0 \\ 0 & 0 & 0 & 0 & -147 & 315 & 0 \\ 0 & 0 & 0 & 0 & -692 & 168 & -63 \\ 0 & 0 & 0 & 0 & -594 & -189 & 252 \end{pmatrix} \rightarrow_6$$

$$\begin{pmatrix} -63 & 0 & 0 & 0 & 231 & 0 & 0 \\ 0 & -63 & 0 & 0 & 98 & 21 & 0 \\ 0 & 0 & -63 & 0 & -149 & -21 & 0 \\ 0 & 0 & 0 & -63 & 135 & 0 & 0 \\ 0 & 0 & 0 & 0 & -3068 & 0 & -315 \\ 0 & 0 & 0 & 0 & 0 & -3068 & -147 \\ 0 & 0 & 0 & 0 & -594 & -189 & 252 \end{pmatrix} \rightarrow_7$$

$$\begin{pmatrix} -63 & 0 & 0 & 0 & 231 & 0 & 0 \\ 0 & -63 & 0 & 0 & 98 & 21 & 0 \\ 0 & 0 & -63 & 0 & -149 & -21 & 0 \\ 0 & 0 & 0 & -63 & 135 & 0 & 0 \\ 0 & 0 & 0 & 0 & -3068 & 0 & -315 \\ 0 & 0 & 0 & 0 & 0 & -3068 & -147 \\ 0 & 0 & 0 & 0 & 0 & 0 & 15683 \end{pmatrix}$$

The obtained polynomial remainder sequence is incomplete and we only have the remainder $-63x + 135$ of degree 1 after the thired step and remainder 15683 of degree 0 after the seventh step.

Matrix operations in the each step:

1. $\delta_2 = \begin{vmatrix} 3 & 3 \\ 2 & 5 \end{vmatrix} = 9; \begin{vmatrix} 3 & 3 \\ 2 & 5 \end{vmatrix} = 9; \begin{vmatrix} 3 & -4 \\ 2 & -2 \end{vmatrix} = 2; \begin{vmatrix} 3 & 0 \\ 2 & 1 \end{vmatrix} = 3; \begin{vmatrix} 3 & 0 \\ 2 & 0 \end{vmatrix} = 0;$

$\begin{vmatrix} 3 & 0 \\ 2 & 0 \end{vmatrix} = 0; \begin{vmatrix} 3 & 3 \\ 5 & 5 \end{vmatrix} = 0; \begin{vmatrix} -4 & 3 \\ -2 & 5 \end{vmatrix} = -14; \begin{vmatrix} 0 & 3 \\ 1 & 5 \end{vmatrix} = -3; \begin{vmatrix} 0 & 3 \\ 0 & 5 \end{vmatrix} = 0; \begin{vmatrix} 0 & 3 \\ 0 & 5 \end{vmatrix} = 0.$

2. $9\begin{pmatrix} 3 & 3 & -4 & 0 & 0 \\ 5 & 5 & -2 & 1 & 0 \end{pmatrix} - \begin{pmatrix} 0 & 3 \\ 0 & 2 \end{pmatrix}\begin{pmatrix} 0 & -14 & -3 & 0 & 0 \\ 9 & 2 & 3 & 0 & 0 \end{pmatrix} =$

$\begin{pmatrix} 0 & 21 & 45 & 0 & 0 \\ 27 & 41 & -24 & 9 & 0 \end{pmatrix}.$

3. $\delta_4 = \begin{vmatrix} 0 & 21 \\ 27 & 41 \end{vmatrix} \frac{1}{9} = -63; \begin{vmatrix} 0 & -45 \\ 27 & -24 \end{vmatrix} \frac{1}{9} = 135; \begin{vmatrix} 0 & 0 \\ 27 & 9 \end{vmatrix} \frac{1}{9} = 0; \begin{vmatrix} 0 & 0 \\ 27 & 0 \end{vmatrix} \frac{1}{9} = 0;$

$\begin{vmatrix} -45 & 21 \\ -24 & 41 \end{vmatrix} \frac{1}{9} = -149; \begin{vmatrix} 0 & 21 \\ 9 & 41 \end{vmatrix} \frac{1}{9} = -21; \begin{vmatrix} 0 & 21 \\ 0 & 41 \end{vmatrix} \frac{1}{9} = 0.$

4. $\left((-63)\begin{pmatrix}-3 & 0 & 0\\ 3 & 0 & 0\end{pmatrix} - \begin{pmatrix}0 & -14\\ 9 & 2\end{pmatrix}\begin{pmatrix}-149 & -21 & 0\\ 135 & 0 & 0\end{pmatrix}\right)\frac{1}{9} = \begin{pmatrix}231 & 0 & 0\\ 98 & 21 & 0\end{pmatrix}.$

5. $(-63)\begin{pmatrix}3 & -4 & 0\\ 5 & -2 & 1\\ 3 & 3 & -4\end{pmatrix} - \begin{pmatrix}0 & 0 & 3 & 3\\ 0 & 0 & 2 & 5\\ 0 & 0 & 0 & 3\end{pmatrix}\begin{pmatrix}231 & 0 & 0\\ 98 & 21 & 0\\ -149 & -21 & 0\\ 135 & 0 & 0\end{pmatrix} =$

$$\begin{pmatrix}-147 & 315 & 0\\ -692 & 168 & -63\\ -594 & -189 & 252\end{pmatrix}.$$

6. $\delta_6 = \frac{\begin{vmatrix}-147 & 315\\ -692 & 168\end{vmatrix}}{(-63)} = -3068; \quad \frac{\begin{vmatrix}-147 & 0\\ -692 & -63\end{vmatrix}}{(-63)} = -147; \quad \frac{\begin{vmatrix}0 & 315\\ -63 & 168\end{vmatrix}}{(-63)} = -315;$

7. $\delta_7 = (252(-3068) - (-315)(-594) - (147)(-189))/(-63) = 15683.$

References

1. Akritas, A.G., Akritas, E.K. and Gennadi I. Malaschonok. *Matrix computation of subresultant polynomial remainder sequences in integral domains.* Reliable Computing **1** (1995), pp. 375–381.
2. Akritas, A. G. *Elements of Computer Algebra with Applications.* J. Wiley Interscience, New York, 1989.
3. Akritas, A. G. A new method for computing polynomial greatest common divisors and polynomial remainder sequences. Numerische Mathematik **52**, 119–127, 1988.
4. Akritas, A. G. Exact algorithms for the matrix-triangularization subresultant prs method. Proceedings of the Conference on Computers and Mathematics, Boston, Massachusetts, 145–155, June 1989.
5. Bareiss, E. H. Sylvester's identity and multistep integer-preserving Gaussian elimination. Mathematics of Computation **22**, 565–578, 1968.
6. Coppersmith, D. and S. Winograd. in Proc. 19th Annu ACM Symp. on Theory of Comput., 1987, 1–6.
7. Dodgson, C. L. Condensation of determinants. Proceedings of the Royal Society of London **15**. 150–155, 1866.
8. Habicht, W. Eine Verallgemeinerung des Sturmschen Wurzelzählverfahrens. Commentarii Mathematici Helvetici **21**, 99–116, 1948.
9. Loos, R. Generalized polynomial remainder sequences. In *Computer Algebra Symbolic and Algebraic Computations*, ed B. Buchberger, G.E. Collins and R. Loos. Springer Verlag, New York, *Computing Supplement* **4**, 115–138, 1982.
10. Malaschonok, G. I. Solution of a system of linear equations in an integral domain. USSR Journal of Computational Mathematics and Mathematical Physics **23**, 1497–1500, 1983 (in Russian).
11. Malaschonok, G.I. Algorithms for the solution of systems of linear equations in commutative rings. In *Effective Methods in Algebraic Geometry*, Edited by T. Mora and C. Traverso, Progress in Mathematics 94, Birkhauser, Boston-Basel-Berlin, 1991, 289–298.
12. Malaschonok, G.I. Recursive Method for the Solution of systems of Linear Equations,*Computational Mathematics* (A. Sydow Ed, Proceedings of the 15th IMACS World Congress, Vol. I, Berlin, August 1997), Wissenschaft & Technik Verlag, Berlin 1997, 475–480.

13. Strassen, V. Gaussian Elimination is not optimal. *Numerische Mathematik* **13**, 354–356, 1969.
14. Sylvester, J. J. On a theory of the syzygetic relations of two rational integral functions, comprising an application to the theory of Sturm's functions, and that of the greatest common measure. Philoshophical Transactions **143**, 407–548, 1853.
15. Van Vleck, E. B. On the determination of a series of Sturm's functions by the calculation of a single determinant. Annals of Mathematics **1**, Second Series, 1–13, 1899–1900.
16. Waugh, F. V. and P. S. Dwyer Compact computation of the inverse of a matrix. Annals of Mathematical Statistics **16**, 259–271, 1945.

About Simultaneous Representation of Two Natural Numbers by a Sum of Three Primes

I.A. Allakov and M.I. Israilov

Termez State University, Samarkand State University
Uzbekistan

Let a_{ij} $(i = 1, 2;\ j = 1, 2, 3)$, b_1, b_2 be integer numbers and p_1, p_2, p_3 be primes. Let us consider the problem of solvability of the system

$$b_i = a_{i1}p_1 + a_{i2}p_2 + a_{i3}p_3,\ i = 1, 2 \tag{1}$$

under conditions:

a) for arbitrary prime p there exist such integers l_1, l_2, l_3, $1 \leq l_1, l_2, l_3 \leq p - 1$, which satisfy the system of the linear congruence:

$$a_{i1}l_1 + a_{i2}l_2 + a_{i3}l_3 \equiv b_i (\mathrm{mod} p),\ i = 1, 2;$$

b) there exist such real positive numbers y_1, y_2, y_3 for which the equalities:

$$a_{i1}y_1 + a_{i2}y_2 + a_{i3}y_3 \equiv b_i (\mathrm{mod} p),\ i = 1, 2;$$

are satisfied.

Set $B = 3 \max |a_{ij}|$, $i = 1, 2;\ j = 1, 2, 3$. Let $U(X)$ be the set of pairs $\bar{b}(b_1, b_2)$, $1 \leq b_1, b_2 \leq X$, which satisfy the conditions a) and b). Let $E(x)$ be the quantity pairs $\bar{b} = (b_1, b_2) \in U(X)$, which cannot be represented in the form (1).

In this paper the following theorem is proved:

Theorem 1. *If X is a sufficiently large positive integer, then*

$$E(X) < X^{\gamma},\ \ \gamma < 1.9999833. \tag{2}$$

Let further a_1, a_2 be relatively prime integers, which are not simultaneously negative, and let $E_1(X)$ be the quantity of positive integers $n \leq X$ with conditions $(n, a_1, a_2) = 1$, $n \equiv a_1 + a_2 (mod 2)$, $n \neq a_1p_1 + a_2p_2$. Then, using the estimation (2) we can prove:

Corollary. *For sufficiently large X the estimate*

$$E_1(X) < X^{\gamma},\ \ \gamma < 1.9999834 \tag{3}$$

is true.

The result of this theorem strengthens the corresponding results of paper [1] (see theorems 1, 2, and B in [1]).

Note that the estimates of type (3), in particular case of $a_1 = a_2 = 1$ (for exceptional set in Goldbach's binary problem) were obtained in [2,3] corresponding with the right-hand side $O(X^{0,99})$ and $X^{0,982}$, but in this paper, the problem statement is more general.

The conditions a) and b) correspond to the conditions of congruence solvability and of the positive solvability in Tarry's problem (see [4], Ch. 5.). A suitable selection of values of parameters in the proof of estimate (2) enable us to improve the numerical value of γ with respect to its value calculated directly according to paper [1]. In particular, we succeeded here to replace the factor $\exp(-c_3\delta^{-\frac{1}{2}})$ by $\exp(-c_3\delta^{-1})$ (see Lemma 2 below). In the proof of theorem, the scheme of proof of the paper [1] and the numerical results of [3] are used with respect to the Dirichlet's L-function boundaries of the zeros-free domain.

For computation of the value of γ, a programm was written in the algorithmical language PASCAL and realized on PC IBM AT-386.

Proof of the Theorem. Here we will follow the scheme of Theorem 2 of [1].

Put

$$\left.\begin{array}{lll} X \geq B^{\exp(10\delta^{-2})}, & & N = 18B^3X, \\ Q = N^{\delta}, & L = NQ^{-c_1}, & T = Q^{c_2}, \end{array}\right\} \tag{4}$$

where $c_1 = \frac{1}{6} + 0,1\delta$; $c_2 = \frac{100}{3} + 4\varepsilon$. From (4) it follows that

$$B < Q^{0.1\delta} \tag{5}$$

We will use the following lemma on the zeros of the Dirichlet's L-function:

Lemma 1. *a) The functions $L(S,X)$, $S = \sigma$. It can have the unique real zero $\tilde{\delta} = 1 - \tilde{\beta}$ for one real primitive character $\tilde{\kappa}$ modulo $\tilde{r} \leq Q$ in the domain*

$$\sigma \geq 1 - 0.0109986((1 + c_2^{-1})\ln T)^{-1}, \ |t| \leq T; \tag{6}$$

b) if $L(S,X)$ has such an (exceptional) zero, then the domain (6) can be replaced with the domain

$$\sigma > 1 - \frac{1}{81\ln Q}\ln(\frac{1}{200\tilde{\delta}\ln Q}, \ |t| \leq^{\varepsilon}, \ \tilde{\delta}\ln Q \leq \frac{1}{200l}; \tag{7}$$

c) this exceptional zero satisfies the inequalities

$$0.4961\tilde{r}^{-\frac{1}{2}}\ln^{-2}\tilde{r} < 1 - \tilde{\beta} < 0.3\ln^{-1}\tilde{r}, \tag{8}$$

As a matter of fact, this is a well known theorem on zeros of the L function. About numerical values of the constants see theorem 1.1 of [3]. Note that in the proof of conclusion a) it is necessary to use the condition $q \leq Q$, $|t| \leq T$ instead of $q \leq Q$, $|t| \leq p^{\frac{19}{4}}$ used in [3].

Let $\Omega = (1 - \tilde{\beta}\ln T$ and let $\sum\limits_{\kappa(\mathrm{mod}\, q)}{}^{*}$, $\sum\limits_{|\gamma|<T}{}'$, respectively, denote summing up by all primitive characters and zeros $p = \beta + i\gamma$ in the rectangle

$$\frac{1}{2} \geq \beta \geq 1 - 0.0109986\left((1 + c_2^{-1})\ln T\right)^{-2}, \ |\gamma| \geq T.$$

$E_{\tilde{\beta}}$ equals 1 or 0 according to the existence or non-existence of the exceptional zero indicated in lemma 1. Using these notations we formulate another lemma.

Lemma 2. *For arbitrary real y and small $\delta > 0$ for $N > y \geq L$ the inequalities*

$$\sum_{q\leq Q}\ \sum_{\kappa(mod_q)}{}^{*},\ \sum_{|\gamma|<T}{}' y^{\beta-1} < \begin{cases} c_3\exp(-c_4\delta^{-1}),\ if\ E_{\tilde{\beta}}=0 \\ c_5\exp(-c_6\delta^{-1}),\ if\ E_{\tilde{\beta}}=1 \end{cases}$$

were $c_3 = 1.333734,\ c_4 = 2.4026066,\ c_5 = 5793.5692,\ c_6 = \frac{1}{108}$ *are true.*

Proof. The proof of this lemma is similar to the proof of lemma 2.1 of [6]. Instead of the Selber's density estimation used in [6] it is necessary to use the Juttila's density estimation in the variant of Chen (see lemma 4 in [2]). According to this lemma 4 [2], for $\varepsilon > 0$ and $T \geq y^{2\varepsilon}$ we have:

$$\sum_{q\leq Q}\ \sum_{\kappa(\mathrm{mod}\,q)}{}^{*},\ \sum_{|\gamma|<T}{}' N_\kappa(\alpha,T) < \begin{cases} \left(T^3y^{-2\varepsilon}\right)^{2(1-\alpha)},\ if\ \alpha\geq 1-\varepsilon \\ \left(T^3y^{-2\varepsilon}\right)^{4(1-\alpha)},\ if\ \alpha\geq 1-\varepsilon \end{cases}$$

Choosing $\varepsilon < c_2\frac{\delta}{2}$, put $T^{12}y^{-8\varepsilon} \geq y^{\frac{1}{4}}$. It follows that δ has to satisfy the condition

$$\delta \leq (1+32\varepsilon)(48c_2 + c_1(+32\varepsilon))^{-1} = \delta(\varepsilon). \tag{9}$$

Reasoning further as in proof of lemma 2.1 from [6] and using Lemma 1 we obtain the assertion of the lemma.

If $E_{\tilde{\beta}} = 1$ and $\tilde{r} \leq Q^{\frac{1}{8}}$, then the constants c_5 and c_6 in Lemma 2 can be replaced with $c_7 = 11.438538$ and $c_8 = \frac{8}{27}$, respectively.

1. Notations and estimation in complementary squares.

Let

$$\tau = N^{-1}T^{\frac{1}{4}},\ e(\alpha) = e^{2\pi i\alpha},\ e_q(\alpha) = e(\frac{\alpha}{q}),$$

$$S(\alpha) = \sum_{L<p\leq N} \ln pe(p\alpha) \text{ and } I(\bar{b}) = \sum_{p_1,p_2,p_3} \ln p_1 \ln p_2 \ln p_3.$$

The summation in the right-hand side of the last equality is carried out over all prime numbers with conditions

$$L < p_1, p_2, p_3 \geq N,\ \sum_{1\leq j\leq 3} a_{ij}p_j = b_i,\ i = 1,2.$$

By representing $I(\bar{b})$ in the form of double integral within the limits $[\tau, 1+\tau]^2$ and dividing the square $[\tau, 1+\tau]^2$ into the main M'_1 and complementary M'_2 small squares (for details see §4, [1]) we obtain

$$I(\bar{b}) = I_1(\bar{b}) + I_2(\bar{b}), \tag{2}$$

where

$$I_i(\bar{b}) = \int\int_{M'_1} e(-\bar{x}_b)\prod_{j=1}^{3} S(\bar{x}_j)dx_1dx_2,\ i = 1,2;$$

$$\bar{x}_b = b_1x_1 + b_2x_2,\ \bar{x}_j = a_{ij}x_1 + a_{2j}x_j,\ j = 1,2,3.$$

For $I_2(\bar{b})$, by virtue of lemma 5.2 and estimate (5.3) of [1], we have

$$\sum_{b_1=-\infty}^{\infty}\sum_{b_2=-\infty}^{\infty} |I_2(\bar{b})|^2 << N^4BQ^{-\frac{1}{2}}\ln^{10} N.$$

It follows that, excepting at most

$$E^{(1)}(X) << N^2BQ^{-frac12+2\lambda_1}\ln^{10} N$$

pairs (b_1, b_2) from $V(X)$, the inequality

$$|I_2(\bar{b})| < NQ^{-\lambda_1} \tag{11}$$

is satisfied.

Put $\lambda_1 = \frac{1}{6}$ if $E_\beta = 0$, and $\lambda_1 = \frac{11}{48} - 0.3\delta$ if $E_{\bar{\beta}} = 1$.

2. **The main squares.**

Now let us consider $I_1(\bar{b})$. Denote

$$C_\kappa(m) = \sum_{1\le l\le q} \kappa(l)l_q(ml),\ C_q(m) = C_{\kappa_0}(m),$$

$$I(y) = \int_L^N l(xy)\sum_{|\gamma|\le T}' x^{p-1}dx,\ G_j(\bar{h},q,\bar{\eta}) = \sum_{\kappa(\text{mod}\, q)} C_\kappa(\bar{h}_j)I_\kappa(\bar{\eta}_i),$$

$$\bar{h}_b = b_1h_1 + b_2h_2,\ \bar{h}_j = a_{ij}h_1 + a_{2j}h_j,$$

$$\bar{\eta}_b = b_1\eta_1 + b_2\eta_2,\ \bar{h}_j = a_{ij}\eta_1 + a_{2j}\eta_j,\ j = 1,2,3.$$

27 terms appear under multiplication in product

$$\prod_{j=1}^{3}(C_q(\bar{h}_j)I(\bar{\eta}_j) - \delta_q C_{\bar{\kappa}\kappa_0}(h_j)\bar{I}(\bar{\eta}_j) - G_j(\bar{h},q,\bar{\eta})).$$

Let us divide these terms into three groups: the first group contains only the term $\prod_{j=1}^{3} c_q(\bar{h}_j)I(\bar{\eta}_j)$; the second group contains the term having at least one factor $G_j(\bar{h},q,\bar{\eta})$, there are 19 such terms; the third group contains the remaining terms. Here $\delta_q = 1$, if $\kappa(q) = \kappa_0(q)$ is the principal character. Denote by K_j ($j = 1,2,3$) the sum of terms from the ith group.

If, now, in proof of the estimate (6.7) [1] we exclude from the consideration at most

$$E^{(2)}(X) < X^2Q^{-\frac{1}{6}+1.2\delta}$$

pairs $(b_1, b-2) \in U(X)$, then for the pairs left instead of (8.2) in [1] we will have:

$$I_1(\bar{b}) = M_1 + M_2 + M_3 + O(NQ^{-\frac{1}{3}}), \tag{12}$$

where

$$M_i = \sum_{q \le Q} \frac{1}{\varphi^3(q)} \sum\nolimits'_{\bar{h}} l_q(-\bar{h}_b) \int\int_{R^2} K_i l(-\bar{\eta}_b) d\eta_1 d\eta_2.$$

Let us consider M_i. Put

$$A(q) = \frac{1}{\varphi^3(q)} \sum\nolimits_h l_q(-\bar{h}_b) \prod_{j=1}^{3} C_q(\bar{h}_j),\ S(p) = 1 + A(p)$$

and

$$M_0 = N|\det_{23}|^{-1} \prod_p S(p) \int_D dx_1, \tag{3}$$

where

$$d_{23} = \det \begin{vmatrix} a_{12}\ a_{13} \\ a_{22}\ a_{23} \end{vmatrix},$$

$$D = \{X_1 :\ LN^{-1} \le X_1, X_2, X_3 \le 1\},$$

A reasoning used in proof of lemma 7.4 [1] in our choice of parameters (see (4)) gives that, if $c_9 < c_1 - 0.2\delta$, then

$$\int_D dX_1 > 0.99Q^{-c_9} \tag{14}$$

with the exception of at most $E^{(3)}(X) < X^2 Q^{-c_9+0.6\delta}$ of pairs $(b_1, b_2) \in U(X)$.

Put

$$c_9 = \frac{1}{6} - 0.2\delta \text{ if } E_{\tilde{\beta}} = 0 \text{ and } c_9 = \frac{1}{28} - 0.7\delta \text{ if } E_{\tilde{\beta}} = 1.$$

Since, by virtue of lemma 7.2 c) of [1], $\prod_p S(p) > c_{10} > 0$, then using (14) from (13) we have:

$$M_0 > c_{11} N Q^{-c_9} B^{-2} \tag{15}$$

with the exception of at most $E^{(3)}(X)$ of pairs $(b_1, b_2) \in U(X)$.

Further we will use the following lemmas.

Lemma 3. *For all pairs $(b_1, b_2) \in U(X)$ we have:*

$$M_1 + M_2 \ge (c_{12}\Omega)^3 M_0 - O(N\tilde{r}Q^{-\frac{4}{5}}),$$

where

$$c_{12} = (1 - c_1\delta)c_2^{-1}\delta^{-1} \exp(-0.0109986(1 - c_1\delta)(1 + c_2)^{-1}\delta^{-1}).$$

Lemma 4. *If $\frac{1}{2} < \gamma < 1$ then the estimate*

$$M_3 << N\tilde{r}{-}1 + \gamma_1 B^3 (\ln\ln N)^2$$

holds for all pairs $(b_1, b_2) \in [1, X]^2$ with the exception of at most $E^{(4)} \le X^2 \tilde{r}^{1-2\gamma}$ of pairs (b_1, b_2) from them

These lemmas are, as a matter of fact, lemmas 8.4 and 8.5 of [1], respectively.

Here we note the following cases: in proof of the lemmas 3 it is necessary to take into account that, by (4) $1 - L^{-\tilde{\beta}-1} \geq c_{12}\Omega$, and lemma 4 in [1] is proved when $\gamma_1 = \frac{3}{4}$.

Let $\kappa_j(\mathrm{mod} r_j)$ be the primitive characters, $r = [r_1, r_2, r_3]$ is the least common multiple of integer numbers r_1, r_2, r_3, and

$$Z(q) = \sum_h l_q(-\bar{h}_b) \prod_{j=1}^{3} C_{\kappa_j \kappa_0}(\bar{h}_j)$$

Lemma 5. *The inequality*

$$|\sum_{q\leq Q, r|q} Z(q)\varphi^{-3}(q)| < 2.3617415 \prod_p S(p)$$

is true.

Proof. Conducting the arguments just as in proof of lemma 4.6 [6] we find:

$$|\sum_{q\leq Q, r|q} Z(q)\varphi^{-3}(q)| \leq \prod_{p|r} S(p) \prod_{\neq|r} (1 + A(p)).$$

If $A(p) > 0$ then the following statement of the lemma immediately follows from here. If $A(p) < 0$ then

$$\prod_{\neq|r}(1 + |A(p)|) = \prod_{\neq|r}(1 + A(p)) \prod_{\neq|r}((1 - 2A(p))(1 + A(p))^{-1})$$

Denote the second product of the right-hand side of the last equality by Π_1. Then it is enough to show that $\Pi_1 < 2.361741\delta$. By (7.5) and lemma 7.1 d) of work [1], we have $S(p) = 1 + A(p) = p^2\varphi^{-3}(p)N(p) > 0$, where $N(p)$ is the number of triples (l_1, l_2, l_3) with $1 \leq l_j \leq p$, $(l_j, p) = 1$, $j = 1, 2, 3$ and

$$\sum_{1\leq j\leq 3} a_{rj} l_j \equiv b_j (\mathrm{mod}\, p),\ i - 1, 2.$$

It is evident that $N(2) = 1$ and $A(2) = 3 > 0$; $N(3)$ is equal to 1 or 2; if $N(3) = 1$, then $A(3) = \frac{9}{8} - 1$ and if $N(3) = 2$ then $A(3) = \frac{9\cdot}{8} - 1 > 0$. Therefore, we can suppose that $p \geq 5$. From (7.7) and (7.8) of [1] it follows that $p - 3 \leq N(p) \leq p - 1$.

Therefore, $-A(p)(1 + A(p))^{-1} \leq (3p - 1)p^{-2}(p - 3)^{-1}$ and

$$\Pi_1 \leq \prod_{p\neq|r,\, p\geq 5} (1 + 2(3p - 1)p^{-2}(p - 3)^{-1})$$

Denote $\alpha(p) = 1 + 2(3p - 1)p^{-2}(p - 3)^{-1}$, then

$$\Pi_1 \leq 2.3133894 \prod_{p\geq 103} \alpha(p). \tag{16}$$

Taking into account the fact that

$$\prod_{p\geq 103} \alpha(p) < \prod_{p\geq 103} (1+6.16p^{-2}) = \exp(\sum_{p\geq 103} \ln(1+\frac{6.16}{p^2})) < 1.020901,$$

from (16) we get $\Pi_1 \leq 2.3617415$.

Now reasoning just as in the proof of lemma 8.6 [1] and using the results of lemmas 2 and 5, we obtain

Lemma 6. *For all pairs $(b_1, b-2) \in U(X)$ the following inequalities are valid:*

$$M_2 < \begin{cases} c_{13}M_0\exp(-c_4\delta^{-1}) & if\ E_{\tilde\beta}=0, \\ c_{13}M_0\Omega^3\exp(-c_8\delta^{-1}) & if\ E_{\tilde\beta}=1,\ and\ \tilde r \leq Q^{\frac{1}{8}} \\ c_{15}M_0\Omega^3\exp(-c_6\delta^{-1}) & if\ E_{\tilde\beta}=1,\ and\ \tilde r > Q^{\frac{1}{8}}, \end{cases}$$

where

$$c_{13} = 2.3617415c_3(3+3c_3\exp(-c_4\delta^{-1}) + (c_3\exp(-c_4\delta^{-1}))^{-2}),$$
$$c_{14} = c_{14}(c_7, c_8) = 2.3617415c_7\{3(1+L^{2(\tilde\beta-1)} + 2L^{(\tilde\beta-1)}$$
$$+ (1+L^{(\tilde\beta-1)})c_7\Omega^3\exp(-c_8\delta^{-1}) + (c_7\Omega^3\exp(-c_8\delta^{-1}))^2\},$$

$c_{15} = c_{14}(c_5, c_6)$, *i.e. in order to obtain an expression for c_{15} it is enough to replace c_7 and c_8 with c_5 and c_6, respectively, in the expression for c_{14}. The values c_i $(i=\overline{3,8})$ are defined in Lemma 2.*

3. **Completion of the proof of the theorem.**

At first let $E_{\tilde\beta=0}$. Put

$$\lambda_1 = \frac{1}{16},\ c_9 = \frac{1}{6} - 0.2\delta,\ c_2 = \frac{100}{3},\ c_1 = \frac{1}{6} + 0.1\delta,\ \delta < \delta_1 = c_4(\ln c_{13} - \ln 0.99)^{-1}.$$

Then by virtue of (10), (11), (12), (15) and Lemma 6 and $M_1 = M_0 + O(NQ^{-\frac{4}{5}})$ (see lemma 8.1 [1]) we have:

$$I(\bar b) \geq I_1(\bar b) - |I_2(\bar b)| > M_0(1 - c_{13}\exp(-c_4\delta^{-1})) - O(NQ^{-\frac{1}{3}}) - NQ^{-\lambda_1} >$$
$$> 0.02c_{11}NQ^{-c_9}B^{-2} > N^{1-0.5\delta} \tag{17}$$

with the exception of

$$E_1(X) = E^{(1)}(X) + E^{(2)} + E^{(3)}(X) < X^2Q^{-\frac{1}{6}+1.2\delta}$$

of pairs $(b_1, b_2) \in U(X)$.

Let now $E_{\tilde\beta=1}$ and $\tilde r \leq Q^\lambda$, $0 < \lambda \geq 1$. Put

$$\lambda_1 = \frac{11}{48} - 0.3\delta,\ c_9 == \frac{1}{24} - 0.7\delta,\ \lambda = \frac{1}{8},\ c_1 = \frac{1}{6} + 0.1\delta,$$

$$c_2 = \frac{100}{3}, \ \delta < \delta_2 = (108\lambda(\ln c_{14} - \ln(0.99c_{12})))^{-1}.$$

Then taking into account the fact that $Q > 0.4961\lambda^{-2}Q^{-\frac{\lambda}{2}}\ln^{-1} Q$ (see Lemma 1), from (10), (11), (12), (15) and Lemmas 3 and 6 we obtain:

$$I(\bar{b}) >> NQ^{-c_9-\frac{3\lambda}{2}-0.3\delta} - \lambda Q^{-\lambda_1} > N^{1-\frac{11}{48}\delta} + 0.3\delta^2 > N^{1-0.5\delta} \tag{18}$$

with the exception of $E_2(X) < X^2Q^{-\frac{1}{24}} + 1.4\delta$ of pairs $(b_1, b_2) \in U(X)$.

Finally let $E_{\tilde{\beta}} = 1$ and $\tilde{r} > Q^\delta$, $0 < \delta < 1$. In this case we put

$$\lambda_1 = \frac{1}{28}, \ \lambda = \frac{1}{8}, \ \gamma_1 = \frac{3}{4}, \ c_9 = \frac{1}{28} - 0.7\delta,$$
$$\delta < \delta_3 = c_6(\ln c_{15} - \ln((6l)^3)0.99))^{-1}.$$

Then using (10), (11), (12), (15) and Lemmas 4 and 6, we find:

$$I(\bar{b}) >> NQ^{-c_9}B^{-2}(1 - c_{15}\Omega^3\exp(-c_6\delta^{-1}) - O(NQ^{-\lambda+\lambda\gamma+0.4\delta})-$$
$$O(NQ^{-\frac{1}{3}}) - NQ^{-\lambda_1} > NQ^{-\frac{1}{24}+0.3\delta} = N^{1-\frac{\delta}{24}+0.3\delta^2} \tag{19}$$

with the exception of $E_3(X) << X^2Q^{-\frac{1}{4}+1.4\delta}$ pairs $(b_1, b_2) \in U(X)$.

Note that each of δ_1, δ_2, δ_3 and $\delta(\varepsilon)$ contain its own δ. This matter makes difficult the computation of the values of δ. Therefore, in way of searching for value of δ_1 it is necessary to give δ the values satisfying all of the conditions according to δ mentioned above. For determination of values of δ_2, δ_3 and $\delta(\varepsilon)$ as an initial value of δ we can take δ_1, δ_2 and δ_3 respectively. Here it is necessary to try to assign to δ a value as big as possible.

The computation made by using the program written in the algorithmical language PASCAL and realized on PC IBM XT 386 shows that if $0 < \varepsilon < 10^{-8}$ then

$$\delta = \min(\delta_1, \ \delta_2, \ \delta_3, \ \delta(\varepsilon)) = 1.0054445 \cdot 10^{-4}.$$

Then all conditions with respect to δ are satisfied, and from (17)–(19) for sufficiently big natural numbers N we obtain $N(\bar{b}) > N^{0.9}$ with the exception of $E(X) < X^{1,9999833}$ of pairs $\bar{b} = (b_1, b_2) \in U(X)$.

The proof of the corollary is similar to the proof of theorem B of [1], only instead of theorem 2 of [1], it is necessary to use the estimate (2) of the present work.

References

1. Liu, M.C., Tsang, E.M.: *J. reine und angew. Math.* **399** (1989) 109–136.
2. Chen, J.R., Pan, C.D.: *Sci. Sinica* **23** (1980) 416–430.
3. Allakov, I.A.: *Exceptional Set of a Sum of Two Prime Numbers*, Candidate Thesis, Leningrad, 1983, pp. 1–148.
4. Hua, L.K.: *Additive Theory of Prime Numbers*, AMS, Providence, Rhode Island, 1965.
5. Rosser, J.B., Schoenfeld, L.: *Illinois J. Math.* No. 6 (1962) 64–94.
6. Liu, M.C., Tsang, E.M.: *Proc. Intern. Number Th. Conf. 1987 Laval University*, Canad. Math. Soc., Berlin, New York, 1989, pp. 595–624.

On a Description of Irreducible Component in the Set of Nilpotent Leibniz Algebras Containing the Algebra of Maximal Nilindex, and Classification of Graded Filiform Leibniz Algebras

Sh.A. Ayupov and B.A. Omirov

Samarkand State University
15, University bld.,Samarkand, Uzbekistan

Abstract. This paper is devoted to the study of Leibniz algebras introduced by Loday in [1-2] as an analogue of zero "noncommutative" Lie algebras. We define the notion of zero-filiform Leibniz algebras and study their properties. There is a notion of p-filiform Lie algebras for $p \geq 1$ [3], which loses a sense in case $p = 0$, since Lie algebra has at least two generators. In the case of Leibniz algebras for $p = 0$ this notion substantial, and thereby, introduction of a zero-filiform algebra is quite natural. We also investigate the complex non–Lie filiform Leibniz algebras. In particular, we give some equivalent conditions of filiformity of Leibniz algebras and describe complex Leibniz algebras, which were graded in natural way.

1 The Description of Irreducible Component Containing the Nilpotent Leinbiz Algebra of Maximal Nilindex

Definition 1. *An algebra L over a field F is called Leibniz algebra, if for any $x, y, z \in L$ it satisfies the Leibniz identity:*

$$[x,[y,z]] = [[x,y],z] - [[x,z],y]$$

where $[,]$ is the multiplication in L.

Note that if in L the identity $[x,x] = 0$ is satisfied, then the Leibniz identity coincides with the Jacobi identity. Therefore, Leibniz algebras are a "noncommutative" analogue of Lie algebras.

For arbitrary algebra L we will define lower central sequence

$$L^{<1>} = L, \qquad L^{<n+1>} = [L^{<n>}, L];$$

Definition 2. *An algebra L is called nilpotent, if there exists $n \in N$ such, that $L^{<n>} = 0$.*

It is easy to see that the index of nilpotency of an arbitrary n-dimensional nilpotent algebra does not exceed $n+1$.

Definition 3. *An n-dimensional Leibniz algebra L is called zero-filifirm, if $\dim L^i = (n+1)-i, \quad 1 \leq i \leq n+1$.*

It is obvious that the definition of a zero-filiform algebra L is equivalent to the fact that algebra L has maximal index of nilpotency.

Lemma 1. *In any n-dimensional zero-filiform Leibniz algebra there is a basis with the following multiplication:*

$$[x_i, x_1] = x_{i+1} \ \ for \ \ 1 \leq i \leq n-1, \qquad [x_i, x_j] = 0, \ \ for \ \ j \geq 2 \tag{1}$$

Proof. Let L be a n-dimensional zero-filiform Leibniz algebra and let $\{e_1, e_2, \ldots, e_n\}$ be the basis of algebra L such that $e_1 \in L^1 \backslash L^2, \ \ e_2 \in L^2 \backslash L^3, \cdots, e_n \in L^n$ (such basis obviously exists). Since $e_2 \in L^2$, for some elements a_{2p}, b_{2q} of algebra L we have the following:

$$e_2 = \sum [a_{2p}, b_{2q}] = \sum \alpha_{ij}^{(2)} [e_i, e_j] = \alpha_{11}^{(2)} [e_1, e_1] + k_3,$$

where $k_3 \in L^3$, i.e. $e_2 = \alpha_{11}^{(2)} [e_1, e_1] + k_3$. Note that $\alpha_{11}^{(2)} [e_1, e_1] \neq 0$ (otherwise $e_2 \in L^3$). We obtain in a similar way:

$$e_3 = \sum [[a_{3p}, b_{3q}], c_{3s}] = \sum \alpha_{ijk}^{(3)} [[e_i, e_j], e_k] = \alpha_{111}^{(3)} [[e_1, e_1], e_1] + k_4,$$

where $k_4 \in L^4$, i.e.

$$e_3 = \alpha_{111}^{(3)} [[e_1, e_1], e_1] + k_4.$$

Also $\alpha_{111}^{(3)} [[e_1, e_1], e_1] \neq 0$ (otherwise $e_3 \in L^4$). Continuing this procedure we obtain that the elements

$$x_1 := e_1, \ \ x_2 := [e_1, e_1], \ \ x_3 := [[e_1, e_1], e_1], \cdots, x_n := [[[e_1, e_1], e_1], \cdots, e_1]$$

are different from zero. Linear independence of these elements is easy to check and, consequently, they form a basis of algebra L. We have thereby: $[x_i, x_1] = x_{i+1}$, for $1 \leq i \leq n-1$, besides $[x_i, x_j] = 0$, for $j \geq 2$. Indeed, if $j = 2$, then

$$[x_i, x_2] = [x_i, [x_1, x_1]] = [[x_i, x_1], x_1] - [[x_i, x_1], x_1] = 0.$$

Suppose that it was proved for all $j > 2$. Inductive suggestion and the following equality:

$$[x_i, x_{j+1}] = [x_i, [x_j, x_1]] = [[x_i, x_j], x_1] - [[x_i, x_1], x_j] = 0$$

give the proof for $j+1$.
Lemma is proved.

Further we will denote by L_0 the algebra with the multiplication (1).

Let x be an element of $L\backslash[L,L]$. For nilpotent operator of right multiplication R_x we shall define the decreasing sequence $C(x) = (n_1, n_2, \cdots, n_k)$ consisting of the dimensions of Jordan blocks of operator R_x. On the set of such sequences, we define a lexicographic order, i.e.

$$C(x) = (n_1, n_2, \cdots, n_k) \leq C(y) = (m_1, m_2, \cdots, m_s) \Leftrightarrow \exists i \in N$$

such that $n_j = m_j \ \forall j < i$ and $n_i < m_i$.

Definition 4. *The sequence* $C(L) = \max\limits_{x \in L\backslash[L,L]} C(x)$ *is called the characteristsc sequence of algebra* L.

Definition 5. *The set* $Z(L) = \{x \in L : \ [y,x] = 0, \ \ \forall y \in L\}$ *is called the right annihilator.*

Example 1. Let L be an arbitrary algebra and $C(L) = (1, \cdots, 1)$, then L is abelian.

Example 2. Let L be a n-dimensional Leibniz algebra. Then by Lemma 1 L is the zero-filiform Leibniz algebra if and only if $C(L) = (n, 0)$.

We consider an arbitrary algebra L from the set of n-dimensional Leibniz algebras over a field F. Let $\{e_1, \cdots, e_n\}$ be a basis of algebra L, then algebra L is defined up to isomorphism by the multiplication of basis elements as follows: $[e_i, e_j] = \sum\limits_{k=1}^{n} \gamma_{ij}^k e_k$, where γ_{ij}^k are structural constants.

So, each n-dimensional algebra over a field F with a fixed basis can be considered as a point in n^3-dimensional space of structural constants with the Zariski topology. Then a natural action of the group $GL_n(F)$ on F^{n^3} corresponds to the change of basis, by this action the orbit of point is a set of all isomorphic algebras.

Let $\Im_n(F)$ be the set of structural constants of all n-dimensional Leibniz algebras over a field F and N_n be a subset of $\Im_n(F)$ consisting of structural constants of all n-dimensional nilpotent Leibniz algebras over the field F.

From the Leibniz identity we have the polinomial identities:

$$\sum_{l=1}^{n} \left(\gamma_{jk}^l \gamma_{il}^m - \gamma_{ij}^l \gamma_{lk}^m + \gamma_{ik}^l \gamma_{lj}^m\right) = 0$$

for structural constants, and so the set $\Im_n(F)$ in F^{n^3} is an affine variety.

Definition 6. *We define the action of group* $GL_n(F)$ *on the set* $\Im_n(F)$ *as follows:* $[x,y]_g := g[g^{-1}x, g^{-1}y]$, *where* $g \in GL_n(F)$ *and* $x, y \in L$. *By* $Orb_n(L)$ *we dcnotc thc orbit* $GL_n^* L$ *of algebra* L.

The $Orb_n(L)$ obviously consists of all algebras isomophic to L (stabilizer of algebra L is group $Aut(L)$, i.e. $Orb_n(F) = GL_n(F)/Aut(L)$). The closure of the orbit $Orb_n(L)$ is understood in the case of the arbitrary field F in the sense of Zariski topology, but when $F = C$ it coincides with the closure with respect to Euclidean topology.

It is easy to see that the scalar matrices from the $GL_n(F)$ act on $\Im_n(F)$ as scalars, so the orbits $Orb_n(L)$ are the cones without the top $\{0\}$ corresponding to the Abelian algebra a_n. So a_n belongs to $\overline{Orb_n(L)}$ $\forall L \in \Im_n(F)$. In particular, $Orb_n(L)$ contains only one closed orbit of a_n ($a_n is$ abelian).

Since from [4] we have that the set $\{L \in \Im_n(F) : \quad dimZ(L) \geq n-1\}$ is closed in Zariski topology, and therefore

$$\overline{Orb_n(L_0)} \subseteq N_n \cap \{L \in \Im_n(F) : \dim Z(L) \geq n-1\}.$$

For convenience we will denote by

$$N_nZ := N_n \cap \{L \in \Im_n(F) : \dim Z(L) = n-1\},$$

the case $\dim Z(L) = n$ is not interesting for us since then L is abelian.

Lemma 2. *Let L be an algebra from the set N_nZ with characteristic sequence $C(L) = (m, n-m)$. Then for $m = \frac{n}{2}$ it is isomorphic to the algebra*
$[e_1, e_n] = 0, \quad [e_2, e_n] = e_1, \cdots, [e_m, e_n] = e_{m-1}, [e_{m+1}, e_n] = 0,$
$[e_{m+2}, e_n] = e_{m+1}, [e_{m+3}, e_n] = e_{m+2}, \cdots, [e_n, e_n] = e_{n-1},$
and for $m > \frac{n}{2}$ it is isomorphic to the one of two following non-isomorphic algebras:
$[e_1, e_m] = 0, \quad [e_2, e_m] = e_1, \cdots, [e_m, e_m] = e_{m-1}, [e_{m+1}, e_m] = 0,$
$[e_{m+2}, e_m] = [e_{m+1}[e_{m+3}, e_m] = e_{m+2}, \cdots, [e_n, e_m] = e_{n-1};$
or $[e_1, e_n] = 0, \quad [e_2, e_n] = e_1, \cdots, [e_m, e_n] = e_{m-1}, [e_{m+1}, e_n] = 0,$
$[e_{m+2}, e_n] = e_{m+1}, [e_{m+3}, e_n] = e_{m+2}, \cdots, [e_n, e_n] = e_{n-1}.$

Proof. Let $\{e_1, \cdots, e_n\}$ be a basis of algebra L and $L \in N_nZ$ and $C(L) = (m, n-m)$, then $\exists x \in L \backslash [L, L]$ such that

$$R_x = \begin{pmatrix} J_m & 0 \\ 0 & J_{n-m} \end{pmatrix},$$

i.e.

$$[e_1, x] = 0, \quad [e_2, x] = e_1, \cdots, [e_m, x] = e_{m-1}, [e_{m+1}, x] = 0,$$

$$[e_{m+2}, x] = e_{m+1}, \quad [e_{m+3}, x] = e_{m+2}, \cdots, [e_n, x] = e_{n-1}.$$

We may assume that x is a basis element (it is possible since $\dim Z(L) = n-1$). Since $\dim Z(L) = n-1$, then $[L, L] \subseteq Z(L)$, and so x does not belong to the linear envelop of vectors $\{e_1, \cdots, e_{m-1}, e_{m+1}, \cdots, e_{n-1}\} \subseteq Z(L)$, therefore, $x = e_m$ or $x = e_n$. In the case $m = \frac{n}{2}$, the following change

$$\bar{e}_1 = e_{m+1}, \quad \bar{e}_2 = e_{m+2}, \cdots, \bar{e}_m = e_n, \quad \bar{e}_{m+1} = e_1, \quad \bar{e}_{m+2} = e_2, \cdots, \bar{e}_n = e_m$$

gives us that the following algebras:

$$[e_1,e_m]=0,\quad [e_2,e_m]=e_1,\cdots,[e_m,e_m]=e_{m-1},[e_{m+1},e_m]=0,$$

$$[e_{m+2},e_m]=e_{m+1},[e_{m+3},e_m]=e_{m+2},\cdots,[e_n,e_m]=e_{n-1};$$

and

$$[e_1,e_n]=0,\quad [e_2,e_n]=e_1,\cdots,[e_m,e_n]=e_{m-1},[e_{m+1},e_n]=0,$$

$$[e_{m+2},e_n]=e_{m+1},[e_{m+3},e_n]=e_{m+2},\cdots,[e_n,e_n]=e_{n-1}$$

are isomorphic.

In the case $m>\frac{n}{2}$ suppose that above algebras are isomorphic, i.e. $\exists\varphi$ isomorphism between them. Then $\varphi(e_m)=\alpha_1e_1+\alpha_2e_2+\cdots+\alpha_ne_n$, where $\alpha_n\neq 0$. It is known that generating elements are mapped by isomorphism onto generating elements. Thus $\varphi(e_m),\varphi(e_m)]=\varphi(e_{m-1}),\cdots,[\varphi(e_2),\varphi(e_m)]=0$ (since $m>n-m$), i.e. we come to contradiction.
Lemma is proved.

Further, if $\dim Z(L)=n-1$, then we can define the algebra L by means of right multiplication operator of an element x, where $x\in Z(L)$.

Corollary 1. *Let $L\in N_nZ$ and $C(L)=(n_1,\cdots,n_s)$. Then L is isomorphic to one of the algebras:*

$$R_{e_{n_1}}=\begin{pmatrix} J_{n_1}\cdots\cdots 0\\ 0J_{n_2}\cdots 0\\ \cdots\cdots\cdots\cdots\\ 0\cdots\cdots 0J_{n_s}\end{pmatrix},\cdots,R_{e_{n_1+\cdots+n_s}}=\begin{pmatrix} J_{n_1}\cdots\cdots 0\\ 0J_{n_2}\cdots 0\\ \cdots\cdots\cdots\cdots\\ 0\cdots\cdots 0J_{n_s}\end{pmatrix},$$

where $J_{n_1},\cdots,J_{n_s}$ are Jordan blocks having the dimensions sizes $n_1,\cdots,n_s$ respectively. In particular $R_{e_{n_1+\cdots+n_{i-1}}}\cong R_{e_{n_1+\cdots+n_{i-1}+n_i}}$ if and only if $n_{i-1}=n_i$.

Proof. Let L satisfy the conditions of Lemma 2, then by arguments similar to Lemma 2 it follows that L is one of the above mentioned algebras specified above. Let $n_{i-1}=n_i$, where $2\leq i\leq s$, then the following change of basis:

$$\bar{e}_{n_1+\cdots+n_{i-2}+1}:=e_{n_1+\cdots+n_{i-2}+n_{i-1}+1},$$
$$\bar{e}_{n_1+\cdots+n_{i-2}+2}:=e_{n_1+\cdots+n_{i-2}+n_{i-1}+2},\cdots,$$
$$\bar{e}_{n_1+\cdots+n_{i-1}}:=e_{n_1+\cdots+n_i},\quad \bar{e}_{n_1+\cdots+n_{i-2}+n_{i-1}+1}:=e_{n_1+\cdots+n_{i-2}+1},$$
$$\bar{e}_{n_1+\cdots+n_{i-2}+n_{i-1}+2}:=e_{n_1+\cdots+n_{i-2}+2},\cdots,\quad \bar{e}_{n_1+\cdots+n_i}:=e_{n_1+\cdots+n_{i-1}},$$
$$\bar{e}_i=e_i$$

for other elements, gives the isomorphism of following algebras: $R_{e_{n_1+\cdots+n_{i-1}}}$, $R_{e_{n_1+\cdots+n_i}}$. As in the lemma 2 one can see that the algebra $R_{e_{n_1+\cdots+n_{i-1}}}$ is not isomorphic to algebra $R_{e_{n_1}+\cdots+n_i}$ if $n_{i-1}\neq n_i$ for some i.
Corollary is proved.

Under the conditions of corollary one has the following

Corollary 2. *The number of non-isomorphic algebras in the set N_nZ is equal to the cardinality of the set $\{n_1, \cdots, n_s\}$.*

Lemma 3. *Let L be an algebra with basis $\{e_1, \cdots, e_n\}$, which belongs to the set N_nZ. Then $L \in \overline{Orb_n(L_0)}$ if and only if $C(L) = C(e_n)$.*

Proof. Putting $\bar{e}_i := e_{n+1}$ for $1 \leq i \leq n$, we obtain $C(L_0) = C(e_n)$, i.e. $L_0 \cong R_{\bar{e}_n} = J_n$. Let L satisfy the conditions of lemma i.e.

$$L \cong R_{e_n} = \begin{pmatrix} J_{n_1} & \cdots & \cdots & 0 \\ 0 & J_{n_2} & \cdots & 0 \\ \cdots & \cdots & \cdots & \cdots \\ 0 & \cdots & 0 & J_{n_s} \end{pmatrix}.$$

Let us take the family of matrices $(g_{\lambda_1})_{\lambda_1 \in R \setminus \{0\}}$ defined as follows: $g_{\lambda_1}(e_i) = \lambda_1^{-1} e_i$ for $1 \leq i \leq n_1$, $g_{\lambda_1}(e_i) = e_i$ for $n_1 + 1 \leq i \leq n$. Taking limit by this family for λ_1 tending to zero, i.e. $\lim\limits_{\lambda_1 \to 0} g_{\lambda_1}^{-1}[g_{\lambda_1}(e_i), g_{\lambda_1}(e_j)]$ we obtain that

$$L_0 \to R_{e_n} = \begin{pmatrix} J_{n_1} & 0 \\ 0 & J_{n-n_1} \end{pmatrix}, \quad \text{for } \lambda_1 \to 0.$$

We will further take the family of matrices $(g_{\lambda_2})_{\lambda_2 \in R \setminus \{0\}}$ defined as $g_{\lambda_2}(e_i) = \lambda_2^{-1} e_i$ for $n_1 + 1 \leq i \leq n_1 + n_2$, $g_{\lambda_2}(e_i) = e_i$ for $1 \leq i \leq n_1$ and $n_1 + n_2 + 1 \leq i \leq n$. Taking limit by this family for λ_2 tending to zero i.e. $\lim\limits_{\lambda_2 \to 0} g_{\lambda_2}^{-1}[g_{\lambda_2}(e_i), g_{\lambda_2}(e_j)]$ we obtain that

$$L_0 \to R_{e_n} = \begin{pmatrix} J_{n_1} & 0 & 0 \\ 0 & J_{n_2} & 0 \\ 0 & 0 & J_{n-n_1-n_2} \end{pmatrix}, \quad \text{for } \lambda_2 \to 0.$$

Continuing this process s times we obtain that the algebra defined by operator

$$R_{e_n} = \begin{pmatrix} J_{n_1} & \cdots & \cdots & 0 \\ 0 & J_{n_2} & \cdots & 0 \\ \cdots & \cdots & \cdots & \cdots \\ 0 & \cdots & 0 & J_{n_s} \end{pmatrix}$$

belongs to $\overline{Orb_n(L_0)}$.

Let $L \in \overline{Orb_n(L_0)}$, then the multiplication in it is given by multiplication in L_0 as follows: $[e_i, e_j] = \lim\limits_{\lambda \to 0} g_\lambda^{-1}[g_\lambda e_i, g_\lambda e_j]$. Since for any $\lambda \neq 0$,

$$g_\lambda(lin(e_1, \cdots, e_{n-1})). \subseteq lin(e_1, \cdots, e_{n-1})$$

we have equalities $[e_i, e_j] = 0$ for $1 \leq j \leq n-1$. So L is defined by operator R_{e_n}. Let $Q^{-1} R_{e_n} Q = J$ (J is the Jordan form of operator R_{e_n}), then taking the family $(g_\lambda Q)_{\lambda \in R \setminus \{0\}}$ we can assume that operator R_{e_n} has the Jordan form, i.e. $C(L) = C(e_n)$.

Lemma is proved.

Since orbit of zero-filiform algebra is open in the affine variety N_n, it follows from [5] that its closure coincides with irreducible component of N_n and follows theorem is true.

Theorem 1. *The irreducible component of the variety N_n, containing the zero-filiform algebra consists up to isomorphizm of the following algebras*

$$R_{e_n} = \begin{pmatrix} J_{n_1} & \cdots & \cdots & 0 \\ 0 & J_{n_2} & \cdots & 0 \\ \cdots & \cdots & \cdots & \cdots \\ 0 & \cdots & 0 & J_{n_s} \end{pmatrix}, \quad \textit{where} \quad n_1 + \cdots + n_s = n.$$

Proof. Follows from Lemma 3 and Corollary 1.

Remark 1. From Theorem 1 it follows that the number of non isomorphic algebras belonging to irreducible component of variety N_n containing the algebra L_0 is equal to the number $p(n)$ of solutions in integers of the equation

$$x_1 + x_2 + \cdots + x_n = n, \quad x_1 \geq x_2 \geq \cdots \geq x_n \geq 0.$$

An asymptotic value of $p(n)$, which is given in [6] by $p(n) \approx \frac{1}{4n\sqrt{3}} e^{A\sqrt{n}}$, where $A = \pi\sqrt{\frac{2}{3}}$ ($p(n) \approx g(n)$ means that $\lim\limits_{n\to\infty} \frac{p(n)}{g(n)} = 1$) shows that the set of non-isomorphic Leibniz algebras, from the irreducible component of N_n, containing the algebra L_0 is not large, i.e. the number of orbits in this component is finite for any n.

2 Classification of Complex Naturally Graded Filiform Leibniz Algebras

Definition 7. *A Leibniz algebra L is called filiform, if $\dim L^i = n - i$, where $n = \dim L$ and $2 \leq i \leq n$.*

Lemma 4. *Let L be a n-dimensional Leibniz algebra. Then the following conditions are equivalent:*
a) $C(L) = (n-1, 1)$;
b) L is filiform Leibniz algebra;
c) $L^{n-1} \neq 0$ and $L^n = 0$.

Proof. Implications *a)* $\Rightarrow$ *b)* $\Rightarrow$ *c)* are obvious.
b) $\Rightarrow$ **a)**. Let $\{e_1, \cdots, e_n\}$ be a basis of the filiform algebra L such that

$$\{e_3, \cdots, e_n\} \subseteq L^2, \quad \{e_4, \cdots, e_n\} \subseteq L^3, \cdots, \{e_n\} \subseteq L^{n-1}.$$

We consider the multiplications:

$$[x, e_1 + \alpha e_2] = \gamma_1 e_3 + \alpha\beta_1 e_3, \quad [e_3, e_1 + \alpha e_2] = \gamma_2 e_4 + \alpha\beta_2 e_4,$$
$$[e_4, e_1 + \alpha e_2] = \gamma_3 e_5 + \alpha\beta_3 e_5, \cdots, [e_n, e_1 + \alpha e_2] = 0,$$

where x is an arbitrary element of L and $\forall i\ \gamma_i^2+\beta_i^2\neq 0$. We choose α such that $\forall i\ \ \gamma_i+\alpha\beta_i\neq 0$. Then $z:=e_1+\alpha e_2\in L\backslash[L,L]$ and $C(z)=(n-1,1)$.

c) $\Rightarrow$ b). Let $L^n=0$, then we obtain a decreasing chain of subalgebras:

$$L\supset L^2\supset L^3\supset\cdots\supset L^{n-1}\supset L^n=0$$

of the length n. It is clear that either $dimL^2=n-1$ or $dimL^2=n-2$ (otherwise $L^{n-1}=0$). Suppose that $dimL^2=n-1$. Then we choose basis $\{e_1,\cdots,e_n\}$ of the algebra L, corresponding to the filtration

$$L\supset L^2\supset L^3\supset\cdots\supset L^{n-1}\supset L^n=0.$$

Suppose that $\dim L^s/L^{s+1}=2$ $(s\neq 1)$, i.e. $\{e_s,e_{s+1}\}\in L^s\backslash L^{s+1}$. By arguments as in proof of Lemma 1 and making corresponding changes of variables we may assume that $e_s=[[[e_1,e_1],e_1],\cdots,e_1]+k_{s+1}$, ($s$ times product and $k_{s+1}\in L^{s+1}$) and $e_{s+1}=[[[e_1,e_1],e_1],\cdots,e_1]+l_{s+1}$ (s times product and $l_{s+1}\in L^{s+1}$). Therefore, $e_s-e_{s+1}\in L^{s+1}$. Thus, we obtain a contradiction with the assumption that $\dim L^s/L^{s+1}=2$. Hence, $\dim L^k/L^{k+1}=1$ $(1\leq i\leq n-1)$. But then the basis of the n-dimensional algebra L will consist of $n-1$ elements, which is a contradiction. This contradiction shows that the assumption $\dim L^2=n-1$ is false. Thus, we have $\dim L^i=n-1$, where $n=\dim L$ and $2\leq i\leq n$, i.e. L is filiform algebra.

Lemma is proved.

Further we shall consider the algebra L as a pair (V,μ), where V is the underlying vector space and μ is the multiplication on V, which defines the algebra L.

Let (V,μ) be an $n+1-$dimensional complex filiform Leibniz algebra. We define a natural graduation of algebra (V,μ) as follows. Take $V_1(\mu)=V$, $V_{i+1}(\mu):=\mu(V_i(\mu),V)$ and $W_i:=V_i(\mu)/V_{i+1}(\mu)$, then $V=W_1+W_2+\cdots+W_n$, where $\dim W_1=2$, $\dim W_i=1$, $2\leq i\leq n$. By [7, Lemma 1] we have inclusions: $\mu(W_i,W_j)\subseteq W_{i+j}$. Thus, we have graduation, which is called a natural graduation.

Similarly to [8], over an infinite field we can find the basis $e_0,e_1\in W_1$, $e_i\in W_i$ $(i\geq 2)$ in space V with a bilinear map μ such that $\mu(e_i,e_0)=e_{i+1}$ and $\mu(e_n,e_0)=0$.

Also for convenience denote $\mu(x,y)$ by $[x,y]$.

Case 1. $[e_0,e_0]=\alpha e_2$ $(\alpha\neq 0)$. Then $e_2\in Z(\mu)$ (where $Z(\mu)$ is a right annihilator of the algebra L), i.e. $e_3,\cdots,e_n\in Z(\mu)$. By changing:

$$\bar{e}_1=\frac{1}{\alpha}e_1,\quad \bar{e}_2=\alpha e_2,\quad \bar{e}_3=\frac{1}{\alpha}e_3,\cdots,\quad \bar{e}_n=\frac{1}{\alpha}e_n,$$

we can suppose that α is equal to zero. Thus, $[e_0,e_0]=e_2$, $[e_i,e_0]=e_{i+1}$ and $[e_n,e_0]=0$. Suppose that $[e_0,e_1]=\beta e_2$ and $[e_1,e_1]=\gamma e_2$. Then we have:

$$[e_0,[e_1,e_0]]=[[e_0,e_1],e_0]-[[e_0,e_0],e_1],$$

i.e.

$$\beta e_3 = [e_2, e_1] \text{ and } [e_1, [e_0, e_1]] = [[e_1, e_0], e_1] - [[e_1, e_1], e_0],$$

i.e. $\gamma e_3 = [e_2, e_1]$,
hence $\beta = \gamma$. By induction for number of basis elements and using equality $[e_i, [e_0, e_1]] = [[e_i, e_0], e_1] - [[e_i, e_1], e_0]$, it is easy to prove that $[e_i, e_1] = \beta e_{i+1}$, i.e. in case 1 we obtain the algebra:

$$[e_0, e_0] = e_2, [e_i, e_0] = e_{i+1}, \quad [e_1, e_1] = \beta e_2, [e_i, e_1] = \beta e_{i+1}, \quad [e_0, e_1] = \beta e_2.$$

Case 2. $[e_0, e_0] = 0$ & $[e_1, e_1] = \alpha e_2$ $(\alpha \neq 0)$. Then $e_2 \in Z(\mu) \Rightarrow e_3, \cdots, e_n \in Z(\mu)$. Putting $\bar{e}_0 = \alpha e_0$, $\bar{e}_2 = \alpha e_2$, $\bar{e}_3 = \alpha^2 e_3, \cdots, \bar{e}_n = \alpha^{n-1} e_n$, we can suppose $\alpha = 1$, i.e. $[e_1, e_1] = e_2$, $[e_i, e_0] = e_{i+1}$. We denote $[e_0, e_1] = \beta e_2$. Then one has:

$$[e_0, [e_1, e_0]] = [[e_0, e_1], e_0] - [[e_0, e_0], e_1] \Rightarrow [[e_0, e_1], e_0] = 0,$$

i.e. $\beta[e_2, e_0] = \beta e_3 = 0 \Rightarrow \beta = 0$. By induction for number of basis elements and using equality $[e_i, [e_0, e_1]] = [e_{i+1}, e_1] - [[e_i, e_1], e_0]$, it is easy to show that $[e_i, e_1] = e_{i+1}$, i.e. in Case 2 we obtain the algebra:

$$[e_i, e_0] = e_{i+1}, \quad [e_i, e_1] = e_{i+1} \quad (i \geq 1).$$

By the following change of variables: $\bar{e}_0 := e_0 - e_1$, $\bar{e}_i := e_i$, we obtain the algebra $[\bar{e}_i, \bar{e}_1] = \bar{e}_{i+1}$. It is easy to see that the obtained algebra is isomorphic to algebra in Case 1 for $\beta = 1$ $(e'_0 := e_0 - e_1, \quad e'_i := e_i)$.

Case 3. $[e_0, e_0] = 0$ & $[e_1, e_1] = 0$. We denote $[e_0, e_1] = \alpha e_2$.

Subcase 1. $[e_0, e_1] = \alpha e_2$ $(\alpha \neq -1)$. Then $e_2 \in Z(\mu) \Rightarrow e_3, \cdots, e_n \in Z(\mu)$. Since $\alpha \neq -1$, then putting $\bar{e}_1 = e_1 + e_0$ we obtain $\bar{e}_1^2 = (\alpha + 1)$ and $[\bar{e}_1, e_0] = e_2$, i.e. we are in Case 2.

Subcase 2. $[e_0, e_1] = -e_2$. To consider this subcase we need the following lemma

Lemma 5. *Let (V, μ) be the $n+1$-dimensional naturally graded filiform Leibniz algebras with basis $\{e_0, e_1, \cdots, e_n\}$ such that $[e_1, e_1] = [e_0, e_0] = 0$, $[e_0, e_1] = -e_2, [e_i, e_0] = e_{i+1}$. Then (V, μ) is a Lie algebra.*

Proof. By induction for number of basis variables and using the equality

$$[e_0, [e_i, e_0]] = [[e_0, e_i], e_0] - [[e_0, e_0], e_i]$$

it is easy to show that

$$[e_0, e_i] = -[e_i, e_0] \quad (1 \leq i \leq n).$$

From $[e_1, [e_1, e_0]] = [[e_1, e_1], e_0] - [[e_1, e_0], e_1]$ we have $[e_1, e_2] = -[e_2, e_1]$. Using the equalities

$$\begin{aligned}
[e_1, e_{i+1}] &= [e_1, [e_i, e_0]] = [[e_1, e_i], e_0] - [[e_1, e_0], e_i] \\
&= -[[e_i, e_1], e_0] - [e_2, e_i] = [e_0, [e_i, e_1]] - [e_2, e_i] \\
&= [[e_0, e_i], e_1] - [[e_0, e_1], e_i] - [e_2, e_1] \\
&= [[e_0, e_i], e_1] + [e_2, e_i] - [e_2, e_i] = -[[e_i, e_0], e_1] = -[e_{i+1}, e_1]
\end{aligned}$$

and using the initial assumption of induction we get that $[e_1, e_i] = -[e_i, e_1]$ $(1 \leq i \leq n)$. So we have $[e_1, e_i] = -[e_i, e_1]$ and $[e_0, e_i] = -[e_i, e_0]$ $(0 \leq i \leq n)$. Let's prove the equality $[e_i, e_j] = -[e_j, e_i]$ $\forall i, j$. We'll prove by induction by i and for any value j. Notice that we can suppose that j is more than 1. Using the chain of equalities

$$\begin{aligned}
&[e_{i+1}, e_j] = [[e_i, e_0], [e_{j-1}, e_0] = [[[e_i, e_0], e_{j-1}], e_0] - [[[e_i, e_0], e_0], e_{j-1}] \\
&= -[[e_0, [e_i, e_0], e_{j-1}] + [[e_0, [e_i, e_0]], e_{j-1}] \\
&= [e_0, [[e_0, e_i], e_{j-1}]] - [[e_0, [e_0, e_i]], e_{j-1}] \\
&= [[e_0, [e_0, e_i]], e_{j-1}] - [[e_0, e_{j-1}], [e_0, e_i]] \\
&- [[[e_0, e_0], e_i], e_{j-1}] + [[e_0, e_i], e_0], e_{j-1}] \\
&= [[[e_0, e_0], e_i], e_{j-1}] - [[[e_0, e_i], e_0], e_{j-1}] \\
&- [[[e_0, e_0], e_i], e_{j-1}] - [[e_{j-1}, e_0], [e_i, e_0], \\
&\quad [e_i, e_0]] + [[[e_0, e_i], e_0], e_{j-1}] = -[e_j, e_{i+1}]
\end{aligned}$$

we have anti-commutativity for basis elements of algebra (V, μ).
Lemma is proved.

Thus, we have that the only possible non-Lie naturally graded filiform Leibniz algebras are the following ones:

$$[e_0, e_0] = e_2, \quad [e_i, e_0] = e_{i+1}, \quad [e_i, e_1] = \beta e_{i+1}, \quad [e_0, e_1] = \beta e_2$$

Let $\beta \neq 1$, then by the following change:

$$\bar{e}_0 = (1-\beta)e_0, \bar{e}_1 = -\beta e_0 + e_1, \; \bar{e}_2 = (1-\beta)^2 e_2, \cdots, \bar{e}_n = (1-\beta)^n e_n,$$

we can suppose that $\beta = 0$.

Now consider the case when $\beta = 1$, i.e.

$$[e_0, e_0] = e_2, \quad [e_i, e_0] = e_{i+1}, \quad [e_i, e_1] = e_{i+1}, \quad [e_0, e_1] = e_2 \quad (1 \leq i \leq n).$$

Making the change $\bar{e}_1 = e_1 - e_0$, we have:

$$[e_0, e_0] = e_2, \quad [e_i, e_0] = e_{i+1} \quad (1 \leq i \leq n).$$

Let us show that the algebras:

$$[e_0, e_0] = e_2, [e_i, e_0] = e_{i+1}, \; (1 \leq i \leq n-1)$$

and

$$[e_0, e_0] = e_2, [e_i, e_0] = e_{i+1}, \quad (2 \leq i \leq n-1)$$

are not isomorphic.

Assume the opposite, and let φ be an isomorphism between these algebras, i.e. $\varphi : L_1 \to L_2$, and let $\varphi(e_i) = \sum\limits_{j=0}^{n} \alpha_{ij} e_j$.

Consider the product

$$[\varphi(e_0), \varphi(e_0)] = \left[\sum_{j=0}^{n} \alpha_{0j} e_j, \alpha_{00} e_0\right] = \alpha_{00}\left(\alpha_{00} e_2 + \alpha_{02} e_3 + \quad \cdots + \alpha_{0,n-1} e_n\right).$$

On the other hand, $\varphi([e_0, e_0]) = \varphi(e_2) = \sum_{j=0}^{n} \alpha_{2j} e_j$. Comparing these equalities we obtain:

$$\alpha_{20} = \alpha_{21} = 0, \quad \alpha_{22} = \alpha_{00}^2, \quad \alpha_{2,k} = \alpha_{00}\alpha_{0,k-1} \quad \text{for} \quad 3 \leq k \leq n \tag{2}$$

Consider the product

$$[\varphi(e_i), \varphi(e_0)] = \left[\sum_{j=0}^{n} \alpha_{ij} e_j, \alpha_{00} e_0\right]$$
$$= \alpha_{00} \sum_{j=0}^{n} \alpha_{ij} [e_j, e_0] = \alpha_{00}(a_{i,0} e_2 + \alpha_{i,2} e_3 + \cdots + \alpha_{i,n-1} e_n).$$

On the other hand, $\varphi([e_i, e_0]) = \varphi(e_{i+1}) = \sum_{j=0}^{n} \alpha_{i+1,j} x_j$ for $1 \leq i \leq n-1$.

Therefore we obtain: $\alpha_{i+1,0} = \alpha_{i+1,1} = 0$,

$$\alpha_{i+1,2} = \alpha_{00}\alpha_{i,0}, \alpha_{i+1,k} = \alpha_{00}\alpha_{i,k-1} \text{ for } 3 \leq k \leq n, 1 \leq i \leq n-1 \tag{3}$$

From (3) we have $\alpha_{22} = \alpha_{00}\alpha_{10}$ and since $\alpha_{00} \neq 0$ (otherwise φ is degenerated) it follows from (2) $\Rightarrow \alpha_{00} = \alpha_{10}$.

Consider $\varphi([e_0, e_1]) = \varphi(0) = 0$.

On the other hand,

$$[\varphi(e_0), \varphi(e_1)] = \left[\sum_{j=0}^{n} \alpha_{0,j} e_j, \alpha_{10} e_0\right]$$
$$= \alpha_{10} \sum_{j=0}^{n} \alpha_{0j} [e_j, e_0] = \alpha_{10}(\alpha_{00} e_0 + \alpha_{02} e_3 + \cdots + \alpha_{0,n-1} e_n) = 0.$$

Hence $\alpha_{10}\alpha_{00} = 0$ and so $\alpha_{10} = 0$, i.e. the first column of the matrix of isomorphism $[\varphi]$ is zero $\Rightarrow \varphi$ is degenerated.

Thus, the following theorem is proved.

Theorem 2. *There exist only two non isomorphic naturally graded complex $n+1$-dimensional non-Lie filiform Leibniz algebras μ_0^n and μ_1^n, namely:*

$$\mu_0^n: \quad \mu_0^n(e_0.e_0) = e_2, \qquad \mu_0^n(e_i, e_0) = e_{i+1}, \ \textit{for } 1 \leq i \leq n-1.$$

$$\mu_1^n: \quad \mu_1^n(e_0.e_0) = e_2, \qquad \mu_1^n(e_i, e_0) = e_{i+1}, \ \textit{for } 2 \leq i \leq n-1.$$

Omitted products are zero.

Remark 2. Complex naturally graded filiform Lie algebras were classified in [8]. Thus, one has a classification of complex naturally graded Leibniz algebras.

Corollary 3. *Any $n+1$- dimensional complex non-Lie filiform Leibniz algebra is isomorphic to one of the following algebras:*
$\mu(e_0.e_0)=e_2, \mu(e_i,e_0)=e_{i+1}, \mu(e_0,e_1)=\alpha_3e_3+\alpha_4e_4+\cdots+\alpha_{n-1}e_{n-1}+\Theta e_n,$
$\mu(e_i,e_1)=\alpha_3e_{i+2}+\alpha_4e_{i+3}+\cdots+\alpha_{n+1-i}e_n$ *for* $1\le i\le n$.

$\mu(e_0,e_0)=e_2, \mu(e_i,e_0)=e_{i+1}, \mu(e_0,e_1)=\beta_3e_3+\beta_4e_4+\cdots+\beta_ne_n, \mu(e_1,e_1)=\gamma e_n,$
$\mu(e_i,e_1)=\beta_3e_{i+2}+\beta_4e_{i+3}+\cdots+\beta_{n+1-i}e_n$ *for* $2\le i\le n$.
Omitted products are zero.

Proof. Direct calculation shows that the above algebras are Leibniz algebras. From Theorem 2 we have that any $n+1$-dimensional complex non-Lie filiform Leibniz algebra μ is isomorphic to one of the following two algebras: $\mu_0^n+\beta$, where $\beta(e_0,e_0)=0, \beta(e_i,e_0)=0$, for $1\le i\le n-1$ and $\beta(e_i,e_j)\in lin(e_{i+j+1},\cdots,e_n)$ for $i\neq 0$ and $\beta(e_0,e_j)\in lin(e_{j+2},\cdots,e_n)$ for $1\le j\le n-2$;
$\mu_1^n+\beta$, where $\beta(e_0,e_0)=0, \beta(e_i,e_0)=0$, for $2\le i\le n-1$ and $\beta(e_i,e_j)\in lin(e_{i+j+1},\cdots,e_n)$ for $i,j\neq 0$ and $\beta(e_0,e_j)\in lin(e_{j+2},\cdots,e_n)$ for $1\le j\le n-2$.

Case 1. $\mu\cong\mu_0^n+\beta$. Then $\mu(e_0,e_0)=\mu_0^n(e_0,e_0)=x_2$ and $\mu(e_i,e_0)=\mu_0^n(e_i,e_0)=e_{i+1}$ for $1\le i\le n-1$, i.e. $e_2,e_3,\cdots,e_n\in Z(\mu)$, i.e. $\mu(e_i,e_j)=0$ for $2\le j\le n$, $0\le i\le n$.

Put $\mu(e_1,e_1)=\alpha_3e_3+\alpha_4e_4+\cdots+\alpha_ne_n$.

Consider $\mu(e_i,\mu(e_0,e_1))=\mu(\mu(e_i,e_0),e_1)-\mu(\mu(e_i,e_1),e_0)$. Since $\mu(e_0,e_1)\in Z(\mu)$, then $\mu(e_i,\mu(e_0,e_1))=0$ and therefore $\mu(\mu(e_i,e_0),e_1)=\mu(\mu(e_i,e_1),e_0)$ for any $i\ge 1$. Thus,

$$\mu(e_i,e_1)=\alpha_3e_{i+2}+\alpha_4e_{i+3}+\cdots+\alpha_{n+1-i}e_n \text{ for } 1\le i\le n.$$

Put $\mu(e_0,e_1)=\Theta_3e_3+\Theta_4e_4+\cdots+\Theta_ne_n$.

Consider $\mu(e_0,\mu(e_1,e_0))=\mu(\mu(e_0,e_1),e_0)-\mu(\mu(e_0,e_0),e_1)$. We have similarly the equality $\mu(\mu(e_0,e_1),e_0)=\mu(\mu(e_0,e_0),e_1)$. Since $\mu(e_0,e_0)=e_2$ and $\mu(e_i,e_0)=e_{i+1}$, it follows that

$$\Theta_3e_4+\Theta_4e_5+\cdots+\Theta_{n-1}e_n=\alpha_3e_4+\alpha_4e_5+\cdots+\alpha_{n-1}e_n.$$

Hence

$$\mu(e_0,e_1)=\alpha_3e_3+\alpha_4e_4+\cdots+\alpha_{n-1}e_{n-1}+\Theta_ne_n.$$

So in Case 1 we obtain the following class:

$$\begin{aligned}&\mu(e_0,e_0)=e_2, \mu(e_i,e_0)=e_{i+1},\\&\mu(e_0,e_1)=\alpha_3e_3+\alpha_4e_4+\cdots+\alpha_{n-1}e_{n-1}+\Theta_ne_n,\\&\mu(e_i,e_1)=\alpha_3e_{i+2}+\alpha_4e_{i+3}+\cdots+\alpha_{n+1-i}e_n \text{ for } 1\le i\le n.\end{aligned}$$

Case 2. $\mu\cong\mu_1^n+\beta$. Then

$$\mu(e_0, e_0) = \mu_1^n(e_0, e_0) = x_2 \text{ and } \mu(e_i, e_0) = \mu_1^n(e_i, e_0) = e_{i+1} \text{ for } 2 \le i \le n-1$$
$$\Rightarrow e_2, e_3, \cdots, e_n \in Z(\mu), \text{ i.e. } \mu(e_i, e_j) = 0 \text{ for } 2 \le j \le n, \quad 0 \le i \le n.$$

Put $\beta(e_1, e_0) = \alpha_3 e_3 + \alpha_4 e_4 + \cdots + \alpha_n e_n$, then by making the change

$$\bar{e}_1 := e_1 - \alpha_3 e_2 - \alpha_4 e_3 - \cdots + \alpha_n e_{n-1}$$

we obtain:

$$\mu(\bar{e}_1, e_0) = \mu_1^n(\bar{e}_1, e_0) = \mu_1^n(-\alpha_3 e_2 - \alpha_4 e_3 - \cdots + \alpha_n e_{n-1}, e_0) + \beta(e_1, e_0) = 0.$$

So we can assume that $\mu(e_1, e_0) = 0$.

Put $\mu(e_0, e_1) = \beta_3 e_3 + \beta_4 e_4 + \cdots + \beta_n e_n$.

Consider

$$\mu(e_1, \mu(e_0, e_1)) = \mu(\mu(e_1, e_0), e_0) - \mu(\mu(e_0, e_0), e_1),$$

since $\mu(e_0, e_1) \in Z(\mu)$ and $\mu(e_1, e_0) = 0$ one has $\mu(\mu(e_1, e_1), e_0) = 0$, but e_0 on the left-hand side annihilate only e_n, then $\mu(e_1, e_1) = \gamma e_n$.

Consider

$$\mu(e_i, \mu(e_0, e_1)) = \mu(\mu(e_i, e_0), e_1) - \mu(\mu(e_i, e_1), e_0) \text{ for } 2 \le i \le n-1.$$

Since $\mu(e_0, e_1) \in Z(\mu)$, we have:

$$\mu(\mu(e_i, e_0), e_1) = \mu(\mu(e_i, e_1), e_0), \quad \text{i.e. } \mu(e_{i+1}, e_1) = \mu(\mu(e_i, e_1), e_0),$$

i.e.

$$\mu(e_i, e_1) = \beta_3 e_{i+2} + \beta_4 e_{i+3} + \cdots + \beta_{n+1-i} e_n \text{ for } 2 \le i \le n-1.$$

So in case 2 we obtain the following class:

$$\begin{aligned} \mu(e_0, e_0) &= e_2, \mu(e_i, e_0) = e_{i+1}, \mu(e_0, e_1) \\ &= \beta_3 e_3 + \beta_4 e_4 + \cdots + \beta_n e_n, \mu(e_1, e_1) = \gamma e_n; \\ \mu(e_i, e_1) &= \beta_3 e_{i+2} + \beta_4 e_{i+3} + \cdots + \beta_{n-1+i} \text{ for } 2 \le i \le n-1. \end{aligned}$$

Corollary is proved.

Remark 2. It is easy to see that the classes of algebras from Corollary 3 are disjoint but the question on isomorphism inside classes remains open.

References

1. Loday, J.-L., Pirashvili, T.: Universal enveloping algebras of Leibniz algebras and (co)homology. *Vath. Ann.* **296** (1993) 139–158.
2. Loday, J.-L.: Une version non commutative des algèbres de Lie: les algèbres de Leibniz. *L'Ens. Math.* **39** (1993) 269–293.

3. Cabezas, J.M., Gomez, J.R., Jimenez-Merchan, A.: Family of p-filiform Lie algebras. In: *Algebra and Operators Theory, Proceedings of the Colloquium in Tashkent, 1997.* Kluwer Academic Publishers, 1998, pp. 93–102.
4. Omirov, B.A.: *Degeneration of Jordan Algebras of Small Dimensions*, Thesis. Nobosibirsk, 1993, pp. 1–21.
5. Shafarevich, I.R.: *Foundations of Algebraic Geometry*, Vol.1, Nauka, Moscow, 1988.
6. Holl, M.: *Combinatorial Analysis*, Mir, Moscow, 1970.
7. Ayupov, Sh.A., Omirov, B.A.: On Leibniz algebras. In: *Algebra and Operators Theory, Proceedings of the Colloquium in Tashkent, 1997.* Kluwer Academic Publishers, 1998, pp. 1–13.
8. Vergne, M.: Cohomologie des algèbres de Lie nilpotentes. Application à l'étude de la varieté des algèbres de Lie nilpotentes. *Bull. Soc. Math. France* **98** (1970) 81–116.

Application of Computer Algebra in Problems on Stabilization of Gyroscopic Systems

Banshchikov Andrey and Bourlakova Larissa[1]

Institute of Systems Dynamics and Control Theory SB RAS, 134, Lermontov str., Irkutsk, 664033, Russia, E-mail : irteg@icc.ru

Abstract. The paper discusses some problems of stabilization of potential systems at the expense of gyroscopic forces. The algorithms described are employed in the software "STABILITY" [1]. This software is based on the computer algebra system "Mathematica" [2] and is designed to investigate stability and stabilization of mechanical systems in symbolic and numeric-symbolic forms.

1 Problem Statement. Lyapunov Matrix Equation

The problem of stability for the trivial solution of the differential equation is discussed :

$$M\ddot{x} + G\dot{x} + Kx = Q(x,\dot{x}) \,, \qquad (1)$$

where $M = M^T > 0$, $G = -G^T$, $K = K^T$ are, respectively, $(n \times n)$ matrices of gyroscopic, potential forces; x, $\dot{x}$ are column matrices of coordinates and velocities; $Q(x,\dot{x})$ is the column matrix of nonlinear forces, $Q(0,0) = 0$.

System (1) is critical in the Lyapunov sense. Especially topical for such systems is the problem of stabilization of potential systems with the aid of gyroscopic forces when the degree of instability is even, i.e. when $\det K > 0$ [3], [4]. This problem is quite important in the aspect of practical applications, still it has not yet been solved completely [5]. Substantial attention has been given to the problem of stability and stabilization of gyroscopic systems in the monograph [6]. A brief survey of the results obtained for this problem by the method of Lyapunov functions can be found in [7].

For the function V :

$$V = \dot{x}^T N\dot{x} + x^T Lx + x^T B^T\dot{x} + \dot{x}^T Bx + F(x,\dot{x}) \,, \quad L = L^T \,, \quad N = N^T \qquad (2)$$

calculate the derivative due to eq. (1) and require that this derivative be represented in the form :

$$\dot{V} = \dot{x}^T W_1\dot{x} + x^T W_2 x + \dot{x}^T W_3 x + x^T W_3^T \dot{x} + \Phi(x,\dot{x}) \,, \quad W_i = W_i^T \quad (i = 1, 2),$$

where $F(x,\dot{x})$, $\Phi(x,\dot{x})$ are nonlinear functions of order larger than two. Hence, we have the system of equations:

$$\begin{cases} B+B^T+GM^{-1}N-NM^{-1}G=W_1\ ; \quad W_2+KM^{-1}B+B^TM^{-1}K=0\ ; \\ KM^{-1}N - NM^{-1}K + B^TM^{-1}G + GM^{-1}B = W_3-W_3^T\ ; \\ 2L = KM^{-1}N + NM^{-1}K + B^TM^{-1}G - GM^{-1}B + W_3 + W_3^T\ ; \end{cases} \qquad (3)$$

$$Q^TM^{-1}(N\dot{x}+Bx) + (\dot{x}^TN+x^TB^T)M^{-1}Q + \frac{\partial F}{\partial x}\dot{x} - \frac{\partial F}{\partial \dot{x}}M^{-1}(G\dot{x}+Kx-Q) = \Phi(x,\dot{x})\ .$$

The system (3) is an extended version of Lyapunov's matrix equation. In the standard situation, such Lyapunov's equation is used for obtaining the Lyapunov function for investigation of asymptotic stability. In our case, for linear systems, only stability is possible. And the most effective Lyapunov function is the sign-definite first integral of system (1). Note that in the general case it is impossible to represent the general solution of (3) in matrix form [8].

Consider some versions of solutions for the system (3).

Possibility 1. Let $N=0$. If $B=-B^T$, then necessarily $W_1=0$. If $B=B^T$, then $W_1=2B$ and for given W_1 there exist W_2, W_3.
Let matrix B have some arbitrary structure. Then $B=1/2\,W_1+\Gamma$, where $\Gamma=-\Gamma^T$ is an arbitrary skew-symmetric matrix, and for the given matrix W_1 there exist $W_3-W_3^T = (1/2\,W_1-\Gamma)M^{-1}G+GM^{-1}(1/2\,W_1+\Gamma)$, $W_2 = -KM^{-1}(1/2\,W_1+\Gamma) - (1/2\,W_1-\Gamma)M^{-1}K$.

We set $W_1=2M$. Then for the system (3) we have the solution : $N=0$, $B=M$, $W_3=G+\epsilon G^*$, $L=\epsilon G^*$. Consequently, for eq. (1) it is possible to write the relation :

$$\frac{d}{dt}(x^TM\dot{x}+\epsilon x^TG^*x) = (\dot{x}+M^{-1}Px)^TM(\dot{x}+M^{-1}Px) - x^TA_1x + x^TQ\ ,$$
$$(\,2P=\epsilon G^*+\epsilon G^{*T}+G\ ,\quad A_1=K+P^TM^{-1}P\,)\ ,$$

where ϵ is an arbitrary number (>0 or ≤ 0) ; G^* is a matrix such that $G=G^*-G^{*T}$, whereas all elements in G^*, which lie below the diagonal, are zero.

Having applied Lyapunov's theorem on instability to these relations it is possible to formulate

Theorem 1. *If* $A_1 = K+1/4(\epsilon G^{*T}+\epsilon G^*+G)^TM^{-1}(\epsilon G^{*T}+\epsilon G^*+G) < 0$, *(where* ϵ *is an arbitrary number), then the solution* $\dot{x}=x=0$ *of system (1) is unstable for any nonlinear forces.*

If $\epsilon = 0$ then the following known result [9] follows from Theorem 1 : the trivial solution of system (1) is unstable for $4K - GM^{-1}G < 0$.

If $4K - GM^{-1}G = 0$ then for the linear eq. (1) we can write the following chain of relations :

$$\frac{d}{dt}(x^T Mx)/2 = x^T M\dot{x} = \tilde{V}_1 ,$$

$$\dot{\tilde{V}}_1 = \frac{d}{dt}(x^T M\dot{x}) = \dot{x}^T M\dot{x} - x^T G\dot{x} - 1/4x^T GM^{-1}Gx = \tilde{V}_2 , \qquad \dot{\tilde{V}}_2 = 0 .$$

The system assumes the first integral $\tilde{V}_2 = c_1$, and from the above sequence of relations it follows that $x^T Mx = ct^2 + c_2 t + c_3$, i.e. there takes place instability with respect to $\dot{x}$, x . Hence, the following theorem holds.

Theorem 2. *If* $4K - GM^{-1}G \leq 0$ *then the linear system (1) is unstable.*

From Theorem 1 and Theorem 2 we can deduce impossibility conditions of gyroscopic stabilization for an unstable potential system. In particular, if a linear potential system has odd degree of instability ($\det K < 0$) then gyroscopic stabilization is impossible [3].

Possibility 2. Let $N \neq 0$. If $B = 0$ then $W_2 = 0$. If $B = -B^T$ then for the first equation in (3) for the given W_1 the solution for N always exists. The system corresponding to the third equation in (3) can be incompatible. If $B = B^T$ then from given matrices W_1 , W_2 we can determine N , B , W_3 . If B is a matrix of arbitrary structure then from the given matrix W_2 we always can find a (non-unique) solution for the matrix B. Let us substitute this solution into the first equation in (3) and find a solution for N. The remaining equations of (3) can be used for choosing W_3 .
If we choose $W_3 = KM^{-1}N + B^T M^{-1}G$, we have $L = W_3 + W_3^T$.

2 First Integrals and Theorems on Stability

Let $W_i = 0$ $(i = 1,2,3)$. Then from (3) we obtain conditions of existence for the first integral of the linear system (1). For the nonlinear system (1) we can add to (3) the following equations:

$$\dot{x}^T\left(\frac{\partial F}{\partial x} + 2NM^{-1}Q + GM^{-1}\frac{\partial F}{\partial \dot{x}}\right) \equiv 0 ;$$

$$x^T(2B^T M^{-1}Q - KM^{-1}\frac{\partial F}{\partial \dot{x}}) \equiv 0 ; \qquad Q^T M^{-1}\frac{\partial F}{\partial \dot{x}} \equiv 0 . \tag{4}$$

These equations have been obtained as sufficient conditions so that $\Phi(x,\dot{x}) \equiv 0$. The system (3) ($W_i = 0$) always has a trivial solution $N = 0$, $B = 0$. As any uniform linear system the system (3) ($W_i = 0$) can have (i) only trivial solution or (ii) a set of solutions. If there exists nontrivial solution N^* , B^* then
(i) $N_1^* = N^* \pm N^{*T}$, $B_1^* = B^* \pm B^{*T}$,
(ii) $N^{**} = \mathcal{P}N^*\mathcal{Q}^T + \mathcal{Q}N^*\mathcal{P}^T$, $B^{**} = \mathcal{P}^T B^*\mathcal{Q} + \mathcal{Q}^T B^*\mathcal{P}$ (where $\mathcal{P}$ and $\mathcal{Q}$ are matrices which commute with $M^{-1}G$, $M^{-1}K$) are solutions too. From eqs. (3) ($W_i = 0$) it follows that the following statement holds.

Statement. *The linear system (1) has* $n_1 \geq n$ *independent quadratic first integrals.*

Eqs.(4) are used to construct $F(x, \dot{x})$ (if it exists) or to find such Q that the function (2) is a first integral.

Let us write some solutions of (3) and the first integrals in matrix form since the solution of specific problems may be complicated by the cumbersome character of computations.

It is known that for the system (3) ($W_i = 0$) there exist the solutions :

a) $N = M$, $B = 0$, $L = K$, and hence, the linear system (1) has the energy integral

$$H = V_1 = \dot{x}^T M \dot{x} + x^T K x = const\ ; \tag{5}$$

For the nonlinear system (1) the energy integral $H_1 = H - U(x) = \text{const}$ exists if nonlinear forces are gyroscopic and/or potential, that is $Q_p = -\partial U(x)/\partial x$. This follows from eqs. (4).

b) $N = K - GM^{-1}G$, $B = G^T M^{-1} K$, $L = KM^{-1}K + GM^{-1}G$; the integral of the linear system

$$V_2 = \dot{x}^T K \dot{x} + (Kx + G\dot{x})^T M^{-1}(Kx + G\dot{x}) = \text{const} \tag{6}$$

corresponds to this solution.

c) $N = MK^{-1}M$, $B = MK^{-1}G$; the first integral of the linear system is :

$$V_3 = \dot{x}^T MK^{-1}M\dot{x} + x^T(M - GK^{-1}G)x + 2\dot{x}TMK^{-1}Gx = \text{const}.$$

Consider some other solutions for the system (3) ($W_i = 0$).

2.1. $B = 0$, $N = N^*$, where N^* satisfies the equations

$$GM^{-1}N - NM^{-1}G = 0\,, \qquad KM^{-1}N - NM^{-1}K = 0\,.$$

Let one of the following restriction be imposed on the matrices of (1)

$$GM^{-1}K = KM^{-1}G\ ; \tag{7}$$
$$GM^{-1}K = -KM^{-1}G\ ; \tag{8}$$
$$K(M^{-1}G)^2 = (GM^{-1})^2 K\ , \tag{9}$$

and the conditions $\det G = 0, \det K = 0$ fail to be satisfied simultaneously.

2.1.1. $N^* = (GM^{-1})^{p-1}GM^{-1}G(M^{-1}G)^{p-1}$ $(p = 0, 1, \dots)$; the solution takes place in case where any of the conditions (7)-(9) is satisfied. The following first integral of the linear system corresponds to this solution:

$$V_4 = \dot{x}^T(GM^{-1})^{p-1}GM^{-1}G(M^{-1}G)^{p-1}\dot{x} + x^T K(M^{-1}G)^{2p}x = \text{const};$$

2.1.2. $N^* = (GM^{-1})^p KM^{-1}K(M^{-1}G)^p$; the solution takes place in the case where any of the conditions (7) or (8) is satisfied. The following first integral of the linear system corresponds to this solution:

$$V_5 = \dot{x}^T(GM^{-1})^p KM^{-1}K(M^{-1}G)^p \dot{x}$$
$$+ x^T(GM^{-1})^p KM^{-1}KM^{-1}K(M^{-1}G)^p x = \text{const} \quad (\text{ when } (7)\ ) ;$$
$$V_5 = \dot{x}^T(GM^{-1})^p KM^{-1}K(M^{-1}G)^p \dot{x}$$
$$+ (-1)^p\, x^T(GM^{-1})^p KM^{-1}KM^{-1}K(M^{-1}G)^p x = \text{const} \quad (\text{ when } (8)\ ) ;$$

2.1.3. $N^* = (GM^{-1})^p(KM^{-1})^m K(M^{-1}K)^m(M^{-1}G)^p$ $(p, m = 0, 1, \dots)$; the solution exists if $(GM^{-1})^{2p+1}K(M^{-1}K)^{2m} = (KM^{-1})^{2m}K(M^{-1}G)^{2p+1}$ and if the condition (9) holds for an even p, and (7) or (8) for an odd p. The following first integral of the linear system corresponds to this solution:

$$V_6 = \dot{x}^T(GM^{-1})^p(KM^{-1})^m K(M^{-1}K)^m(M^{-1}G)^p \dot{x}$$
$$+ x^T(GM^{-1})^p(KM^{-1})^m KM^{-1}K(M^{-1}K)^m(M^{-1}G)^p x = \text{const.}$$

2.2. Let $N = 0$. Then the system (3) has solutions $B = -B^T$:

2.2.1. $B = (GM^{-1})^p G(M^{-1}G)^p$ $(p = 0, 1, \dots)$; the solution takes place in the case where condition (7) is satisfied. The following first integral of the linear system corresponds to this solution:

$$V_7 = 2\dot{x}^T(GM^{-1})^p G(M^{-1}G)^p x + x^T G(M^{-1}G)^{2p+1} x = \text{const}; \quad (10)$$

2.2.2. $B = (KM^{-1})^m G(M^{-1}K)^m$ $(m = 0, 1, \dots)$; the solution takes place in case where the condition (7) is satisfied. The following first integral of the linear system corresponds to this solution:

$$V_8 = 2\dot{x}^T(KM^{-1})^m G(M^{-1}K)^m x$$
$$+ x^T GM^{-1}(KM^{-1})^m G(M^{-1}K)^m x = \text{const};$$

2.2.3. $B = GM^{-1}K$; the solution takes place in case where the condition (7) is satisfied. The following first integral of the linear system corresponds to this solution:

$$V_9 = 2\dot{x}^T GM^{-1}Kx - x^T GM^{-1}KM^{-1}Gx = \text{const};$$

2.2.4. $B = GK^{-1}M$ $(\det K \neq 0\ )$; the solution takes place in case where the condition (7) is satisfied. The following first integral of the linear system corresponds to this solution:

$$V_{10} = 2\dot{x}^T MK^{-1}Gx - x^T GK^{-1}Gx = \text{const};$$

2.2.5. $B = MK^{-1}GM^{-1}GM^{-1}G$; the solution takes place in case where the conditions $\det K \neq 0$, $\det G \neq 0$, and either (i) (7) or (ii) conditions (9) and $(GM^{-1})^3 K = K(M^{-1}G)^3$ hold. The following first integral of the linear system

$$V_{11} = 2\dot{x}^T MK^{-1}GM^{-1}GM^{-1}Gx - x^T GK^{-1}GM^{-1}GM^{-1}Gx = \text{const};$$

corresponds to this solution.

2.2.6. $B = MG^{-1}K$; the solution takes place in case where the conditions (7), $\det G \neq 0$ are satisfied. The following first integral of the linear system corresponds to this solution:

$$V_{12} = 2\dot{x}^T KG^{-1}Mx - x^T Kx = \text{const.}$$

Solutions for the group (2.2) are of interest since they suggest integrals of the linear system, which hold also for the nonlinear system (1) if nonlinear forces $Q(x)$ satisfy the second equation in (4): $xB^T M^{-1}Q = 0$. For the integrals (10) and for V_8, V_9 , the expression $Q = f'(x^T Mx)Mx$ (where $f'(x^T Mx)$ is a scalar function) satisfies the equation indicated above; while for the integrals V_7 (for $p = 0$) and V_{10}, V_{11}, V_{12} the expression $Q = f'(xKx)Kx$ (where $f'(x^T Kx)$ is a scalar function) satisfies the same equation.

2.3. If $\det G \neq 0$ and $KM^{-1}GM^{-1}KG^{-1}M = MG^{-1}KM^{-1}GM^{-1}K$, $(GM^{-1})^2K = K(M^{-1}G)^2$ then the linear system (1) assumes the first integral:

$$V_{13} = \dot{x}^T G^T M^{-1}KG^{-1}M\dot{x} + 2x^T KM^{-1}G\dot{x}$$
$$+ x^T (KM^{-1}GM^{-1}G - KM^{-1}GM^{-1}KG^{-1}M)x = \text{const.}$$

For $M = E$ (E is the identity matrix) this integral coincides with the one indicated in [7].

The first integrals (or their bundles) obtained above may be used for constructing Lyapunov functions in the process of proving stability of the trivial solutions for (1) when Sylvester inequalities for sign-definiteness of the quadratic part of the integral (2) with respect to all the variables hold.

From the Lyapunov second method theorems it follows : if $K > 0$ then the trivial solution of the linear (as well as for potential and/or gyroscopic nonlinear forces) system (1) is stable with respect to the variables $\dot{x}$, x [6]. In this case, the energy integral (5) has been chosen in the capacity of the Lyapunov function. When the matrix K is not positive definite we construct a bundle of integrals. The conditions of sign-definiteness for this bundle are sufficient stability conditions for the considered critical system as well as conditions of gyroscopic stabilization.

Consider the Lyapunov function

$$V = V_2 - \alpha H = \dot{x}^T(K - GM^{-1}G - \alpha M)\dot{x} - \dot{x}^T GM^{-1}Kx$$
$$+ x^T KM^{-1}G\dot{x} + x^T(KM^{-1}K - \alpha K)x .$$

The Sylvester conditions of positive definiteness for this form are

$$KM^{-1}K - \alpha K > 0 , \quad K - \alpha M - GM^{-1}G + GM^{-1}K(K - \alpha M)^{-1}G > 0 .$$

Well-known formulas for the block matrix have been used above. The inequalities obtained contain the indeterminate multiplier α . If α is assigned arbitrarily then the stability conditions can be "rough". Note (due to [4]) that the conditions of existence of α needed to satisfy the above inequalities are the sufficient conditions of stability for the trivial solution of the linear system (1). Hence, Theorem 3 holds.

Theorem 3. *If there exists an* α *such that* $KM^{-1}(K-\alpha M) > 0$, $(K-\alpha M) - GM^{-1}G + GM^{-1}K(K-\alpha M)^{-1}G > 0$ *then the linear system (1) is stable with respect to* x, $\dot{x}$.

As it will be obvious from Example 3.1 (below), Theorem 3 suggests stability conditions close to necessary ones.

If the bundle of integrals (5) and (10) (for $p = 0$) itself is used in Lyapunov's stability theorem and next Theorem 2 is applied then it is possible to prove that Theorem 4 holds.

Theorem 4. *If* $GM^{-1}K - KM^{-1}G = 0$ *then the condition* $4K - GM^{-1}G > 0$ *is sufficient for the stability of the linear system (1) with respect to the variables* x, $\dot{x}$. *This condition is also necessary [10].*

Conditions of Theorem 4 are sufficient for the stability of the trivial solution of nonlinear system (1) with the restriction (7) when $Q = f'(y)Mx$, where $y = x^TMx$, $f'(y) = \partial f/\partial y$ or $Q = f'(z)Kx$, where $z = x^TKx$, $f'(z) = \partial f/\partial z$.

If the first integral obtained is not sign-definite then it is possible to consider the combination $V = \sum V_i^2$ $(i = 1, 2, \ldots, p)$ as the Lyapunov function. Hence it is necessary to verify the compatibility of the equations $V_i = 0$. The conditions, for which these equations are incompatible, are sufficient for the stability of the trivial solution of (1).

3 Examples

3.1 Stabilization of the System with Two Degrees of Freedom

Theorem 1 in [7] suggests simple but very "rough" stability conditions. For proving Theorem 1 in [7] the bundle of integrals (5) and (6) with an assigned coefficient has been employed. But the bundle of integrals (5) and (6) allows to obtain conditions more close to the necessary ones. Let us demonstrate this by an example from [7]:

$$\begin{aligned} \ddot{x}_1 - 4\dot{x}_2 + (k_1 - 7)x_1 &= 0, \\ \ddot{x}_2 + 4\dot{x}_1 + (k_2 - 7)x_2 &= 0, \quad k_i < 7. \end{aligned} \tag{11}$$

Construct the bundle

$$\begin{aligned} V = V_2 - \alpha H = (9 - \alpha + k_1)\dot{x}_1^2 + (9 - \alpha + k_2)\dot{x}_2^2 + (7 - k_1)(7 + \alpha - k_1)x_1^2 \\ + (7 - k_2)(7 + \alpha - k_2)x_2^2 + 8(k_2 - 7)x_2\dot{x}_1 - 8(k_1 - 7)x_1\dot{x}_2, \end{aligned} \tag{12}$$

where α is an indeterminate constant coefficient.

Signdefiniteness conditions for (12) may be considered as polynomials with respect to α :

$$f_1(\alpha) = (7+\alpha-k_1) > 0 , \quad f_2(\alpha) = (7+\alpha-k_2) > 0 ,$$
$$f_3(\alpha) = \alpha^2 - (2+k_1+k_2)\alpha + k_1k_2 - 7(k_1+k_2) + 49 < 0 . \qquad (13)$$

Investigate the possibility of choosing α for the satisfaction of the conditions (13). For the first two linear polynomials for $\alpha > \alpha^* = \max\{(k_1-7),(k_2-7)\}$ we have $f_1 > 0$, $f_2 > 0$. The real value of $\alpha > 0$, for which $f_3 < 0$, exists when the conditions

$$2+k_1+k_2 > 0 , \quad f_4(k_1,k_2) = (2+k_1+k_2)^2 - 4(k_1-7)(k_2-7) > 0 \quad (14)$$

hold. These conditions are sufficient for stability since they ensure the existence of the coefficient α for which (13) holds. Conditions (14) coincide with the necessary stability conditions (without the boundary $f_4(k_1,k_2) = 0$) obtained as the conditions that the roots of the characteristic equation of (11) are purely imaginary. Now let us estimate the domain of parameters k_1, k_2 satisfying (14). From Theorem 2 it follows that when (i) $k_1 = k_2 = 3$ or (ii) both $k_i < 3$ ($i = 1,2$), the system (11) is unstable. Consequently, such points (k_1, k_2) do not occur in the stability domain. Theorem 1 in [7] suggests the following values of the parameters under which the system is stable : $7 > k_i > 3$, ($i = 1,2$). If one selects (similarly to [7]) $\alpha = 7 - k_1$ then from (13) it follows that $14 > k_1 + k_2 > 6$. Let $k_1 + k_2 = 6$. Having substituted $k_2 = 6 - k_1$ in (13), we have:

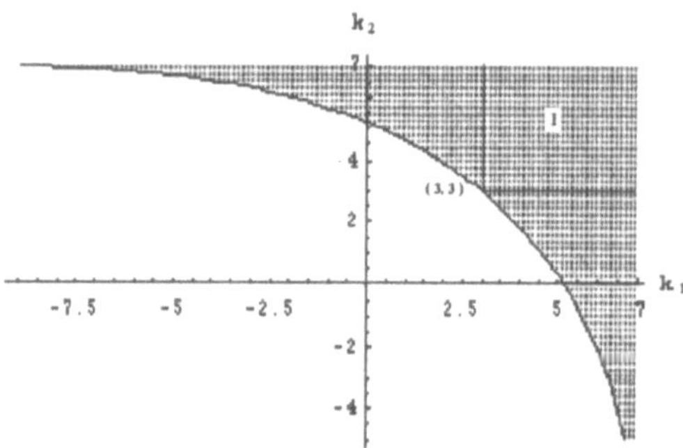

Fig. 1. Shaded area is domain of gyroscopic stabilization.

$f_1 = 7+\alpha-k_1 > 0$, $f_2 = 1+\alpha+k_2 > 0$, $f_3 = 7-8\alpha+\alpha^2+6k_1-k_1^2 < 0$. The roots of the polynomial $f_3(\alpha) = 0$ are $\alpha_{1,2} = 4 \pm (3-k_1)$. Now, choose $\alpha = 4$. Hence $f_3 = -(3-k_1)^2 < 0$ for any $k_1 < 7$, $k_2 < 7$, $k_1 + k_2 \geq 6$. Fig.1 shows the domain of the parameters k_1 , k_2 from conditions (14). The domain has been constructed by the software **Region** intended for graphic solving of systems of 2-parameter algebraic inequalities [11]. Shaded area (without boundaries) corresponds to the domain of gyroscopic stabilization. The index "I" denotes the result obtained in [7].

3.2 On Satellite Stabilization

Consider parametric analysis of gyroscopic stabilization conditions for the equilibrium position of a satellite with a gravitational stabilizer on the circular orbit.

The linear equations of motion are subdivided into two subsystems [12]: with respect to the pitch angle (θ)

$$M_1\,\ddot{x}_1 + K_1\,x_1 = 0\,, \tag{15}$$

$$K_1 = 3\omega^2\begin{pmatrix} b-a & f \\ f & f \end{pmatrix}; \quad M_1 = \begin{pmatrix} c & f \\ f & d \end{pmatrix} > 0\,; \quad x_1 = \begin{pmatrix} \theta \\ \alpha \end{pmatrix}$$

and with respect to the yaw and bank angles (ψ , φ)

$$M_2\,\ddot{x}_2 + G\,\dot{x}_2 + K_2\,x_2 = 0\,, \tag{16}$$

$$K_2 = \omega^2\begin{pmatrix} c-b & 0 & 0 \\ 0 & 4(c-a) & 4f \\ 0 & 4f & 3f+d \end{pmatrix}; \quad M_2 = \begin{pmatrix} a & 0 & 0 \\ 0 & b & f \\ 0 & f & d \end{pmatrix} > 0\,;$$

$$G = \omega\begin{pmatrix} 0 & \mathrm{g} & 0 \\ -\mathrm{g} & 0 & 0 \\ 0 & 0 & 0 \end{pmatrix}; \quad x_2 = \begin{pmatrix} \psi \\ \varphi \\ \beta \end{pmatrix},$$

where ω is an absolute value of orbital angular velocity ;

$$a = J_y\,; \quad b = J_x + m\,r\,(r+l) + \frac{1}{3}m\,l^2 + m_0\,(r+l)^2\,; \quad c = b + J_z - J_x\,;$$

$$d = (\frac{1}{3}m + m_0)\,l^2\,; \quad f = (\frac{1}{2}m + m_0)\,r\,l\ +\ d\,; \quad \mathrm{g} = J_z - J_x - J_y\,;$$

m , m_0 are the masses of the rod and the weight on its end, respectively; $l > 0$ is the rod length; $r > 0$ is the distance between the mass center of the system and attachment point of the rod; J_x , J_y , J_z are principal moments of inertia of the satellite ; α, β are the rotation angles of the rod with respect to the satellite body.

The research has been carried out on IBM PC with the aid of the computer algebra system "Mathematica" in symbolic-numeric form.

The system of eqs. (15)-(16) is critical in Lyapunov sense. From equations (3) the first integrals of differential equations of motion have been obtained. For the subsystem (15) there exists the energy integral (5). The condition of definite positiveness for this integral writes $b - a - f > 0$. If it holds then the subsystem (15) is characterized by secular stability. The subsystem (16) is under the influence of potential and gyroscopic forces. For (16) the general solution of eqs. (3) depends on 3 free parameters n_{11} , b_{21} , b_{31} (they are elements of the matrices N , B). This integral writes :

$$V = \mathrm{n}_{11}H + \mathrm{b}_{21}V_{13} + \mathrm{b}_{31}V_{14} - x^T S\,x\,, \quad \text{where} \quad x^T = (\,\dot{\psi},\ \dot{\varphi},\ \dot{\beta},\ \psi,\ \varphi,\ \beta\,).$$

The non-zero coefficients of the quadratic form V are

$$S_{11} = \mathrm{n}_{11} \; ; \qquad S_{24} = 2\,\mathrm{b}_{21} \; ; \qquad S_{34} = 2\,\mathrm{b}_{31} \; ;$$

$$S_{22} = \mathrm{n}_{11}\,\frac{b}{a} + \frac{\mathrm{b}_{31}\,f\,h}{(f-d)\,\omega}\big(\,b\,(d-f)\,\mathrm{g} + 4\,a\,(f^2 - b\,d)\,\big)$$
$$+ \frac{\mathrm{b}_{21}\,h}{\mathrm{g}\,\omega}\,\big(\,a\,(c-a)\,(3\,f^2 - 4\,b\,d) - f^2\,\mathrm{g}\,(2\,b - c) + (c-b)\,b^2\,d\,\big)\;;$$

$$S_{23} = \mathrm{n}_{11}\,\frac{2\,f}{a} + \frac{2\,\mathrm{b}_{21}\,f\,h}{\mathrm{g}\,\omega}\,\big(\,\mathrm{g}\,(d\,(c-a) - f^2\,) + 3\,a\,(f^2 - b\,d)\,\big)$$
$$+ \frac{2\,\mathrm{b}_{31}\,h}{(f-d)\,\omega}\,\big(\,\mathrm{g}\,b\,d\,(d-f) + 4\,a\,f\,(f^2 - b\,d)\,\big)\;;$$

$$S_{33} = \mathrm{n}_{11}\,\frac{d}{a} + \frac{\mathrm{b}_{21}\,h}{\mathrm{g}\,\omega}\,\big(\,3\,a\,f\,(f^2 - b\,d) - d\,\mathrm{g}\,(\,f^2 - d\,(c-a))\,\big)$$
$$+ \frac{\mathrm{b}_{31}\,h}{\mathrm{g}\,(d-f)\,f\,\omega}\,\big(\,3\,a\,f\,(\,f^3\,(b+d-f) + b\,d\,(\,d\,(c-a) - f\,(b+d)\,)\,)$$
$$+ \mathrm{g}\,(\,(f-d)\,(\,f^2\,(f^2 - b\,d) + b\,d\,(c\,d - f^2)\,) - 4\,f^3\,d\,a + b\,d^3\,a\,)\,\big)\;;$$

$$S_{44} = \mathrm{n}_{11}\,\frac{(c-b)\,\omega^2}{a} + \frac{(\,\mathrm{b}_{21}\,d - \mathrm{b}_{31}\,f\,)\,\mathrm{g}\,\omega}{f^2 - b\,d}\;;$$

$$S_{15} = -\,8\,a\,h\,(\,\mathrm{b}_{21}\,(f^2 + a\,d - c\,d) + \mathrm{b}_{31}\,\mathrm{g}\,f\,)\;;$$

$$S_{55} = \mathrm{n}_{11}\,\frac{4\,(c-a)\,\omega^2}{a} + \frac{4\,\mathrm{b}_{21}\,h\,\omega}{\mathrm{g}}\,\big(\,(\,(c - 8\,a)\,(c\,d - f^2\,)\,\mathrm{g}$$
$$+ 3\,a\,(\,b\,(f^2 - c\,d) + a\,d\,(c-a)\,) + 2\,a\,d\,(c\,(c-b) - a\,(b+a)\,)\,\big)$$
$$+ \frac{4\,\mathrm{b}_{31}\,h\,f\,\omega}{(f-d)}\,\big(\,(5\,a - c)\,(f-d)\,\mathrm{g} + 4\,a\,(f^2 - b\,d)\,\big)\;;$$

$$S_{16} = 2\,a\,h\,(\,3\,\mathrm{b}_{21}\,f\,(d-f) + \mathrm{b}_{31}\,(3\,b\,f - 4\,f^2 + b\,d)\,)\;;$$

$$S_{56} = \mathrm{n}_{11}\,\frac{8\,f\,\omega^2}{a} + \frac{8\,\mathrm{b}_{21}\,h\,f\,\omega}{\mathrm{g}}\,\big(\,\mathrm{g}\,(c\,d - f^2 + a\,(3\,f - 4\,d)) + 3\,a\,(f^2 - b\,d)\,\big)$$
$$+ \frac{8\,\mathrm{b}_{31}\,h\,f\,\omega}{(f-d)}\,\big(\,\mathrm{g}\,f\,(d-f) - a\,(\,b\,(3\,f + d) - f\,(7\,f - 3\,d)\,)\,\big)\;;$$

$$S_{66} = n_{11}\,\frac{(3f+d)\,\omega^2}{a} + \frac{b_{21}\,h\,\omega}{g}\,\big(\,3\,(8a+b-c)\,f^3 + 3\,(c\,g + a\,(a-c))\,f\,d$$
$$+\,(c-a)\,g\,d^2 - f^2\,(g\,d + 9\,a\,(b+d))\,\big)$$
$$+\,\frac{b_{31}\,h\,\omega}{(f-d)\,f\,g}\,\big(\,4\,(4a+b-c)\,f^5 + f^3\,(2\,d\,g\,(b+2\,a) + 9\,a\,b\,(b+2\,d))$$
$$+\,f^4\,(4\,d\,g + 12\,a\,(c-d-a) + 3\,b\,(c-b-12\,a)) + b\,(c-a)\,g\,d^3$$
$$+\,b\,f^2\,d\,(3\,a\,(d-2\,b) - g\,(5\,d + 9\,a + 3\,c)) + b\,f\,d^2\,(2\,g\,(c-3\,a) - 3\,a\,b)\,\big)\,,$$

where $$h = \frac{1}{(b-c)\,(b\,d - f^2)}\,.$$

The structure of the matrix of quadratic form V is

$$S = \begin{pmatrix} S_{11} & 0 & 0 & 0 & S_{15} & S_{16} \\ 0 & S_{22} & S_{23} & S_{24} & 0 & 0 \\ 0 & S_{23} & S_{33} & S_{34} & 0 & 0 \\ 0 & S_{24} & S_{34} & S_{44} & 0 & 0 \\ S_{15} & 0 & 0 & 0 & S_{55} & S_{56} \\ S_{16} & 0 & 0 & 0 & S_{56} & S_{66} \end{pmatrix}.$$

Sylvester's conditions of definite positiveness for the matrix S are :

$$\begin{cases} S_{11} > 0\,, \quad S_{22} > 0\,, \quad S_{22}\,S_{33} - S_{23}^2 > 0\,, \quad S_{11}\,S_{55} - S_{15}^2 > 0\,, \\ 2\,S_{23}\,S_{24}\,S_{34} - S_{24}^2\,S_{33} - S_{22}\,S_{34}^2 - S_{23}^2\,S_{44} + S_{22}\,S_{33}\,S_{44} > 0\,, \\ 2\,S_{15}\,S_{16}\,S_{56} - S_{16}^2\,S_{55} - S_{11}\,S_{56}^2 - S_{15}^2\,S_{66} + S_{11}\,S_{55}\,S_{66} > 0\,. \end{cases} \tag{17}$$

If the 5th and 6th conditions (17) are written in explicit form via the mechanical system's parameters then one can readily see that they coincide with the accuracy up to positive factors. So, further only one of them will be used.

Suppose that the potential subsystem (16) is unstable but has an even degree of instability when the following conditions are satisfied:

$$b > c > a\,, \qquad (c - a)\,(3f + d) - 4f^2 < 0\,. \tag{18}$$

Then due to Thomson-Tait's theorem [3] gyroscopic stabilization of the system is possible. Consider this stabilization problem. Let the parameters of the system be

$l = 12$ m ; $r = 2.5$ m ; $m_0 = 1.531$ kg ; $m = 0.0143$ kg ;
$\omega = 0.0011$ rad/s. ; $J_y = 75\ \mathrm{kg\,m^2}$; $J_x = 75\ \mathrm{kg\,m^2}$; $J_z = 28\ \mathrm{kg\,m^2}$.
For these numerical values conditions (18) are satisfied.

The numeric-symbolic Sylvester's conditions (17) have the form :

condition[[1]] : $n_{11} > 0$;
condition[[2]] : $1.\, b_{21} - 3.11379\, b_{31} + 0.000894071\, n_{11} > 0$;
condition[[3]] : $1.\, b_{21}{}^2 + 0.514611\, b_{21}\, b_{31} - 13.4219\, b_{31}{}^2 + 0.008646\, b_{21}\, n_{11} - 0.0267956\, b_{31}\, n_{11} + 2.83076 * 10^{-6}\, n_{11}{}^2 > 0$;
condition[[4]] : $-1.\, b_{21}{}^2 + 6.27861\, b_{21}\, b_{31} - 9.85523\, b_{31}{}^2 + 1.1155 * 10^{-6}\, n_{11}{}^2 - 0.0104645\, b_{31}\, n_{11} + 0.00307247\, b_{21}\, n_{11} > 0$;
condition[[5]] : $-1.\, b_{21}{}^3 + 6.93492\, b_{21}{}^2\, b_{31} + 9.75785\, b_{31}{}^3 + 0.01315\, b_{31}{}^2\, n_{11} + 0.00247856\, b_{21}{}^2\, n_{11} - 0.0125568\, b_{21}\, b_{31}\, n_{11} - 14.8825\, b_{21}\, b_{31}{}^2 + 2.96312 * 10^{-6}\, b_{31}\, n_{11}{}^2 - 3.11525 * 10^{-7}\, b_{21}\, n_{11}{}^2 - 4.47952 * 10^{-10}\, n_{11}{}^3 > 0$.

Present the graphic solution of this system of algebraic inequalities. For a fixed value of the parameter $n_{11} = 1/\omega > 0$, the result of operation of the program written in the programming language "Mathematica"

Region[{ *condition*[[2]] > 0 , *condition*[[3]] > 0 , *condition*[[4]] > 0 , *condition*[[5]] > 0 } , { b_{31} , −1.5 , 2 } , { b_{21} , −2 , 8 }]

is shown in Fig. 2.

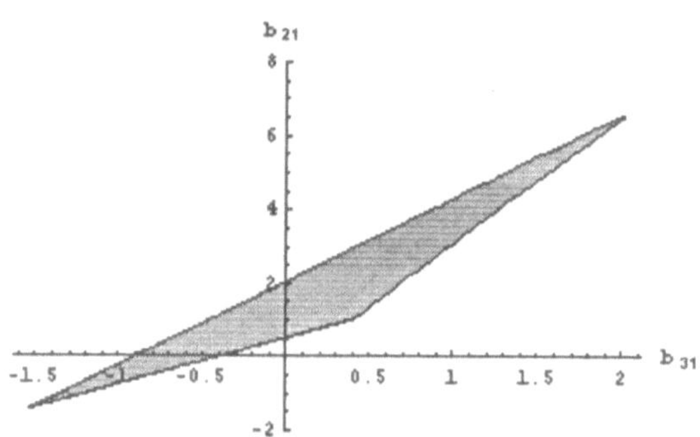

Fig. 2. Shaded area is the domain for choosing free parameters.

The figure shows the area where the parameters n_{11}, b_{21}, b_{31} can be chosen from the conditions of definite positiveness for the integral V . The area is not empty. Hence, system is stable.

References

1. Banshchikov, A., Bourlakova, L.: Information and Research System "Stability". Izvestia RAN. Teoria i sistemi upravlenia. No. 2 (1996) 13 - 20 (Russian). English transl. in Journal of Computer and Systems Sciences International. **35**, 2 (1996) 177-184.

2. Wolfram S. : Mathematica: A System for Doing Mathematics by Computer. Addison-Wesley Publ. Co., Redwood City, 1988.
3. Thomson, W., Tait, P.: Treatise on Natural Phylosophy. Part 1. Cambridge University Press, Cambridge, 1879.
4. Chetayev, N.G.: Stability of Motion. Works on Analytical Mechanics. Izd. AN SSSR, Moscow, 1962 (Russian).
5. Kuzmin, P.A.: Small Oscillations and Stability of Motion. Nauka, Moscow, 1973 (Russian).
6. Merkin D.: Gyroscopic Systems. Nauka, Moscow, 1974 (Russian).
7. Bulatovic R.: On stability of linear potential gyroscopic systems in the cases of maximum potential energy. Prikladnaya Matematika i Mekhanika. **61**, 3 (1997) 385-389 (Russian); English transl. in Journal of Applied Mathematics and Mechanics. **61**, 3 (1997).
8. Lankaster, P.: Theory of Matrices. Academic Press, New York, London, 1969.
9. Hagedorn, P.: Über die Instabilität konservativer Systeme mit gyroskopischen Kräften. Arch. Rat. Mech. Anal. **58**, 1 (1975) 1 - 9.
10. Huscyin, K., Hagedorn, P., Teschner, W.: On the stability of linear conservative gyroscopic systems. Journal of Applied Mathematics and Physics. **34**, 6 (1983) 807-815.
11. Banshchikov, A., Bourlakova, L.: Algorithms of Symbolic Computation Used in Stability Analysis. Programmirovanie. No. 3 (1997) 72 - 80 (Russian). English transl. in Programming and Computer Software. **23**, 3 (1997) 173 - 179.
12. Potapenko, E.: Dynamics of a spacecraft with line active control by a gravitational stabilizer. Kosmicheskie issledovania. **26**, 5 (1988) 699 - 708. English transl. in Cosmic Research. **26**, 5 (1988).

Implementing Computational Services Based on OpenMath*

Matthias Berth[1], Frank-Michael Moser[1], and Arrigo Triulzi[2]

[1] Institut für Mathematik/Informatik, Universität Greifswald, 17487 Greifswald, Germany berth@mail.uni-greifswald.de, moser@mail.uni-greifswald.de

[2] School of Mathematical Sciences, Queen Mary and Westfield College, Mile End Rd, London E1 4NS, UK
arrigo@maths.qmw.ac.uk

Abstract. OpenMath is a standard for representing mathematical objects. This report describes our experiences in implementing a facility for OpenMath-based symbolic computation services to be made available over the Internet. Services can be implemented in REDUCE, *Mathematica* or Maple, they are accessible via a Java client or using shell scripts. Our experimental client/server system is publicly available under an Open Source license. We address some issues in the development of OpenMath-based services such as implementing Phrasebooks and different choices of input syntax.

1 Introduction

Our need for communicating mathematical expressions amongst algebra packages arose in the context of the CATHODE II Esprit working group[1]. In the framework of the CATHODE II working group, software packages are developed which implement a wide spectrum of algorithmic solution methods for ordinary differential equations (ODEs). They cover approximate and exact methods, ranging from a single first order ODE to systems of ODEs and PDEs, both linear and nonlinear. The packages are based on different Computer Algebra systems (Maple, REDUCE, and Aldor, among others), and usually they represent a substantial amount of development time and expertise. Typically, a package is implemented by a group which also develops the algorithms contained in it. In this setting, the need for communication between different Computer Algebra (CA) systems arises naturally, as some packages can make good use of the tools provided by others. A number of prototypes that translate on a syntactical level were developed on a case-by-case basis, in particular between Aldor and Maple [6].

A more precise requirement is to have a means of sending mathematical objects from one package to another such that its mathematical meaning is

* This work was developed partly under the CATHODE II Esprit Working Group 24490 [9].

[1] CATHODE is an acronym for "**C**omputer **A**lgebra **T**ools for **H**andling **O**rdinary **D**ifferential **E**quations"

preserved. We chose to build on top of the OpenMath standard because it is extensible and able to capture a wide range of mathematical concepts[2]. Our aim was not to have the fastest system possible[3] but to prove that it was possible to integrate diverse CA systems using OpenMath as a common platform.

A number of proposals for intercommunicating mathematical computation systems have been put forward in [18, 2, 29, 3, 4] and earlier in [15–17]. In particular [2] deals with the problem of rendering and input of formulae and [29] introduces a protocol for data exchange for "Internet Accessible Mathematical Computation". The paper by Le and Howlett [18] describes instead an environment in which MathML [7], OpenMath and VRML are all used to provide a computational server on the Internet. The papers on integrating MuPad and Singular [3, 4] discuss the use of the MP protocol between these two systems. In this particular instance the focus was on the performance rather than the ability to be open to other CA systems.

This document describes our experiences in using OpenMath tools for a prototype, involving three CA systems (Maple, REDUCE and *Mathematica*) and a Java program. The system described herein is available at [5]. Our work would not have been possible without a number of software components developed by others, in particular the NAOMI [20] OpenMath implementation and Maple parser/printer from the PolyMath libraries [27] along with experimental versions of REDUCE OpenMath support by Winfried Neun and Marc Gaëtano [21]. We had to develop some tools ourselves, so we give a suitably detailed account of how they were implemented.

Our aim is to provide an infrastructure for computational *"services"*, similar in concept to the ideas expressed by Kajler in [15] and Fateman in [11]. The goal behind our design is that the user should not need to know the network location or indeed the CA system which eventually carries out the computation (or "service").

2 OMWS/OMD: a Client and Server for Mathematical Computation

Our prototype uses a client/server model of computation. A server makes available computational *services* like evaluation, simplification, or solving ordinary differential equations, to a number of clients. The server then talks to the CA systems, invoking them via shell scripts.

We provide a simple client (see figure 2), a Java application/applet, having means to input and display OpenMath objects, and sending them to a server. It was built as a first stepping stone in gaining experience with OpenMath by implementing the concept of a "mathematical workbench", albeit with restricted functionality.

---

[2] Another influencing factor was that at the time we believed that Maple support was forthcoming [22].

[3] Nevertheless our experience shows that the system is quite responsive when used over the internet.

A typical transaction might look as follows.

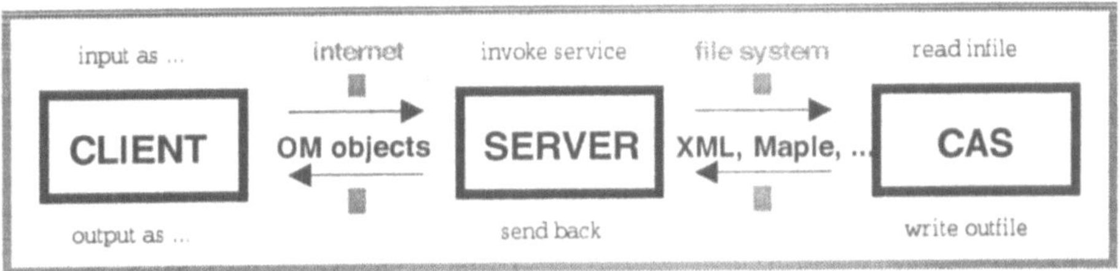

Fig. 1. An OMWS transaction

The user enters an OpenMath object using, say, Maple notation and, together with the object, the user specifies a service that is to be applied to that object:

```
Services::odesolve([diff(y(x), x) = y(x) * y(x) * x, y(x), x])
```

Upon pressing the Query button, the object is sent to the server which then invokes a shell script to start the appropriate CA system. The conversion from/to the native representation of the CA system is done using a Phrasebook[4]. The service itself usually consists of a package written in the CA system's native programming language.

When the service has finished, the output is converted to an OpenMath object before being sent back to the client. The client maintains a list of OpenMath objects that are generated either by user input or as a result of a computation.

2.1 Client – the OpenMath Work Sheet

The client is called OpenMath Worksheet (OMWS). It is programmed in Java, based on the OpenMath library from the PolyMath development group [27]. An OpenMath object can be input in Maple-like syntax or in XML representation. Additionally, both *Mathematica* and OpenMath expression tree formats are available for output.

Most of the time, Maple syntax will be used for both display and input. The conversion from Maple syntax to an OpenMath object is done by the class `MapleOMCodec` provided in the PolyMath package. We have extended version 0.4 of this package and added some fixes.

The Content Dictionary `Basic`[5] is supported, symbols from other CD's can be input as `OtherCD::symbol`.

To invoke a service, the user applies a symbol from the CD `Services` to an OpenMath object. It is possible to nest service calls, for example

[4] A brief introduction to OpenMath terminology is provided as an appendix.

[5] In the recent version of the OpenMath standard, `Basic` was split into a group of smaller CD's. We used the older version because it was supported by the PolyMath libraries and the Reduce/OpenMathinterface

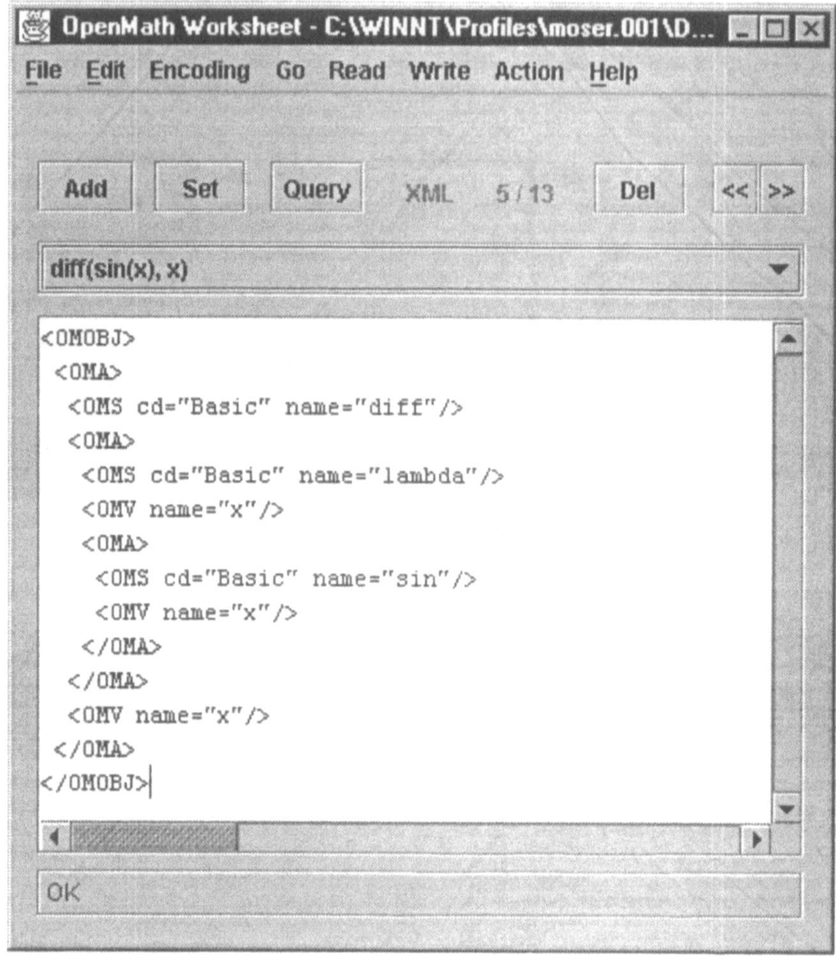

Fig. 2. The OMWS client

```
Services::subst([Services::mapledsolve([diff(y(x),x) * y(x) =
sin(x), y(x), x]), x, 0])
```

will ultimately call **dsolve** from Maple to solve an ordinary differential equation, and replace x by 0 in the solution.

After pressing the Query button, the client will send the complete object to the server. The server will then examine the service symbols and dispatch the object (or its parts) to the appropriate service implementation. When all the services are finished, the resulting OpenMath object is sent back to the OMWS where it will be inserted into the list.

The OMWS can be invoked as a command line filter, making all the conversions available to other tools. Furthermore, the command line version can be used to access services using files as input and output. This file based mechanism is a primitive way to allow other applications, e.g. CA systems, to act as a client.

2.2 Server – the OpenMath Dispatcher

The OpenMath Dispatcher (OMD) dispatches incoming OpenMath objects to services and sends replies back to clients. The server is programmed in Java, it talks to clients by sending serialised Java objects over network sockets. All the communication between client and server is stateless.

OMD is designed such that it does not have to know anything about the services internals. A request is an OpenMath object. The server looks for symbols from the `Services` Content Dictionary and dispatches the subtree to the corresponding service for processing. The correspondence between service symbols and services is described in a directory hierarchy that contains configuration files for each service. It is possible to add new services on the fly by adding a new directory with configuration information.

In our current implementation, any service must be invoked via a shell command. The server puts a representation of the incoming object in a file. The service may choose the representation (e.g. XML, binary) and say in which representation the output will be. When the service is done, the server converts the result from the output file to an OpenMath object which is sent back to the client.

There is a simple facility to build a chain of two or more servers, using the OMWS command line version to send OpenMath objects to other servers. A command line flag makes it possible to wrap a service symbol around the object before sending. Our current installation uses this method to access Maple which is not available on the machine where the principal server runs. Figure 3 shows how multiple servers might cooperate to offer a common set of services. Note that each client only needs to know one of the servers.

2.3 Services Providing a Common Interface

There are some desirable features that would make using services more convenient. One area is a choice in input syntax and more attractive formula display. A user should be able to choose a familiar, more or less natural, notation. In addition to Maple syntax, our prototype allows the input OpenMath objects in *Mathematica* syntax. Instead of integrating a *Mathematica* parser into OMWS, we chose to provide a service that converts a *Mathematica* string to an OpenMath object. On the server side, the string is parsed and converted to an OpenMath object by *Mathematica*. The OMWS receives this object, and the conversion is done. Clearly, this conversion process can be made entirely transparent to the user, allowing the addition of more notations as the need arises. It is quite feasible to have dedicated servers for conversion from/to different CA languages and two-dimensional formula display using graphics or MathML.

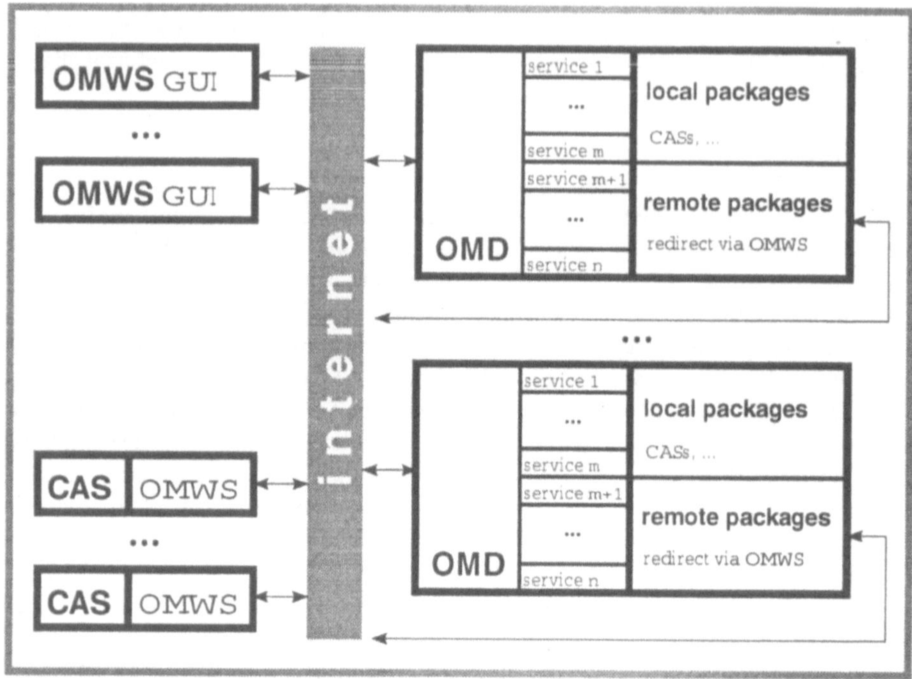

Fig. 3. A system with multiple clients and servers, some CA systems acting as clients

3 Implementing Phrasebooks

It might initially appear as though the implementation of a Phrasebook concerns itself exclusively with the translation of symbols such as `atan` into `arctan`. However, as Content Dictionaries describe abstract mathematical concepts, sometimes a more substantial effort might be necessary. In particular the implementation aspect also covers the details of actually reading and writing an OpenMath object.

Three different approaches were used to address these problems:

Link the OpenMath library. This seems to be the method that is encouraged by the OpenMath consortium. The CA system has direct access to OpenMath objects via the library's data structures. This approach is very efficient but requires some low level programming. It is applied in the Reduce Phrasebook we used.

Use Java to do the entire conversion. This requires a parser for the CA system language, and one for an OpenMath representation. Since a Maple parser/printer was already present in the Polymath libraries, we used this method for Maple.

Use the CA system parser. This method starts from a valid CA system expression that is in one-to-one correspondence to the OpenMath object. Then Phrasebooks are implemented completely using the CA system language.

We used the last method for *Mathematica*. The OpenMath object is translated into an intermediate form which is a valid *Mathematica* expression representing the XML tree. The OMWS is able to generate this intermediate form.

Using the *Mathematica* means for manipulation of expressions, it was easy to implement a Phrasebook for the `Basic` Content Dictionary, and some general support for implementers of other Phrasebooks. We introduced a canonical representation for OpenMath objects in *Mathematica* using heads like `OMApplication`, `OMSymbol`, etc. This canonical representation makes it possible for other packages to handle "raw" OpenMath objects. At a later stage, it may serve as a target for an OpenMath library that uses MathLink for efficiency.

We think that the approach described here can also be useful for connecting other CA system to OpenMath. It uses the CA system language as much as possible, these languages are specialised in processing expression trees, such as an OpenMath tree. It uses a language that is known by the implementors of packages, who will often be the first users of the OpenMath implementation, too. Furthermore, implementing almost anything inside the CA system makes OpenMath objects accessible to users and other packages.

4 OpenMath support for REDUCE

As of writing the status of OpenMath support under REDUCE is in a state of flux. There are 1998 implementations in PSL by Gaëtano, Huchet and Neun ([14] and [13]) one directly in the underlying PSL Lisp and the second linking in the OpenMath C library [12]. Both were experimental "proof of concept" versions designed to implement an early draft of the OpenMath standard. During the development phase of REDUCE version 3.7 a module was written [1] implementing MathML which might seem a suitable stepping stone towards full OpenMath availabilty. This is not the case as MathML is intended for formula display using tools such as web browsers[6], e.g. Netscape, and not as a means of inter-CAS communication although recently there have been suggestions to the contrary. There is also an experimental module by Neun for REDUCE 3.7 presented at the April 2000 Marseille CATHODE II Workshop implementing OpenMath support for PSL [19]. Unfortunately there is no generic implementation for both Standard Lisp versions of REDUCE as yet.

For demonstration purposes we decided to use a version of REDUCE with OpenMath support kindly supplied by Winfried Neun [14]. The aim was to be able to call the REDUCE module `odesolve` [31] from *Mathematica* via OpenMath. There were a number of available interfaces: an early version of OMWS, CGI scripts and a mixture of Perl and UNIX shell scripts. During the development of the demonstration all the above were implemented, in particular the approach finally taken for the link-up between REDUCE and *Mathematica* was via Java and OMD.

[6] Currently MathML is still very much a "markup" language but we cannot fail to notice that the standard is moving more towards semantics with each revision.

On the REDUCE side the CA system was invoked by a UNIX shell script[7] and given as input a simple REDUCE file which would pick up the XML-encoded OpenMath description of the problem from a second file placing the solution, if any in a third file.

Since this experiment it became clear that a more recent OpenMath implementation was required. The obvious solution was to link the OpenMath C Library [12] with REDUCE. In particular this could be made as simple as possible by making use of the CSL [23–25] implementation of REDUCE which is based on the C programming language. At the time the PSL re-implementation by Neun was not yet available hence the decision to choose CSL for ease of development.

The first layer of the implementation creates a one-to-one mapping between the functions defined in the C library and Lisp equivalents by the same name. This means that it is then possible to reproduce any C examples entirely within the Lisp layer (with the exception of remote server invocation which is not portable without substantial support by the underlying Lisp). An intermediate layer acting as an interface between the Lisp and REDUCE algebraic mode would then allow to make use of the OpenMath functionality in a way similar to the one available in the early "proof of concept" implementations [14, 13]. A small number of functions were not implemented due to the portability requirements of CSL and the only extension is the RLisp-visible procedure `omlibversion` returning a list with information on the implementation compiled into REDUCE and the underlying C Library.

The different types available in C and Lisp forced a number of design decisions which are described in more detail in [28].

Following is a short example run using the REDUCE Development System[8]:

```
REDUCE Development Version, Mon Mar 13 17:02:12 2000 ...

1: symbolic$
2* omlibversion();
 ("CSL/OpenMath interface $Revision: 1.3 $" "Mar 13 2000"
  "1.2" "debug is on")
3* dev := ommakedevice("/tmp/test","XML","out");
 1
4* omputobject(dev)$
5* omputint32(dev,45627)$
6* omputendobject(dev)$
7* omclosedevice(dev);
 t
```

[7] We actually used the REDUCE OpenMath client, instead of the server version as this listens on a TCP port and is used when invoking REDUCE via a remote shell over the network.

[8] This is a CSL REDUCE Development System, compiled from C sources to include the additional OpenMath support module and linked with the OpenMath C Library. Please note that the output is slightly abridged by the removal of empty lines.

```
8* bye$
```

and the contents of the test file are then:

```
<OMOBJ>
  <OMI>45627</OMI>
</OMOBJ>
```

The above is intended to demonstrate a session from a low-level point of view. Clearly it is not expected that a REDUCE user would be working at the Lisp level with OpenMath although the option is always available. An example of the C/Lisp "translation layer" can be found in [28] where the function `oMputint32` used above is described in detail.

Although not a complete OpenMath implementation due to the lack of high-level support such as content dictionaries and phrasebooks it is the author's hope that this portable implementation of a C/Lisp "translation layer" for the OpenMath C Library will allow easier support of OpenMath under REDUCE in the future.

Future work will include a test for the package under REDUCE and if possible proper integration into the REDUCE Development System.

5 Conclusion

We have presented the OMD/OMWS client-server system which can be used to provide the necessary infrastructure for active collaboration between CA packages. It achieves this by integrating OpenMath-related software components from various sources.

We strongly believe that an established communication mechanism allowing CA systems to share their strengths is long overdue. It is quite noticeable how the advancement of OpenMath is impaired by the lack of implementations for common CA systems, even though they initially claimed interest. Users need to be encouraged by a strong fixed standard and possibly reference implementations for the major CA systems. Furthermore the availability of both public code releases and a test suite, to validate OpenMath implementations, would surely help in this direction ensuring that multiple incompatible variants of the standard do not appear.

The authors have expended some unnecessary coding effort due to the lack of publicly available information during the course of the development which has led to re-implementing code which had already been written but not made public. Understandably the OpenMath community has been waiting to perfect their current standard before releasing it onto the public but we believe that it would perhaps be better to offer what is available so that work can commence. A good solution now is surely better than a perfect solution later when other, probably incompatible, systems have developed. The academic user base can often be counted upon to provide fixes, suggestions and bring to light issues which might have been overlooked at the drawing board stage.

It is definitely thanks to a small number of people who have gone out of their way to help us that we have been able to put together our experimental system. Perhaps an Open Source approach for implementations would be a step in the right direction but we feel that there is a strong need for both the standard itself and available software implementations to be propagated as much as possible. This would have allowed us to concentrate on the overall CATHODE II package integration aspect instead of on the development of the tools.

Nevertheless we believe that OpenMath is an excellent foundation for establishing computational services. The flexible and extensible design of OpenMath would permit true communication of mathematical objects using a uniform mechanism to access services.

As a prototype, the OWMS server lacks some important features such as strong server side security but this is clearly a deficiency which will need to be addressed in future releases. We plan to use the tools we have developed to turn CATHODE II packages into "OpenMath-enabled services", among others the REDUCE package CRACK [30] and the NODES Maple library [10]. This would allow transparent access to these packages from both CA systems and OMWS or indeed any other OpenMath-compliant system.

It is our hope that from this first step other CATHODE II packages may well benefit from the ability to easily call each other. It is clear that there is great potential in a system comprising all these packages, co-operating to provide the user as much help as possible. Experiences within CATHODE II have shown that it is infeasible to integrate the available packages into one monolithic "super-package" that is implemented on top of a single CA system: the amount of work necessary for re-implementation and coordination would by far exceed the available manpower. We believe that a well structured distributed system with parts implemented in different CA systems might be preferable as well as being more flexible.

A development of this would be a more "dynamic" environment where the actual CA packages not only communicated via OpenMath but actually posed the problem without specifying the solution method. This would not only free the packages from the constraint of knowing apriori which other package to call but also allows transparent introduction of new solvers which adhere to the given calling convention as specified by a suitable Content Dictionary.

Appendix: OpenMath basics [MB]

This section is meant as a brief introduction to basic OpenMath concepts.

From the OpenMath web site [26]:

> OpenMath is an emerging standard for representing mathematical objects with their semantics, allowing them to be exchanged between computer programs, stored in databases, or published on the worldwide web.

OpenMath objects are expressions (expression trees) built up from atomic objects such as symbols, variables, numbers and strings. The atomic objects are combined into larger expressions by applying symbols or variables to smaller expressions. The OpenMath standard places a strong emphasis on preserving the meaning (semantics) of an object. Thus, every symbol has to be defined in a so-called *Content Dictionary (CD)* where its meaning and properties are fixed. A CD groups symbols in a meaningful way (e.g. there are content dictionaries for linear algebra and polynomials). The translation from the OpenMath object into a "native" CA system object is done by a piece of software called a *Phrasebook*. A Phrasebook is highly specific to the chosen CA system, whereas a Content Dictionary tries to describe a platform-independent representation of some mathematical concepts. Usually, a Content Dictionary will come with many Phrasebooks, one for each CA system that is supposed to communicate using objects from the CD.

Extending OpenMath to a new area of mathematics amounts to defining a new Content Dictionary and implementing or extending corresponding Phrasebooks.

The full, current, standard can be found in [26] and in particular [8].

References

1. Luis Alvarez-Sobreviela. *REDUCE-MathML Interface*. Konrad-Zuse-Zentrum für Informationstechnik Berlin, Nov 1998. Part of the REDUCE 3.7 manual.
2. Olivier Arsac, Stéphane Dalmas, and Marc Gaëtano. The design of a customizable component to display and edit formulas. In *Proc. of ISSAC '99*, pages 283–290. ACM Press, 1999.
3. Olaf Bachmann, Hans Schönemann, and Andreas Sorgatz. Connecting MuPAD and Singular with MP. *mathPAD*, 8(1):8–17, Mar 1998. ISSN 0941-9187.
4. Olaf Bachmann, Hans Schönemann, and Andreas Sorgatz. Connecting MuPAD and Singular with MP. *MapleTech*, 5(2/3):117–121, Jul 1999.
5. Matthias Berth and Frank-Michael Moser. OMWS web site, 2000. `http://paul.math-inf.uni-greifswald.de/Cathode2/omws/`.
6. Manuel Bronstein. CATHODE II demonstration of the maple-aldor package for solving 2nd order homogeneous linear ordinary differential equations, 1999. Available online at: `http://www-lmc.imag.fr/lmc-cf/Manuel.Bronstein/kovacic_demo.html`.
7. S. Buswell, S. Devitt, A. Diaz, N. Poppelier, B. Smith, N. Soiffer, R. Sutor, and S. Watt. Mathematical markup language (MathML) 1.0 specification. Technical report, World-Wide-Web Consortium, 1998. Available online at: `http://www.w3.org/TR/1998/REC-MathML-19980407`.
8. Olga Caprotti and Arjeh M. Cohen. The OpenMath standard. Technical report, RIACA, Eindhoven, 1999. Available online at: `http://www.nag.co.uk/projects/OpenMatha/omstd/`.
9. CATHODE II web site, 1996. `http://www-lmc.imag.fr/cathode2`.
10. Marco Codutti. NODES: Non linear Ordinary Differential Equations Solver. In *Proc. of ISSAC '92*. ACM Press, Jul 1992. Software available at: `http://cso.ulb.ac.be/~nodes`.
11. Richard Fateman. Network servers for symbolic mathematics. In *Proc. of ISSAC '97*, pages 249–256. ACM Press, 1997.

12. Marc Gaëtano and Stéphane Dalmas. The INRIA OpenMath library. Technical report, Projet Safir, INRIA Sopia Antipolis, 1998.
13. Marc Gaëtano and Claude Huchet. Second OpenMath–REDUCE version. Technical report, Konrad–Zuse–Zentrum für Informationstechnik Berlin, 1998.
14. Marc Gaëtano, Claude Huchet, and Winfried Neun. The realisation of an OpenMath server for REDUCE. Technical report, Konrad–Zuse–Zentrum für Informationstechnik Berlin, 1998.
15. N. Kajler. Building a computer algebra environment by composition of collaborative tools. In *LNCS 721, Proc. of DISCO '92*, pages 85–94. Springer-Verlag, 1992.
16. N. Kajler. CAS/PI: A portable and extensible interface for computer algebra systems. In *Proc. of ISSAC '92*, pages 376–386. ACM Press, 1992.
17. N. Kajler. *Environnment Graphique Distribué pour le Calcul Formel.* PhD thesis, Université de Nice Sophia-Antipolis, Mar 1993.
18. Ha Le and Chris Howlett. Client-server communication standards for mathematical computation. In *Proc. of ISSAC '99*, pages 299–305. ACM Press, 1999.
19. Herbert Melenk and Winfried Neun. *Portable Standard Lisp.* Konrad–Zuse–Zentrum für Informationstechnik Berlin, Berlin, version 4.2 edition, 1997.
20. NAOMI web site, 1999. `http://naomi.math.ca/`.
21. Winfried Neun. Notes on PSL REDUCE/OpenMath interface. Private Communication, 1998. Detailed information on how to install and operate the PSL REDUCE/OpenMath prototype system.
22. Y. N.Lakshman, Bruce Char, and Jeremy Johnson. Software components using symbolic computation for problem solving environments. In *Proc. of ISSAC '98*, pages 46–53. ACM Press, 1998.
23. Arthur Norman. Codemist Standard Lisp (CSL): technical overview and details. Technical report, Codemist Ltd., Jul 1991.
24. Arthur Norman. Compact delivery support for REDUCE. In *LNCS 1128, Proc. of DISCO '93*, pages 331–330. Springer-Verlag, 1993.
25. Arthur Norman. Compact delivery support for REDUCE. *Journal of Symbolic Computation*, 19:133–143, 1995.
26. OpenMath web site, 1999. `http://www.openmath.org/`.
27. PolyMath Development Team, 1999. `http://pdg.cecm.sfu.edu/`.
28. Arrigo Triulzi. OpenMath support under CSL-based REDUCE. Technical report, Queen Mary and Westfield College, Apr 2000.
29. Paul S. Wang. Design and protocol for internet accessible mathematical computation. In *Proc. of ISSAC '99*, pages 291–306. ACM Press, 1999.
30. Thomas Wolf and Andreas Brand. *The Computer Algebra package CRACK for investigating PDEs.* Queen Mary and Westfield College, Nov 1998. Part of the REDUCE 3.7 manual and also:
`http://reduce.maths.qmw.ac.uk/packages/crack/`.
31. Francis Wright. *ODEsolve.* Queen Mary and Westfield College, Nov 1998. Part of the REDUCE 3.7 manual, also online at:
`http://reduce.maths.qmw.ac.uk/packages/odesolv1/`.

Group Classification of the Navier-Stokes Equations for Compressible Viscous Heat-Conducting Gas *

Vasiliy V. Bublik

Institute of Theoretical and Applied Mechanics, Russian Academy of Sciences, Siberian Branch, Institutskaya str., 4/1, Novosibirsk 630090, Russia
bublik@itam.nsc.ru

Abstract. We consider the problems of using computer algebra systems for group classification of partial differential equations. The presence of arbitrary elements makes the solution of determining equations more difficult. It offers necessity for active human control over this process by using dialog regime. Methods of computer algebra for problems of group classification are illustrated by the example of a system of the Navier-Stokes equations for compressible viscous heat-conducting gas. This system has five arbitrary elements. The solution of the problem of finding equivalence transformations is presented. The group classification is carried out in some particular cases.

1 Introduction

The methods of classical group analysis provide the most widely applicable technique to find closed form solutions of differential equations [1, 2]. Group classification of physical models gives insight into their properties and allows one to identify important particular cases.

The basic procedure for finding the symmetry group allowed by a given differential equation is well known. The size of the determining system, however, grows dramatically with the order and the number of variables in the equation. As a specific real-world application, the equations of viscous, heat-conducting gas dynamics will be studied in the paper. Here one obtains an overdetermined system with about 200,000 equations. Because of this, computer algebra systems such as *Mathematica*, *MACSYMA*, *Maple*, *REDUCE*, *AXIOM*, and *MuPAD*, are extremely useful aids in such computations. A review of the purpose, methods, algorithms, and literature was done in [3].

However, in group classifications the problem becomes even worse due to the presence of arbitrary functions (modelling the equations of state or the dependence of the viscosity on thermodynamical parameters, etc.). Thus, the determining equations divide into two types. The first one named classifying

* This research was financially supported by INTAS (grant 99-1222) and the Russian Foundation for Basic Research (grant 99-01-00515).

equations contains arbitrary elements. The second one contains no arbitrary elements. Even the second type equations cannot generally be solved automatically. There are examples when the determining system for the classical Lie symmetry generators cannot be explicitly integrated by the known methods and algorithms. The first type equations cannot be trusted totally to computer because the automatic completion to involution of this system may lead to a loss in some particular cases. It is necessary to resolve the determining system under active human control.

The problem of group classifications can be split into two stages. In the first stage, the admissible equivalence transformations must be found. This requires the involutive analysis of an intermediate highly overdetermined system. Here the arbitrary elements are treated as additional dependent variables. This stage can be performed automatically with the use of slightly modified software packages.

In the second stage, the group classification itself is performed. This classification is based on the equivalence transformations. In this stage, the computer algebra system is used only for computing determining equations. Choosing a specific equation and its solving cannot be automatized. Because of this, problems of group classification are very laborious. The process of solving can be made effectively in dialog regime.

2 General Statement of the Problem

For the purpose of illustrating the use of computer algebra for problems of group classification we consider a system of the Navier-Stokes equations for compressible viscous heat-conducting gas. The governing equations are

$$\rho \frac{d\boldsymbol{v}}{dt} = -\nabla p + \nabla(\lambda \operatorname{div} \boldsymbol{v}) + \operatorname{div}(2\mu \mathrm{D}) \ , \tag{1}$$

$$\frac{d\rho}{dt} + \rho \operatorname{div} \boldsymbol{v} = 0 \ , \tag{2}$$

$$\rho T \frac{dS}{dt} - \operatorname{div}(\varkappa \nabla T) + \lambda (\operatorname{div} \boldsymbol{v})^2 + 2\mu \mathrm{D} : \mathrm{D} \ . \tag{3}$$

Here, $\boldsymbol{v} = (u, v, w)$ is the velocity, D is the strain tensor, ρ is the density, T is the temperature, $\mu = \mu(\rho, T)$ is the first viscosity coefficient, $\lambda = \lambda(\rho, T)$ is the second viscosity coefficient, $\varkappa = \varkappa(\rho, T)$ is the heat conductivity coefficient, $p = p(\rho, T)$ is the pressure, $S = S(\rho, T)$ is the entropy,

$$\frac{d}{dt} = \frac{\partial}{\partial t} + u\frac{\partial}{\partial x} + v\frac{\partial}{\partial y} + w\frac{\partial}{\partial z} \ , \quad \nabla = \left(\frac{\partial}{\partial x}, \frac{\partial}{\partial y}, \frac{\partial}{\partial z}\right) \ , \quad \mathrm{D} : \mathrm{D} = \sum_{i,j=1}^{3} \mathrm{D}_{ij}^2 \ .$$

We assume that

$$3\lambda + 2\mu \geq 0 \ , \quad \frac{\partial S}{\partial T} \neq 0 \ .$$

We solve the group classification problem for system (1)–(3) with the arbitrary elements μ, λ, $\varkappa$, p, and S [1].

For a particular case of system (1)–(3) the group classification was made in the previous papers:

- group classification of the gas dynamics equations ($\mu = \lambda = \varkappa = 0$) was done in [4];
- group classification of the radiation hydrodynamics equations ($\mu = \lambda = 0$) was done in [5];
- group classification of the two-dimensional equations for a motion of a viscous, heat-conducting perfect gas ($\mu = \mu(T), 3\lambda + 2\mu = 0, \varkappa = \varkappa(T), p = \rho T$) was done in [6];
- group classification of the three-dimensional equations for a motion of a viscous, heat-conducting gas with constant viscosity and heat conductivity coefficients ($\mu \equiv 1$, $3\lambda + 2\mu = 0$, $\varkappa \equiv 1$) was done in [7];
- group classification of the two-dimensional equations for a stable flow of a viscous, heat-conducting perfect gas ($\mu = \mu(T)$, $3\lambda + 2\mu = 0$, $\varkappa = \varkappa(T)$, $p = \rho T$) was done in [8].

3 Equivalence Transformations

To perform the group classification one needs to find equivalence transformations [1, 2]. The nondegenerate transformation of the space of dependent and independent variables, and arbitrary functions, which acts only on arbitrary functions and saves structure of the differential equations, is called the equivalence transformation of these equations. We use a generalization of the equivalence transformations proposed by S. V. Meleshko [9, 10]. See all necessary formulae below.

The vector notations used are as follows:

$$
\begin{gathered}
\boldsymbol{q} = (t, x, y, z) \ , \quad \boldsymbol{\phi} = (u, v, w, \rho, T) \ , \\
\boldsymbol{\tau} = (\mu, \lambda, \varkappa, p, S) \ , \ \boldsymbol{h} = (t, x, y, z, u, v, w, \rho, T) \ , \\
\phi_i^k = \frac{\partial \phi^k}{\partial q^i} \ , \quad \phi_{ij}^k = \frac{\partial \phi_i^k}{\partial q^j} \ , \quad \tau_m^l = \frac{\partial \tau^l}{\partial h^m} \\
(k = 1, \dots, 5 \ , \quad i, j = 1, \dots, 4 \ , \quad l = 1, \dots, 5 \ , \quad m = 1, \dots, 9) \ .
\end{gathered}
$$

Then we add the conditions for μ, λ, $\varkappa$, p, and S to the system (1)–(3)

$$
\tau_i^k = 0 \quad (k = 1, \dots, 5 \ , \quad i = 1, \dots, 7) \ . \tag{4}
$$

For the system (1)–(3), (4) we find the full group of transformations for the space of variables t, x, y, z, u, v, w, ρ, T, μ, λ, $\varkappa$, p, and S with admissible operator

$$
X^e = \sum_i \xi^i \frac{\partial}{\partial q^i} + \sum_j \eta^j \frac{\partial}{\partial \phi^j} + \sum_k \alpha^k \frac{\partial}{\partial \tau^k} \ .
$$

The functions ξ^i, η^j, α^k depend on t, x, y, z, u, v, w, ρ, T, μ, λ, $\varkappa$, p, and S. Taking into account the equations (4), we calculate the components of prolonged operator

$$X_p^e = X^e + \sum_{i,j} \zeta_j^i \frac{\partial}{\partial \phi_j^i} + \sum_{i,j,k} \zeta_{jk}^i \frac{\partial}{\partial \phi_{jk}^i} + \sum_{i,j} \beta_j^i \frac{\partial}{\partial \tau_j^i}$$

from the formulae

$$\zeta_j^i = D_j^e \eta^i - \sum_k \phi_k^i D_j^e \xi^k \ , \quad \zeta_{jk}^i = D_k^e \zeta_j^i - \sum_l \phi_{jl}^i D_k^e \xi^l \ ,$$

$$\beta_j^i = D_j' \alpha^i - \tau_8^i D_j' \eta^4 - \tau_9^i D_j' \eta^5 \ ,$$

$$D_j^e = \frac{\partial}{\partial q^j} + \sum_k \phi_j^k \frac{\partial}{\partial \phi^k} + \sum_{i,k} \phi_{ij}^k \frac{\partial}{\partial \phi_i^k} + \left(\phi_j^8 \tau_8^l + \phi_j^9 \tau_9^l\right) \frac{\partial}{\partial \tau^l} \quad (j = 1, \ldots, 4) \ ,$$

$$D_j' = \frac{\partial}{\partial h^j} \quad (j = 1, \ldots, 7) \ ,$$

$$D_8' = \frac{\partial}{\partial h^8} + \sum_i \tau_8^i \frac{\partial}{\partial \tau^i} \ , \quad D_9' = \frac{\partial}{\partial h^9} + \sum_i \tau_9^i \frac{\partial}{\partial \tau^i} \ .$$

The difference in calculations between the component ζ_j^i and β_j^i arises because the functions $u(t,x,y,z)$, $v(t,x,y,z)$, $w(t,x,y,z)$, $\rho(t,x,y,z)$, $T(t,x,y,z)$ and $\mu(\rho,T)$, $\lambda(\rho,T)$, $\varkappa(\rho,T)$, $p(\rho,T)$, $S(\rho,T)$ are from the different spaces.

After applying the operator X_p^e to each of the equations (1)–(3), (4) we go over to the manifold of (1)–(3), (4) by eliminating the derivatives $\frac{\partial u}{\partial t}$, $\frac{\partial v}{\partial t}$, $\frac{\partial w}{\partial t}$, $\frac{\partial \rho}{\partial t}$, $\frac{\partial T}{\partial t}$, $\frac{\partial \mu}{\partial t}$, $\frac{\partial \mu}{\partial x}$, $\frac{\partial \mu}{\partial y}$, $\frac{\partial \mu}{\partial z}$, $\frac{\partial \mu}{\partial u}$, $\frac{\partial \mu}{\partial v}$, $\frac{\partial \mu}{\partial w}$, $\frac{\partial \lambda}{\partial t}$, $\frac{\partial \lambda}{\partial x}$, $\frac{\partial \lambda}{\partial y}$, $\frac{\partial \lambda}{\partial z}$, $\frac{\partial \lambda}{\partial u}$, $\frac{\partial \lambda}{\partial v}$, $\frac{\partial \lambda}{\partial w}$, $\frac{\partial \varkappa}{\partial t}$, $\frac{\partial \varkappa}{\partial x}$, $\frac{\partial \varkappa}{\partial y}$, $\frac{\partial \varkappa}{\partial z}$, $\frac{\partial \varkappa}{\partial u}$, $\frac{\partial \varkappa}{\partial v}$, $\frac{\partial \varkappa}{\partial w}$, $\frac{\partial p}{\partial t}$, $\frac{\partial p}{\partial x}$, $\frac{\partial p}{\partial y}$, $\frac{\partial p}{\partial z}$, $\frac{\partial p}{\partial u}$, $\frac{\partial p}{\partial v}$, $\frac{\partial p}{\partial w}$, $\frac{\partial S}{\partial t}$, $\frac{\partial S}{\partial x}$, $\frac{\partial S}{\partial y}$, $\frac{\partial S}{\partial z}$, $\frac{\partial S}{\partial u}$, $\frac{\partial S}{\partial v}$, and $\frac{\partial S}{\partial w}$. After this we split up the resulting equations with respect to the remaining derivatives and solve the derived determining equations.

We can conveniently begin to investigate the determining equations which follow from (4). We obtain the functions η^4, η^5, α^1, ..., α^5 depending on ρ, T, μ, λ, $\varkappa$, p, and S. Thereafter the determining equations resulting from (2) are studied. These equations have the first order, therefore, they can be more readily solved. Then the equations are solved, which follow from (1) and (3).

Here and in what follows, all necessary calculations were carried out on the computer using the computer algebra system *REDUCE* [11]. We will provide a more detailed description in the next section.

As a result of computer calculations, we obtain the system (1)–(3), (4) in the space of variables t, x, y, z, u, v, w, ρ, T, μ, λ, $\varkappa$, p, and S admits the equivalent transformation corresponding to the operators

$$X_1 = \frac{\partial}{\partial x} \ , \quad X_2 = \frac{\partial}{\partial y} \ , \quad X_3 = \frac{\partial}{\partial z} \ , \tag{5}$$

$$X_4 = t\frac{\partial}{\partial x} + \frac{\partial}{\partial u} \ , \quad X_5 = t\frac{\partial}{\partial y} + \frac{\partial}{\partial v} \ , \quad X_6 = t\frac{\partial}{\partial z} + \frac{\partial}{\partial w} \ , \tag{6}$$

$$X_7 = z\frac{\partial}{\partial y} - y\frac{\partial}{\partial z} + w\frac{\partial}{\partial v} - v\frac{\partial}{\partial w} \ , \quad X_8 = x\frac{\partial}{\partial z} - z\frac{\partial}{\partial x} + u\frac{\partial}{\partial w} - w\frac{\partial}{\partial u} \ , \tag{7}$$

$$X_9 = y\frac{\partial}{\partial x} - x\frac{\partial}{\partial y} + v\frac{\partial}{\partial u} - u\frac{\partial}{\partial v} \ , \quad X_{10} = \frac{\partial}{\partial t} \ , \tag{8}$$

$$R_1 = t\frac{\partial}{\partial t} + x\frac{\partial}{\partial x} + y\frac{\partial}{\partial y} + z\frac{\partial}{\partial z} + \mu\frac{\partial}{\partial \mu} + \lambda\frac{\partial}{\partial \lambda} + \varkappa\frac{\partial}{\partial \varkappa} \ , \tag{9}$$

$$R_2 = \rho\frac{\partial}{\partial \rho} + \mu\frac{\partial}{\partial \mu} + \lambda\frac{\partial}{\partial \lambda} + \varkappa\frac{\partial}{\partial \varkappa} + p\frac{\partial}{\partial p} \ , \tag{10}$$

$$R_3 = x\frac{\partial}{\partial x} + y\frac{\partial}{\partial y} + z\frac{\partial}{\partial z} + u\frac{\partial}{\partial u} + v\frac{\partial}{\partial v} + w\frac{\partial}{\partial w} + \\ +2\mu\frac{\partial}{\partial \mu} + 2\lambda\frac{\partial}{\partial \lambda} + 4\varkappa\frac{\partial}{\partial \varkappa} + 2p\frac{\partial}{\partial p} + 2S\frac{\partial}{\partial S} \ , \tag{11}$$

$$R_4 = T\frac{\partial}{\partial T} - \varkappa\frac{\partial}{\partial \varkappa} - S\frac{\partial}{\partial S} \ , \quad \Sigma_p = \frac{\partial}{\partial p} \ , \quad \Sigma_S = \frac{\partial}{\partial S} \ . \tag{12}$$

4 Group Classification

Let us describe the group classification techniques for (1)–(3).

We consider the system (1)–(3) as a system of second order differential equations for five unknown functions u, v, w, ρ, and T. We seek the operator admitted by the equations in the form

$$X = \sum_i \xi^i \frac{\partial}{\partial q^i} + \sum_j \eta^j \frac{\partial}{\partial \phi^j} \ .$$

We calculate the components of prolonged operator

$$X_p = X + \sum_{i,j} \zeta^i_j \frac{\partial}{\partial \phi^i_j} + \sum_{i,j,k} \zeta^i_{jk} \frac{\partial}{\partial \phi^i_{jk}}$$

from the formulae

$$\zeta^i_j = D_j \eta^i - \sum_k \phi^i_k D_j \xi^k \ , \quad \zeta^i_{jk} = D_k \zeta^i_j - \sum_l \phi^i_{jl} D_k \xi^l \ ,$$

$$D_j = \frac{\partial}{\partial q^j} + \sum_k \phi^k_j \frac{\partial}{\partial \phi^k} + \sum_{i,k} \phi^k_{ij} \frac{\partial}{\partial \phi^k_i} \quad (j = 1, \dots, 4) \ .$$

After applying the operator X_p to each of the equations (1)–(3) we go over to the manifold of (1)–(3) by eliminating the derivatives $\frac{\partial u}{\partial t}$, $\frac{\partial v}{\partial t}$, $\frac{\partial w}{\partial t}$, $\frac{\partial \rho}{\partial t}$, and $\frac{\partial T}{\partial t}$. After this we split up the resulting equations with respect to the remaining derivatives and solve the obtained determining equations.

The computer algebra system *REDUCE* was used for calculating the prolonged operator X_p by applying the X_p to every equation from (1)–(3), going over to the manifold of (1)–(3) and splitting up the resulting equations with respect to derivatives. For efficient allocation of machine resources we will begin to study the determining equations resulting from (2). The splitting process is accomplished by differentiation. Each of the determining equations is resolved manually. The refined representations for ξ^i and η^j are input into the *REDUCE* program for derivation of the next determining equations. In deciding on specific determining equation it is necessary to consider first of all those that add no more complexity to inner machine representation of ξ^i and η^j. For problems of group classification it is also necessary to study all determining equations, which contain no arbitrary functions and only then we resolve other determining equations (which are classifying equations).

According to the general theory [1] the group of equivalence transformations contains the kernel of the fundamental Lie algebra, that allows a group under all arbitrary functions. In the case of specific type of arbitrary functions, the admissible group may be expanded.

As a result of symbolic computations, we have obtained that the basis of kernel of the fundamental Lie algebra allowed by equations (1)–(3) consists of operators $X_1, \ldots, X_{10}$ defined by (5)–(8). The analysis of particular form of arbitrary elements is now performed [7].

Table 1. Group classification for the case of perfect gas

μ	λ	$\varkappa$	S	Extensions
$f(T)$	$g(T)$	$h(T)$	$\varphi(T) + s_0 \log \rho$	Y_1
T^ω	$l_0 T^\omega$	$k_0 T^\omega$	$s_0 \log T + s_{01} \log \rho$	Y_1, Y_2
T^ω	$l_0 T^\omega$	$k_0 T^\omega$	$s_0 \log T + \varphi(\rho)$	Y_3
T^ω	$l_0 T^\omega$	$k_0 T^\omega$	$\varphi(\rho T^{1-\omega}) + s_0 \log \rho$	Y_2
T^ω	$l_0 T^\omega$	$k_0 T^\omega$	$\varphi(\rho T^{1/2-\omega}) + s_0 \log \rho$	Y_4

In Table 1, the result of group classification for the perfect gas case ($p = \rho T$) is given. The functions f, g, h, and φ are the arbitrary functions of indicated arguments. The quantities ω, l_0, k_0, s_0, and s_{01} are arbitrary constants. Extensions of kernel of the fundamental Lie algebra are presented in the last column. The operators Y_1, Y_2, Y_3, and Y_4 are determined by the formulae

$$Y_1 = t\frac{\partial}{\partial t} + x\frac{\partial}{\partial x} + y\frac{\partial}{\partial y} + z\frac{\partial}{\partial z} - \rho\frac{\partial}{\partial \rho} ,$$

$$Y_2 = x\frac{\partial}{\partial x} + y\frac{\partial}{\partial y} + z\frac{\partial}{\partial z} + u\frac{\partial}{\partial u} + v\frac{\partial}{\partial v} + w\frac{\partial}{\partial w} + 2(\omega - 1)\rho\frac{\partial}{\partial \rho} + 2T\frac{\partial}{\partial T} ,$$

$$Y_3 = 2(\omega-1)t\frac{\partial}{\partial t} + (2\omega-1)x\frac{\partial}{\partial x} + (2\omega-1)y\frac{\partial}{\partial y} + (2\omega-1)z\frac{\partial}{\partial z} + \\ +u\frac{\partial}{\partial u} + v\frac{\partial}{\partial v} + w\frac{\partial}{\partial w} + 2T\frac{\partial}{\partial T} ,$$

$$Y_4 = t\frac{\partial}{\partial t} - u\frac{\partial}{\partial u} - v\frac{\partial}{\partial v} - w\frac{\partial}{\partial w} + (1-2\omega)\rho\frac{\partial}{\partial \rho} + 2T\frac{\partial}{\partial T} .$$

The research of other particular cases is in progress now.

References

1. Ovsiannikov, L. V.: Group analysis of differential equations. Nauka, Moscow, 1978. (English translation published by Academic Press, New York, 1982).
2. CRC handbook of Lie group analysis of differential equations. Vol. 1: Symmetries, exact solutions, and conservation laws. CRC Press, Boca Raton, New York, London, Tokyo, 1994.
3. CRC handbook of Lie group analysis of differential equations. Vol. 3: New trends in theoretical developments and computational methods. CRC Press, Boca Raton, New York, London, Tokyo, 1996.
4. Ovsiannikov, L. V.: Group properties of differential equations. Academy of Sciences, Siberian Branch, U.S.S.R., Novosibirsk, 1962. (English translation by Bluman, G. W., 1972, unpublished).
5. Coggeshall, S. V., Axford, R. A.: Lie group invariance properties of radiation hydrodynamics equations and their associated similarity solutions. Phys. Fluids. **29** No. 8 (1986) 2398–2420
6. Bublik, V. V.: Group classification of the two-dimentional equations of motion of a viscous heat conducting perfect gas. Prikl. Mekh. Tekhn. Fiz. **37** No. 2 (1996) 27–34. (English translation in Appl. Mech. and Techn. Phys. by Plenum Publishing Corp.).
7. Bublik, V. V.: Group classification of equations for dynamics of a viscous heat conducting gas. Dinamika Sploshnoi Sredy. Novosibirsk. **113** (1998) 27–31. (In Russian).
8. Meleshko, S. V.: Group classification of two-dimensional stable viscous gas equations. Int. J. Nonlin. Mech. **34** No. 3 (1998) 449–456.
9. Meleshko, S. V.: Group classification of the equations of two-dimensional motions of a gas. J. Appl. Math. Mech. **58** No. 4 (1995) 56–62.
10. Meleshko, S. V.: Generalization of the equivalence transformations. Nonlin. Math. Phys. **3** No. 1–2 (1996) 170–174.
11. Hearn, A. C.: REDUCE User's Manual. Version 3.6. Santa Monica, 1995.

Plotting Functions and Singularities with Maple and Java on a Component-based Web Architecture

Dieter Bühler, Corinne Chauvin, and Wolfgang Küchlin

University of Tübingen
Symbolic Computation Group
Sand 13,
D-72076 Tübingen, Germany
{buehler, chauvin, kuechlin}@informatik.uni-tuebingen.de,
WWW home page: http://www-sr.informatik.uni-tuebingen.de

Abstract. In this paper we present a client/server plotting system for analytical functions that features a reliable computation and visualization of singularities. The system is part of a Web-based course on Mathematics for first-year students at the University of Tübingen. In an educational context where students of the first or second semesters are addressed, it is crucial that the plotted graphs are correct, since the students may not yet be able to interpret faulty graphs in an appropriate way. The singularities are computed by a computer algebra system which is transparently accessible through a Web interface. Our implementation of the singularity computations is more complete than the existing ones that we found as parts of various algebra systems. The rendering of the graphs and the visualization of the singularities is performed by a Java applet which can be executed within almost all of the current Java-aware Web browsers. No additional software, like plug-ins or proprietary browser extensions, has to be installed on the client machine.

1 Introduction

With the rapid development of the World Wide Web, activities for teleteaching, distance learning, and entire virtual universities are gaining ground. The German state of Baden-Württemberg has launched several research projects to develop and evaluate those efforts. One of them develops a Web-based course of study for Computer Science in the Life Sciences[1] at the University of Tübingen. Our research group is involved in the subproject[2] that develops the introductory course in Mathematics for the first-year students. The Web site provides access to the lecture notes, which are available in HTML format, a guest book, a chat

[1] German: *Bio-Informatik*. See http://www-ra.informatik.uni-tuebingen.de/-bioinformatik/

[2] See http://mfb.informatik.uni-tuebingen.de - Please use login *guest* and password *BIOINFORMATIK*

and full text search facility, and several interactive Java applets which visualize mathematical concepts.

We developed a simple Java package for numerical evaluations, but whenever symbolic computation is needed the applets delegate the computation to a computer algebra system like Maple or MuPAD. This delegation is completely transparent for the user who uses her or his well-known Web browser as the single interface to our system. The Java-aware Web browser and the connection to the Internet is the only software requirement on the client side. This approach is also used in a related research project dealing with Web-based teaching of automation engineering concepts and telematics where the algebra system at the back-end is replaced by industrial devices [4, 3].

Though the functions we have to plot are quite simple analytic functions, we have to be able to plot functions which contain tricky singularities, as in 0 for $sin(x)/x$, where the function can be extended to a continuous function. Such functions are traditional examples studied in an analysis course. Furthermore, we would like to be able to deal with complex singularities, such as the ones around 0 in $1/sin(1/x)$, where 0 is an accumulation point for singularities.

Having looked at some of the most common mathematical software systems, we came to the conclusion that they were not really dealing with singularities plotting, so that it was not feasible to reuse their plotting functions. In most of the cases, the singularities were represented by a vertical line, without any indications. Maple proposes a *discont* option for its plotter which allows to suppress those vertical lines. It remains difficult to understand that there is a singularity at this place, and furthermore, this option may cause a computation error, for instance for some logarithmic functions. Anyway, all of the plot functions of these software systems ignore singularities like in $sin(x)/x$, cf. Section 2.

There exist other sophisticated plotting systems [6, 1] which feature a lot of techniques like supersampling and interval arithmetic in order to produce accurate plots. None of them explicitly computes and visualizes the singularities in the way our system does.

Implementing a function that determines the singularities in Java, so that it could be used directly in a Java applet would have been a major effort, whereas most of the functionalities and data types we need still exist in mathematical software systems. As the general setting is client-server based, our approach to solving this problem is to use an architecture for Web-based algebraic components, which allows us to implement the computation of singularities using a computer algebra system. Such architectures have been described and used e.g. in [9, 5] and [7].

We use a hybrid setting which delegates the difficult singularities computation to a pool of computer algebra servers and which delegates the rendering of the graphs to a Java applet. The applet uses the results of the server program to plot singularities as hatched vertical regions.

Our first hope was that we could use one of the standard algebra programs like Maple or Mathematica or a numeric program like Matlab as a Java wrapped server in order to directly reuse one of their functions for computing singularities.

Unfortunately, these programs do not have a method for computing singularities that would really fit our purpose. Therefore we had to look for an algorithm of our own that could be used on our server. We implemented this algorithm in Maple, as we could reuse some quite strong functionalities of the Maple library.

2 Plotting Singularities in Existing Systems

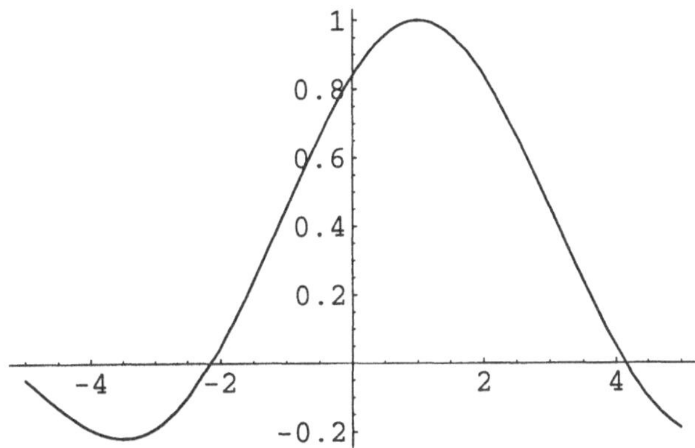

Fig. 1. Mathematica - Plot of the function $sin(x-1)/(x-1)$

The existing systems only plot a vertical line in case the computation of the value of the function in a region give as a result a huge numerical value, or an error. This produces mistakes of two types: first, that some singularities are ignored, second, that the function may be plotted with more singularities than it really has, originating either from the computation or from the visualization. The following examples of plotting show first a plot where the singularity is missed because the limits are bounded (there is a removable discontinuity in $x = 1$) (cf. Figure 1). The second one is the plot of a function with an infinite number of singularities and the last one shows a highly-oscillating function with only one singularity (cf. Figure 2). These last cases show that it is not easy to learn from the plot which function presents singularities and which one does not, and to determine where they are, which is highly desirable in our educational context.

3 Plotter Requirements

We want the students to start the plotter from the mathematics course Web page with their Java-aware browser. They should just type in the function, press the *Plot* button and possibly specify the visible area.

We defined the following requirements for the plotter: first of all, a clear representation of the singularities so that they are distinct from points where the

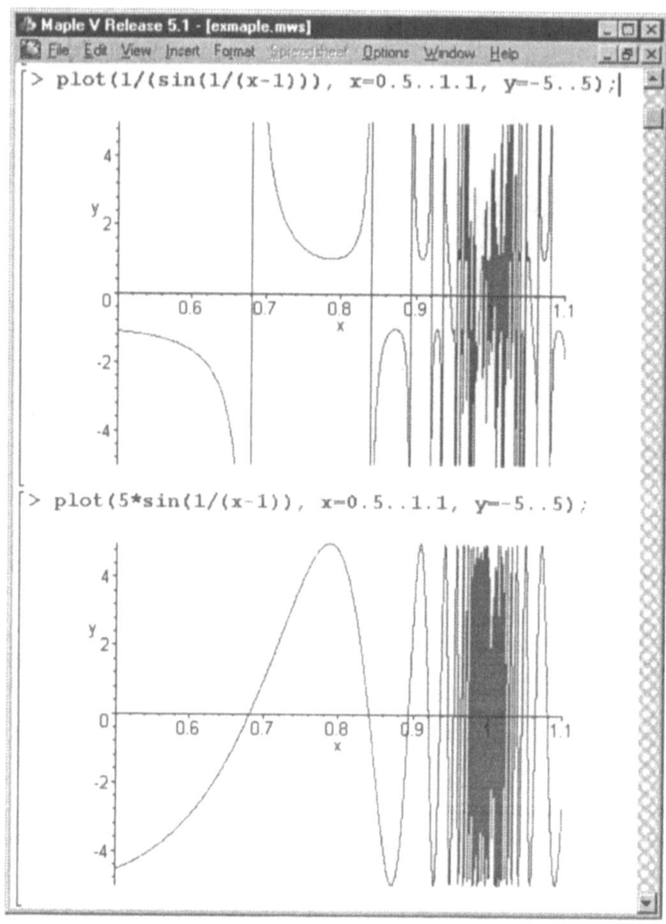

Fig. 2. Plots of the functions $1/sin(1/(x-1))$ and $5sin(1/(x-1))$ - Maple

function is highly oscillating; second, an indication that more than one singularity is present in a region, in the case of two or more singularities which are confounded because of the precision of the plotter, but which can be plotted correctly with an appropriate change of scale; third, an indication that there is an infinite number of singularities around a point (accumulation point), which implies that no scale change would allow all the singularities to be plotted. This leads to the following specifications:

The required data for the computation of the singularities are:

- the function itself,
- the interval where the function is to be plotted
- the precision.

To be able to plot the possible singularities in a way that satisfies our requirement, we need the following data from the server system, i.e. our computer algebra software:

- an array of intervals of doubles, which may be empty, indicating where singularities are to be found, i.e where the function should not be plotted. Example: the output [0.726..0.728, 1.512..1.514] means that there is at least one singularity between 0.726 and 0.728, and at least one between 1.512 and 1.514.
- an array of integers, possibly empty, which are pointers into the first array, pointing to the intervals of the first array which include accumulation points for singularities.

4 Computation of the Singularities

The case of functions like $sin(x)/x$ leads us to use a computer algebra system for the computations of the singularities; this kind of singularities (removable discontinuities) can not be detected with numerical methods because the function and its derivative are bounded and continuous around the singularities, and that the function can be extended to a continuous function.

Among the most common computer algebra software systems, Maple proposes two powerful functions to compute singularities that we could reuse. The *discont* function is an algebraic procedure giving all the existing singularities over the Reals in the form of one or more algebraic expressions or numbers; the *fdiscont* function is a numeric procedure, providing the singularities of a function within a given interval as a list of intervals of doubles.

4.1 The *fdiscont* function

The Maple function *fdiscont* is a numeric function computing the singularities of a function within a given interval looking for discontinuities of the function or of its first derivative. Optionally, *fdiscont* could be applied to certain functions applying the Newton method to its inverse. *fdiscont* is a very powerful tool in case of functions which are not ill-conditioned.

The first case where *fdiscont* fails is for singularities which are not really discontinuities (removable singularities), such as for $sin(x)/x$. The second problem is that, in the case of tricky functions, *fdiscont* detects as singularities points where the growth of the function or of its derivatives are very high. It can be only partly resolved augmenting the precision of the computation. The problem is thus the following: *fdiscont* can compute either too few, or too many singularities.

4.2 The *discont* function

The Maple function *discont* computes singularities of a function over the reals, and gives the result in the form of one or more algebraic expressions or numbers.

These expressions can be relatively complex, using more than one variable, which can be over the reals or the integers. Being an algebraic expression, the result of *discont* provides all the results and no false point.

Unfortunately, this result is not always directly exploitable. Extracting the numeric information that we need out of these algebraic expressions is very often a rather difficult task. Depending on the type of the algebraic expression, it may be possible to find all its roots, and then check which ones are in the plotting interval. Some other expressions returned by *discont* (e.g. some *RootOf* numbers) can be easily converted to reals. In some other cases, it may be possible to find some of the roots modulo an expression, and then to try to find which of these roots are of interest for the plotting. Unfortunately, this algebraic solving is not always possible, so that this algebraic solution wouldn't help in a lot of cases.

4.3 Collecting Numerical Solutions of *discont* and *fdiscont*

For some functions, *discont* returns numerical values or the values can be extracted easily. In case the computation of *discont* shows that there exist singularities and that they are not given in a numerical form, this algorithm then makes a call to *fdiscont* and merges its solution with the subset of numerical solutions given by *discont*.

This allows to find some removable singularities that *fdiscont* cannot find. This allows too, in most of the cases, to suppress the points returned by *fdiscont* which are not singularities, as these points are not kept when *discont* returns only numerical values.

4.4 Computation of the Accumulation Points

As *discont* provides an algebraic expression that gives all the singularities of the given function, it is possible in some cases to check if this function will present an accumulation point for singularities. Given the algebraic expression $A(\alpha)$, if A has a finite limit l approaching $+\infty$ or $-\infty$, then l will be an accumulation point for singularities.

For instance, a call to *discont* for the function $1/sin(1/x)$ gives as result $[0, 1/(\Pi Z)]$. That means that 0 is a singularity, and that for all integers Z, $1/(\Pi Z)$ is a singularity. The limit for $1/(\Pi Z)$ in ∞ is 0, so that there will be an infinite number of singularities around 0.

4.5 Solving *discont* Algebraic Expressions

The following can be proposed as an improvement for a further implementation. As our goal is to plot simple functions, but to plot them with an adequate representation of the singularities, it may be a good solution to apply *discont*, which can theoretically provide us with all the solutions, and, taking into account possibly modulo part of the expression, to look for the roots of this expression using numerical procedures such as *fsolve*.

This would be a simplification of the numerical computation, like the Newton search of zeros implemented in *fdiscont*, because the algebraic expression resulting from *discont* may be simpler than the first function.

5 The Plotting System

The function plotter[3] (FP) visualizes the graphs and singularities of analytical functions with one unknown. It is realized with a Java applet which can be executed by all Web browsers featuring a Java 1.1 compliant virtual machine (cf. Figure 3).

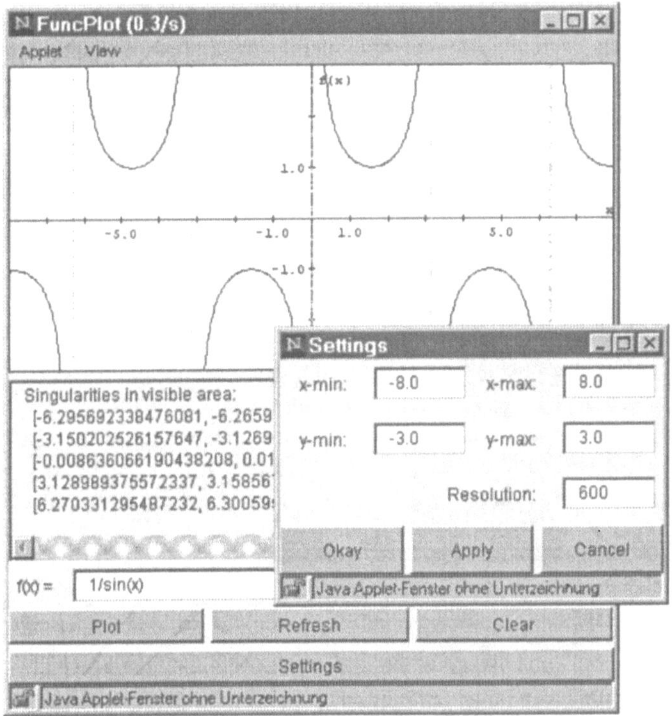

Fig. 3. The function plotter executed by Netscape Navigator

No additional software, like plug-ins or proprietary browser extensions, have to be installed at the client machine. The plotter is used in an introductory course in Mathematics at the University of Tübingen which is available online[4].

[3] http://mfb.informatik.uni-tuebingen.de/FuncPlot.html

[4] http://mfb.informatik.uni-tuebingen.de - Please use login *guest* with password *BIOINFORMATIK*

The Web site features a number of other interactive Java applets to demonstrate various mathematical concepts.

In order to compute the singularities of the plotted functions, the FP applet acts as a client to a Java RMI [8] server application providing access to Maple functionality. We use an adoption of the client/server architecture presented in [9] which wraps the computer algebra system into a Web component.

5.1 The Client

The client applet visualizes arbitrary real functions with one unknown. The user is free to choose the extension of the visible area and the scaling in the x and the y direction. The function expressions are parsed and numerically evaluated by our *MathExpression* Java package. The resulting data structures are instances of the class *Evaluation* that contains all information relevant for the plot, i.e. the specification of the visible area, the number of values to be computed, the computed value pairs and the singularities which are encapsulated by the *Singularities* class.

The class *Singularities* provides convenient access to the intervals that contain single or complex singularities. The actual instances of this class are created by our FP server. The client requests a *Singularities* object by a remote invocation of the *getSingularities()* method which is declared in our *BasicCA* Java interface. The *getSingularities()* method is parameterized by a textual representation of the function expression, the minimum x value, the maximum x value and the precision.

The client window displays the information about detected singularities as hatched regions in the plotting area and as textual information in the text area at the lower end of the window (cf. Figure 4). The distinction between single singularities and accumulation points of singularities is achieved by using different colors for the hatched regions and corresponding labels in the text area, respectively.

5.2 The Server

The server is a Java application that runs on a pool of Solaris multi-processor workstations. The communication interface to the server consists of the Java interface *BasicCA* derived from *java.rmi.Remote* and the class *BasicCAImpl* which implements the methods declared in *BasicCA*. A Java client requests a remote reference to this implementation via the Java *registry* and uses the corresponding object as if it would be local.

When the client (remotely) invokes the *getSingularities()* method the *BasicCAImpl* object creates a new Maple process on the workstation which computes the singularities in the way described in Section 4. The result of the computation is translated to a *Singularities* object which is returned via the network. The access to the operating system is done through the *java.lang.Runtime* programming interface.

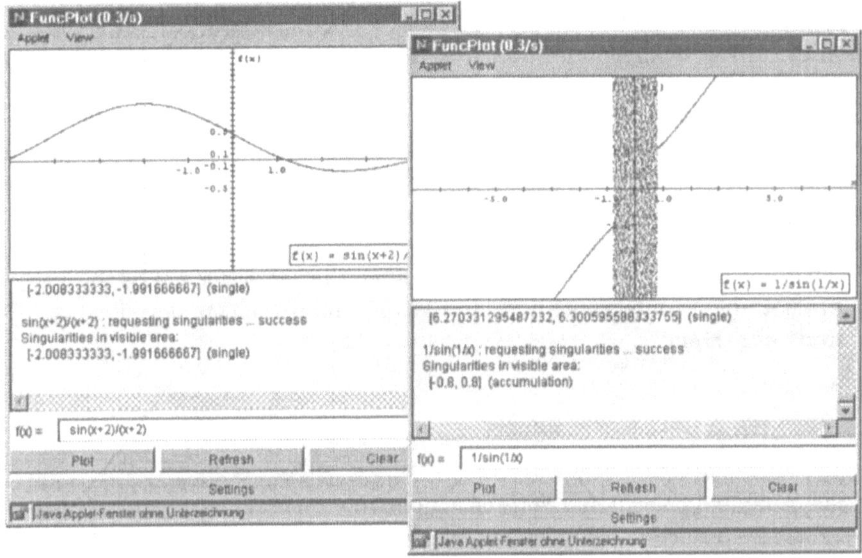

Fig. 4. A single and a complex singularity

Since it is not yet clear how we will deal with Maple licensing issues we integrated a limit for the number of concurrent Maple users into the server. Therefore a client may have to wait in a queue until the corresponding server regains free capacity according to the purchased Maple licenses. All client requests are registered in a detailed log file which is analyzed for evaluation purposes.

6 Discussion

Our plotting system is a hybrid system consisting of a fat server side with a sophisticated component-based architecture featuring parallelism, load balancing and easy extensibility [9, 5] and a Java client applet that performs the numerical part of the evaluation and the rendering of the graph. The server side only computes the singularities. The client is executable in an ordinary Web browser without any further software installation.

The fact that we render the graphs on the client side and do not just transfer bitmaps produced by a CAS or any other sophisticated plotting system has several advantages. The rendering can easily be adjusted and extended. For example, a further applet in our system using the same Java framework provides an animation of the differentiation of a given function[5]. This animation is only possible because we do have available the graphical data on the client side. Furthermore, the user can interact with the graphics. For example the plotter

[5] for instance: http://mfb.informatik.uni-tuebingen.de/private/Mathe-Tools/-examples/example22.html

displays the current mouse pointer position in world coordinates. Thus, the user can query the x and y value of any displayed point just by positioning the mouse pointer over it.

A further benefit is the reduction of network load since we only transfer small serialized *Singularities* objects over the network. This is especially important since we want the students to really *work* with the plotter, trying lots of different parameter settings rather than spending their time waiting for a new plotting result.

The drawback of this approach is that, at the current stage of the project, we are not able to integrate all of the plotting improvements described in [6] or [1] into our system.

7 Summary

In this paper we presented a Web-based plotting system for analytical functions that features a reliable computation and visualization of singularities. This is an improvement over the plotting facilities provided by most of the current computer algebra systems. In an educational context where students of the first or second semesters are addressed, it is crucial that the plotted graphs are correct, since the students may not yet be able to interpret faulty graphs in an appropriate way.

A further improvement is the easy use of the graphical user interface of the plotter because the students do not have to learn the dedicated syntax of a given computer algebra system. Since the rendering of the graph is done in Java on the client side, the client can directly use the graphical data to let the user query the x and y value of any displayed point just by positioning the mouse pointer over it, for example.

The singularities are computed using Maple functionalities. The algorithm we implemented presents an improvement of the existing functions, since it provides numerical results more accurate than the ones of $fdiscont$, and since it allows to compute accumulation points for singularities.

We believe that interactive experiences of mathematical concepts play an important role in education. The presented plotting system offers easy access to those experiences since no additional software has to be installed at the client side and no new syntax has to be learned by the user. First evaluation results [2] show that the students appreciate this new kind of Web-based teaching experience.

References

1. R. Avitzur, O. Bachmann, and N. Kajler. From honest to intelligent plotting. In A.H.M. Levelt, editor, *Proc. of the International Symposium on Symbolic and Algebraic Computation (ISSAC '95)*, Montreal, Kanada, 1995. ACM Press.
2. B. Barquero, U. Creß , and F. W. Hesse. Evaluation der multimedialen Lehrveranstaltungen im Wintersemester 99-00. Internal Report, 2000.

3. D. Bühler, W. Küchlin, G. Gruhler, and G. Nusser. The Virtual Automation Lab - Web based teaching of automation engineering concepts. In *Proc. of the 7th IEEE International Conference and Workshop on the Engineering of Computer Based Systems (ECBS 2000)*. IEEE Computer Society Press, April 2000.
4. D. Bühler, G. Nusser, G. Gruhler, and W. Küchlin. A Java client/server system for accessing arbitrary CANopen fieldbus devices via the Internet. *South African Computer Journal*, (24):239–243, November 1999.
5. M. El Kahoui and A. Weber. Deciding Hopf bifurcations by quantifier elimination in a software-component architecture. *Journal of Symbolic Computation*, 2000. To appear.
6. R. Fateman. Honest plotting, global extrema, and interval arithmetic. In P. Wang, editor, *Proc. of the International Symposium on Symbolic and Algebraic Computation (ISSAC '92)*, Berkeley, USA, July 1992. ACM Press.
7. Y. N. Lakshman, B. Char, and J. Johnson. Software components using symbolic computation for problem solving environments. In O. Gloor, editor, *Proc. of the International Symposium on Symbolic and Algebraic Computation (ISSAC '98)*, Rostock, Germany, 1998. ACM Press.
8. Sun Microsystems, http://java.sun.com/products/jdk/1.1/docs/guide/rmi/spec/-rmiTOC.doc.html. *Remote Method Invocation Specification.*
9. A. Weber, W. Küchlin, and B. Eggers. Parallel computer algebra software as a Web component. *Concurrency: Practice and Experience*, 10(11–13):1179–1188, 1998.

Acknowledgements

We would like to thank Andreas Weber for his useful comments and Wolfgang Westje for his help in setting up various parts of the system.

Stable Self-Oscillatory Regimes in Volterra Models of Three Populations

T.E. Buriev and V.E. Ergashev

Department of differential equations and mathematical physics
Samarkand State University
lyceum@samuni.silk.org

Abstract. For two models of the quantitative dynamics of a predator-prey system such as the generalized three-dimensional Lotka-Voltterra models the existence of the stable self-oscillatory regimes of behavior is investigated basing on qualitative and bifurcation theories as well as on computer experiment.

The models being consideration are described by systems of three ordinary differential equations (ODEs):

$$\dot{x}_i = a_i x_i + \sum a_{ij} x_i x_j, \ (i,j = \overline{1,3}). \tag{1}$$

The systems of type (1) describe the quantitative dynamics (densities of quantity) of three populations interacting in predator-prey regime, where $x_i(t)$ are the quantities of predator (or prey's) populations, the coefficients a_i are the constant parameters characterizing autotrophics (if $a_i > 0$) or heteretrophics (if $a_i < 0$) of the ith populations, the a_{ij} characterize the interactions between the populations i and j (when $i \neq j$), a_{ii} are the coefficients of intraspecific competition.

The first model under consideration is described by the following system of ODEs:

$$\left.\begin{aligned} \dot{x}_1 &= c_1 x_1 - c_{11} x^2 - c_{12} x_1 x_2 + c_{13} x_1 x_3 \\ \dot{x}_2 &= c_2 x_2 + c_{21} x_1 x_2 - c_{22} x_2^2 - c_{23} x_2 x_3 \\ \dot{x}_3 &= c_3 x_3 - c_{31} x_1 x_3 + c_{32} x_2 x_3 - c_{33} x_3^2 \end{aligned}\right\} \tag{2}$$

where all $c_{ij} \geq 0$, $c_i > 0$. The coefficients C_{ij} are the modules of coefficients a_{ij}. Denote by $C = (c_{ij}) \quad (i,j = 1,\bar{}3)$ the matrix of coefficients of the system (2).

The system (2) is obtained from the initial system (1), which obviously demonstrates the directions of populations flows.

In system (2), every population is as well predator as prey.

For system (2) without intraspecific competitions (at $c_{ii} = 0$) and with symmetrical matrix there is the theorem of Volterra [1], according to which any initial quantities of populations unboundedly increase with time.

We consider here the effects of violation of the Volterra's conditions and prove the existence of stable self-oscillatory regimes behavior.

Let us introduce the new variables in system (2) by substitutions

$$x_1 = c_1 x/c_{11}, \quad x_2 = c_2 y/c_{22}, \quad x_3 = c_3 z/c_{33}, \quad t = \tau/c_1. \tag{3}$$

We can then rewrite system (2) as

$$\left.\begin{aligned} \dot{x} &= x(1 - x - b_1 y + d_1 z) \\ \dot{y} &= \gamma_1 y(1 - y - b_2 z + d_2 x) \\ \dot{z} &= \gamma_2 z(1 - z - b_3 x + d_3 y) \end{aligned}\right\} \tag{4}$$

where

$$\begin{aligned} &b_1 = c_2 c_{12}/c_1 c_{22}, \quad d_1 = c_3 c_{13}/c_1 c_{33} \\ &b_2 = c_{22} c_3/c_2 c_{33}, \quad d_2 = c_1 c_{21}/c_2 c_{11} \\ &b_3 = c_{31} c_1/c_3 c_{11}, \quad d_3 = c_{32} c_1/c_3 c_{22} \\ &\gamma_1 = c_2/c_1 \quad \gamma_2 = c_3/c_1 \end{aligned} \tag{5}$$

and $(\cdot)$ denotes τ-derivatives.

The second system has the following form:

$$\left.\begin{aligned} \dot{x}_1 &= c_1 x_1 - c_{12} x_1 x_2 - c_{11} x_1^2 \\ \dot{x}_2 &= -c_2 x_2 + c_{21} x_1 x_2 - c_{22} x^2 - c_{23} x_2 x_3 \\ \dot{x}_3 &= -c_3 x_3 + c_{32} x_2 x_3 - c_{33} x_3^2 \end{aligned}\right\} \tag{6}$$

System (4) describes the dynamics of the quantity of community of three populations interacting by scheme $1 \Rightarrow 2 \Rightarrow 3$ ("producent"-"consument"-"predator"). This system in the absence of intraspecific competition in second and third populations (at $c_{22} = c_{33} = 0$) was investigated in [2], where it was shown that populations coexist in a stable stationary equilibrium.

We consider the system (6) at $c_{22} > 0$, $c_{33} > 0$ and prove the absence of closed integral surface.

After substitution (3) the system (6) can be written as

$$\left.\begin{aligned} \dot{x} &= x(1 - x - b_1 y) \\ \dot{y} &= \gamma_1 y(1 - y - b_2 z + d_2 x) \\ \dot{z} &= \gamma_2 z(1 - z + d_3 y) \end{aligned}\right\} \tag{7}$$

where the coefficients b_i, d_i, γ_i are defined by (5).

II. Consider the system (4). It has the following equilibrium points: $O(0,0,0)$, the points lying on the coordinate axes, and coordinate planes. The system also has the equilibrium point $B(x^*, y^*, z^*)$ in the first coordinate octant of space. The coordinates x^*, y^*, and z^* are determined by the following system of algebraic equations:

$$\left.\begin{aligned} 1 - x - b_1 y + d_1 z = 0 \\ 1 - y - b_2 z + d_2 x = 0 \\ 1 - z - b_3 x + d_3 y = 0 \end{aligned}\right\} \tag{8}$$

Solving this system for coordinates of the equilibrium point B, we obtain:

$$z^* = \frac{1 - b_3 - d_3 + b_1 d_2 + b_1 b_3 - d_2 d_3}{(d_1 d_2 + b_2)(d_3 - b_1 b_3) + (b_1 d_2 + 1)(d_1 b_3 - 1)},$$

$$x^* = \frac{1 - b_1 + (d_1 - b_1 b_2) z^*}{1 + b_1 b_2}, \quad y^* = \frac{1 + d_2 + (d_1 d_2 + b_2) z^*}{1 + b_1 d_2}.$$

Now we will show the existence of the closed integral surface around equilibrium $B(x^*, y^*, z^*)$. Let us consider the system (4) with the following coefficients: $c_1 = c_2 = c_3 = 1$ and matrix C of special form:

$$C = \begin{pmatrix} 1+d & -2 & -d \\ -d & 1+d & -2 \\ -2 & -d & 1+d \end{pmatrix}, \tag{9}$$

where d is a constant parameter and $d \leq -1$. Then the coordinates of equilibrium B are $x^* = y^* = z^* = 1$.

The eigenvalues of linearization matrix of system (4) at point B are determined from the following third-degree algebraic equation:

$$\lambda^3 - 3(d+1)\lambda^2 + 3(1+d^2)\lambda + 3d^2 + 3d + 7 = 0. \tag{10}$$

An analysis of equation (10) shows that one of the eigenvalues is real and negative for all values of parameter d and two eigenvalues are with negative real parts at $d < -\frac{4}{3}$, and positive real parts at $d > -\frac{4}{3}$. At $d = -\frac{4}{3}$ equation (10) has the form:

$$(\lambda + 1)(\lambda^2 + \frac{25}{3}) = 0$$

and its eigenvalues are $\lambda_1 = -1$, $\lambda_{2,3} = \pm\frac{5}{\sqrt{3}}i$. Thus, we have two purely imaginary eigenvalues.

Therefore, the equilibrium $B(1,1,1)$ is a complex node–focus with one stable invariant manifold $W^u(B)$. The first Lyapunov value L_2 (third focus value) of equilibrium $B(1,1,1)$ (calculated by formula from [3]) is negative at $d = -\frac{4}{3}$.

That is why, according to Andronov-Hopf theorem [3], from stable complex node-focus equilibrium point $B(1,1,1)$ at $d > -\frac{4}{3}$, a stable limit cycle arround unstable equilibrium point B appears.

For the proof of the existence of limit cycle in the system at other values of parameters we have made numeral investigation on PC. Calculations were carried out by using the Interactive Integrator TraX and LOCBIF developed at the Institute of Mathematical Problems of Biology of the Russian Academy of Sciences (Pushchino, Moscow region). [4, 5, 6]. The results of investigations of the system show that, in parameter space $\{b_1, b_2, b_3, d_1, d_2, d_3\}$ there is a region (D), to the points of which the existence of a stable limit cycle around equilibrium point $B(x^*, y^*, z^*)$ corresponds. It means that in region (D) of parametric space in community populations, the stable coexistence of all populations in the stable self oscillatory regimes behavior is realized around the equilibrium point $B(x^*, y^*, z^*)$.

III. Investigations of the system (7). The spatial equilibrium point of the system $B(x^*, y^*, z^*)$ has the following coordinates:

$$x^* = \frac{1 - b_1 + b_1 b_2 + b_2 d_3}{1 + b_1 d_2 + b_2 d_3}, \quad y^* = \frac{1 + b_1 + d_2}{1 + b_1 d_2 + b_2 d_3}, \quad z^* = \frac{d_3 + d_2 d_3 - b_1 b_2 - 1}{1 + b_1 d_2 + b_2 d_3}. \tag{11}$$

In the first octant of the phase space the equilibrium B is stable. An analysis of the behavior of the solutions of the system at infinity by generalized substitutions of Poincaré;

$x = v/u, y = w/u, z = 1/u$ for ended axis OZ

$x = 1/u, y = v/u, z = w/u$ for ended axis OX

$x = v/u, y = 1/u, z = w/u$ for ended axis OY

showed that all of the equilibrium states on the Poincaré's sphere are unstable. Consequently the infinitely far part of the positive octant is unstable.

Now we will prove the absence of the closed integral surface for system (7) by using the generalized criteria of Bendixon-Dulac about absence of limit cycle for three-dimensional system of ODEs.

The sufficient condition of absence of the closed integral surface in some region (D) of the phase space is the existence of the continuously differentiable function $\Phi(,,z)$, such that the expression

$$D = \frac{\partial}{\partial x}(P\Phi) + \frac{\partial}{\partial y}(Q\Phi) + \frac{\partial}{\partial z}(R\Phi)$$

keeps its sign in region (D), where

$$P = xP_1(x_1, y_1, z_1), \; Q = yQ_1(x_1, y_1, z_1), \; R = zR_1(x_1, y_1, z_1)$$

are the right-hand sides of system (7).

We will seek the function $\Phi(,,z)$ in the form:

$$\Phi(x, y, z) = x^{k-1} y^{m-1} z^{n-1}, \tag{12}$$

where k, m, n are the constants which have to be determined.

Define expression D:

$$\begin{aligned} D &= \Phi\frac{\partial P}{\partial x} + P\frac{\partial \Phi}{\partial x} + \Phi\frac{\partial Q}{\partial y} + Q\frac{\partial \Phi}{\partial y} + \Phi\frac{\partial R}{\partial z} + R\frac{\partial \Phi}{\partial z} \\ &= \Phi(\frac{\partial P}{\partial x} + \frac{\partial Q}{\partial y} + \frac{\partial R}{\partial z}) + P\frac{\partial \Phi}{\partial x} + Q\frac{\partial \Phi}{\partial y} + R\frac{\partial \Phi}{\partial z}; \end{aligned}$$

$$\begin{aligned} \frac{\partial P}{\partial x} &= 1 - 2x - b_1 y, \\ \frac{\partial Q}{\partial y} &= \gamma_1 - 2\gamma_1 y - \gamma_1 b_2 z + \gamma_1 d_2 x, \\ \frac{\partial R}{\partial z} &= \gamma_2(1 - z + d_3 y) - \gamma_2 z, \end{aligned}$$

$$\frac{\partial \Phi}{\partial x} = (k-1)x^{k-2}y^{m-1}z^{n-1},$$
$$\frac{\partial \Phi}{\partial y} = (m-1)x^{k-1}y^{m-2}z^{n-1},$$
$$\frac{\partial \Phi}{\partial z} = (n-1)x^{k-1}y^{m-1}z^{n-2}$$

$$\begin{aligned} D &= x^{k-1}y^{m-1}z^{n-1}(1-2x-b_1y+\gamma_1-2\gamma_1y-\gamma_1b_2z+\gamma_1d_2z \\ &+ \gamma_2(1-z+d_3y)-\gamma_2z)+x(1-x-b_1y)(k-1)x^{k-2}y^{m-1}z^{n-1} \\ &+ \gamma_1y(1-y-b_2z+d_2x)(m-1)x^{k-1}y^{m-2}z^{n-1} \\ &+ \gamma_2z(1-z+d_3y)(n-1)x^{k-1}y^{m-1}z^{n-2} \\ &= x^{k-1}y^{m-1}z^{n-1}[k+\gamma_1m+\gamma_2n+(-k+\gamma_1d_2m-1)x \\ &+ (-b_1k-\gamma_1m+\gamma_2d_3n-\gamma_1)y+(-\gamma_1b_2m-\gamma_2n-\gamma_2)z]. \end{aligned} \tag{13}$$

The values of constants k, m, n are defined by the system of the algebraic equations

$$\left.\begin{aligned} -k+\gamma_1d_2m-1=0 \\ -b_1k-\gamma_1m+\gamma_2d_3n-\gamma_1=0 \\ -\gamma_1b_2m-\gamma_2n-\gamma_2=0 \end{aligned}\right\} \tag{14}$$

Then we obtain:

$$D = (k+\gamma_1m+\gamma_2n)x^{k-1}y^{m-1}z^{n-1}$$

At

$$k+\gamma_1m+\gamma_2n \neq 0, \tag{15}$$

$D=0$ only on coordinate planes $x=0, y=0, z=0$, which are the integrals of the system. Therefore, if the condition (15) is satisfied in the first octant of phase space, the system has no closed integral surface.

References

1. Volterra, V.: *Mathematical Theory of Struggle for Existence*, Nauka, Moscow, 1978 (in Russian).
2. Bazykin, A.D.: *Mathematical Biophysics of Interacting Populations*, Nauka, Moscow, 1985 (in Russian).
3. Bautin, N.N, Leontovich, E.A.: *Methods of Qualitative Analyses of Dynamic Systems in the Plane*, Nauka, Moscow, 1976 (in Russian).
4. Khibnik, A.I.: *Using TraX: Tutorial to Accompany TraX, A Program for Simulation and Analyses of Dynamical Systems (Version 1.1)*, Exeter Software, New York, 1990.
5. Khibnik, A.I., Kuznetsov, Yu.A., Levitin, V.N. and Nikolaev, E.V.: *LOCBIF: Interactive Local Bifurcation Analyser (Version 2)*, Computer Algebra Netherlands Expertise Center, Amsterdam, 1993.
6. Levitin, V.N.: *TraX: Simulation and Analyses of Dynamical Systems (Version 1.1)*, Exeter Software, New York, 1989.

Computing "Small" 1–Homological Models for Commutative Differential Graded Algebras*

CHATA group**

Depto. de Matemática Aplicada I, Universidad de Sevilla
Avda. Reina Mercedes s/n, 41012 Sevilla, Spain,
e-mail: real@cica.es

Abstract. We use homological perturbation machinery specific for the algebra category [13] to give an algorithm for computing the differential structure of a small 1–homological model for commutative differential graded algebras (briefly, CDGAs). The complexity of the procedure is studied and a computer package in Mathematica is described for determining such models.

1 Introduction

The description of efficient algorithms for homological computation can be considered to be a very important topic in Homological Algebra. These algorithms can be used mainly in the resolution of problems in Algebraic Topology; but this subject also impinges directly on the development of diverse areas such as Combinatorial Designs, Code Theory, Concurrency Theory or Cohomological Physics.

Starting from a finite CDGA A, we establish an algorithm for obtaining an "economical" 1–homological model hBA, in the sense that the number of algebra generators of hBA is less than that of the reduced bar construction $\bar{B}(A)$. In order to get the 1–homology of A, we would need to compute the homology groups of the model hBA. This computation can be reduced to a simple problem of Linear Algebra (see [11] for a complete explanation of this method).

Our main technique is homological perturbation machinery [6, 7, 9]. Homological Perturbation Theory is often used to replace given chain complexes by other smaller, homotopic chain complexes which are more readily computable. An essential notion in this theory is that of contraction. A contraction $c = (f, g, \phi)$ between two differential graded modules (N, d_N) and (M, d_M) is a special homotopy equivalence between both modules such that the corresponding homology groups are isomorphic. The morphisms f, g and ϕ are called *projection*, *inclusion* and *homotopy* of the contraction, respectively. The Basic Perturbation Lemma

* Authors are partially supported by the PAICYT research project FQM–0143 from Junta de Andalucía and the DGES–SEUID research projects PB98–1621–C02–02 from Education and Science Ministry (Spain).

** Álvarez V., Armario J.A., Frau M.D., González–Díaz R., Jiménez M.J., Real P. and Silva B.

is the heart of this theory and states that given a contraction $c = (f, g, \phi)$, and a perturbation δ of d_N (that is, $(d_N + \delta)^2 = 0$), then there exists a new contraction $c_\delta = (f_\delta, g_\delta, \phi_\delta)$ from $(N, d_N + \delta)$ to $(M, d_M + d_\delta)$, satisfying

$$f_\delta = f(1 - \delta\Sigma_c^\delta\phi)\,, \quad g_\delta = \Sigma_c^\delta g\,, \quad \phi_\delta = \Sigma_c^\delta\phi$$

$$d_\delta = f\delta\Sigma_c^\delta g\,, \tag{1}$$

where $\Sigma_c^\delta = \sum_{i\geq 0}(-1)^i\,(\phi\delta)^i = 1 - \phi\delta + \phi\delta\phi\delta - \cdots + (-1)^i(\phi\delta)^i + \cdots$.

It is necessary to emphasize that a nilpotent condition for the composition $\delta\phi$ is required for guaranteeing the finiteness of the formulas.

The basic idea we use in this paper is the establishment (via composition, tensor product or perturbation of contractions) of an explicit contraction from an initial differential graded module N to a free differential graded module M of finite type, so that the homology of N is computable from that of M.

This "modus operandi" has been used by the authors in previous works [1, 8, 3].

Working in the context of CDGAs, Homological Perturbation Theory immediately supplies a general algorithm computing the 1–homology of these objects at graded module level. Nevertheless, this procedure, already presented by Lambe in [12], bears, in general, high computational charges and actually restricts its application to the low dimensional homological calculus.

This algorithm is refined, taking advantage of the multiplicative structures, in [2]. More precisely, the Semifull Algebra Perturbation Lemma [13, Sec. 4] is used for designing the algorithm **Alg1**. The input of this method is a CDGA A given in the form of a "twisted" tensor product of n exterior and polynomial algebras, and the output is a contraction c_δ (produced via perturbation) from the reduced bar construction $\bar{B}(A)$ to a smaller differential graded algebra hBA, which is free and of finite type as a graded module. In this case, we say that the pair $\{c_\delta, hBA\}$ (or, simply, hBA) is a 1–homological model of A. Taking advantage of the fact that the differential d_{hBA} of hBA is a derivation, that is, a morphism compatible with the product of the 1–homological model, it is only necessary to know the value of this morphism applied to the generators of the model (let us observe that there are n algebra generators). This implies a substantial improvement in the computation of the differential on the small model hBA.

In spite of this improvement, the computational cost for determining the morphism d_{hBA} applied to an algebra generator of hBA is enormous, since the differential d_{hBA} follows the formula (1) and the homotopy ϕ of c has an essentially exponential nature not only in time but also in space.

We develop some techniques, which comprise what we call *Inversion Theory* and which first appears in [13]. In consequence, we refine the formula for ϕ, which is involved in the description of d_{hBA}. This study is based on the observation that the the projection f applied to certain elements (those "with inversions") is always null. It follows that a not insignificant number of terms in the formula of the morphism ϕ can be eliminated in the composition $f\delta(\phi\delta)^i g$, which appears in the formula (1) of d_δ. In such a way, we derive an upgraded algorithm **Alg2**.

The article is organized as follows: Notation and terminology are introduced in Section 2. In Section 3 the algorithm **Alg1**, which was described in [2] is recalled. Our contribution starts in Section 4 which is devoted to explaining Inversion Theory and describing the algorithm **Alg2**. An analysis of the complexity of **Alg1** and **Alg2** for computing the differential structure of the small 1–homological model hBA is carried out in Section 5 and a comparison between both algorithms is given. Finally, in Section 6 we also give several examples illustrating the implementation of **Alg2** carried out using Mathematica 3.0.

2 Preliminaries

Although relevant notions of Homological Algebra are explained through the exposition of this paper, most common concepts are not explicitly given (they can be found, for instance, in [10] or [15]).

Let Λ be a commutative ring with the non zero unit, which will be considered to be the ground ring. A *DGA–module* (M, d_M, ξ_M, η_M) is a module endowed with:

- A graduation, that is, $M = \oplus_{n \in \mathbb{N}} M_n$.
- A differential, $d_M : M \to M$, which decreases the degree by one and satisfies $d_M^2 = 0$.
- An augmentation, $\xi_M : M \to \Lambda$, with $\xi_M d_1 = 0$.
- A coaugmentation, $\eta : \Lambda \to M$, with $\xi_M \eta_M = 1_\Lambda$.

We will respect Koszul conventions. The *homology* of a differential graded module M, is a graded module $H_*(M)$, where $H_n(M) = \mathrm{Ker}\, d_n / \mathrm{Im}\, d_{n+1}$. We are specially interested on CDGAs, $(A, d_A, *_A, \xi_A, \eta_A)$ which are differential graded modules endowed with a product, $*_A$, that is commutative in a graded sense. A morphism $\delta : A \to A$ which decreases the degree by one, is a *derivation* if $\delta *_A = *_A(1 \otimes \delta + \delta \otimes 1)$.

Three particular algebras are of special interest in the development of this paper: exterior, polynomial and divided power algebras. Let n be a fixed non–negative integer.

- The *exterior algebra* $E(x, 2n+1)$ is the graded algebra with generators 1 and x of degrees 0 and $2n+1$, respectively, and the trivial product, that is, $x \cdot x = 0$ and $x \cdot 1 = x$.
- The *polynomial algebra* $P(y, 2n)$ consists in the graded algebra with generators 1 of degree 0 and y of degree $2n$. The product is the usual one in polynomials, i.e.: $y^i \cdot y^j = y^{i+j}$, for non negative integers i and j.
- Finally, the *divided power algebra* $\Gamma(y, 2n)$ is the graded algebra with generators 1 and y ($y = y^{(1)}$) of respective degrees 0 and $2n$. The product is defined by the rules $y^{(i)} \cdot y^{(j)} = \binom{i+j}{i} y^{(i+j)}$, i and j being non–negative integers.

Each one of these three types of algebras can be considered as a CDGA with the trivial differential.

Now, we shall recall a standard algebraic tool which allows us to preserve the product structure of the initial CDGA through the procedure of homological computation. The *reduced bar construction* [10] associated to a CDGA A is defined as the differential graded module $\bar{B}(A)$:

$$\bar{B}(A) = \Lambda \oplus \text{Ker } \xi_A \oplus (\text{Ker } \xi_A \otimes \text{Ker } \xi_A) \oplus \cdots \oplus (\text{Ker } \xi_A \otimes \cdots \otimes \text{Ker } \xi_A) \oplus \cdots .$$

An element from $\bar{B}(A)$ is denoted by $\bar{a} = [a_1|\cdots|a_n]$. There is a *tensor graduation* $|\ |_t$ given by $|[a_1|\cdots|a_n]|_t = \sum_{i=1}^n |a_i|$, as well as a *simplicial graduation* $|\ |_s$, which is defined by $|\bar{a}|_s = |[a_1|\cdots|a_n]\,|_s = n$. The total degree of $\bar{a}$ is given by $|\bar{a}| = |\bar{a}|_t + |\bar{a}|_s$.

The total differential is given by the sum of the tensor and simplicial differentials. The *tensor differential* is defined by:

$$d_t[a_1|\cdots|a_n] = -\sum_{i=1}^{n}(-1)^{|[a_1|\cdots|a_{i-1}]|}[a_1|\cdots|d_A a_i|\cdots|a_n]\,.$$

The *simplicial differential* acts by cutting down the simplicial degree by using the product given in A.

When the algebra A is commutative, it is possible to define a multiplicative structure on $\bar{B}(A)$ (via an operator called the *shuffle product*), so that the reduced bar construction also becomes a CDGA.

Given two non–negative integers p and q , a *(p,q)–shuffle* is defined as a permutation π of the set $\{0,\ldots,p+q-1\}$, such that $\pi(i) < \pi(j)$ when $0 \le i < j \le p-1$ or $p \le i < j \le p+q-1$ is the case.

Let us observe that there are $\binom{p+q}{p}$ different (p,q)–shuffles.

So, given a CDGA A, the *shuffle product* $\star : \bar{B}(A) \otimes \bar{B}(A) \longrightarrow \bar{B}(A)$, is defined by:

$$[a_1|\cdots|a_p] \star [b_1|\cdots|b_q] = \sum_{\pi\in\{(p,q)\text{–shuffles}\}} (-1)^{\varepsilon(\pi,a,b)}[c_{\pi(0)}|\cdots|c_{\pi(p+q-1)}]\,;$$

where $(c_0,\ldots,c_{p-1},c_p,\ldots,c_{p+q-1}) = (a_1,\ldots,a_p,b_1,\ldots b_q)$ and

$$\varepsilon(\pi,a,b) = \sum_{\pi(i)>\pi(p+j)} |[a_i]||[b_j]|\,.$$

Let n be a non–negative integer. The n–homology of a CDGA A (see [10]) consists in the homology groups of the iterated reduced bar construction $\bar{B}^n(A) = \bar{B}(\bar{B}^{n-1}(A))$, being $\bar{B}^0(A) = A$.

Let $\{A_i\}_{i\in I}$ be a set of CDGAs. A *twisted tensor product* $\tilde{\otimes}^{\rho}_{i\in I}A_i$ is a CDGA satisfying the following conditions:

i) $\tilde{\otimes}^{\rho}_{i\in I}A_i$ coincides with the tensor product $\otimes_{i\in I}A_i$ as a graded algebra.

ii) The differential operator consists in the sum of the differential of the banal tensor product and a derivation ρ.

A *contraction* $c : \{N, M, f, g, \phi\}$ [4,9], also denoted by $(f, g, \phi) : N \overset{c}{\Rightarrow} M$, from a differential graded module (N, d_N) to a differential graded module (M, d_M) consists in a homotopy equivalence determined by three morphisms f, g and ϕ; $f : N_* \to M_*$ (projection) and $g : M_* \to N_*$ (inclusion) being two differential graded module morphisms and $\phi : N_* \to N_{*+1}$ a homotopy operator. Moreover, these data are required to satisfy the following rules:

$$fg = 1_M\,, \quad \phi d_N + d_N\phi + gf = 1_N\,, \quad f\phi = 0\,, \quad \phi g = 0\,, \qquad \phi\phi = 0\,.$$

There are two basic operations between contractions which give place to new contractions: tensor product and composition of contractions.

In this paper we use a particular type of contraction between CDGAs. Given two CDGAs A and A', a *semifull algebra contraction* $(f, g, \phi) : A \Rightarrow A'$ [13] consists of an inclusion g that is a morphism of CDGAs, a quasi–algebra projection f and a quasi–algebra homotopy ϕ. We recall that

1. The projection f is said to be a *quasi–algebra projection* whenever the following conditions hold:

$$f(\phi *_A \phi) = 0, \quad f(\phi *_A g) = 0, \quad f(g *_A \phi) = 0\,. \qquad (2)$$

2. The homotopy operator ϕ is said to be a *quasi–algebra homotopy* if

$$\phi(\phi *_A \phi) = 0, \quad \phi(\phi *_A g) = 0, \quad \phi(g *_A \phi) = 0\,. \qquad (3)$$

The class of all semifull algebra contractions is closed under composition and tensor product of contractions. Moreover, this class is closed under perturbation.

Theorem 1. *[13]*
Let $c : \{N, M, f, g, \phi\}$ be a semifull algebra contraction and $\delta : N \to N$ be a perturbation–derivation of d_N. Then, the perturbed contraction c_δ, is a new semifull algebra contraction.

3 Computability of the 1–Homology of CDGAs. First Algorithm

Here we recall the algorithm described in [2] for the computation of a 1–homological model of a CDGA.

It is commonly known that every CDGA A "factors", up to homotopy equivalence, into a tensor product of exterior and polynomial algebras endowed with a differential–derivation; in the sense that there exists a homomorphism connecting both structures, which induces an isomorphism in homology.

In fact, our input is a twisted tensor product of algebras $A = \tilde{\otimes}^{\rho}_{i \in I} A_i$ where I denotes a finite set of indices, ρ is a differential–derivation and A_i an exterior or a polynomial algebra, for every i. In our algorithmic approach, we encode A by

1. a sequence of non–negative integers $n_1 \leq n_2 \leq \cdots \leq n_k$, such that n_i represents the degree of the algebra generator x_i of A_i;
2. a k–vector $\bar{v} = (v_1, v_2, \ldots, v_k)$, such that v_i is $\rho(x_i)$ for all i.

The principal goal is to obtain a "chain" of semifull algebra contractions starting at the reduced bar construction $\bar{B}(A)$ and ending up at a smaller free (as a module) CDGA. In that way, we determine a 1–homological model for A.

Now, we consider the following three **semifull** algebra contractions which are used, firstly, to find the structure of a graded module of a 1–homological model for a CDGA:

– The contraction defined in [4, 5] from $\bar{B}(A \otimes A')$ to $\bar{B}(A) \otimes \bar{B}(A')$, where A and A' are two CDGAs.

$$C_{\bar{B}\otimes} : \{\bar{B}(A \otimes A'), \bar{B}(A) \otimes \bar{B}(A'), f_{\bar{B}\otimes}, g_{\bar{B}\otimes}, \phi_{\bar{B}\otimes}\};$$

- $f_{\bar{B}\otimes}[a_1 \otimes a'_1|\cdots|a_n \otimes a'_n]$

$$= \sum_{i=0}^{n} \xi_A(a_{i+1} *_A \cdots a_n)\xi_{A'}(a'_1 *_{A'} \cdots a'_i)[a_1|\cdots|a_i] \otimes [a'_{i+1}|\cdots|a'_n]$$

- $g_{\bar{B}\otimes}([a_1|\cdots|a_n] \otimes [a'_1|\cdots|a'_m])$

$$= [a_1 \otimes \theta'|\cdots|a_n \otimes \theta'] \star [\theta \otimes a'_1|\cdots|\theta \otimes a'_n],$$

  where θ and θ' are the units in A and A' respectively.

- up to sign, $\phi_{\bar{B}\otimes}([a_1 \otimes a'_1|\cdots|a_n \otimes a'_n])$

$$= \sum \pm\xi_A(a_{n-q+1} *_A \cdots a_n)[a_1 \otimes a'_1|\cdots|a_{\bar{n}-1} \otimes a'_{\bar{n}-1} \\ |a'_{\bar{n}} *_{A'} \cdots a'_{n-q}|c_{\pi(0)}|\cdots|c_{\pi(p+q)}],$$

  where $\bar{n} = n-p-q$, $(c_0, \ldots, c_{p+q}) = (a_{\bar{n}}, \ldots, a_{n-q}, a'_{n-q+1}, \ldots a'_n)$ and the sum is taken over all the $(p+1, q)$–shuffles π and $0 \leq p \leq n-q-1 \leq n-1$.

  Let us note that the complexity of $g_{\bar{B}\otimes}$ and $\phi_{\bar{B}\otimes}$ is exponential since shuffles are involved in both formulas.

  Given a tensor product $\otimes_{i\in I} A_i$ of CDGAs, a contraction from $\bar{B}(\otimes_{i\in I} A_i)$ to $\otimes_{i\in I} \bar{B}(A_i)$ is easily determined by applying $C_{\bar{B}\otimes}$ several times in a suitable way. This new contraction is also denoted by $C_{\bar{B}\otimes}$.

– The isomorphism of differential graded algebras (therefore, a contraction)

$$C_{\bar{B}E} : \{\bar{B}(E(u, 2n+1)), \Gamma(\underline{u}, 2n+2), f_{\bar{B}E}, g_{\bar{B}E}, 0\}$$

  described in [5], where

$$f_{\bar{B}E}([u|\overset{m \text{ times}}{\cdots}|u]) = \underline{u}^{(m)}; \quad g_{\bar{B}E}(\underline{u}^{(m)}) = [u|\overset{m \text{ times}}{\cdots}|u].$$

– The contraction

$$C_{\bar{B}P} : \{\bar{B}(P(v,2n)), E(\underline{v}, 2n+1), f_{\bar{B}P}, g_{\bar{B}P}, \phi_{\bar{B}P}\}$$

stated in [5], where

$$f_{\bar{B}P}([v^r]) = \begin{cases} 0 \text{ if } r \neq 1 \\ \underline{v} \text{ if } r = 1 \end{cases}, \quad f_{\bar{B}P}([v^{r_1}|\cdots|v^{r_m}]) = 0\,;$$

$$g_{\bar{B}P}(\underline{v}) = [v] \quad \text{and} \quad \phi_{\bar{B}P}([v^{r_1}|\cdots|v^{r_m}]) = [v|v^{r_1-1}|\cdots|v^{r_m}]\,.$$

Thanks to these three contractions, it is possible to establish, by composition and tensor product of contractions, the following semifull algebra contraction $C = (f, g, \phi)$:

$$\bar{B}(\otimes_{i\in I} A_i) \;\Rightarrow\; \otimes_{i\in I}\bar{B}(A_i) \;\Rightarrow\; \otimes_{i\in I} hBA_i\,,$$

where hBA_i represents an exterior or a divided power algebra with a generator $\underline{x}_i$, depending on whether A_i is a polynomial or an exterior algebra with a generator x_i.

In order to obtain the differential structure of the 1–homological model for the twisted tensor product $\tilde{\otimes}^{\rho}_{i\in I} A_i$, the next step is to perturb C. The perturbation ρ produces a perturbation–derivation δ on the tensor differential of $\bar{B}(\otimes_{i\in I} A_i)$:

$$\delta([a_1|\cdots|a_n]) = \sum_{i=1}^{n} (-1)^{|[a_1|\cdots|a_{i-1}]|}[a_1|\cdots|\rho(a_i)|\cdots|a_n]\,.$$

Now, by applying Theorem 1, a new semifull algebra contraction $(f_\delta, g_\delta, \phi_\delta)$ is constructed:

$$\bar{B}(\tilde{\otimes}^{\rho}_{i\in I} A_i) \overset{(C)_\delta}{\Rightarrow} (\otimes_{i\in I} hBA_i, d_\delta)\,,$$

where the differential d_δ is determined by the perturbation procedure (Basic Perturbation Lemma). That means that $hBA = \otimes_{i\in I}(hBA_i, d_\delta)$ is a 1–**homological model** of $A = \tilde{\otimes}_{i\in I} A_i$. Let us emphasize that the Basic Perturbation Lemma provides finite formulas. Indeed, this is a consequence of two facts: the perturbation δ does not change the simplicial degree and ϕ increases this degree.

Procedure 1 *Algorithm* **Alg1**.

Input: *A finite CDGA A:* $((n_1, \ldots, n_k), (v_1, \ldots, v_k))$.

Output: $((n_1+1, \ldots, n_k+1), (w_1, \ldots, w_k))$
a 1–homological model of the CDGA $\otimes^{\rho}_{i=1,\ldots,k} A_i$, *$A_i$ being the exterior algebra $E(x_i, n_i)$, if n_i is odd and $P(x_i, n_i)$ if n_i is even.*

$w_1 = 0$,
for $i = 2$ **to** k
  $w_i = d_\delta(\underline{x}_i)$, *where* x_i *is the algebra generator of degree* n_i
endfor

Naturally, the first components of the vector $\bar{v}$ must be zero, because they correspond to the image of the algebra generators with the lowest degree under ρ.

Moreover, a general algorithm for computing the 1–homology of CDGAs can be described . Clearly, the homology of the 1–homological model obtained can be computed using an algorithm based on the establishment of Smith's normal form of the matrices representing the differentials at each degree [14, 11].

The computational cost of constructing the contraction $(C)_\delta$ is high. Let us note that both the inclusion and homotopy operators of the contraction $C_{B\otimes}$ give an answer in exponential time. In fact, the formula of the differential operator d_δ produced by the homological perturbation machinery is given by:

$$d_\delta \;=\; f\,\delta(1 - \phi\,\delta + \phi\,\delta\,\phi\,\delta - \cdots)\,g\,.$$

With regard to the previous remarks, a first impression is that obtaining d_δ generally becomes a procedure of exponential nature.

It is possible to take advantage of d_δ being a derivation. Indeed, the fact that d_δ is a derivation implies that it is only necessary to know this morphism applied to the generators of the model (let us observe that there are as many generators as the cardinal of the set of indices I indicated). This is an enormous improvement in the computation of the differential on the small model. In spite of this, computing d_δ on an algebra generator is extremely time–consuming.

4 Inversion Theory

In this section, we go further in the simplification of the computation of the differential d_δ. For clarity, we begin this work considering only two algebras.

As we have seen before, obtaining d_δ is an extremely expensive procedure. The morphism responsible for this is the homotopy operator, ϕ, due, essentially, to the shuffles that are involved in the formulas of $\phi_{B\otimes}$ and $g_{B\otimes}$. We intend to eliminate these shuffles, and, with this aim in mind, we define the concept of inversion.

Definition 1. Let A and A' be CDGAs and let us consider a homogeneous element $[a_1 \otimes a'_1 | a_2 \otimes a'_2 | \cdots | a_n \otimes a'_n]$ from $\bar{B}(A \otimes A')$. We say that a component $\theta \otimes a'_i$ from that element, is responsible for an *inversion*, if there exists an index $j > i$ with $a_j \neq \theta$ (where θ is the unit of A). In this sense, such an element presents k inversions if there exist k components responsible for an inversion.

We will say that an element from $\bar{B}(A \otimes A')$ has k inversions, if it is a sum of elements which each have, as a minimum, k inversions.

Let us consider the contraction

$$(f_{B\otimes}, g_{B\otimes}, \phi_{B\otimes}) : \bar{B}(A \otimes A') \Rightarrow \bar{B}(A) \otimes \bar{B}(A')$$

described in the previous section. We analyze the behaviour of the component morphisms of this contraction with respect to inversions. For this purpose, we do not take into account the signs in the formulas referred to.

- The image of an element with at least one inversion under $f_{B\otimes}$, is null.
- The injection $g_{B\otimes}$, applied to $[a_1|\cdots|a_n] \otimes [a'_1|\cdots|a'_m]$, produces:
  - a unique term with no inversions (that one which comes from juxtaposition),
  - n terms with one inversion,
  - $\binom{n+m}{n} - n - 1$ terms with more than one inversion.
- As for the homotopy operator $\phi_{B\otimes}$, we can state that the image of a homogenous element under $\phi_{B\otimes}$ gives rise to a sum of elements which, if non null, have at least one more inversion than the original one. Let us note that an inversion is produced by the component $a'_{\bar{n}} *_{A'} \cdots a'_{n-q}$, which is always on the left side of those components $a_{\bar{n}}, \ldots, a_{n-q}$ of each summand in the formula of $\phi_{B\otimes}$.

Let us consider the contraction which provides us with a 1–homological model for the tensor product of two CDGAs, A and A':

$$(f, g, \phi) : \bar{B}(A \otimes A') \Rightarrow \bar{B}(A) \otimes \bar{B}(A') \Rightarrow hBA \otimes hBA' \qquad (4)$$

where

$$\begin{aligned} f &= (f_{BA} \otimes f_{BA'})f_{B\otimes}, \\ g &= g_{B\otimes}(g_{BA} \otimes g_{BA'}) \\ \phi &= \phi_{B\otimes} + g_{B\otimes}(\phi_{BA} \otimes g_{BA'}f_{BA'} + 1_{BA} \otimes \phi_{BA'})f_{B\otimes} \end{aligned}$$

Let us note that the image of an element with an inversion under f is also null, since the first morphism applied is $f_{B\otimes}$.

Now we assume that there is a perturbation ρ of the tensor product of the algebras A and A'. This perturbation induces, in a natural way, a perturbation δ on $\bar{B}(A \otimes A')$. Let us analyze the behaviour of such a morphism with respect to inversions.

Lemma 1. *Let us consider a perturbation δ for $\bar{B}(A \otimes A')$ induced by a perturbation–derivation ρ for $A \otimes A'$ such that $\rho(A) \subset A$. The image of a homogeneous element with k inversions under δ, is a sum of elements with at least $k - 1$ inversions.*

Proof. Let us point out that a component of a homogeneous element from $\bar{B}(A \otimes A')$ is responsible for, at most, one inversion and that δ acts only on a component of the element at each term of the resultant sum. □

Attending to the Basic Perturbation Lemma, one can obtain from the contraction (4), a new contraction:

$$(f_\delta, g_\delta, \phi_\delta) : \bar{B}(A\tilde{\otimes}^\rho A') \Rightarrow (hBA \otimes hBA', d_\delta)\,.$$

We recall the formula for d_δ:

$$d_\delta \;=\; f\,\delta\,(1 - \phi\,\delta + \phi\,\delta\,\phi\,\delta - \cdots)\,g.$$

We can observe that f is the last morphism applied in the formula. If at any stage, an element y obtained by applying ϕ, has more than one inversion, then $\delta(y)$ will have at least one inversion. In this way, each time we apply $\delta\,\phi$, we obtain a sum of homogeneous elements with at least one inversion, and, therefore, the image of these elements under f is null. This means that we only have to consider the summands of ϕ having, at most, one inversion.

In consequence, we can establish the following theorem where we considerably reduce the complexity of the computation of d_δ.

Theorem 2. *The formula for ϕ, that is involved in the definition of d_δ, is the following:*

$$\phi = \bar{\phi}_{B\otimes} + \bar{g}_{B\otimes}(\phi_{BA} \otimes g_{BA'}f_{BA'} + 1 \otimes \phi_{BA'})f_{B\otimes},$$

where

– $\bar{\phi}_{B\otimes}([a_1 \otimes a'_1|\cdots|a_n \otimes a'_n])$

$$= \sum_{0\le p\le n-q-1\le n-1} (-1)^{\varphi(n,p,q)}\xi_A(a_{n-q+1} *_A \cdots a_n)[a_1 \otimes a'_1|\cdots|a_{\bar{n}-1} \otimes a'_{\bar{n}-1} \\ |a'_{\bar{n}} *_{A'} \cdots a'_{n-q}|a_{\bar{n}}|\cdots|a_{n-q}|a'_{n-q+1}|\cdots|a'_n]$$

being $\bar{n} = n - p - q$ and

$$\varphi(n,p,q) = \bar{n} - 1 + |[a_1|\cdots|a_{\bar{n}-1}]|_t + |[a'_1|\cdots|a'_{n-q}]|_t \\ + \sum_{k=0}^{p}\sum_{\ell=0}^{k} |a_{n-q-k}|\;|a'_{n-q-\ell}|\,.$$

– $\bar{g}_{B\otimes}([a_1|\cdots|a_n] \otimes [a'_1|\cdots|a'_m])$

$$= [a_1|\cdots|a_n|a'_1|\cdots|a'_m]$$

$$+\sum_{i=1}^{n-1}(-1)^{|[a_{i+1}|\cdots|a_n]||a'_1|}\,[a_1|\cdots|a_i|a'_1|a_{i+1}|\cdots|a_n|a'_2|\cdots|a'_m]$$

$$+(-1)^{|a'_1||[a_1|\cdots|a_n]|}\,[a'_1|a_1|\cdots|a_n|a'_2|\cdots|a'_m]\,.$$

Let us note that now the number of summands in the formula above for $\phi_{B\otimes}$ is

$$\sum_{q=0}^{n-1}\sum_{p=0}^{n-q-1} 1 = \frac{n^2+n}{2},$$

in contrast to the original number of summands:

$$\sum_{q=0}^{n-1}\sum_{p=0}^{n-q-1} \binom{p+q+1}{q} = 2^{n+1} - n - 2.$$

On the other hand, the formula for $g_{B\otimes}$ is reduced to n summands, instead of $\binom{m+n}{n}$.

This theorem is easy to generalize, by induction, to the general case of a twisted tensor product of CDGAs $\tilde{\otimes}^{\rho}_{i\in I}A_i$ with $I = \{1,\ldots,n\}$, where A_i is an exterior or a polynomial algebra with generator x_i and $|x_i| \leq |x_{i+1}|$. Therefore, as we saw in Section 3, the following semifull algebra contraction can be established:

$$\bar{B}(\otimes_{i=1}^{n}A_i) \Rightarrow \otimes_{i=1}^{n}\bar{B}(A_i) \Rightarrow \otimes_{i=1}^{n}hBA_i$$

where hBA_i is a polynomial or a divided power algebra.

The key to understanding the generalization is the fact that the inversions in $\bar{B}(\otimes_{i=1}^{n}A_i)$ are those of the last tensor product (as they were defined at the beginning of the Section, with $A = \otimes_{i=1}^{n-1}A_i$ and $A' = A_n$) along with those of $\otimes_{i=1}^{n-1}A_i = (\otimes_{i=1}^{n-2}A_i)\otimes A_{n-1}$ with respect to the last tensor product, and so on.

Summing up, we obtain an algorithm **Alg2** having the same input and output as the Algorithm **Alg1** of the last section, but speeding up the steps concerning the image of the algebra generators under d_δ.

Procedure 2 *Algorithm* **Alg2**.

Input *and* **Output***: the same as in* **Alg1**.

$w_1 = 0$,
for $i = 2$ **to** k
  $w_i = d_\delta(\underline{x}_i)$, *where* x_i *is the algebra generator of degree* n_i
  (using Theorem 4)
endfor

5 Complexity

In this section we give a comparison of the algorithms **Alg1** and **Alg2** from the point of view of their complexity. We are mainly interested in measuring the efficiency of the corresponding steps concerning the obtention of the differential d_δ. We consider the degree of the algebra generator as the size of an instance. We take as elementary operations those ones generating each homogeneous term produced by the different morphisms and a worst–case analysis of the algorithms is carried out.

We calculate the total number of elementary operations needed for computing d_δ on a generator $\underline{x}_k$ of degree k, for both **Alg1** and **Alg2**. We hold that this number for **Alg1** is

$$\sum_{i=0}^{\lfloor k/k_0+1 \rfloor} (i!\,r^i + ((i+1)! + (i+2)!\,)r^{i+1}) \left(\prod_{j=1}^{i} 2^{j+1} - j - 2 \right),$$

and for **Alg2** is

$$\sum_{i=0}^{\lfloor k/k_0+1 \rfloor} (r^i + (i+3)r^{i+1}) \left(\prod_{j=1}^{i} \frac{j^2+j}{2} \right),$$

where $k_0 = \min_{1 \leq i \leq n} |x_i|$ and r is the maximum number of summands given by $\rho(x_i)$, where x_i ranges over all the algebra generators of A.

In the following table, the required time for computing $d_\delta(\underline{x}_k)$ is showed, supposing that our computer carries out 10^6 elementary operations per second. Let us denote $s = \lfloor k/k_0 + 1 \rfloor$. Note that, for example, that $s = 5$ means that if $k_0 = 10$, the degrees of the algebra generators range over the set $\{10, 11, \ldots, 54, 55\}$.

Table 1. Required time

	in **Alg1**	in **Alg2**
$s = 3;\ r = 2$	0.1 sec.	0.002 sec.
$s = 3;\ r = 3$	0.5 sec.	0.009 sec.
$s = 4;\ r = 3$	31.20 sec.	0.04 sec.
$s = 5;\ r = 3$	3.19 days	17.6 sec.
$s = 6;\ r = 2$	1.45 years	1.17 min
$s = 6;\ r = 3$	24.75 years	19.56 min

6 Implementation Performance

The algorithm **Alg2** has been implemented. The user supplies an encoding of a finite CDGA in the form of a twisted tensor product of exterior and polynomial algebras to the program, which computes an encoding of a small 1–homological model of this algebra.

This program is written in Mathematica 3.0, consisting in 300 lines of code and 10 basic functions.

In order to give some indication of the implementation, we report on the time taken to compute 1–homological models for certain CDGAs:

1. $E(x_1, x_2, x_3; 1) \otimes P(x_4, x_5; 2) \otimes E(x_6; 3) \otimes P(x_7; 4) \otimes P(x_8; 6)$.

   Input: $((1,1,1,2,2,3,4,6),$
   $(0,0,0,x_1 - x_2, x_2, x_1x_2, x_2x_4 + x_1x_5 + x_1x_2x_3, x_1x_2x_6))$.

   Output: $((2,2,2,3,3,4,5,7),(0,0,0,\underline{x}_1 - \underline{x}_2, \underline{x}_2, 0, \underline{x}_1\underline{x}_2 - 2\underline{x}_2^2, 0))$

   Time: $d_\delta(\underline{x}_4)$ in 0.55 sec.
   $d_\delta(\underline{x}_5)$ in 0.27 sec.
   $d_\delta(\underline{x}_6)$ in 0.33 sec.
   $d_\delta(\underline{x}_7)$ in 2.14 sec.
   $d_\delta(\underline{x}_8)$ in 0.28 sec.

2. $E(x_1; 1) \otimes P(x_2; 2) \otimes P(x_3; 6) \otimes P(x_4; 10) \otimes P(x_5; 26)$.

   Input: $((1,2,6,10,26),(0,-2x_1, x_1x_2^2, 3x_1x_2x_3, 8x_1x_2x_3^2x_4))$.

   Output: $((2,3,7,11,27),(0,-2\underline{x}_1, -8\underline{x}_1^3, -192\underline{x}_1^5, 21799895040\underline{x}_1^{13}))$

   Time: $d_\delta(\underline{x}_2)$ in 0.16 sec.
   $d_\delta(\underline{x}_3)$ in 0.60 sec.
   $d_\delta(\underline{x}_4)$ in 2.36 sec.
   $d_\delta(\underline{x}_5)$ in 1571.47 sec.

3. $E(x_1; 1) \otimes P(x_2; 2) \otimes P(x_3; 4) \otimes E(x_4; 5) \otimes E(x_5; 7) \otimes P(x_6; 14)$.

   Input: $((1,2,4,5,7,14),(0,-x_1, x_1x_2, 0, 2x_1x_4, -x_1x_4x_5))$.

   Output: $((2,3,5,6,8,15),(0,-\underline{x}_1, -\underline{x}_1^2, 0, 0))$

   Time: $d_\delta(\underline{x}_2)$ in 0.17 sec.
   $d_\delta(\underline{x}_3)$ in 0.33 sec.
   $d_\delta(\underline{x}_4)$ in 0.26 sec.
   $d_\delta(\underline{x}_5)$ in 0.17 sec.

For each algebra, we summarize the results and the time taken to compute a description of the model. All CPU times are in seconds and calculations were carried out on a Pentium III, 128Mb RAM, 7.2Gb Hard disk space.

This program produces as output an encoding of a certain differential graded algebra which could be introduced into another program in order to calculate the homology of such objects. We intend to tackle this task in the near future.

References

1. Álvarez V., Armario J.A., González–Díaz R., Real P.: *Algorithms in Algebraic Topology and Homological Algebra. The Problem of the Complexity.* Extended Abstracts of the First Workshop on Computer Algebra and Scientific Computing, CASC'98 (1998) Saint Petersburg.
2. Álvarez V., Armario J.A., Real P., Silva B.: *HPT and Computability of the Homology of CDGAs.* Conference on Secondary Calculus and Cohomological Physics (1997) Moscow. Electronic Proceedings of the EMS (http://www.emis.de/proceedings).
3. Armario J.A., Real P., Silva B.: *On p–Minimal Homological Models of Twisted Tensor Products of Elementary Complexes Localized over a Prime.* Contemporary Mathematics **227** (1999) 303–314.
4. Eilenberg S., Mac Lane S.: *On the Groups* $H(\pi, n)$*, I.* Annals of Math. **58** (1953) 55–139.
5. Eilenberg S., MacLane S.: *On the Groups* $H(\pi, n)$*, II.* Annals of Math. **60** (1954) 49–139.
6. Gugenheim V.K.A.M., Lambe L.A.: *Perturbation Theory in Differential Homological Algebra, I.* Illinois J. Math. **33** (1989) 556–582.
7. Gugenheim V.K.A.M., Lambe L.A., Stasheff J.D.: *Perturbation Theory in Differential Homological Algebra, II.* Illinois J. Math. **35** n. 3 (1991) 357–373.
8. González–Díaz R., Real P.: *Computing Cocycles on Simplicial Complexes.* Proceedings of the Second Workshop on Computer Algebra and Scientific Computing, CASC'99 (1999) Springer–Verlag, 177–190.
9. Huebschmann J., Kadeishvili T.: *Small Models for Chain Algebras.* Math. Zeit. **207** (1991) 245–280.
10. MacLane S.: *Homology.* Classics in Mathematics (1995) Springer.
11. Munkres, J.R.: *Elements of Algebraic Topology.* Addison-Wesley Publishing Company (1984).
12. Lambe L.A.: *Homological Perturbation Theory, Hochschild Homology and Formal Groups.* Contemp. Math. **134** (1992) 183–218.
13. Real P.: *Homological Perturbation Theory and Associativity.* To appear in Homology, Homotopy and Applications.
14. Veblen O.: *Analysis Situs.* A.M.S. Publications **5** (1931).
15. Weibel C.A.: *An Introduction to Homological Algebra.* Cambridge Studies in Advanced Mathematics **38**, Cambridge University Press (1994).

Effective computation of algebra of derivations of Lie algebras

L.M. Camacho, J.R. Gómez, R.M. Navarro, and I. Rodríguez

Dpto. Matemática Aplicada I. Univ. Sevilla. Avda. Reina Mercedes S.N. 41012 Sevilla (Spain). E-mail: lcamacho@cica.es - jrgomez@cica.es
Dpto. de Matemáticas. Univ. de Extremadura. Avda. de la Universidad S.N. 10071 Cáceres (Spain). E-mail: rnavarro@unex.es
Departamento Matemáticas. Escuela Politécnica Superior, Universidad de Huelva, Huelva (Spain). Email: rodgar@uhu.es

Abstract. By the software *Mathematica 3.0* we design a program that allows calculate the algebra of derivations of any Lie algebra. As an example we use this program in a special class of Lie algebras, $(n-4)$-filiform Lie algebras of dimension n.

1 Introduction

The spaces of derivations, viewed into the theory of Lie algebras are an invaluable tool to resolve many problems. Examples of these can be the classification of solvable Lie algebras from the nilpotent ones; determination of the first space of cohomology [6]; the calcule of dimension of orbits. Moreover the knowledge of the space of derivations has an increasing importance in the study of other problems next to Physics. Specifically to the interaction of particles [9] as the Gelfand-Kirillov's conjecture about the field of fractions of enveloping algebras of Lie algebras [1], [2].

Till now, if we wanted obtain the space of derivations of any Lie algebra it was needed a suitable gradation for the algebra. That is, a gradation with homogeneous subspaces of dimension as low as possible [3], [5]. But there are many cases (many types of algebras) in which to lead to any gradation like above is specially difficult. These cases correspond to all the algebras with nilindex low.

In this paper we are going to describe a process that allows to determine the space of derivations $\mathcal{D}er(\mathfrak{g})$ for a given Lie algebra $\mathfrak{g}$ without any restriction about the gradation.

So it will be particulary important with algebras of low nilindex, that up to now they were very difficult. Note that the process bellow is also valid to determine the algebra or space of derivations for high nilindex or, even, for non-nilpotent Lie algebras.

The *descending central sequence* of a Lie algebra $\mathfrak{g}$ of dimension n is defined by $\mathcal{C}^0(\mathfrak{g}) = \mathfrak{g}$, $\mathcal{C}^i(\mathfrak{g}) = [\mathfrak{g}, \mathcal{C}^{i-1}(\mathfrak{g})]$. If $\mathcal{C}^k(\mathfrak{g}) = 0$, for some k, the corresponding Lie algebra is called *nilpotent*. The smallest integer k such that the equality $\mathcal{C}^k(\mathfrak{g}) = 0$ holds is called the *nilindex* of $\mathfrak{g}$. In general, the nilindex ranges from 1 (*abelian*) to $n-1$ (*filiform Lie algebra*).

In this way, the *derived sequence* $(\mathcal{D}^n(\mathfrak{g}))$ of ideals in $\mathfrak{g}$ is defined inductively by $\mathcal{D}^1(\mathfrak{g}) = \mathfrak{g}$, and $\mathcal{D}^n(\mathfrak{g}) = [\mathcal{D}^{n-1}(\mathfrak{g}), \mathcal{D}^{n-1}(\mathfrak{g})]$ for $n > 1$. If there exists an integer n such that $\mathcal{D}^n(\mathfrak{g}) = \{0\}$, then $\mathfrak{g}$ is said to be a *solvable Lie algebra*, for more details see [10].

Let $\mathfrak{g}$ be a Lie algebra over $\mathbf{K}$. A linear endomorphism d of $\mathfrak{g}$ is called a derivation of $\mathfrak{g}$ if it satisfies

$$d([X,Y]) = [d(X),Y] + [X,d(Y)] \qquad \forall X,Y \in \mathfrak{g}$$

It is easy to see that the set $\mathcal{D}er(\mathfrak{g})$ of all derivations of $\mathfrak{g}$ is a vector subspace of End$\mathfrak{g}$, furthermore $\mathcal{D}er(\mathfrak{g})$ is a Lie algebra over $\mathbf{K}$ for the bracket $[d,d'] = d \circ d' - d' \circ d \qquad \forall d,d' \in \mathcal{D}er(\mathfrak{g})$

A $\mathbf{Z}$-gradation for a Lie algebra $\mathfrak{g}$ consists in a decomposition as direct sum into vectorial spaces $\mathfrak{g} = \bigoplus_{i \in \mathbf{Z}} \mathfrak{g}_i$ such that $[\mathfrak{g}_i, \mathfrak{g}_j] \subset \mathfrak{g}_{i+j}$.

Let $\mathfrak{g}$ be a Lie algebra on a field $\mathbf{K}$. It is often convenient to use the language of modules along with the (equivalent) language of representations. As in other algebraic theories, there is a natural definition (see [7]). A vector space V over $\mathbf{K}$, endowed with a bilinear operation $\mathfrak{g} \times V \to V$ (denoted $(x,v) \mapsto x.v$ or just xv) is called a **$\mathfrak{g}$-module** if the following condition is satisfied

$$[x,y].v = x.y.v - y.x.v \quad \forall x,y \in \mathfrak{g},\ v \in V$$

For example, if $\phi : \mathfrak{g} \to End(\mathrm{V})$ is a representation of $\mathfrak{g}$ (i.e., an homomorphism of Lie algebras), then V may be viewed as a $\mathfrak{g}$-module via the action $x.v = \phi(x)(v)$. Conversely, given a $\mathfrak{g}$-module V, this equation defines a representation $\phi : \mathfrak{g} \to End(\mathrm{V})$.

Throughout the following sections, $\mathbf{K}$ is a field of characteristic 0, and all algebras and modules are finite dimensional over $\mathbf{K}$.

Let V be a $\mathfrak{g}$-module, and $\phi : \mathfrak{g} \to End(\mathrm{V})$ the corresponding representation.

V (or ϕ) is called *simple* (or *irreducible*) if $V \neq \{0\}$ and V has no submodules other than (0) and V.

V (or ϕ) is called *semisimple* (or *completely reducible*) if V is the direct sum of simple submodules.

Note that $\mathfrak{g}$ may be semisimple as a $\mathfrak{g}$-module without being a semisimple Lie algebra; for example, $\mathfrak{g} = \mathbf{K}$.

Theorem 1. *[9] (H. Weyl). If $\mathfrak{g}$ is semisimple, all $\mathfrak{g}$-modules (of finite dimension) are semisimple.*

It is possible to use the foregoing results to obtain some general results concerning semisimple representations of arbitrary Lie algebras and the structure of reductive Lie algebras.

A Lie algebra $\mathfrak{g}$ is called *reductive* if its radical (rad $\mathfrak{g}$) coincides with its center (cf. Koszul [8]), and in this case, $\mathfrak{g}$ is the direct sum of its center (that is an abelian algebra) and $\mathcal{D}(\mathfrak{g})$ (that is semisimple) [11].

We show now the following decisive criterion for the semisimplicity of a representation of an arbitrary Lie algebra.

Theorem 2. *[11] Let $\mathfrak{g}$ be a reductive Lie algebra over* $\mathbf{K}$*, and let ϕ be a finite-dimensional representation of $\mathfrak{g}$. Then ϕ is completely reducible if and only if for every element X in the radical of $\mathfrak{g}$, $\phi(X)$ is a semisimple endomorphism (i.e., diagonalizable over the algebraic closure of* $\mathbf{K}$*)*

Furthermore, let V be the space on which ϕ acts, so if the hypothesis of the theorem are satisfied we can then find distinct elements $\lambda_1, \ldots, \lambda_p$ on the dual space (rad $\mathfrak{g}$)* and subspaces $V_1, \ldots, V_p$ of V such that V is the direct sum of the V_i and $V_i = \{v : v \in V, Xv = \lambda_i(X)v \ \forall X \in rad \ \mathfrak{g}\}$, obtaining with that direct sum the decomposition into simple $\mathfrak{g}$-submodules of the semisimple $\mathfrak{g}$-module V.

The below result holds if we restrict to the algebras $\mathfrak{g} = \mathfrak{a}$ with $\mathfrak{a}$ abelian instead of consider general reductive Lie algebras $\mathfrak{g} = \mathfrak{g}_1 \oplus \mathfrak{a}$ with $\mathfrak{g}_1$ semisimple and $\mathfrak{a}$ abelian Lie algebras. By applying the preceding theorem to a particular abelian Lie algebra (which always coincides with its radical) we are going to obtain a new method that simplifies the computations involved in the determination of the algebra of derivations of any Lie algebra.

2 Toral-Method

Let $\mathfrak{g}$ be an arbitrary Lie algebra, and $\mathcal{D}er(\mathfrak{g})$ the corresponding algebra of derivations that is also a Lie algebra. We are now consider a special subalgebra of $\mathcal{D}er(\mathfrak{g})$ called T that it is constituted by the set, the largest that can be found, of derivations that are simultaneously diagonalizable. This algebra is abelian, coincides with its radical and it is formed by diagonal endomorphisms X, so any endomorphism of the form $\phi(X)$ there will be a semisimple endomorphism. Thus, by applying the precedent Theorem every T-module will be semisimple via the correspondence between representations completely reducible and semisimple modules.

An easy calculation shows that $\mathfrak{g}$ and $\mathcal{D}er(\mathfrak{g})$ are both structure of T-modules by considering the operations or "products" $t.g = t(g), \forall t \in T, g \in \mathfrak{g}$ and $t.d = [t, d], \forall t \in T, d \in \mathcal{D}er(\mathfrak{g})$ respectively. So $\mathfrak{g}$ and $\mathcal{D}er(\mathfrak{g})$ are semisimple T-modules, obtaining from that fact the following decompositions into direct sums, in the sense of vectorial spaces, of simple T-modules

- $\mathfrak{g} = \mathfrak{g}_{\alpha_1} \oplus \mathfrak{g}_{\alpha_2} \oplus \cdots$
  $\mathfrak{g}_{\alpha_i} = \{g \ / \ t.g = \alpha_i(t)g, \ \forall \ t \in T\}$
  with $\alpha_i \in$ T*, such that $\alpha_i(T_j) = \delta_{i,j}$ with $\{T_i\}$ a basis of T, and the subspaces $\mathfrak{g}_{\alpha_i}$ verifies that $[\mathfrak{g}_{\alpha_i}, \mathfrak{g}_{\alpha_j}] \subset \mathfrak{g}_{\alpha_i+\alpha_j}$, i. e., if there exist $\mathfrak{g}_{\alpha_i}$, $\mathfrak{g}_{\alpha_j}$ with $[\mathfrak{g}_{\alpha_i}, \mathfrak{g}_{\alpha_j}] \neq 0$ then the space $\mathfrak{g}_{\alpha_i+\alpha_j}$ also exists.
- $\mathcal{D}er(\mathfrak{g}) = \mathcal{D}_{\lambda_1} \oplus \mathcal{D}_{\lambda_2} \oplus \cdots$
  $\mathcal{D}_{\lambda_i} = \{d \in \mathcal{D}er(\mathfrak{g}) \ / \ t.d = \lambda_i(t) \cdot d, \ \forall \ t \in T\}$
  $\lambda_i \in$ T*, and the subspaces $\mathcal{D}_{\lambda_i}$ verify similar conditions to the precedents $\mathfrak{g}_{\alpha_i}$.

It remains to determine which is the form of each space $\mathcal{D}_{\lambda_i}$ belonging to the gradation obtained for $\mathcal{D}er(\mathfrak{g})$. In the other hand, if we use the previous gradation for $\mathfrak{g}$, any derivation $d \in \mathcal{D}_{\lambda_i}$ verifies $d(\mathfrak{g}_{\alpha_i}) \subset \mathfrak{g}_{\alpha_j}$, from that, with an easy calculation we lead to the following form to λ_i, that is $\lambda_i(t) = \alpha_j(t) - \alpha_i(t)$. This last fact determines the type of derivations that form part of every $\mathcal{D}_{\alpha_j - \alpha_i}$.

The steps that can be followed to calculate the algebra of derivations of a Lie algebra are the following:

Step 1:

The first step consists in determining into $\mathcal{D}er(\mathfrak{g})$ a particular set, the biggest possible, constituted by diagonal derivations. For that, we choose a gradation for $\mathfrak{g}$ ($\mathfrak{g} = \bigoplus_{i \in \mathbf{Z}} \mathfrak{g}_i$) in the sense that this gradation must contain as many homogeneous subspaces of dimension one as possible.

An easy way to find diagonal derivations is to determine the subspace d_0 of $\mathcal{D}er(\mathfrak{g})$ such that

$$d_0(\mathfrak{g}_i) \subset \mathfrak{g}_i, \quad i \in \mathbf{Z}$$

continuously, we selected those elements of a certain basis in witch they are diagonals.

Step 2:

Let T be a subalgebra form by the diagonal derivations. Thus,

$$t \in T \Longrightarrow t = \sum_{i \in \mathbf{Z}} \alpha_i(t) d_0^i$$

where the $\alpha_i' s$ are the dual basis of the d_0^i's, these last one are the diagonal derivations.

Step 3:

In this step, we are going to determine the differences of the "weights" [3] of the subspaces. Then, imposing to a generic derivations in each subspaces the conditions that actual derivation must hold, we determine all possible derivations. In this step, we have used as an invaluable tool the software *Mathematica* given the huge amount of data involved in these computations.

Note 1. The method essentially reduces the computations comparing with the methods that have been used up to now, being specially efficient for that algebras of low nilindex whose treatment results very complicated by the precedent methods. A method used usually [3] need to find a suitable gradation for the considered algebra. This gradation would be as large as possible. This is, we will look a gradation containing homogeneous subspaces of dimension as low as possible. The difficulty for to obtain this gradations increasing when the nilindex of these algebras is low. In theorem 4 we obtain the space of derivations, $\mathcal{D}er(\mathfrak{g})$, for a family of Lie algebras of dimension $n \geq 8$ and nilindex 4.

---

[3] The notion of weight is usually used in the theory of semisimple algebras although we used it for arbitrary algebras using an analogous meaning with the semisimple case.

Note 2. If the subalgebra T is one-dimensional, our method is reduced to the one used in [3], so the new method (called *Toral-method*) generalizes the precedent. Usually, the subalgebra T obtained is most large when the nilindex is low. Thus, the method exposed in this paper is most interesting when we are working with Lie algebra of nilindex low. In this case is most difficult to obtain a large gradation of nilpotent Lie algebra considered. For filiform Lie algebra both ways are, usually, similar.

Note 3. The Toral-method can be used to determine $\mathcal{D}er(\mathfrak{g})$ even if $\mathfrak{g}$ is not nilpotent. Essentially we not used the condition of nilpotency of $\mathfrak{g}$ for the Toral-method.

3 Algebra of derivations of Lie algebras

3.1 Algebra of derivations with *Mathematica*

The program that we explicit allows to obtain the space of derivations of any Lie algebra for concrete dimensions. It is very useful in order to guess the dimension and a basis of the aforementioned space for the generic case, i.e. generic dimension of the Lie algebra (see the following section). The program can be separated in four parts.

1. Using T (the subalgebra form by diagonal derivations) we calculate all of the possible differences between weights of subspaces `t[i]` (that we will given in each case) by the function `Dif[i_,j_]`, as follows

```
Dif[i_, j_] := Simplify[t[i][[1]]-t[j][[1]]];
   pesos[n_] :=Module[{i, j, k, l, m}, dim = n;
       For[i = 1, i <= dim, i++,
         For[j = 1, j <= dim, j++, p[i, j] = {};
           If[Dif[i, j] =!= v, p[i, j] = {i, j};

             For[k = j + 1, k <= dim, k++,
               If[Dif[i, j] === Dif[i, k],p[i, j] =
                 Join[p[i, j], {i, k}];
                  Dif[i, k] = v, {}]];

             For[l = i + 1, l <= dim, l++,
               For[m = 1, m <= dim, m++,
                 If[Dif[i, j] === Dif[l, m],
                   p[i, j] = Join[p[i, j], {l, m}];
                     Dif[l, m] = v, {}]]]
                      , {}]]]];
```

As it shows by the list of lists `p[i,j]` (associated to each `Dif[i,j]`) we collect all the possible pair of index `{l,m}` such as weight of `t[m]` less the weight of `t[l]` is exactly `Dif[i,j]`.

2. Now we create the generic space of derivations d[i,j] associated to each p[i,j] that verifies d[i,j](t[m]) is contained in the space t[l] for all the pairs {l,m} of p[i,j].

```
derpesos[m_] :=
     Module[{i, j, q, r, k},  dim = m;
     d[i_, j_][0] := 0;
     d[i_, j_][a_ x_] :=  a d[i, j][x];
     d[i_, j_][x_ + y_] := d[i, j][x] + d[i, j][y];
     d[i_, j_][x[k_]] := 0;
       For[i = 1,  i <= dim, i++,
         For[j = 1, j <= dim, j++,
                     If[p[i, j] =!= {},
             For[q = 1, q < (Length[p[i, j]]/2) + 1, q++,
                For[r = 1,
                 r < Length[t[p[i, j][[2  q]]][[2]]] + 1, r++,
                 d[i, j][t[p[i, j][[2  q]]][[2, r]]] =
                   Sum[a[p[i, j][[2  q]], r, s]
                       t[p[i, j][[2  q - 1]]][[2, s]], {s,
                       1, Length[t[p[i, j][[2  q - 1]]][[2]]]} ]
                     ]], {}]]]];
```

3. We solve the equations ec[i,j,l,p,m] that results by imposing that the below spaces d[i,j] are effectively spaces of derivations of the algebra. We substitute the solutions obtained into the generic expression of d[i,j] leading to der[i,j]. In particular we work with the image of each vector x[k] by der[i,j], which will be noted der[i,j,k].

```
ec[i_, j_, l_, p_, m_] :=
      ec[i, j, l, p, m] =
        Coefficient [
          Expand[Plus[-d[p1, m1][mu[x[i], x[j]]],
              mu[d[p1, m1][x[i]], x[j]], mu[x[i], d[p1, m1][x[j]]
                   ]]],x[l]];
    ecder[p_, m_, n_] :=
      Module[{i, j, l, q, s, k, h, g, t, u}, dim = n; p1 = p;
         m1 = m;
        Lec[h_, g_] := {};
        For[i = 0, i < dim - 1, i++,
          For[j = i + 1, j < dim, j++,
            For[l = 0, l < dim , l++,
              Lec[p1, m1] = Union[Lec[p1, m1], {ec[i, j, l, p1,
                m1]}]]]];
        der[h_, g_, k_] :=
          d[h, g][x[k]]  /. Solve[Lec[h, g] == 0][[1]];
```

4. Finally we arrive to the dimension DimBaseDer and a basis ElBaseDer of the algebra of derivations as sum and union respectively of the dimension and particulary basis of each space der[i,j].

```
parder[h_, g_, t_] :=
         Select[Variables[Table[der[h, g, u], {u, 0, t - 1}]],
           FreeQ[#, x]&];
       d2[p1, m1, q_][x[i_]] :=
         der[p1, m1, i]  /.
           Dispatch[
             Join[{parder[p1, m1, dim][[q]] -> 1},
               Table[parder[p1, m1, dim][[s]] -> 0,
                  {s, 1, Length[parder[p1, m1, dim]]}]]];
       Length[parder[p1, m1, dim]]

       DimBaseDer[m_, n_] :=
     Module[{i, j, k}, dim = n; base = 0; dimtoro = m;
       For[i = 1, i <= dimtoro, i++,
           For[j = 1, j <= dimtoro, j++,
             If[(p[i, j] =!= {})  && (ecder[i, j, dim] != 0),
               base = base + \ Length[parder[i, j, dim]], {}]]]
         Print[base]]

    ElBaseDer[m_, n_] :=
     Module[{i, j, k, q, s},  dimtoro = m;  dim = n; s = 1;
       For[i = 1, i <= dimtoro, i++,
         For[j = 1, j <= dimtoro, j++,
           If[(p[i, j] =!= {})  && (ecder[i, j, dim] != 0),
             For[q = 1, q < Length[parder[i, j, dim]] + 1,q++,
             For[k = 0, k < dim, k++,
             If[d2[i, j, q][x[k]] === 0, {},
                   Print["d" , s , "[x[", k , "]]:=",
                     d2[i, j, q][x[k]]]]];
                Print["---------------"]; s = s + 1], {}]]]]
```

3.2 Applications

Next, we effect a concrete application of the toral-method by computing the algebra of derivations of one concrete family of $(n-4)$-filiform Lie algebras.

Theorem 3. *[4] In dimension n, $n \geq 8$, there are exactly $6n-29$, pairwise non-isomorphic, $(n-4)$-filiform Lie algebras and such that, being $\{X_0, X_1, X_2, X_3, X_4, Y_1, Y_2, \ldots, Y_{n-5}\}$ an adapted basis, their laws can be given by*

$$
\begin{array}{llll}
\mathfrak{g}_n^{1,r},\ 1 \leq r \leq \left\lfloor \frac{n-3}{2} \right\rfloor, & \mathfrak{g}_n^{2,r},\ 1 \leq r \leq \left\lfloor \frac{n-3}{2} \right\rfloor, & \mathfrak{g}_n^{3,r},\ 1 \leq r \leq \left\lfloor \frac{n-3}{2} \right\rfloor, & \mathfrak{g}_n^{4,r},\ 1 \leq r \leq \left\lfloor \frac{n-3}{2} \right\rfloor, \\
\mathfrak{g}_n^{5,r},\ 1 \leq r \leq \left[\frac{n-3}{2} \right], & \mathfrak{g}_n^{6,r},\ 1 \leq r \leq \left[\frac{n-5}{2} \right] & \mathfrak{g}_n^{7,r},\ 1 \leq r \leq \left[\frac{n-5}{2} \right] & \mathfrak{g}_n^{8,r},\ 1 \leq r \leq \left[\frac{n-5}{2} \right] \\
\mathfrak{g}_n^{9,r},\ 1 \leq r \leq \left[\frac{n-5}{2} \right] & \mathfrak{g}_n^{10,r},\ 1 \leq r \leq \left[\frac{n-4}{2} \right] & \mathfrak{g}_n^{11,r},\ 1 < r < \left[\frac{n-5}{2} \right] & \mathfrak{g}_n^{12,r},\ 1 \leq r \leq \left[\frac{n-4}{2} \right].
\end{array}
$$

The definition of laws of the defined algebras into the theorem can be found in [4]. Thus, the multiplication table of $\mathfrak{g}_n^{1,r}$ is

$$\mathfrak{g}_n^{1,r} = \begin{cases} [X_0, X_i] = X_{i+1}, & 1 \leq i \leq 3, \\ [Y_{2i-1}, Y_{2i}] = X_4, & 1 \leq i \leq r-1. \end{cases} \quad 1 \leq r \leq \left\lfloor \frac{n-3}{2} \right\rfloor$$

Lemma 1. *The linear applications* $ad(X_0)$, $ad(X_1)$, $ad(X_2)$, $ad(X_3)$, $ad(Y_{2i})$, $ad(Y_{2i-1})$, d_0^1, d_0^2, d_0^{i+2}, d_0^{j-r+3}, $d_{\alpha_1-\alpha}$, $d_{\delta_i-\alpha}$, $d_{2\alpha+\alpha_1-\delta_i}$, $d_{\beta_j-\alpha}$, $d_{2\alpha}$, $d_{3\alpha}$, $d_{\beta_j-\alpha_1}$, $d_{3\alpha+\alpha_1-2\delta_i}$, $d_{\delta_i-\delta_k}$, $d_{\beta_j-\delta_k}$ $d_{2\delta_i-3\alpha-\alpha_1}$, $d_{\beta_j-3\alpha-\alpha_1+\delta_i}$, $d_{3\alpha+\alpha_1-\beta_j}$, $d_{\beta_j-\beta_l}$, $1 \leq i, k \leq r-1$, $i \neq k$ *and* $1 \leq j, l \leq n-2r-3$, $l \neq j$ *of the space* $\mathfrak{g}_n^{1,r}$, *where*

$$\begin{cases} d_0^1(X_0) = X_0 \\ d_0^1(X_2) = X_2 \\ d_0^1(X_3) = 2X_3 \\ d_0^1(X_4) = 3X_4 \\ d_0^1(Y_{2i}) = 2Y_{2i} \end{cases} \quad \begin{cases} d_0^2(X_1) = X_1 \\ d_0^2(X_2) = X_2 \\ d_0^2(X_3) = X_3 \\ d_0^2(X_4) = 3X_4 \\ d_0^2(Y_{2i}) = Y_{2i} \end{cases} \quad \begin{cases} d_0^{i+2}(Y_{2i-1}) = Y_{2i-1} \\ d_0^{i+2}(Y_{2i}) = -Y_{2i} \end{cases}$$

$$\left\{ d_0^{j-r+3}(Y_j) = Y_j \right. \quad \left\{ d_{\alpha_1-\alpha}(X_0) = X_1 \right. \quad \begin{cases} d_{\delta_k-\alpha}(X_0) = Y_{2i-1} \\ d_{\delta_k-\alpha}(Y_{2i}) = -X_3 \end{cases}$$

$$\begin{cases} d_{2\alpha+\alpha_1-\delta_i}(X_0) = Y_{2i} \\ d_{2\alpha+\alpha_1-\delta_i}(Y_{2i-1}) = X_3 \end{cases} \quad \begin{cases} d_{2\alpha}(X_1) = X_3 \\ d_{2\alpha}(X_2) = X_4 \end{cases} \quad \left\{ d_{3\alpha}(X_1) = X_4 \right.$$

$$\left\{ d_{3\alpha+\alpha_1-2\delta_i}(Y_{2i-1}) = Y_{2i} \right. \quad \left\{ d_{2\delta_i-3\alpha-\alpha_1}(Y_{2i}) = Y_{2i-1} \right. \quad \begin{cases} d_{\delta_i-\delta_k}(Y_{2k-1}) = Y_{2i-1} \\ d_{\delta_i-\delta_k}(Y_{2i}) = Y_{2k} \end{cases}$$

$$\left\{ d_{\beta_j-\alpha}(X_0) = Y_{j+2r-2} \right. \quad \left\{ d_{\beta_j-\alpha_1}(X_1) = Y_{j+2r-2} \right. \quad \left\{ d_{\beta_j-\delta_i}(Y_{2i-1}) = Y_{j+2r-2} \right.$$

$$\left\{ d_{\beta_j-3\alpha-\alpha_1+\delta_i}(Y_{2i}) = Y_{j+2r-2} \right. \quad \left\{ d_{\beta_j-\beta_l}(Y_{l+2r-2}) = Y_{j+2r-2} \right. \quad \left\{ d_{3\alpha+\alpha_1-\beta_j}(Y_{j+2r-2}) = X \right.$$

are derivations over $\mathfrak{g}_n^{1,r}$.

The proof is obvious.

Theorem 4. *The linear mapping described above constitute a basis of* $Der(\mathfrak{g}_n^{1,r})$

Proof. We consider the following gradation of the algebra $\mathfrak{g}_n^{1,r}$

$$\mathfrak{g}_n^{1,r} = \bigoplus_{i=-r-2}^{n-r-1} \mathfrak{g}_i$$

where

$$\mathfrak{g}_i = \begin{cases} \langle 0 \rangle, & i = -2, -3 \\ \langle X_0 \rangle, & i = -1, \\ \langle X_{4-i} \rangle, & 0 \leq i \leq 3, \\ \langle Y_{2i-7} \rangle, & 4 \leq i \leq r+2, \\ \langle Y_{-2i-6} \rangle, & -r-2 \leq i \leq -4, \\ \langle Y_{i+r-4} \rangle, & r+3 \leq i \leq n-r-1 \end{cases}$$

In order to find diagonal derivations, let d_0 be the linear application such that

$$\left| \begin{array}{l} d_0(X_i)=c_iX_i,\ 0\le i\le 4,\\ d_0(Y_i)=d_iY_i,\ 1\le i\le n-5. \end{array}\right.$$

By imposing d_0 is a derivation, the following restrictions for the parameters are obtained

- $c_{i+1}=c_0+c_i,\ 1\le i\le 3$
- $d_{2k}=3c_0+c_1-d_{2k-1},\ 1\le k\le r-1,\ 1\le r\le \lfloor\frac{n-3}{2}\rfloor$

Therefore, it remaind the following parameters

$$\{c_0,c_1\}\cup\{d_{2i-1},\ 1\le i\le r-1\}\cup\{d_j,\ 2r-1\le j\le n-5\}$$

and it follows that

$$\dim\mathcal{D}_0^{1,r}=2+(r-1)+(n-2r-3)=n-r-2$$

Hence, we have the following basis for $\mathcal{D}_0^{1,r}$

$$\mathcal{B}_0^r=\{d_0^1,\ d_0^2,\}\cup\{d_0^{i+2},\ 1\le i\le r-1\}\cup\{d_0^{j-r+3},\ 2r-1\le j\le n-5\}$$

In this way, let T be a subalgebra formed by the diagonal derivations, that is

$$t\in T\implies t=\sum_{i\in\mathbf{Z}} i(t)d_0^i$$

$$t:\begin{cases} X_0 & \longrightarrow \alpha X_0,\\ X_1 & \longrightarrow \alpha_1 X_1,\\ X_2 & \longrightarrow (\alpha+\alpha_1)X_2,\\ X_3 & \longrightarrow (2\alpha+\alpha_1)X_3,\\ X_4 & \longrightarrow (3\alpha+\alpha_1)X_4,\\ Y_{2i-1} & \longrightarrow \delta_i Y_{2i-1}, \qquad 1\le i\le r-1,\\ Y_{2i} & \longrightarrow (3\alpha+\alpha_1-\delta_i)Y_{2i-1}, \ 1\le i\le r-1,\\ Y_{j+2r-2} & \longrightarrow \beta_j Y_{j+2r-2}, \qquad 1\le j\le n-2r-3. \end{cases}$$

which determine the following gradation

$$\mathfrak{g}_n^{1,r}=\mathfrak{g}_{3\alpha+\alpha_1-\delta_{r-1}}\oplus\ldots\oplus\mathfrak{g}_{3\alpha+\alpha_1-\delta_1}\oplus\{0\}\oplus\{0\}\oplus\mathfrak{g}_\alpha\oplus\mathfrak{g}_{3\alpha+\alpha_1}\oplus\mathfrak{g}_{2\alpha+\alpha_1}\oplus$$
$$\mathfrak{g}_{\alpha+\alpha_1}\oplus\mathfrak{g}_{\alpha_1}\oplus\mathfrak{g}_{\delta_1}\oplus\mathfrak{g}_{\delta_2}\oplus\ldots\oplus\mathfrak{g}_{\delta_{r-1}}\oplus\mathfrak{g}_{\beta_1}\oplus\ldots\oplus\mathfrak{g}_{\beta_{n-2r-3}}$$

Next, we are going to determine the differences of the weights of the subspaces, obtaining

Then

$$\dim Der(\mathfrak{g}_n^{1,r})=n^2-5n-2nr+r^2+8r+10.$$

Note 4. The program that we explicit allows to obtain the space of derivations of any Lie algebra for concrete dimensions. By induction we can be computed the space of derivations of Lie algebras in arbitrary dimension.

In this paper, we has obtained the space $\dim Der(\mathfrak{g})$ for some concrete dimensions with the aid of the program. After, we have induced the expression for the general case. Finally, we are proved the general result.

Table 1. Resume

Differences	Derivations	Dimension
$\alpha_1 - \alpha$	$d_{\alpha_1-\alpha}$	$dim(\mathcal{D}_{\alpha_1-\alpha}) = 1$
α_1	$ad(X_1)$	$dim(\mathcal{D}_{\alpha_1}) = 1$
$\alpha + \alpha_1$	$ad(X_2)$	$dim(\mathcal{D}_{\alpha+\alpha_1}) = 1$
$2\alpha + \alpha_1$	$ad(X_3)$	$dim(\mathcal{D}_{2\alpha+\alpha_1}) = 1$
$\delta_i - \alpha$	$d_{\delta_i-\alpha}$	$dim(\oplus\mathcal{D}_{\delta_i-\alpha}) = r - 1$
$2\alpha + \alpha_1 - \delta_i$	$d_{2\alpha+\alpha_1-\delta_i}$	$dim(\oplus\mathcal{D}_{2\alpha+\alpha_1-\delta_i}) = r - 1$
$\beta_j - \alpha$	$d_{\beta_j-\alpha}$	$dim(\oplus\mathcal{D}_{\beta_j-\alpha}) = n - 2r - 3$
α	$ad(X_0)$	$dim(\mathcal{D}_{\alpha}) = 1$
2α	$d_{2\alpha}$	$dim(\mathcal{D}_{2\alpha}) = 1$
3α	$d_{3\alpha}$	$dim(\mathcal{D}_{3\alpha}) = 1$
$\beta_j - \alpha_1$	$d_{\beta_j-\alpha_1}$	$dim(\oplus\mathcal{D}_{\beta_j-\alpha_1}) = n - 2r - 3$
$3\alpha + \alpha_1 - \delta_i$	$ad(Y_{2i-1})$	$dim(\oplus\mathcal{D}_{3\alpha+\alpha_1-\delta_i}) = r - 1$
δ_i	$ad(Y_{2i})$	$dim(\oplus\mathcal{D}_{\delta_i}) = r - 1$
$3\alpha + \alpha_1 - 2\delta_i$	$d_{3\alpha+\alpha_1-2\delta_i}$	$dim(\oplus\mathcal{D}_{3\alpha+\alpha_1-2\delta_i}) = r - 1$
$2\delta_i - 3\alpha - \alpha_1$	$d_{2\delta_i-3\alpha-\alpha_1}$	$dim(\oplus\mathcal{D}_{2\delta_i-3\alpha-\alpha_1}) = r - 1$
$\delta_i - \delta_k$	$d_{\delta_i-\delta_k}$	$dim(\oplus\mathcal{D}_{\delta_i-\delta_k}) = (r-2)(r-1)$
$3\alpha + \alpha_1 - \beta_j$	$d_{3\alpha+\alpha_1-\beta_j}$	$dim(\oplus\mathcal{D}_{3\alpha+\alpha_1-\beta_j}) = n - 2r - 2$
$\beta_j - \delta_i$	$d_{\beta_j-\delta_i}$	$dim(\oplus\mathcal{D}_{\beta_j-\delta_i}) = (r-1)(n-2r-3)$
$\beta_j - 3\alpha - \alpha_1 + \delta_i$	$d_{\beta_j-3\alpha-\alpha_1+\delta_i}$	$dim(\oplus\mathcal{D}_{\beta_j-3\alpha-\alpha_1+\delta_i}) = (r-1)(n-2r-3)$
$\beta_j - \beta_l$	$d_{\beta_j-\beta_l}$	$dim(\oplus\mathcal{D}_{\beta_j-\beta_l}) = (n-2r-3)(n-2r-4)$

References

1. J. Alev and F. Dumas. *Sur le corps des fractions de certaines algèbres quantiques.* J. Algebra, 170:229-265, 1994.
2. J. Alev, A. Ooms and M. Van der Bergh. *A class of counterexamples to the Gel'fand-Kirillov Conjecture.* T. Am. Math. Soc., 348:1709-1715, 1996.
3. J. M. Cabezas. *Una generalización de las álgebras de Lie filiformes.* PhD thesis, Universidad de Sevilla, 1996.
4. J.M. Cabezas and J.R. Gómez. *$(n-4)-$filiform Lie algebras.* Communications in Algebra, 27(10):4803-4819, 1999.
5. J.R. Gómez and R.M. Navarro. *Espacios de derivaciones de álgebras de Lie con Mathematica.* Proc. of IV Encuentro de Álgebra Computacional y Aplicaciones, (EACA'98). Sigüenza, 1998.
6. M. Goze and Y. Khakimdjanov. *Nilpotent Lie Algebras.* Kluwer Academic Publishers, 1996.
7. J.E. Humphreys. *Introduction to Lie Algebras and Representation Theory.* Springer-Verlag. Graduate Texts in Mathematics 9, 1987.
8. J.L. Koszul. *Homologie et cohomologie des algèbres de Lie.* Bull. Soc. Math. France, 78:65-127, 1950.
9. D.H. Sattinger and O.L. Weaver. *Lie Groups and Algebras with Applications to Physics, Geometry and Mechanics.* Springer-Verlag New York Inc., 1986.
10. J-P. Serre. *Lie Algebras and Lie Groups,* Springer-Verlag. Lecture Notes in Mathematics, 1992.
11. V.S. Varadarajan. *Lie Groups, Lie Algebras and Their Representations.* Springer-Verlag, Graduate Texts in Mathematics, 102, 1984.

An Integration and Reduction for Ordinary Non linear Differential Equations Library

J. Della Dora, F. Richard-Jung

LMC-IMAG, Tour Irma, 51 rue des Mathmatiques
38041 Grenoble Cedex (France)
Jean.Della-Dora@imag.fr, Francoise.Jung@imag.fr

Abstract. In this paper, we review the algorithmic work done in the last years by the Grenoble Computer Algebra team in the domain of differential equations. We organize in a logic manner the studies developed in different directions and we show how the modules obtained from these studies could cooperate in order to build IRONDEL, a Library for the Integration and Reduction of Ordinary Non linear Differential Equations.

1 Introduction

Let $\mathbb{K} = \mathbb{R}$ or $\mathbb{C}$ and consider a map

$$F : \mathbb{K} \times \mathbb{K}^r \times \mathbb{K}^{(d+1)n} \longrightarrow \mathbb{K}^m.$$

We are interested in equations of the type:

$$F(t, X, Y, \dot{Y}, \ldots, Y^{(d)}) = 0, \tag{1}$$

where

- t is the independent variable,
- X is a r-vector of parameters,
- $Y, \dot{Y}, \ldots, Y^{(d)}$ are the dependent variables, Y is a n-vector and $\dot{Y}, \ldots, Y^{(d)}$ are its successive derivatives with respect to t.

This is a general form for an **implicit system** of ordinary differential equations[1].

First of all, we remark that this type of equations arises in many applications,

[1] *ordinary* is to be understood here in opposition to *partial*.

and especially in the modelling of practical problems: modelling of electrical circuits, of chemical reactions, mechanical dynamic with constraints...
In the whole paper, we will distinguish "global studies", which give global information on the equations (invariants,...) and on the solutions (rational solutions, exponential solutions, more generally closed-form solutions,...) and "local methods", which give local information on the solutions in the neighborhood of a point, by various series expansions.
If $\mathbb{D}$ is an integral domain, we use the classical notations $\mathbb{D}[x], \mathbb{D}(x), \mathbb{D}[[x]], \mathbb{D}((x))$ to denote respectively the ring of polynomials in one variable with coefficients in $\mathbb{D}$, its fraction field, the ring of formal power series in one variable with coefficients in $\mathbb{D}$ and its fraction field. The similar notations $\mathbb{D}[X], \mathbb{D}(X), \mathbb{D}[[X]]$ are used in the multi-variate case.

2 Scalar equations: $n = m = 1$

2.1 Algebraic equations: $d = 0$

We consider **algebraic equations** $f(y) = 0$, where f is polynomial in y with coefficients series in $X = (x_1, \dots, x_r)$, i.e. $f \in \mathbb{D}[y]$, $\mathbb{D} = \mathbb{C}[[X]]$.
1. The case $\mathbf{r = 0}$ is a particular case (in general, f is a real function, and not a polynomial) of the numerical method due to Newton [20], approximating a solution y^* of $f(y) = 0$ by the sequence defined by some initial value y_0 and the recurrence formula $y_k = y_{k-1} - \frac{f(y_{k-1})}{f'(y_{k-1})}$.
2. The case $\mathbf{r = 1}$ is well-known too: it is the case of algebraic curves. It led to the definition of the field of Puiseux series: $\mathbb{C}((x))^* = \cup_{k \in \mathbb{N}^*} \mathbb{C}((x^{1/k}))$ and the main result is the Puiseux theorem: this field is algebraically closed [32].
From an algorithmic point of view, we can analyse the solutions of such an equation in two steps:
- produce a square-free factorization of f, to obtain the multiplicity of each solution (global information);
- compute the Puiseux expansion of each solution (local information).
In this case, we will say that f satisfies the *regular case* if $f(0) = 0 \bmod x$ and $f'(0) \neq 0 \bmod x$: then f admits one solution in $x\mathbb{C}[[x]]$, which can be computed by a quadratic Newton iteration (the same formula as in the case $r = 0$).
The Newton algorithm consists in repeating basic transformations of the type $y = cx^\alpha + \tilde{y}$, where α is the slope (in $\mathbb{Q}$) of an edge of the Newton polygon of the current equation and c is a non null root of the characteristic equation associated to the edge, until the new undetermined function $\tilde{y}$ verifies an equation $\tilde{f}$ satisfying the conditions of the regular case.
Note that it is in this context that appears for the first time a *rational* version of the Newton algorithm [10]. The term *rational* means here that if we start with $f \in \mathbb{D}[y], \mathbb{D} = \mathbb{k}[[x]]$, $\mathbb{k}$ a subfield of $\mathbb{C}$, we will collect the solutions in order to express them in $\hat{\mathbb{k}}((x))^*$, where $\hat{\mathbb{k}}$ is an algebraic extension of $\mathbb{k}$, as small as possible.
3. The multi-variate case ($\mathbf{r > 1}$) leads to the definition of the Newton polyhedron of f, which is a polyhedral set in $\mathbb{R}^{r+1}$ and to the introduction of series

with exponents in a cone, which are a generalization of the Puiseux series [18]. Let us consider a strongly convex rational cone $\sigma \subset \mathbb{R}^r$ of dimension r. For each $k \in \mathbb{N}^*$, we define the additiv semigroup $\mathcal{S}_\sigma^k = \sigma \cap \frac{1}{k}\mathbb{Z}^r$. Then we build the semigroup ring $\mathbb{C}[[\mathcal{S}_\sigma^k]]$ of all formal fractional power series of the form $\sum_{\alpha \in \mathcal{S}_\sigma^k} a_\alpha X^\alpha, a_\alpha \in \mathbb{C}$, $\mathbb{C}[[\mathcal{S}_\sigma]] = \cup_{k=1}^{\infty} \mathbb{C}[[\mathcal{S}_\sigma^k]]$, the ring of series with support in σ, and $\mathbb{C}((\mathcal{S}_\sigma)) = \cup_{\alpha \in \mathbb{Q}^r} X^\alpha \mathbb{C}[[\mathcal{S}_\sigma]]$, the ring of series with support in some translate of σ.
In particular, we recognize $\mathbb{C}((x))^* = \mathbb{C}((\mathcal{S}_\sigma))$ when $r = 1$ and $\sigma = \mathbb{R}^+$.
In this context, the scheme of the Newton algorithm is the same as for equations in one variable: it consists in repeating basic transformations of the type $y = cX^\alpha + \tilde{y}$, where α is the slope (in $\mathbb{Q}^r$) of an edge of the Newton polyhedron of the current equation and c is a non zero root of the characteristic equation associated to the edge, until the new undetermined function $\tilde{y}$ satisfies conditions of the *regular case* and can be computed by a quadratic Newton iteration [5].
The result is a finite number of well determined (thanks to the discriminant of f [12]) strongly convex rational cones and for each cone σ among them, as much solutions in $\mathbb{C}((\mathcal{S}_\sigma))$ as the degree of f.

2.2 Linear differential equations

We consider equations of the form:

$$L(y) = \sum_{k=0}^{d} a_k y^{(k)} = 0.$$

For this type of equations, several packages have been produced in the last years:
1. If $a_k \in \mathbb{C}[t]$, **global** methods have been developed for the research of Liouvillian solutions, beginning with the Risch algorithm ($d = 1$), going on with the Kovacic algorithm ($d = 2$), and its generalization to higher degree, passing through the study of the Galois differential group of the equation (two basic papers on the subject, one by M.F. Singer and the other by M. Bronstein, can be found in part 1 of [30]).
Besides, we can also consider as a **global information** the first step of our study, which consists in locating the singularities of the solutions as root of a_d, and possibly the point at infinity. By "locating", we mean "defining the singularity formally": in practice, if $a_k \in \mathbb{k}[t]$, where $\mathbb{k}$ is an algebraic extension of $\mathbb{Q}$, the finite ones are manipulated as algebraic numbers. Then, after translation or change of variable $z = 1/t$, we are working in the neighborhood of the origin.
2. If $a_k \in \mathbb{C}[[t]]$, **local** methods permit to compute a basis of formal solutions of the form:

$$y(z) = e^{Q(1/z)} z^\lambda \tilde{y}(z), \tag{2}$$

where $t = z^q$, $q \in \mathbb{N}^*$, $\lambda \in \mathbb{C}$, Q is a polynomial with coefficients in $\mathbb{C}$ and $\tilde{y} \in \mathbb{C}[[z]][\log z]$.

This is the more general form of an expansion in the neighborhood of an irregular singularity. It covers the case of a regular singularity ($q = 1$, i.e. $z = t$, and $Q = 0$) and the case of a regular point (the previous conditions and $\lambda = 0$, $\tilde{y} \in \mathbb{C}[[t]]$).
For these equations, the Newton algorithm consists in repeating basic transformations of the type $y = \exp(\frac{-c}{\gamma t^\gamma})\tilde{y}$, where γ is the slope of an edge of the Newton polygon of the current equation, and c is a non zero root of the associated characteristic equation, until the new undetermined function $\tilde{y}$ verifies an equation whose Newton polygon has an edge of null slope, and so can be computed by the Frobenius algorithm.
The algorithms for computing these expansions have been first implemented in REDUCE by C. Dicrescenzo, E. Tournier and J. Della Dora, in the package called DESIR[2] [29]. A *rational* version, in the same sense as in the algebraic case, was first proposed by M. Barkatou in [4] and implemented in MAPLE by E. Pfluegel [22].
3. If we suppose that the previous series $a_k, k = 0, \ldots, d$ are convergent in the neighborhood of the origin, the study goes on with a **numerical** and **graphical** package. In this case, we can use results of convergence, k-summability or multi-summability of the series which appear in the formal solutions, to calculate *actual* solutions of the differential equation (see part 2 of [30], by J. Martinet and J.P. Ramis).
First of all, we have to remember that each singularity has been formally placed at the origin: in order to recover numerical information around the singularity, we have to evaluate it... and for this we can refer to the paragraph 2.1, case $r = 0$.
Then, we recall that the power series which appear in the solutions when the origin is a regular singular point, are convergent. Of course, it is well-known that the series which appear in the formal solutions when the origin is an irregular singular point, are generally divergent. But they are always Gevrey of order γ, where γ is the slope of an edge of the Newton polygon of L.
Let $e^{Q(1/z)}z^\lambda\tilde{y}(z)$ a solution of $L(y) = 0$. Denote by $\tilde{L}$, the differential operator obtained from L after extracting the exponential part, such that $\tilde{L}(z^\lambda\tilde{y}) = 0$. The Newton polygon of $\tilde{L}$ has an edge of null slope (of length ≥ 1) and κ edges of slopes $\gamma_1 > \gamma_2 > \ldots > \gamma_\kappa > 0$.
In the particular case $\kappa = 1$ – notice that this hypothesis is verified by all the equations of order 2, and therefore all special functions – the series $\tilde{y}_j$ are γ_1-summable and numerical algorithms for summing the series have been successfully developed [27].
In the general case, the series $\tilde{y}_j$, which appear in $\tilde{y}$, are $(\gamma_1, \gamma_2, \ldots, \gamma_\kappa)$-summable [2] and their sum can be calculated by iterated Borel and Laplace transformations [1]. This theoretical process has been transformed into an effective algorithm, mixing symbolic and numerical parts, to calculate numerical values of the actual solutions [19].
Based on previous experiments of graphical representation of such functions

[2] Regular and Irregular Singularities of Differential Equations

(complex functions of the complex variable are four dimensional objects) [25, 26], a new software called COMPAS has been especially designed to help the user defining his problem, computing the solution by different ways and representing in the complex plane the image of the solution along a path. We use the colour to associate a point on the path in the domain and its image in the co-domain. This representation is very useful for the comparison of calculation methods, for determining the practical regions of validity of various numerical methods (classical numerical integration, classical summation of convergent series, k or multi-summation of divergent series...), for illustrating the Stokes phenomenon...

4. If $a_k \in \mathbb{C}[t]$, an other approach has been studied, which consists in expanding the solution in series of orthogonal polynoms. This is now part of the MAPLE package ORTHOSERIES, which contains all basic procedures needed for the manipulation of series of orthogonal polynoms (differentiation, application of a differential or difference operator, change of basis, fast calculus of partial sums, etc) [24].

5. Scalar linear differential equations with a parameter ϵ are studied in [17]. The coefficients a_k are supposed to belong to $\mathbb{C}[[t]][\epsilon]$: definition of a Newton polygon, algorithm for computing formal solutions, proof of the Gevrey type of the series appearing in the solutions.

2.3 Non linear differential equations: $d \geq 1$

We consider now equations of the form $f(y, y', \ldots, y^{(d)}) = 0$, where f is polynomial in the dependent variables $y, y', \ldots, y^{(d)}$: $f \in \mathbb{D}[y, y', \ldots, y^{(d)}]$.

1. From a **global** view point, it seems natural to have a geometric view of such an equation. This can be done (classically) using methods of Ritt differential algebra or Cartan exterior differential algebra. In this context, the problem of producing the general solution and, if any, the essential and particular singular solutions, has been studied in [15]

2. Local methods for equations $f(y, y', \ldots, y^{(d)}) = 0$, where $\mathbb{D} = \mathbb{C}[[t]]$ are based on a Newton polygon (in $\mathbb{R}^2$) first introduced by Fine in [11].

If we carry out the simple transformation of the algebraic case, that is to say, we put $y = ct^\gamma + \tilde{y}$, we are able to distinguish two types of candidates for γ:

•$(a)\gamma$ is the opposite of the slope of an edge of the Newton polygon ($\gamma \in \mathbb{Q}^*$); we can define a characteristic polynomial, of which c must be a root. **But**, this polynomial has no longer the good properties of the previous characteristic polynomials (degree equal to the length of the edge, and non null constant term) assuring that it has exactly "the length of the edge" non zero roots. In other words, there can be edges, which fails to give a solution of the requisite form. Nevertheless, using this first type of candidates, J. Cano defines *good edges* and proves the existence of at least one Puiseux series solution, for a particular type of equations: equations of order 1 and degree 1 in y' [7, 6]. Moreover, it will be useful for the following to notice that this result has been obtained without assuming that f is polynomial in y ($f \in \mathbb{C}[[t, y]]$).

•$(b)\gamma$ can be a root of the characteristic equation associated to a vertex. If γ

is real, it has to be situated strictly between the opposite of the slope of the adjacent edges. This possibility led to the definition of the set of generalized power series [13]:

$$\Omega = \{\sum_{k=0}^{+\infty} a_k t^{\beta_k}, a_k \in \mathbb{C}, \beta_k \in \mathbb{R}, \beta_0 > \beta_1 > \ldots, \text{ and } \lim_{k\to+\infty} \beta_k = -\infty\}$$

and to the discussion about the possibility of finding a solution of the equation in Ω.
The possibility for γ of being **complex** (as in the Frobenius algorithm, in the linear differential case) has been explored in [9].
In order to illustrate the power and the failing of this method, we give two examples:
Example 1: $t(y')^2 - tyy'' - yy' = 0$. For this equation, the Newton polygon consists in the quadrant $(\mathbb{R}^+)^2$ translated at the point $(0,2)$ and the characteristic polynomial associated to this vertex is... null! So we deduce that $y = ct^\alpha$ is solution, **for all** $(c, \alpha) \in \mathbb{C}^2$.
Example 2: consider the equation $z_1^3 + 12t^3 z_0^2 - 7t^3 z_0 z_1 + t^3 z_1^2 + t^{11} = 0$ ($z_0 = y, z_1 = ty', z_2 = t\frac{dz_1}{dt}$). An exhaustive study of the edges and multiple vertices of the Newton polygon of f gives no candidate for γ, even if we authorize γ to belong to $\mathbb{R}$ or to $\mathbb{C}$. We conclude that this equation does not have any solution beginning by a monomial ct^γ.
Moreover, it is possible that this phenomenon appears at a further step of the algorithm: when we have the first terms of a solution, we are not sure that the process for determining higher order terms does not stop.
The first ideas to overpass this diffculty by introducing logarithmic terms can be found in [9].
3. In the particular case $d = 1$, an other approach can be taken. Consider $f(t, y, y') = 0$, with $f \in \mathbb{C}[[t]][y, y']$ and use the resolution of $f(t, y, p)$ to express p in terms of $X = (t, y)$.
Let $p \in \mathbb{C}((S_\sigma))$ for some strongly convex rational cone σ, $p \in X^\alpha \mathbb{C}[[S_\sigma^k]], \alpha \in \mathbb{Q}^2$. A formal change of variables permits to define $\tilde{X} = (\tilde{t}, \tilde{y})$ such that the initial equation $f(t, y, y') = 0$ is transformed into a quasi-linear one:

$$a(\tilde{t}, \tilde{y})\frac{d\tilde{y}}{d\tilde{t}} + b(\tilde{t}, \tilde{y}) = 0,$$

where a and b are (classical) series, elements of $\mathbb{C}[[\tilde{t}, \tilde{y}]]$.
So we arrive at the quasi-linear case, for which significant results exist yet (see 3(a) below).

3 Systems

For systems, we do a similar review of the different cases.
Without loss of generality, we can of course suppose that $d \leq 1$ and consider

systems

$$F(t, X, Y, \dot{Y}) = 0. \tag{3}$$

As introduction, we propose to resume the first studies done on these equations by numericians. For this purpose, we will consider that $\mathbb{K} = \mathbb{R}$, $r = 0$ (no parameter), $n = m$ (squared systems) and that F is sufficiently differentiable. Under these conditions, it is usual to distinguish two cases:

- if there exists a point $(t_0, Y_0, P_0) \in \mathbb{R} \times \mathbb{R}^n \times \mathbb{R}^n$ for which $\frac{\partial F}{\partial P}(t_0, Y_0, P_0)$ is invertible, locally in the neighborhood of the point (t_0, Y_0, P_0) the system (3) can be rewritten in an **explicit form**[3]: $\dot{Y} = G(t, Y)$.

- on the contrary, if for all $(t, Y, P) \in \mathbb{R} \times \mathbb{R}^n \times \mathbb{R}^n$, $\frac{\partial F}{\partial P}(t, Y, P)$ is singular, the system is called a **Differential Algebraic Equation**.

In order to illustrate some of the basic notions introduced in this study, and the difficulties we encounter, we present the classical problem of a pendulum. The motion of a ball of mass m, suspended at the end of a string of fixed length L is described by the system:

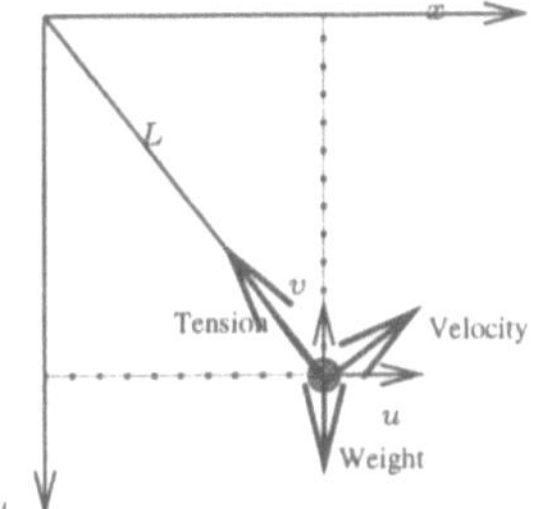

$$\begin{cases} \dot{x} = & u \\ \dot{y} = & v \\ \dot{u} = & -\lambda x \\ \dot{v} = & -\lambda y - g \\ 0 = x^2 + y^2 - L^2 \end{cases},$$

where (x, y) is the position of the ball, (u, v) are the cartesian coordinates of the velocity, and $\lambda = T/mL$ (T is the norm of the tension vector).
For this system $F(t, Y, \dot{Y}) = 0$, we ask a first question: given an initial value, $t_0, Y_0 = (x_0, y_0, u_0, v_0, \lambda_0)$, are the existence and unicity of a solution guaranteed such that $Y(t_0) = Y_0$?
Of course, it is obvious that the initial value must verify the "apparent" constraint $x_0^2 + y_0^2 = L^2$. But this is not sufficient: there are **hidden** constraints too. Indeed,

1. if we derive according to t the last (algebraic) equation, we obtain: $xu + yv = 0$.
2. we derive this new constraint a second time and we eliminate the derivatives, using the other equations: $u^2 + v^2 - \lambda L - gy = 0$.
3. we reiterate this process a third time: $-2xu\lambda - 2v(y\lambda + g) - L^2\dot{\lambda} - vg = 0$, or after simplification: $3vg + L^2\dot{\lambda} = 0$.

[3] also called Ordinary Differential Equation in the literature!

After three steps of differentiation, we can write the system under explicit form:

$$\dot{Y} = (\dot{x}, \dot{y}, \dot{u}, \dot{v}, \dot{\lambda}) = G(Y).$$

If we introduce the ideal generated (in $\mathbb{R}[Y]$) by the three constraints

$$(x^2 + y^2 - L^2, xu + yv, u^2 + v^2 - \lambda L - gy),$$

the initial value, and the solution, must belong to the algebraic manifold (in $\mathbb{R}^5$) defined by this ideal.
Let us come back to a general system $F(t, Y, \dot{Y}) = 0$. The derivatives array of order m is defined as

$$F_m(t, Y, \dot{Y}, \ldots, Y^{(m+1)}) = \begin{bmatrix} F(t, Y, \dot{Y}) \\ \dfrac{dF(t, Y, \dot{Y})}{dt} \\ \vdots \\ \dfrac{d^m F(t, Y, \dot{Y})}{dt^m} \end{bmatrix}.$$

Then the **differential index** of F is defined as the smallest integer ν such that the algebraic system $F_\nu(t, Y, Y_1, \ldots, Y_{\nu+1}) = 0$ determines Y_1 as a continuous function of (t, Y).
For the example of the pendulum, the differential index is 3.
This index *measures* the difficulty of the numerical integration of the system. For systems resulting from modelling, it is generally *known*. From the numerical point of view, it is admitted that systems of index less or equal than 3, can be integrated by generalization or adaptation of methods valid for explicit systems. On the contrary, if the index is greater than 3, this approach is no more possible: at each step of integration, it is necessary to use the derivatives array in order to locally explicit the system.
We will see below what kind of information can give a symbolic approach.
With this background, we can go to our classification of a general implicit system $F(t, X, Y, Y) = 0$.

3.1 Algebraic case

Suppose that $d = 0$ and that $F \in \mathbb{D}[X, Y]$: we have to mention this case, because it is the important domain of **polynomial equations in multiple variables** and a lot of work has been done in this area from the computer algebra point of view: methods of elimination, effective computation of Gröbner basis are **global** methods to give a precise description of the common zeros of polynomial equations.
Besides, it seems to be currently a lack of effective local methods for expanding Y in series of X. This will be a natural generalization of section 2.1.
Moreover, we point out that the problem of studying the eigenvalues and the Jordan structure of a matrix depending from a parameter takes part in this

paragraph. Indeed, the equation $(A(\epsilon) - \lambda I)V = 0$, where A is a matrix depending from the parameter ϵ, λ and V are respectively the eigenvalues and the eigenvectors, can be written in the form $F(Y) = 0$ with $Y = (\lambda, V)$ and $X = \epsilon$. In fact, it is also closely related to paragraph 2.1 and the method based on a Newton diagram provided by the Jordan structure of the matrix $A(0)$, can be seen as a lazy way for obtaining the leading term of the roots of the characteristic polynomial of $A(\epsilon)$ (without computing the whole polynomial) [16].
Without loss of generality, we will then restrict ourself to quasi-linear systems

$$A\dot{Y} = B,$$

where A and B are respectively a matrix of type (m, n) and a m-dimensional vector depending on t, X and Y.
Indeed, considering a system $F(Z, \dot{Z}) = 0$, we put $Y = (Z, \dot{Z})$, $A = \begin{pmatrix} I_n & 0 \\ 0 & 0 \end{pmatrix}$ and $B(Y) = (\dot{Z}, F(Z, \dot{Z}))$, so that the initial system becomes quasi-linear.
Moreover we will treat the case without parameter, i.e. suppose $r = 0$.

3.2 Explicit Systems

If there exists a point $(t_0, Y_0) \in \mathbb{K}^{n+1}$ where the system is of rank n, locally, in the neighborhood of such a point, the implicit system can be rewritten **explicitly** $\dot{Y} = G(Y)$.
1. If $G(Y) = HY$, it is the **linear case**. The studies are organized according to the type of H:
• (a) if $H \in \mathcal{M}_n(\mathbb{k}((t)))$ with $\mathbb{k}$ a subfield of $\mathbb{C}$: the first results in the **local** analysis of such systems are due to A. Hilali [14] and A. Wazner (concretized some years later in [33]). The algorithmic work has been continued by M. Barkatou [3] and E. Pflügel [21] and a complete algorithm is now available for the computation of a formal fundamental matrix solution

$$U(z) = \Phi(z) z^{\Lambda} e^{Q(z)} \tag{4}$$

where $t = z^q$, $q \in \mathbb{N}^*$, $\Phi \in \mathcal{M}_n(\hat{\mathbb{k}}[[t]])$ is a matrix of formal power series with coefficients in an algebraic extension of $\mathbb{k}$, $\Lambda \in \mathcal{M}_n(\hat{\mathbb{k}})$ is a constant matrix and Q is a diagonal matrix with elements in $z^{-1}\hat{\mathbb{k}}[z^{-1}]$.
• (b) if $H \in \mathcal{M}_n(\mathbb{k}(t))$: pertinent information about **global** (or closed-form) solutions can be derived from the previous local analysis in the neighborhood of the singularities of the rational functions coefficients of the matrix H. So it is possible to find the (possibly) rational or exponential solutions, to decide if two systems are rationaly equivalent, etc
All the fundamental functions relative to this section can be found in the MAPLE package ISOLDE[4].
2. If $G \in \mathbb{k}[[Y]]^n$, $\mathbb{k}$ a sub-field of $\mathbb{C}$, the **local** analysis begins with the construction of **normal forms** [8].

[4] Integration of Systems of Ordinary Linear Differential Equations

For this purpose, we suppose that 0 is a singular point (or equilibrium point), i.e. $G(0) = 0$.
We decompose

$$G(Y) = HY + G_2(Y) + G_3(Y) + \ldots + G_k(Y) + \ldots,$$

where $H \in \mathcal{M}_n(\mathbb{k})$ and $G_k \in \mathcal{H}_k$ is the homogeneous part of degre k. The problem is to build a formal change of variables $\varphi \in \mathbb{k}[[Z]]^n$ such that $Y = \varphi(Z)$ and $\dot{Z} = \tilde{G}(Z)$ is as simple as possible.
• first step: the choice of the linear part of φ.
Put $Y = PZ, \quad P \in Gl_n(\mathbb{k})$. If $\dot{Y} = HY + \hat{G}(Y)$, then $\dot{Z} = P^{-1}HPZ + P^{-1}\hat{G}(PZ)$: the linear part of $\tilde{G}$ is similar to the linear part of G. In other words, we can replace H by any representant of its similarity class, under convenience (Frobenius form, Jordan form if the eigenvalues of H belong to $\mathbb{k}\ldots$).
• further steps: the choice of the homogeneous part of degre k of φ, for $k \geq 2$. We put $Y = Z + \varphi_k(Z), \quad \varphi_k \in \mathcal{H}_k$. So $\dot{Z} = (I + D\varphi_k(Z))^{-1}\dot{Y}$ and if

$$\dot{Y} = G(Y) = HY + \ldots + G_{k-1}(Y) + G_k(Y) + \ldots,$$

$$\dot{Z} = (I - D\varphi_k(Z) + M)\dot{Y},$$

where $D\varphi_k(Z) \in \mathcal{M}_n(\mathcal{H}_{k-1})$ and M is a matrix whose coefficients are series of valuation greater than $2k - 2$.

$$\dot{Z} = \tilde{G}(Z) = HZ + \ldots + G_{k-1}(Z) + \underbrace{G_k(Z) + H\varphi_k(Z) - D\varphi_k(Z)HZ}_{\tilde{G}_k(Z)} + \tilde{M},$$

where $\tilde{M}$ is a matrix whose coefficients are series of valuation greater than $k+1$. This means that under this change of variables the part of G of degree less than $k-1$ is unchanged.
Moreover, we introduce the linear operator:

$$\begin{aligned} L_H^k : \mathcal{H}_k &\longrightarrow \mathcal{H}_k \\ \varphi(Z) &\longmapsto D\varphi(Z)HZ - H\varphi(Z), \end{aligned}$$

in order to express the homogeneous part of degree k of $\tilde{G}$:

$$\tilde{G}_k = G_k - L_H^k \varphi_k.$$

The question is now: is it possible to choose φ_k such that $\tilde{G}_k = 0$?
Of course, a sufficient condition is that L_H^k is invertible.
This condition can be specified in terms of the eigenvalues of H in the following way: we suppose that H is in Jordan canonical form and we note $\lambda = (\lambda_1, \ldots, \lambda_n)$ its eigenvalues.
We denote by $e_j, j = 1, \ldots, n$ the canonical basis of $\mathbb{k}^n$.
Then we choose for basis of $\mathcal{H}_k$ the family $\mathcal{B} = \{Y^q e_j, |q| = k, j = 1 \ldots n\}$ and range it in such a way that the matrix of L_H^k in this ordered basis is triangular

inferior, with diagonal elements $< q,\lambda > -\lambda_j$, $j = 1,\dots,n$, $q = (q_1,\dots,q_n) \in \mathbb{N}^n, |q| = k$[5].
This leads to the definition of **resonance**: the eigenvalues of H are resonant if and only if there exists j between 1 and n and $q \in \mathbb{N}^n$, $|q| \geq 2$ such that $\lambda_j =< q,\lambda >$.
So we have the **Poincaré theorem**: if the eigenvalues of H do not present any resonance, the initial system $\dot{Y} = G(Y)$ can be transformed into a linear one $\dot{Z} = HZ$.
To go a little further, we decompose G_k and φ_k in the basis $\mathcal{B}$:

$$G_k = \sum_{\substack{|q| = k,\\ 1 \leq j \leq n}} g_{qj} Y^q e_j, \quad \text{and} \quad \varphi_k = \sum_{\substack{|q| = k,\\ 1 \leq j \leq n}} \varphi_{qj} Y^q e_j.$$

With these notations, the g_{qj} are given and the φ_{qj} are to be determined:

$$\tilde{G}_k = 0 \Longleftrightarrow \forall q \in \mathbb{N}^n, |q| = k, \forall j \in [1,n], (< q,\lambda > -\lambda_j)\varphi_{qj} = g_{qj}.$$

In the general case (presence of resonant eigenvalues), we obtain the **Poincaré-Dulac theorem**:
the initial system $\dot{Y} = G(Y)$ can be transformed into the new one $\dot{Z} = \tilde{G}(Z)$, where

$$\tilde{G}_k = \sum_{|q|=k, 1\leq j \leq n < q,\lambda > -\lambda_j = 0} , g_{qj} Z^q e_j.$$

With this approach, in the new series $\tilde{G}$ are remaining only resonant monomials. From the computer algebra point of view, the main default of this approach is the use of the eigenvalues, which supposes in practice the introduction of algebraic numbers over $\mathbb{k}$. Moreover, the change of variables depends on these algebraic numbers and $\varphi \in \hat{\mathbb{k}}[[Y]]^n$, with $\hat{\mathbb{k}}$ an algebraic extension of $\mathbb{k}$.
A slightly different approach is due to **Takens**: if (for all $k \geq 2$) we suppose given $\mathcal{C}_k$ a supplementary subspace of $Im(L_H^k)$ in $\mathcal{H}_k$, we can transform the initial system $\dot{Y} = G(Y)$ into a new one $\dot{Z} = \tilde{G}(Z)$ such that $\tilde{G}_k \in \mathcal{C}_k$.
The first interest of this new vision is to be *rational*, in the sense that it does not use any algebraic extension of the initial field $\mathbb{k}$.
The second one lies in the possibility of obtaining a better reduction of G. To illustrate this affirmation, let us take an elementary example. Consider $\dot{Y} = G(Y)$, where the linear part is

$$H = \begin{pmatrix} 0 & 1 \\ 0 & 0 \end{pmatrix}.$$

The reduction of Poincar-Dulac leaves the system unchanged.
By calculus, we can see that $Y^q e_1$ and $Y^q e_2$, $|q| = k$ and $q_2 \geq 1$ are in $Im(L_H^k)$. So we can choose for $\mathcal{C}_k$ the subspace generated by $y_1^k e_1$ and $y_1^k e_2$, so that

$$\tilde{G}(Z) = \begin{pmatrix} z_2 + \sum_{k\geq 2} a_k z_1^k \\ \sum_{k \geq 2} b_k z_1^k \end{pmatrix}.$$

[5] $< q,\lambda >= \sum_{k=1}^n q_k\lambda_k$ and $|q| = \sum_{k=1}^n q_k$

3. In the particular case of hamiltonian systems, the previous methods have been used successfully in the formal computation of families of periodic solutions for systems coming from celestial mechanic or biology. The corresponding algorithms have been implemented in the MAPLE package SCANNER [34, 35].

3.3 Differential Algebraic Equations

We treat now the case of a DAE $A\dot{Y} = B$: there exists no point $(t_0, Y_0) \in \mathbb{K}^{n+1}$ where $A(t_0, Y_0)$ is of rank n.
1. If A does not depend on Y and B depends linearly on Y, $B = CY$, we have to study **matrix pencils** and use their **normal forms** (Kronecker normal form) to simplify the system [23].
2. $A \in \mathcal{M}_{mn}(\mathbb{D})$ and $B \in \mathcal{M}_{m,1}(\mathbb{D})$, $\mathbb{D} = \mathbb{C}[t, Y]$.
The contribution of the Grenoble team to the **global** study of such systems can be divided in three parts:
• (a) first step: a theoretical mathematical study [28].
For this we have to manipulate the following non trivial objects:
- the differential ring $\mathcal{D} = \mathbb{C}[t]\{Y\}$, ring of polynomials in infinitely many variables, $y_1, \ldots, y_n$ and all their derivatives, with coefficients in $\mathbb{C}[t]$;
- the differential ideal $\mathcal{I}_d$ generated in the previous ring by the polynomials defining the system $A(t, Y)\dot{Y} = B(t, Y)$;
- the ideal of constraints $\mathcal{I}_c = \mathcal{I}_d \cap \mathbb{C}[t, Y]$;
- the manifold of constraints $\mathcal{V}(\mathcal{I}_c)$, the set of common zeros in $\mathbb{C}^{n+1}$ of the polynomials belonging to $\mathcal{I}_c$.
Of course, the knowledge of $\mathcal{V}(\mathcal{I}_c)$ is of central interest because it contains the valid (or *consistent*) initial values and all the solutions live on it.
A theoretical algorithm permits to obtain a **complete description** of the manifold of constraints, i.e. a decomposition in a finite number of components associated to prime ideals $\mathcal{P}_j$, $\mathcal{V}(\mathcal{I}_c) = \cup_j \mathcal{V}(\mathcal{P}_j))$. On each component, the initial system is equivalent to a new one $M_j(t, Y)\dot{Y} = \beta_j(t, Y)$, obtained by successive differentations (as in the example of the pendulum), whose ideal of constraints is $\mathcal{V}(\mathcal{P}_j)$, and which has "good" properties of integration.
Precisely, we note ρ the rank of M_j considered as element of $\mathbb{C}[t, Y]/\mathcal{P}_j$ and we define as **regular** the points $(t, Y) \in \mathcal{V}(\mathcal{P}_j)$ where $M(t, Y)$ is of rank ρ.
Let (t_0, Y_0) such a regular point. We consider the Cauchy problem:

$$\begin{cases} M_j(t, Y)\dot{Y} = \beta_j(t, Y) \\ \quad Y(t_0) \quad = \quad Y_0 \end{cases} .$$

- if $\rho = n$, this problem has a unique solution, holomorphic in a neighborhood of t_0;
- if $\rho < n$, choose a minor $w(t_0, Y_0)$ of rank ρ of $M_j(t_0, Y_0)$ and $n - r$ holomorphic functions $y_k(t)$, satisfying the initial conditions, where k belongs to the set of column indices not appearing in $w(t_0, Y_0)$, then the solution is unique.

As a conclusion of this study, we can define a *differential multi-index* $(\ldots, \nu_j, \ldots)$, where ν_j is the number of differentiations needed to pass from the initial system to $M_j(t,Y)\dot{Y} = \beta_j(t,Y)$, which resume the global behavior of the D.A.E. Nevertheless, this approach is based on the use of prime ideals, which is very onerous from a practical point of view.

• (b) second step: an effective algorithm [28].

Following the philosophy of D5 [10] for the factorization of polynomials, the idea is to perform a lazy factorization of the ideals, which replaces the prime ideals $\mathcal{P}_j$ by radical ideals $\mathcal{I}_j$, the other properties being conserved. This version of the algorithm, using commutative algebra in non integral rings of polynomials and the representation of the ideals by Gröbner basis, has been implemented in MAPLE.

The example of the pendulum (after putting $L = 1$, $g = 1$ and $\lambda = L$) can be represented by the list:

$$[x^2 + y^2 - 1, x1 - u, y1 - v, u1 - Lx, v1 - Ly + 1],$$

where $x1$ denotes $\dot{x}$, $y1$ denotes $\dot{y}$, etc...

The output is a list containing a description of the ideal of constraints as the ideal generated by a Gröbner basis (here only one componant):

$$[x^2 + y^2 - 1, ux + vy, u^2 + v^2 - L - y, Ly^2 + y^3 + v^2 - L - y, vxy - uy^2,$$

$$Lxy + xy^2 + uv, v^2x - uvy - Lx - xy],$$

the differential index 3 and the equivalent system (here explicit):

$$[x1 - u, y1 - v, u1 - Lx, v1 - Ly + 1, L1 - 3v].$$

Nevertheless, this approach is limited by the use of the Gröbner basis and its inherent difficulties, in particular the problem of ordering the variables. For example, the previous solution has been obtained with L, u, v as preponderant variables, when the program does not terminate with the order $x1 > y1 > u1 > v1 > L1 > x > y > u > v > L$.

• (c) third step: a mixted formal-numerical approach [31].

The algorithm (if it terminates) builds a system $M(t,Y)\dot{Y} = \beta(t,Y)$ such that the matrix M is of rank n in $\mathcal{M}_{mn}(\mathbb{R}[t,Y])$ and a finite set of equations $C_j(t,Y) = 0$ describing locally the manifold of constraints: if (t_0, Y_0) is a point satisfying the constraints $C_j(t_0, Y_0) = 0$ and such that the rank of $M(t_0, Y_0)$ is n, then locally in the neighborhood of (t_0, Y_0) the initial system and the system $M(t,Y)\dot{Y} = \beta(t,Y)$ are equivalent and $\mathcal{V}(\mathcal{I}_c)$ coincides with the common zeros of the equations $C_j(t,Y) = 0$.

We see that

- the termination of the algorithm is no more guaranteed;
- the complete description of the manifold of constraints is replaced by a local description,

but the results provided by this new algorithm seem to be sufficient for numerical applications.

In contrast, it uses only linear algebra in $\mathcal{M}_{mn}(\mathbb{R}[t,Y])$, and becomes by the fact very efficient.

In other words it appears as an efficient formal pre-treatment for numerical methods well adapted to low differential indices.
Proceeding the previous step (b), we must now perform the **local** analysis of systems $A\dot{Y} = B$, where A is of rank n, in the neighborhood of a **singular** point, i.e. a point (t_0, Y_0) such that the matrix $A(t_0, Y_0)$ is of rank $< n$. This study has been tackled in [28]. Here we want only to show one example of a so called "impasse points". For this purpose, let us take the following D.A.E.:

$$\begin{pmatrix} 0 & 1 \\ i^2-1 & 0 \\ 0 & 0 \end{pmatrix} \begin{pmatrix} \frac{di}{dt} \\ \frac{du}{dt} \end{pmatrix} = \begin{pmatrix} -i \\ \frac{-i}{3} \\ u - i^3 + 3i \end{pmatrix}.$$

The matrix defining the system is of rank two, except at the points where $i = \pm 1$. All real solutions $(i(t), u(t))$ live on the graph of the function $y = x^3 - 3x$, which presents a local maximum at the point $(-1, 2)$ and a local minimum at the point $(1, -2)$. If we take into account the sign of $\frac{du}{dt}$, we see that any solution **go to** one of these points. Moreover, a simple integration proves that the solution of the Cauchy problem with initial condition $(t_0, i_0, i_0^3 - 3i_0)$, $i_0 > 1$, reaches the point $(i = 1, u = -2)$ in a finite time, and does not survive (in the real domain) after this time. The point $(1, -2)$ (and its symetric $(-1, 2)$) is an "impasse point". Such a phenomenom can be explained in the complex plane.

4 Conclusion

In this paper, we have described what could be a library for a complete study of general implicit differential systems. We have insisted in more details about three particular points:

- the **Newton algorithms**, because it seems to be the "universal" method in local analysis;
- the computation of **Normal forms** of explicit systems, because it is currently the classical way of simplifying the system, before computing the solutions;
- the study of **D.A.E.s**, because they are coming from very important applied problems and because they clearly show the way of combining formal and numerical computations.

We mentionned that each particular study led to implemented algorithms, but the software integration of all the modules remains to be done.

Acknoledgments This work is partially supported by the research cooperation between France and Marocco AI 181 MA 99.

References

1. W. Balser. Summation of formal power series through iterated laplace integrals. *Math. Scand.*, 70:161–171, 1992.

2. W. Balser, B.L.J. Braaksma, J.P. Ramis, and Y. Sibuya. Multisummability of formal power series solutions of ordinary differential equations. *Asympt. Analysis*, 5(1), 1991.
3. M. Barkatou. Contribution à l'étude algorithmique des systèmes différentiels linéaires. H.D.R., Université Joseph Fourier (Grenoble), Novembre 1998.
4. M.A. Barkatou. Rational Newton algorithm for computing formal solutions of linear differential equations. In *Lecture Notes in Computer Science*, volume 358, proceeding of ISSAC 88, Italy, 1989.
5. F. Béringer and F. Jung. Solving multi-variate algebraic equations. in preparation.
6. J. Cano. An extension of the Newton-Puiseux Polygon construction to give solutions of Pfaffian forms. *Annales de l'Institut Fourier*, 3(1), 1993.
7. J. Cano. On the series defined by differential equations, with an extension of the Puiseux Polygon construction to these equations. *Analysis*, 13:103–119, 1993.
8. G. Chen. Contribution à l'étude algorithmique de systèmes d'équations différentielles ou aux différences. H.D.R., Université de Lille, Novembre 1999.
9. J. Della Dora and F. Richard-Jung. About the Newton Polygon Algorithm for non Linear Ordinary Differential Equations. In *International Symposium on Symbolic and Algebraic Computation*, Maui, Hawaii, July 1997.
10. D. Duval. Diverses questions relatives au calcul formel avec des nombres algébriques. Thèse d'Etat, Université Joseph Fourier (Grenoble), Avril 87.
11. H. Fine. On the Fuctions Defined by Differential Equations, with an Extension of the Puiseux Polygon Construction to these Equations. *Amer. Jour. of Math.*, XI:317–328, 1889.
12. P.D. Gonzáles Pérez. Singularités quasi-ordinaires toriques et polyèdre de Newton du discriminant, preprint 1999.
13. D.Yu Grigorèv and M. Singer. Solving Ordinary Differential Equations in Terms of Series with Real Exponents. *Trans A.M.S.*, 1991.
14. A. Hilali. Solutions formelles de systèmes différentiels linéaires au voisinage d'un point singulier. Thèse d'Etat, Université Joseph Fourier (Grenoble), Juin 1987.
15. E. Hubert. *Etude Algébrique et Algorithmique des Singularités des Équations Différentielles Implicites.* PhD thesis, Institut National Polytechnique de Grenoble, 1997.
16. C.-P. Jeannerod and E. Pflügel. A Reduction Algorithm for Matrices Depending on a Parameter. In *International Symposium on Symbolic and Algebraic Computation*, Vancouver, 1999.
17. Y.O. Macutan. Formal Solutions of Scalar Singularly-Perturbed Linear Differential Equations. In *International Symposium on Symbolic and Algebraic Computation*, Vancouver, 1999.
18. J. McDonald. Fiber polytopes and fractional power series. *Jour. of Pure and Applied Algebra*, 104:213–233, 1995.
19. F. Naegele. *Autour de quelques équations fonctionnelles analytiques.* PhD thesis, Institut National Polytechnique de Grenoble, 1995.
20. I. Newton. *La méthode des fluxions et des suites infinies*, volume XXXII. Librairie Scientifique Albert Blanchard, 1966.
21. E. Pflügel. *Résolution symbolique des systèmes différentiels linéaires. Le logiciel* ISOLDE : ÉTUDE THÉORIQUE ET RÉALISATION. PhD thesis, Université Joseph Fourier (Grenoble), 1995.
22. E. Pflügel. On the latest version of DESIR. *Theoretical Computer Science*, 187:81–86, 1997.

23. M.P. Quéré-Stuchlik. *Algoritmique des faisceaux linéaires de matrices, application à la théorie des systèmes linéaires et à la résolution d'équations algébro-différentielles.* PhD thesis, Institut National Polytechnique de Grenoble, 1997.
24. L. Rebillard. *Etude théorique et algorithmique des séries de Chebyshev solutions d'équations différentielles holonomes.* PhD thesis, Institut National Polytechnique de Grenoble, 1997.
25. F. Richard-Jung. *Représentations graphiques de solutions d'équations différentielles dans le champ complexe.* PhD thesis, Université Louis Pasteur (Strasbourg), 1988.
26. L. Testard. *Calculs et visualisation en nombres complexes.* PhD thesis, Institut National Polytechnique de Grenoble, 1997.
27. J. Thomann. Procédés formels et numériques de sommation de séries solutions d'équations différentielles. *Expo. Math.*, 13:223–246, 1995.
28. G. Thomas. *Contributions Théoriques et Algorithmiques l'Etude des Equations Différentielles-Algébriques. Approche par le calcul formel.* PhD thesis, Institut National Polytechnique de Grenoble, 1997.
29. E. Tournier. Solutions formelles d'équations différentielles. Le logiciel de calcul formel : DESIR. Etude théorique et réalisation. Thèse d'Etat, Université Joseph Fourier (Grenoble), Avril 87.
30. E. Tournier, editor. *Computer Algebra and Differential Equations.* Academic Press, Proceedings of CADE88.
31. J. Visconti. *Résolution numérique des Equations Algébro-Différentielles, Estimation de l'erreur globale et Réduction formelle de l'indice.* PhD thesis, Institut National Polytechnique de Grenoble, 1999.
32. R. J. Walker. *Algebraic curves.* Dover edition, 1950.
33. A. Wazner. *Formes canoniques invariantes d'un système différentiel linéaire homogène, Polygone de Newton, Calcul de la partie exponentielle des solutions formelles.* PhD thesis, Université Joseph Fourier (Grenoble), 1998.
34. F. Zinoun. Méthodes formelles pour la réduction la forme normale de systèmes dynamiques. Doctorat de 3ème cycle, Université de Rabat, Décembre 1997.
35. F. Zinoun. Formes normales pour la construction de solutions formelles de systèmes dynamiques non-linéaires. Doctorat National, Université de Rabat, Mai 2000.

Complexity of Derivatives Generated by Symbolic Differentiation

Herbert Fischer and Hubert Warsitz

Fakultät für Mathematik, Technische Universität München,
80290 München, Germany
e-mail: fischer@mathematik.tu-muenchen.de

Abstract. The computational solution of many mathematical problems involves derivatives. Programs for computing derivatives may be (1) hand–coded, (2) set up via function calls and divided differences, or (3) obtained using symbolic differentiation. In practice, the divided differences approach is still the standard technique. But in many cases, derivatives can be computed cheaper and more accurately by symbolic differentiation. In this paper we investigate the complexity of algorithms for computing derivatives of rational functions. In particular, we deal with the forward mode and the reverse mode of symbolic differentiation. We discuss bounds on the amount of work within the described algorithms in terms of rational operations.

1 Introduction

There has been a rumor for long that the derivative is cheap. In fact, for an explicitly given rational function a function-value and the corresponding derivative-value together cost at most *four* times as much as the function-value alone. Hereby costs are measured in terms of rational operations. By a rational operation we understand addition or subtraction or multiplication or division of two real numbers, or changing the sign of a real number. The cheap derivative can be obtained by *symbolic differentiation*, in particular by the *reverse mode.* Approximating the derivative by divided differences costs at least $n + 1$ times as much as the function-value alone, where n is the number of variables. So, for an explicitly given rational function with $n > 3$ (more than 3 variables) we have the startling situation: The approximate derivative, computed with divided differences, costs more than the exact derivative, computed with reverse symbolic differentiation.

It is difficult to locate the beginning of symbolic differentiation, which sometimes is called *automatic differentiation*, *algorithmic differentiation*, or *computational differentiation.* Investigations lead back to a report written 1959 (in Russian) by Beda, Korolev, Sukkikh, and Frolova [2]. In 1981 Rall's book [15] appeared, which now is a standard reference. An overall presentation of the state of the art can be found in the conference proceedings [8] and [3]. Concerning the reverse mode it seems that Linnainmaa [11] was the first who described this technique (1970, in Finnish). Since then the reverse mode has been rediscovered and restated many times, e.g. 1971 by Ostrowski, Wolin, and Borisow [14], 1972 by Tienari [18], 1974 by Werbos [20], 1980 by Miller and Wrathall [13] and by Speelpenning [17], 1983 by Baur and Strassen [1], 1984 by Iri [9], by Kim, Nesterov, and Cherkasskiĭ [10] and by Sawyer [16].

In the present paper we investigate the complexity of algorithms for computing the derivative of a rational function. In particular we deal with the *forward mode* and the *reverse mode.* We give bounds on the amount of work within the described algorithms in terms of rational operations.

Let f be a rational function of n variables and let $\#(f)$ denote the number of rational operations for evaluating $f(x)$. Let f' denote the derivative of f. Assume that A is an algorithm for computing both $f(x)$ and $f'(x)$, requiring $\#(f, f', A)$ rational operations. We relate $\#(f)$ and $\#(f, f', A)$ by the inequality

$$\#(f, f', A) \leq Q \cdot \#(f).$$

Using the forward mode we can achieve $Q = 3n + 1$. The earlier papers on the reverse mode claimed or showed that the algorithm A can be set up such that Q is a constant, independent of n. Baur and Strassen [1] proved that $Q = 4$ for a suitably designed algorithm A. Surprisingly, later papers state $Q = 5$. Even more surprising, there exist papers describing software for the reverse mode which state $Q = 4$ or $Q = 5$ but the described software actually requires $Q = 7$ or higher. This somewhat dubious situation triggered the present paper. Our algorithm RM.G works with $Q = 4$.

2 The Derivative

For our purpose it is essential to know the form in which the function f is given. $f(x)$ is the result of finitely many rational operations, starting with the components $x_1, x_2, \ldots, x_n$ of x and using real constants. Let us assume that f is available in form of an algorithm FUN.

	FUN
(1)	for $k = 1, \ldots, n$ $y_k \leftarrow x_k = k$-th component of x
(2)	for $k = n+1, \ldots, n+t$ $r \leftarrow \omega(k)$ $i \leftarrow \alpha(k)$ $j \leftarrow \beta(k)$ $y_k \leftarrow \Phi_k(y_1, \ldots, y_{k-1})$
(3)	$f(x) \leftarrow y_{n+t}$

Hereby $r = \omega(k) \in \{\mathtt{A, S, M, D, AV, MV, AVC, SVC, MVC, DVC, ACV, SCV, MCV, DCV, SC}\}$, $i = \alpha(k) \in \{1, \ldots, k-1\}$, $j = \beta(k) \in \{1, \ldots, k-1\}$, and the Φ_k are functions as specified in the following table.

r	$\Phi_k(y_1, \ldots, y_{k-1})$	
A	$y_i + y_j$	with $i \neq j$
S	$y_i - y_j$	with $i \neq j$
M	$y_i \cdot y_j$	with $i \neq j$
D	$y_i \,/\, y_j$	with $i \neq j$
AV	$y_i + y_i$	
MV	$y_i \cdot y_i$	
AVC	$y_i + c_j$	
SVC	$y_i - c_j$	
MVC	$y_i \cdot c_j$	
DVC	$y_i \,/\, c_j$	with $c_j \neq 0$
ACV	$c_i + y_j$	
SCV	$c_i - y_j$	
MCV	$c_i \cdot y_j$	
DCV	$c_i \,/\, y_j$	
SC	$-\, y_j$	

r is a *type* and i, j are *subscripts* indicating which variables $y_1, y_2, \ldots$ resp. which constants $c_1, c_2, \ldots$ are employed for computing y_k.

Now we construct a formula for the derivative f' of the given function f. We represent the t rational operations by t *elementary* functions $E_{n+1}, \ldots, E_{n+t}$ in the following way. Let $k \in \{n+1, \ldots, n+t\}$, $r = \omega(k)$, $i = \alpha(k)$, $j = \beta(k)$. Define the functions

$$E_k : D_{k-1} \subseteq \mathbb{R}^{k-1} \longrightarrow \mathbb{R}^k \quad \text{with} \quad E_k(v) = \begin{bmatrix} v \\ \Phi_k(v) \end{bmatrix},$$

$$B : \mathbb{R}^n \to \mathbb{R}^n \quad \text{with} \quad B(x) = x,$$

$$L : \mathbb{R}^{n+t} \to \mathbb{R} \quad \text{with} \quad L(v) = \text{last component of } v.$$

Then we have

$$f(x) = L(E_{n+t}(E_{n+t-1}(\ldots E_{n+2}(E_{n+1}(B(x)))\ldots))), \tag{1}$$

or, using catenation of functions,

$$f = L \circ E_{n+t} \circ E_{n+t-1} \circ \ldots \circ E_{n+2} \circ E_{n+1} \circ B.$$

Let us interprete the formula (1) for $f(x)$. The evaluation of $f(x)$ involves t rational operations. E_{n+1} represents the first rational operation, E_{n+2} represents the second rational operation, and so forth. The function E_k copies all previous intermediate results and performs just one rational operation. And the function L picks the last intermediate result, which is $f(x)$. Hence, formula (1) expresses the stepwise computation of $f(x)$ with FUN. For shorter notation we set

$$z_k := \begin{bmatrix} y_1 \\ \vdots \\ y_k \end{bmatrix} \qquad \text{for } k = n, n+1, \ldots, n+t.$$

Then we find $z_n = B(x)$, and for $k = n+1, \ldots, n+t$

$$z_k = E_k(E_{k-1}(\ldots E_{n+2}(E_{n+1}(B(x)))\ldots)).$$

Using the chain-rule we obtain from (1)

$$f'(x) = L'(z_{n+t}) \cdot E'_{n+t}(z_{n+t-1}) \cdot \ldots \cdot E'_{n+2}(z_{n+1}) \cdot E'_{n+1}(z_n) \cdot B'(x) \quad (2)$$

So, $f'(x)$ is expressed as a product of Jacobian matrices. Since multiplication of matrices is associative, there are various ways to do the multiplications in formula (2). Working from right to left yields a *forward mode* algorithm, and working from left to right yields a *reverse mode* algorithm. Other ways of exploiting formula (2) yield *mixed mode* algorithms.

The elementary functions $E_{n+1}, \ldots, E_{n+t}$ serve to express $f(x)$ and $f'(x)$ in a concise form. The formula (2) for $f'(x)$ will be used to derive algorithms for computing $f'(x)$. Since this formula (2) involves the derivatives $E'_{n+1}, \ldots, E'_{n+t}$, we have to know how the derivative of an elementary function looks like.
The Jacobian matrix $E'_k(z_{k-1})$ has k rows and $k-1$ columns. The upper square part is an identity matrix, and the last row has at most 2 non-zero entries.

3 The Forward Mode

We use equation (2) and perform the multiplications from right to left. The successive products get names as shown in the diagram.

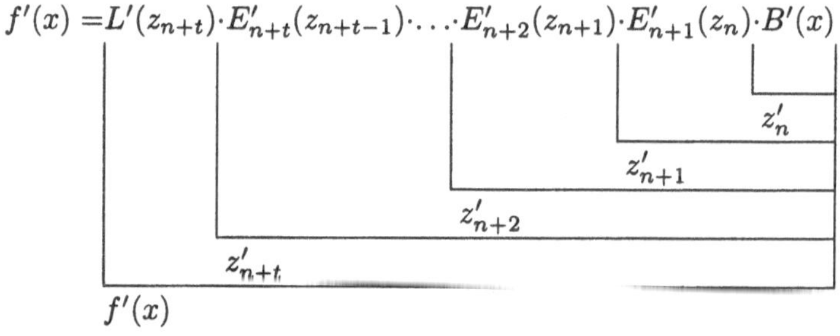

$$f'(x) = L'(z_{n+t}) \cdot E'_{n+t}(z_{n+t-1}) \cdot \ldots \cdot E'_{n+2}(z_{n+1}) \cdot E'_{n+1}(z_n) \cdot B'(x)$$

z'_n

z'_{n+1}

z'_{n+2}

z'_{n+t}

$f'(x)$

Let us outline the agenda of the forward mode.

(1)	set $z_n = B(x)$
	set $z'_n = B'(x)$
(2)	for $k = n+1, \ldots, n+t$ in this order
	compute $z_k = E_k(z_{k-1})$
	compute $z'_k = E'_k(z_{k-1}) \cdot z'_{k-1}$
(3)	$f(x) = L(z_{n+t})$
	$f'(x) = L'(z_{n+t}) \cdot z'_{n+t}$

Now we exploit the special form of the functions $B, E_{n+1}, \ldots, E_{n+t}, L$, and the sparsity of the corresponding Jacobian matrices. Obviously, $z_n = x$. Because of

$$z_k = E_k(z_{k-1}) = \begin{bmatrix} z_{k-1} \\ \Phi_k(z_{k-1}) \end{bmatrix}$$

the computation of z_k from z_{k-1} requires just the computation of

$$y_k = \Phi_k(z_{k-1}).$$

z'_n is an identity-matrix with n rows. z'_k is a matrix with k rows and n columns. The special structure of the Jacobian matrix $E'_k(z_{k-1})$ implies that the first $k-1$ rows of z'_k are the rows of z'_{k-1}. Let us write z'_k row-wise,

$$z'_k = \begin{bmatrix} y'_1 \\ \vdots \\ y'_k \end{bmatrix} \quad \text{for } k = n, n+1, \ldots, n+t.$$

So the computation of z'_k from z'_{k-1} requires just the computation of the last row y'_k of z'_k. For obtaining y'_k we use

$$y'_k = \text{last row of } \{E'_k(z_{k-1}) \cdot z'_{k-1}\} = \{\text{last row of } E'_k(z_{k-1})\} \cdot \begin{bmatrix} y'_1 \\ \vdots \\ y'_{k-1} \end{bmatrix}.$$

r	y_k	y'_k
A	$y_i + y_j$	$y'_i + y'_j$
S	$y_i - y_j$	$y'_i - y'_j$
M	$y_i \cdot y_j$	$y_j \cdot y'_i + y_i \cdot y'_j$
D	y_i / y_j	$(y'_i - y_k \cdot y'_j)/y_j$
AV	$y_i + y_i$	$2 \cdot y'_i$
MV	$y_i \cdot y_i$	$(2 \cdot y_i) \cdot y'_i$
AVC	$y_i + c_j$	y'_i
SVC	$y_i - c_j$	y'_i
MVC	$y_i \cdot c_j$	$c_j \cdot y'_i$
DVC	y_i / c_j	y'_i / c_j
ACV	$c_i + y_j$	y'_j
SCV	$c_i - y_j$	$-y'_j$
MCV	$c_i \cdot y_j$	$c_i \cdot y'_j$
DCV	c_i / y_j	$(-y_k/y_j) \cdot y'_j$
SC	$-y_j$	$-y'_j$

Table 1: Formulas for y_k and y'_k

Finally, $L'(z_{n+t}) = [0, \ldots, 0, 1]$, a matrix with 1 row and $n+t$ columns. Hence, $f'(x) = y'_{n+t}$. Now we state the forward mode algorithm FM.A for computing

$f(x)$ and $f'(x)$. Here it is understood that row $\pm$ row, scalar $\cdot$ row, row/scalar, and sign-change of a row are carried out component-wise.

	forward mode algorithm FM.A
(1)	for $k = 1, \ldots, n$
	$y_k \leftarrow k$-th component of x
	$y'_k \leftarrow k$-th row of the $n \times n$-identity-matrix
(2)	for $k = n+1, \ldots, n+t$
	$r \leftarrow \omega(k)$
	$i \leftarrow \alpha(k)$
	$j \leftarrow \beta(k)$
	$y_k \leftarrow$ according to Table 1
	$y'_k \leftarrow$ according to Table 1
(3)	$f(x) \leftarrow y_{n+t}$
	$f'(x) \leftarrow y'_{n+t}$

Obviously, the algorithm FM.A could have been derived directly from the algorithm FUN, using well-known differentiation rules for rational operations. But then it would have been necessary to start anew for deriving the reverse mode. Our framework employing elementary functions to set up formula (2) seems to be a uniform basis for both the forward mode and the reverse mode, and for mixed modes as well.

Here we mention two modifications of the algorithm FM.A. The first modification FM.B concerns storing the pairs (y_k, y'_k). In FM.A we need space for $n+t$ pairs. We can reduce this storage-requirement considerably by *overwriting* storage locations containing pairs which are no longer relevant. For doing this we need to know at which stage k we can *forget* which pairs. The necessary information is inherent in the algorithm FUN. As soon as a pair is "forgotten", we can use its storage location to store another pair. That, of course, requires some book-keeping.

The second modification of FM.A, we call it FM.C, concerns avoiding rational operations with system-zeros. With a row y'_k we associate an index-set I_k that contains the indices of the significant extries of y'_k. For instance for y'_k we have $I_k = \{k\}$ for $k = 1, ..., n$. The index-sets are built up iteratively. Assume that y'_i, I_i and y'_j, I_j are already available, and that $y_k = y_i + y_j$. Then the *sparse* addition $y'_k = y'_i + y'_j$ is done as follows:

$$y'_{k\mu} = \left\{ \begin{array}{ll} y'_{i\mu} + y'_{j\mu} & \text{for } \mu \in I_i \cap I_j \\ y'_{i\mu} & \text{for } \mu \in I_i \backslash I_j \\ y'_{j\mu} & \text{for } \mu \in I_j \backslash I_i \\ 0 & \text{for } \mu \notin I_i \cup I_j \end{array} \right\} \text{ for } \mu = 1, \ldots, n$$

And for the new index-set we take $I_k = I_i \cup I_j$. The other *sparse* operations with derivatives are defined in an analogeous way. Replacing in FM.A the entry-wise operations by *sparse* operations yields the algorithm FM.C.

4 The Reverse Mode

We use equation (2) and perform the multiplications from left to right. The successive products get names as shown in the diagram.

$$f'(x) = L'(z_{n+t}) \cdot E'_{n+t}(z_{n+t-1}) \cdot \ldots \cdot E'_{n+2}(z_{n+1}) \cdot E'_{n+1}(z_n) \cdot B'(x)$$

p_{n+t+1}

p_{n+t}

p_{n+2}

p_{n+1}

$f'(x)$

Unfortunately, we need the vectors $z_{n+t}, z_{n+t-1}, \ldots, z_n$ in this order. Therefore we first compute these intermediate vectors and $f(x)$, and then we use (2) to obtain $f'(x)$. Let us outline the agenda of the reverse mode.

(1)	compute $y_1, \ldots, y_{n+t}$ and $f(x)$ with FUN
(2.1)	set $p_{n+t+1} = L'(z_{n+t})$
(2.2)	for $k = n+t, \ldots, n+1$ in this order
	compute $p_k = p_{k+1} \cdot E'_k(z_{k-1})$
(2.3)	$f'(x) = p_{n+1} \cdot B'(x)$

Here we illustrate (2.2), which is the basis of the reverse mode.

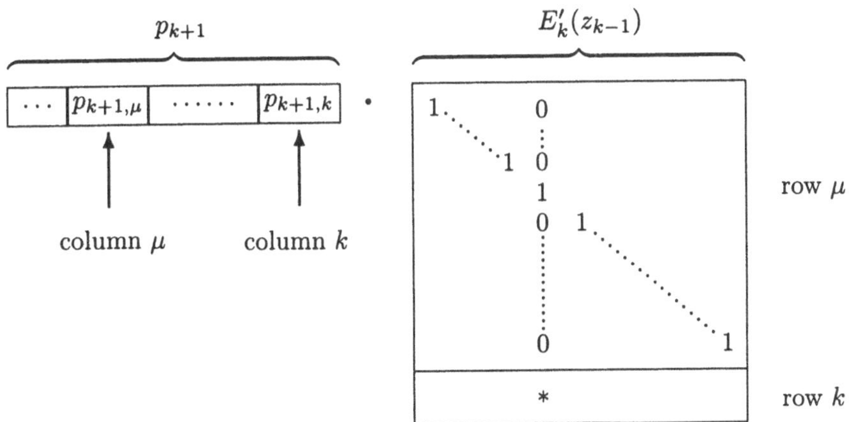

Now we exploit the special form of the functions $B, E_{n+1}, \ldots, E_{n+t}, L$, and the sparsity of the corresponding Jacobian matrices.

$$p_{n+t+1} = L'(z_{n+t}) = [0, \ldots, 0, 1] = \text{ row with } n+t \text{ entries.}$$

Let $k \in \{n+t, \dots, n+1\}$. p_k is a row with $k-1$ entries. The special form of $E'_k(z_{k-1})$ implies that for $\mu = 1, \dots, k-1$

$$p_{k,\mu} = p_{k+1,\mu} + p_{k+1,k} \cdot E'_k(z_{k-1})_{k\mu} \tag{3}$$

Formula (3) suggests overwriting. We introduce a working-row U of length $n+t$,

$$U = \boxed{U_1 \;\cdots\cdots\; U_n \;\cdots\cdots\cdots\cdots\; U_{n+t}}$$

in which we store current values of p_k. (3) then reads for $\mu = 1, \dots, k-1$

$$U_\mu \longleftarrow U_\mu + U_k \cdot E'_k(z_{k-1})_{k\mu} \tag{4}$$

Since at most 2 entries in the last row of $E'_k(z_{k-1})$ are non-zero, the transition from p_{k+1} to p_k causes an *update* of at most 2 entries of U. After completion of block (2.2) the row U contains the values

$$U = \boxed{p_{n+1,1} \cdots p_{n+1,n}} \boxed{p_{n+2,n+1} \cdots p_{n+t+1,n+t}}$$

Finally, since $B'(x)$ is an identity-matrix, we get

$$f'(x) = p_{n+1} = [U_1, \dots, U_n].$$

Now we state several versions of the reverse mode for computing $f(x)$ and $f'(x)$. For shorter notation we set

$$A_{ki} := E'_k(z_{k-1})_{ki} \quad \text{and} \quad A_{kj} := E'_k(z_{k-1})_{kj}.$$

r	A_{ki}	C_{ki}	G_{ki}	A_{kj}	B_{kj}	C_{kj}	D_{kj}	E_{kj}	G_{kj}
A	1	U_k	W_k	1	1	U_k	U_k	U_k	W_k
S	1	U_k	W_k	-1	-1	$-U_k$	$-U_k$	U_k	W_k
M	y_j	$U_k \cdot y_j$	$W_k \cdot y_j$	y_i	y_i	$U_k \cdot y_i$	$U_k \cdot y_i$	$U_k \cdot y_i$	$W_k \cdot y_i$
D	$1/y_j$	U_k/y_j	W_k/y_j	$-y_i/y_j^2$	$-y_k/y_j$	$-U_k \cdot y_k/y_j$	$-C_{ki} \cdot y_k$	$C_{ki} \cdot y_k$	$G_{ki} \cdot y_k$
AV	2	$U_k \cdot 2$	$W_k \cdot 2$						
MV	$2 \cdot y_i$	$U_k \cdot 2 \cdot y_i$	$W_k \cdot 2 \cdot y_i$						
AVC	1	U_k	W_k						
SVC	1	U_k	W_k						
MVC	c_j	$U_k \cdot c_j$	$W_k \cdot c_j$						
DVC	$1/c_j$	U_k/c_j	W_k/c_j						
ACV				1	1	U_k	U_k	U_k	W_k
SCV				-1	-1	$-U_k$	$-U_k$	U_k	W_k
MCV				c_i	c_i	$U_k \cdot c_i$	$U_k \cdot c_i$	$U_k \cdot c_i$	$W_k \cdot c_i$
DCV				$-c_i/y_j^2$	$-y_k/y_j$	$-U_k \cdot y_k/y_j$	$-U_k \cdot y_k/yj$	$U_k \cdot y_k/y_j$	$W_k \cdot y_k/y_j$
SC				-1	-1	$-U_k$	$-U_k$	U_k	W_k

Table 2: Formulas for the reverse mode algorithms

A straightforward use of the update (4) yields the algorithm RM.A.

	reverse mode algorithm RM.A
(1)	compute $y_1, \ldots, y_{n+t}$ and $f(x)$ with FUN
(2.1)	$U \leftarrow [0, \ldots, 0, 1]$ with length $n+t$
(2.2)	for $k = n+t, \ldots, n+1$ in this order $r \leftarrow \omega(k)$ $i \leftarrow \alpha(k)$ $j \leftarrow \beta(k)$
(2.2i)	if $r \in \{$A,S,M,D,AV,MV,AVC,SVC,MVC,DVC$\}$: $A_{ki} \leftarrow$ according to Table 2 $U_i \leftarrow U_i + U_k \cdot A_{ki}$
(2.2j)	if $r \in \{$A,S,M,D,ACV,SCV,MCV,DCV,SC$\}$: $A_{kj} \leftarrow$ according to Table 2 $U_j \leftarrow U_j + U_k \cdot A_{kj}$
(2.3)	$f'(x) \leftarrow [U_1, \ldots, U_n]$

Minor modifications of the algorithm RM.A lead to the algorithms RM.B, RM.C, RM.D, RM.E. We replace in RM.A the update (2.2i) and (2.2j) by cheaper updates.

RM.B update	(2.2i)	if $r \in \{$A,S,M,D,AV,MV,AVC,SVC,MVC,DVC$\}$: $A_{ki} \leftarrow$ according to Table 2 $U_i \leftarrow U_i + U_k \cdot A_{ki}$
	(2.2j)	if $r \in \{$A,S,M,D,ACV,SCV,MCV,DCV,SC$\}$: $B_{kj} \leftarrow$ according to Table 2 $U_j \leftarrow U_j + U_k \cdot B_{kj}$

RM.C update	(2.2i)	if $r \in \{$A,S,M,D,AV,MV,AVC,SVC,MVC,DVC$\}$: $C_{ki} \leftarrow$ according to Table 2 $U_i \leftarrow U_i + C_{ki}$
	(2.2j)	if $r \in \{$A,S,M,D,ACV,SCV,MCV,DCV,SC$\}$: $C_{kj} \leftarrow$ according to Table 2 $U_j \leftarrow U_j + C_{kj}$

RM.D update	(2.2i)	if $r \in \{$A,S,M,D,AV,MV,AVC,SVC,MVC,DVC$\}$: $C_{ki} \leftarrow$ according to Table 2 $U_i \leftarrow U_i + C_{ki}$
	(2.2j)	if $r \in \{$A,S,M,D,ACV,SCV,MCV,DCV,SC$\}$: $D_{kj} \leftarrow$ according to Table 2 $U_j \leftarrow U_j + D_{kj}$

In the algorithm RM.D we recognize that for some types (S,D,SCV,DCV,SC) we first do a sign-change to obtain D_{kj} and then add D_{kj} to U_j. These two operations can be combined to one subtraction by introducing E_{kj}.

RM.E update		
	(2.2i)	if $r \in \{\texttt{A,S,M,D,AV,MV,AVC,SVC,MVC,DVC}\}$: $C_{ki} \leftarrow$ according to Table 2 $U_i \leftarrow U_i + C_{ki}$
	(2.2j+)	if $r \in \{\texttt{A,M,ACV,MCV}\}$: $E_{kj} \leftarrow$ according to Table 2 $U_j \leftarrow U_j + E_{kj}$
	(2.2j−)	if $r \in \{\texttt{S,D,SCV,DCV,SC}\}$: $E_{kj} \leftarrow$ according to Table 2 $U_j \leftarrow U_j - E_{kj}$

In RM.E it may happen that prior to the update one or the other of U_i, U_j is a system-zero. For $k = n + t$ all entries of U except U_{n+t} are system-zeros. And as k runs down from $n + t$ to $n + 1$, these entries remain system-zeros unless they get updated. We want to avoid rational operations with such system-zeros. Therefore we introduce a control-row σ of length $n+t$ with Boolean entries that keeps track of system-zeros. This yields the algorithm RM.F.

	reverse mode algorithm RM.F
(1)	compute $y_1, \ldots, y_{n+t}$ and $f(x)$ with FUN
(2.1)	$U \leftarrow [0, \ldots, 0, 1]$ with length $n + t$ $\sigma \leftarrow [0, \ldots, 0, 1]$ with length $n + t$, Boolean entries
(2.2)	for $k = n + t, \ldots, n + 1$ in this order $r \leftarrow \omega(k),\ i \leftarrow \alpha(k),\ j \leftarrow \beta(k)$
(2.2i)	if $r \in \{\texttt{A, S, M, D, AV, MV, AVC, SVC, MVC, DVC}\}$: $C_{ki} \leftarrow$ according to Table 2 if $\sigma_i = 1 : U_i \leftarrow U_i + C_{ki}$ if $\sigma_i = 0 : U_i \leftarrow \quad C_{ki},\ \sigma_i \leftarrow 1$
(2.2j+)	if $r \in \{\texttt{A, M, ACV, MCV}\}$: $E_{kj} \leftarrow$ according to Table 2 if $\sigma_j = 1 : U_j \leftarrow U_j + E_{kj}$ if $\sigma_j = 0 : U_j \leftarrow \quad E_{kj},\ \sigma_j \leftarrow 1$
(2.2j−)	if $r \in \{\texttt{S, D, SCV, DCV, SC}\}$: $E_{kj} \leftarrow$ according to Table 2 if $\sigma_j = 1 : U_j \leftarrow U_j - E_{kj}$ if $\sigma_j = 0 : U_j \leftarrow \quad -E_{kj},\ \sigma_j \leftarrow 1$
(2.3)	$f'(x) \leftarrow [U_1, \ldots, U_n]$

Finally we turn to the reverse mode algorithm RM.G, which is the cheapest in terms of rational operations in our collection. Let us look at RM.F, in particular at the update $U_j \leftarrow -E_{kj}$. Here some savings seem to be possible. We suspend the sign-change, because in a later update of U_j it may be subsumed by replacing an addition by a subtraction or vice versa, or it may stay pending. Our aim now is to do explicit sign-changes as late as possible. Instead of the row U we use a row W which agrees with U up to signs. And instead of the control-row σ we use a control-row s with entries 1,-1,0 that keeps track of the signs. We update W and s such that $U_\mu = s_\mu \cdot W_\mu$. Since C_{ki} and E_{kj} contain the factor U_k, which is no longer available as a single entity, we set $C_{ki} = s_k \cdot G_{ki}$ and $E_{kj} = s_k \cdot G_{kj}$. Our directive "sign-changes as late as possible" leads from RM.F to the formulas for G_{ki} and G_{kj} in Table 2 and to the algorithm RM.G.

	reverse mode algorithm RM.G
(1)	compute $y_1, \ldots, y_{n+t}$ and $f(x)$ with FUN
(2.1)	$W \leftarrow [0, \ldots, 0, 1]$ with length $n+t$
	$s \leftarrow [0, \ldots, 0, 1]$ with length $n+t$
(2.2)	for $k = n+t, \ldots, n+1$ in this order
	$r \leftarrow \omega(k),\ i \leftarrow \alpha(k),\ j \leftarrow \beta(k)$
(2.2i)	if $r \in \{\mathtt{A}, \mathtt{S}, \mathtt{M}, \mathtt{D}, \mathtt{AV}, \mathtt{MV}, \mathtt{AVC}, \mathtt{SVC}, \mathtt{MVC}, \mathtt{DVC}\}$:
	$G_{ki} \leftarrow$ according to Table 2
	(necessary update $s_i W_i \leftarrow s_i W_i + s_k G_{ki}$)
	if $s_i = 1,\ s_k = 1 : W_i \leftarrow W_i + G_{ki}$
	$s_k = -1 : W_i \leftarrow W_i - G_{ki}$
	if $s_i = -1,\ s_k = 1 : W_i \leftarrow G_{ki} - W_i\ ,\ s_i \leftarrow 1$
	$s_k = -1 : W_i \leftarrow G_{ki} + W_i$
	if $s_i = 0,\ s_k = 1 : W_i \leftarrow G_{ki}\ ,\ s_i \leftarrow 1$
	$s_k = -1 : W_i \leftarrow G_{ki}\ ,\ s_i \leftarrow -1$
(2.2j+)	if $r \in \{\mathtt{A}, \mathtt{M}, \mathtt{ACV}, \mathtt{MCV}\}$:
	$G_{kj} \leftarrow$ according to Table 2
	(necessary update $s_j W_j \leftarrow s_j W_j + s_k G_{kj}$)
	if $s_j = 1,\ s_k = 1 : W_j \leftarrow W_j + G_{kj}$
	$s_k = -1 : W_j \leftarrow W_j - G_{kj}$
	if $s_j = -1,\ s_k = 1 : W_j \leftarrow G_{kj} - W_j\ ,\ s_j \leftarrow 1$
	$s_k = -1 : W_j \leftarrow G_{kj} + W_j$
	if $s_j = 0,\ s_k = 1 : W_j \leftarrow G_{kj}\ ,\ s_j \leftarrow 1$
	$s_k = -1 : W_j \leftarrow G_{kj}\ ,\ s_j \leftarrow -1$
(2.2j−)	if $r \in \{\mathtt{S}, \mathtt{D}, \mathtt{SCV}, \mathtt{DCV}, \mathtt{SC}\}$:
	$G_{kj} \leftarrow$ according to Table 2
	(necessary update $s_j W_j \leftarrow s_j W_j - s_k G_{kj}$)
	if $s_j = 1,\ s_k = 1 : W_j \leftarrow W_j - G_{kj}$
	$s_k = -1 : W_j \leftarrow W_j + G_{kj}$
	if $s_j = -1,\ s_k = 1 : W_j \leftarrow G_{kj} + W_j$
	$s_k = -1 : W_j \leftarrow G_{kj} - W_j\ ,\ s_j \leftarrow 1$
	if $s_j = 0,\ s_k = 1 : W_j \leftarrow G_{kj}\ ,\ s_j \leftarrow -1$
	$s_k = -1 : W_j \leftarrow G_{kj}\ ,\ s_j \leftarrow 1$
(2.2 s)	for $k = 1, \ldots, n$
	if $s_k = 1 : U_k \leftarrow W_k$
	if $s_k = -1 : U_k \leftarrow -W_k$
(2.3)	$f'(x) \leftarrow [U_1, \ldots, U_n]$

5 Complexity

In this section we focus on the number of rational operations for each of the algorithms stated above. Let us agree on the following:

(a) By a rational operation we understand addition, subtraction, multiplication or division of two real numbers, or sign-change of a real number. This point of view agrees with Baur/Strassen [1], Theorem 2.

(b) The function f is given by the algorithm FUN.

(c) $\#(f) :=$ number of rational operations for computing $f(x)$ with FUN.
The definition of $\#(f)$ refers to an algorithm for computing $f(x)$.
Without such a reference, $\#(f)$ would be meaningless.
With our definition we obviously have $\#(f) = t$.

(d) $\#(f, f', \mathrm{A}) :=$ number of rational operations for computing $f(x)$ and $f'(x)$ with algorithm A.

Theorem 1. *For the forward mode algorithms* FM.A, FM.B, *and* FM.C *we have*

$$\begin{aligned}\#(f, f', \mathrm{FM.A}) &\leq (3n+1) \cdot \#(f)\\ \#(f, f', \mathrm{FM.B}) &\leq (3n+1) \cdot \#(f)\\ \#(f, f', \mathrm{FM.C}) &\leq (3n+1) \cdot \#(f)\end{aligned}$$

The upper bounds on $\#(f, f', \mathrm{FM.A})$ and $\#(f, f', \mathrm{FM.B})$ are sharp: If all operations in FUN happen to be of type `M` or `D`, then $\leq$ is in fact $=$.

Example 1. Consider the function $f : \mathbb{R}^n \to \mathbb{R}$ given by the schedule

$$\begin{array}{lll}(1) & y_1 & = x_1\\ & y_2 & = x_2\\ & & \vdots\\ & y_n & = x_n\\ (2) & y_{n+1} & = y_1 \cdot y_2\\ & y_{n+2} & = y_{n+1} \cdot y_3\\ & y_{n+3} & = y_{n+2} \cdot y_4\\ & & \vdots\\ & y_{2n-1} & = y_{2n-2} \cdot y_n\\ (3) & f(x) & = y_{2n-1}\end{array}$$

Hereby we assume $n \geq 3$. Obviously $f(x) = x_1 \cdot x_2 \cdot \ldots \cdot x_n$. We find

$$\#(f, f', \mathrm{FM.A}) = 3n^2 - 2n - 1,$$

$$\#(f, f', \mathrm{FM.C}) = \frac{1}{2}n^2 + \frac{3}{2}n - 2.$$

Hence, *in this example* for large n

$$\#(f, f', \mathrm{FM.C}) \approx \frac{1}{6} \cdot \#(f, f', \mathrm{FM.A}).$$

Transition from algorithm FM.A to algorithm FM.C may save quite a lot of rational operations. Nevertheless, we can not reduce the factor $(3n+1)$ for FM.C in Theorem 1.

Example 2. Consider the function $f : \mathbb{R}^n \to \mathbb{R}$ given by the schedule

$$\begin{array}{lll}
(1) & y_1 & = x_1 \\
 & y_2 & = x_2 \\
 & & \vdots \\
 & y_n & = x_n \\
(2) & y_{n+1} & = y_1 \cdot y_2 \\
 & y_{n+2} & = y_{n+1} \cdot y_3 \\
 & y_{n+3} & = y_{n+2} \cdot y_4 \\
 & & \vdots \\
 & y_{2n-1} & = y_{2n-2} \cdot y_n \\
 & y_{2n} & = y_{2n-1} \cdot y_{2n-1} \\
 & y_{2n+1} & = y_{2n} \cdot y_{2n-1} \\
 & y_{2n+2} & = y_{2n+1} \cdot y_{2n-1} \\
 & & \vdots \\
 & y_{2n+m-2} & = y_{2n+m-3} \cdot y_{2n-1} \\
(3) & f(x) & = y_{2n+m-2}
\end{array}$$

Hereby we assume $n \geq 3$ and $m \geq 3$. Obviously $f(x) = (x_1 \cdot x_2 \cdot \ldots \cdot x_n)^m$. We find

$$\#(f, f', \mathrm{FM.C}) = \left(3n + 1 - \frac{5n^2 - 3n}{2(n+m-2)}\right) \cdot \#(f).$$

The factor on the right hand side can be arbitrarily close to $3n+1$.

Theorem 2. *For the reverse mode algorithms* RM.A *to* RM.G *we have*

$$\begin{array}{ll}
(A) & \#(f, f', \mathrm{RM.A}) \leq 9 \cdot \#(f) \\
(B) & \#(f, f', \mathrm{RM.B}) \leq 8 \cdot \#(f) \\
(C) & \#(f, f', \mathrm{RM.C}) \leq 7 \cdot \#(f) \\
(D) & \#(f, f', \mathrm{RM.D}) \leq 6 \cdot \#(f) \\
(E) & \#(f, f', \mathrm{RM.E}) \leq 5 \cdot \#(f) \\
(F) & \#(f, f', \mathrm{RM.F}) < 5 \cdot \#(f) \\
(G) & \#(f, f', \mathrm{RM.G}) \leq 4 \cdot \#(f)
\end{array}$$

We can not reduce the factors in Theorem 2. In $(A), (B), (C), (D)$ the relation $\leq$ is in fact $=$ if all operations are of type `D`. In (E) the relation $\leq$ is in fact $=$ if all operations are of type `M` or `D`. As for (F), we show with Example 3 that $\#(f, f', \mathrm{FM.F})$ can be arbitrarily close to $5 \cdot \#(f)$. And for (G), we show with Example 4 that the factor 4 can not be replaced by a smaller constant.

Example 3. Consider the function $f : D \subseteq \mathbb{R} \to \mathbb{R}$ given by the schedule:

$$\begin{array}{lll} (1) & y_1 & = x \\ (2) & y_2 & = y_1 + 1 \\ & y_3 & = y_1/y_2 \\ & y_4 & = y_1/y_3 \\ & & \vdots \\ & y_m & = y_1/y_{m-1} \\ (3) & f(x) & = y_m \end{array}$$

Hereby we assume $m \geq 3$. We find

$$\#(f, f', \mathrm{RM.F}) = (5 - \frac{4}{m-1}) \cdot \#(f).$$

The factor on the right hand side can be arbitrarily close to 5.

Example 4. Consider the function $f : D \subseteq \mathbb{R}^2 \to \mathbb{R}$ given by the schedule

$$\begin{array}{lll} (1) & y_1 & = x_1 \\ & y_2 & = x_2 \\ (2) & y_3 & = y_1/y_2 \\ (3) & f(x) & = y_3 \end{array}$$

For this example we have $\#(f, f', \mathrm{RM.G}) = 4 \cdot \#(f)$.

Proofs for Theorem 1 and Theorem 2 can be found in Fischer/Warsitz [6].

6 Implementation

Each of the stated algorithms for computing $f(x)$ and $f'(x)$ has the form of an interpreter

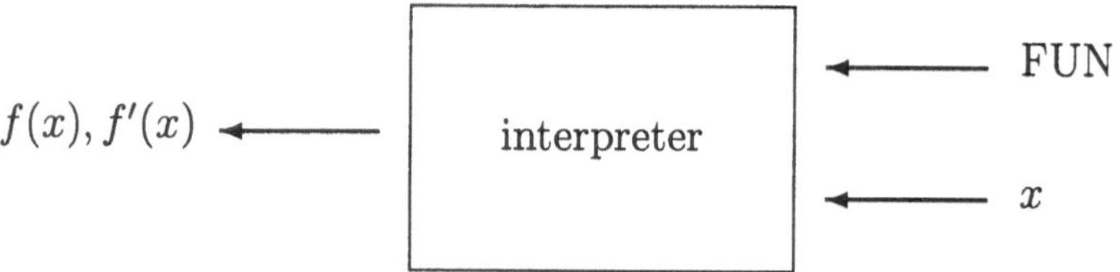

From the algorithm FUN for the function f and a value of x the interpreter computes the values $f(x)$ and $f'(x)$. For each additional value of x the interpreter has to pass through FUN again. Hence, if $f(x)$ and $f'(x)$ have to be computed for several values of x, it may be advantageous to do more or less of the logistic work (looking for types of operations, observing the control-row) *just once.*
Each of the stated algorithms describes in detail how to convert the algorithm FUN into a program P. We get the diagram

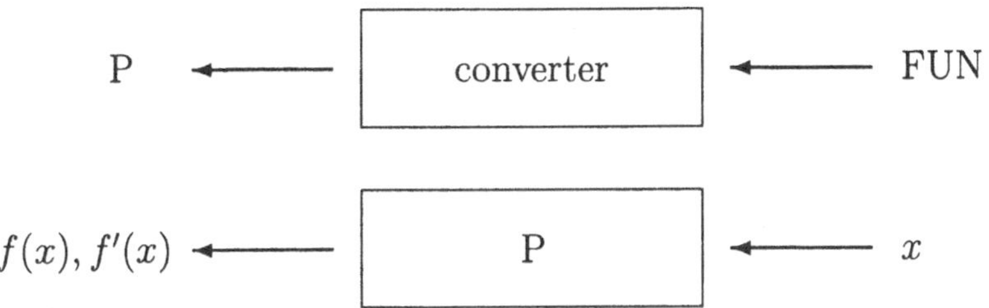

From the algorithm FUN for the function f the converter produces a program P. This program P accepts a value of x and computes the values $f(x)$ and $f'(x)$. So, each of the stated algorithms, or the corresponding converter, performs symbolic differentiation.

7 Conclusion

We described three forward mode algorithms FM.A, FM.B, FM.C, and seven reverse mode algorithms RM.A to RM.G for computing the derivative of rational functions. The cheapest algorithm in terms of rational operations is RM.G, where function and derivative together cost at most 4 times as much as the function alone. Concerning the reverse mode it is surprising that the symbolic differentiation community mainly refers to RM.D, where the relevant factor is 6 (reduced to 5 by some authors ignoring sign-changes). If the factor Q in

$$\#(f, f', \mathrm{A}) \leq Q \cdot \#(f)$$

is decisive, we would like to stress that our algorithm RM.G with factor $Q = 4$ is preferable to all the other algorithms.

References

1. Baur, W., Strassen, V.: *The complexity of partial derivatives.* Theoretical Computer Science 22, 1983, 317–330.
2. Beda, L.M., Korolev, L.N., Sukkikh, N.V., Frolova, T.S.: *Programs for automatic differentiation for the machine BESM.* (in Russian) Technical Report, Institute for Precise Mechanics and Computation Techniques, Academy of Science, USSR, Moscow, 1959.
3. Berz, M., Bischof, Ch., Corliss, G.F., Griewank, A. (editors): *Computational Differentiation: Techniques, Applications, and Tools.* SIAM, Philadelphia, 1996.
4. Fischer, H.: *Automatisches Differenzieren.* In: Wissenschaftliches Rechnen – Eine Einführung in das Scientific Computing, edited by J.Herzberger, Akademie Verlag, Berlin, 1995, 53–104.
5. Fischer, H., Flanders, H.: *A minimal code list.* Theoretical Computer Science 215, 1999, 345–348.
6. Fischer, H., Warsitz, H.: *Complexity investigations concerning derivatives generated by symbolic differentiation.* Technical Report 168, Fakultät für Mathematik, Technische Universität München, 2000

7. Griewank, A.: *Some bounds on the complexity of gradients, Jacobians and Hessians.* In: Complexity in Numerical Optimization, edited by P.M. Pardalos, World Scientific Publishing Company, Singapore, 1993, 128–162.
8. Griewank, A., Corliss, G.F. (editors): *Automatic Differentiation of Algorithms: Theory, Implementation, and Application.* SIAM, Philadelphia, 1991.
9. Iri, M.: *Simultaneous computation of functions, partial derivatives and estimates of rounding errors, complexity and practicality.* Japan J. Appl. Math. 1, 1984, 223–252.
10. Kim, K.V., Nesterov, Yu.E., Cherkasskiĭ, B.V.: *An estimate of the effort in computing the gradient.* Soviet Math. Dokl. 29, 1984, 384–387.
11. Linnainmaa, S.: *The representation of the cumulative rounding error of an algorithm as a Taylor expansion of the local rounding errors.* (in Finnish) Master's Thesis, Department of Computer Science, University of Helsinki, 1970.
12. Linnainmaa, S.: *Taylor expansion of the accumulated rounding error.* BIT 16, 1976, 146–160.
13. Miller, W., Wrathall, C.: *Software for Roundoff Analysis of Matrix Algorithms.* Academic Press, New York, 1980.
14. Ostrowski, G.M., Wolin, Yu.M., Borisow, W.W.: *Über die Berechnung von Ableitungen.* Wissenschaftliche Zeitschrift der Technischen Hochschule für Chemie, Leuna-Merseburg, Vol. 13, 1971, 382–384.
15. Rall, L.B.: *Automatic Differentiation: Techniques and Applications.* Lecture Notes in Computer Science 120, Springer-Verlag, Berlin, 1981.
16. Sawyer, J.W.: *First partial differentiation by computer with an application to categorial data analysis.* The American Statistician 38, 1984, 300–308.
17. Speelpenning, B.: *Compiling fast partial derivatives of functions given by algorithms.* Ph.D. Thesis, Department of Computer Science, University of Illinois, Urbana-Champaign, 1980.
18. Tienari, M.: *On some topological properties of numerical algorithms.* BIT 12, 1972, 409–433.
19. Ulbrich, M., Ulbrich, S.: *Automatic differentiation: A structure-exploiting forward mode with almost optimal complexity for Kantorovič trees.* In: Applied Mathematics and Parallel Computing, edited by H.Fischer, B.Riedmüller, S.Schäffler, Physica-Verlag, Heidelberg, 1996, 327–357.
20. Werbos, P.: *Beyond Regression: New Tools for Prediction and Analysis in the Behavioral Sciences.* Ph.D. Thesis, Harvard University, Cambridge, MA, 1974.

An Assessment of the Efficiency of Computer Algebra Systems in the Solution of Scientific Computing Problems

Victor G. Ganzha[1], Evgenii V. Vorozhtsov[2*] and Michael Wester[3]

[1] Institute of Informatics, Technical University of Munich, Munich 80290, Arcisstr. 21, Germany; e-mail: ganzha@informatik.tu-muenchen.de

[2] Institute of Theoretical and Applied Mechanics, Russian Academy of Sciences, Novosibirsk 630090, Russia; e-mail: vorozh@itam.nsc.ru

[3] Cotopaxi, 1801 Quincy, SE, Albuquerque, New Mexico, USA 87108-4427; e-mail: wester@math.unm.edu

Abstract. Computer algebra systems (CASs) have become an important tool for the solution of scientific computing problems. With the increasing number of general purpose CASs, there is now a need for an assessment of the efficiency of these systems. We discuss some peculiarities associated with the analysis of CPU time efficiency in CASs, and then present results from three specific systems (Maple Vr5, Mathematica 4.0 and MuPAD 1.4) on a sample of intermediate size problems. These results show that Maple Vr5 is generally the speediest on our examples. Finally, we formulate some requirements for developing a comprehensive test suite for analyzing the efficiency of CASs.

1 Introduction

Computer algebra systems (CASs) have become an important tool of scientific computing. Their importance has increased dramatically in the last decade as a result of the enhancement of existing general purpose CASs such as Derive, Macsyma, Maple, Mathematica and REDUCE, and the development of new systems such as Axiom, MuPAD, the TI-92, etc. These CASs enable their users to perform the following important tasks:

- symbolic manipulations,
- numerical computations,
- visualization.

These new possibilities have aroused considerable interest for many researchers and engineers in CASs.

In modern CASs, the availability of many built-in functions enables one to perform not only symbolic manipulations, but also fast and accurate numerical computations as well as visualization of the numerical results with the aid

* The work of this author was supported by Russian Foundation for Basic Research, grant No. 99-01-00573.

of computer graphics. These capabilities provide support for all the important stages of scientific computing methods:

- model formulation,
- implementation,
- numerical treatment,
- visualization,
- integration with existing software.

A further development is the possibility of parallel computations, putting CASs into the foreground of scientific computing and enabling the practical solution of many complex applied problems in the domains of natural science and engineering.

With an increase in the efficiency of the latest CASs in performing numerical computations by a factor of several dozens compared with their earlier versions (as in Mathematica 4.0), it has now become possible not only to solve the relatively simple tasks of scientific computing with the aid of CASs, but also to develop software packages for the solution of some of the more complex problems in this domain.

With the increasing number of general purpose CASs, there is now a need to investigate the comparative efficiency of these systems. One of the criteria, which is now widely accepted for the efficiency assessment of various software systems, is the speed of both numerical and symbolic computations. In the present paper we discuss (in Section 2) some peculiarities associated with the analysis of the efficiency of numerical or symbolic algorithms implemented in a CAS environment.

In Sections 3–6, we compare the capabilities of some specific CASs on the following four applied problems of scientific computing:

(1) plotting the isolines of a grid function of two variables, which is given by the nodes of a curvilinear grid in the plane;
(2) numerical quadrature after Archimedes;
(3) symbolic computation of definite integrals;
(4) approximation of functions with the aid of hierarchical bases.

In Section 7, we formulate some general conclusions about the capabilities of the chosen CASs and outline some directions for future investigations.

2 CASs versus CPU Times

Since CPU time is now widely accepted as an assessment of the efficiency of algorithm implementation, we would like to discuss in this section some peculiarities associated with the analysis of CPU time efficiency for an algorithm implemented in a CAS environment.

The problems of scientific computing can be subdivided into the following two large groups:

(a) Computationally intensive problems which can be solved only on supercomputers with a parallel architecture. An example of such a problem is LES (Large Eddy Simulation) in turbulence modeling; this problem requires several months of computer time running on a Cray supercomputer.

(b) Scientific computing problems of smaller size which can be solved successfully on personal desktop computers.

Parallelization of computations is not yet efficient in the current CAS languages. Therefore, we will limit our discussion to peculiarities of CPU time estimation when performing serial computations on desktop computers.

It is well known that given a specific computer hardware platform, the same numerical algorithm may be executed with different efficiencies depending on the implementation language (e.g., C, Fortran 90, etc.) as well as the particular version of the language used (e.g., Borland C, Watcom C 10.0, Watcom C 11.0, etc.) and the compilation options chosen (for compiled languages). The variety in code speed proves to be even more striking when we try to write a program in the language of a chosen CAS. The speed in these cases depends on the available memory, coding style, function choice, etc.

As an example, let us consider the task of summing a series of numbers using Maple. For users coming from other languages like Fortran or C, using a procedural construct like a `for` loop is tempting. However, in Maple one can also use the `add` function or maybe even `sum` (although the help page does say that this is mainly for symbolic sums). In Figure 1, we compare the CPU time efficiency of using the Maple constructs `for`, `add` and `sum`.

`s:=0:`
`for i from 1`
`to 1000000 do`
`s:=s+i^2`
`od:`
 	
`time = 75.41,`
`bytes = 84868422`
	`add(i^2,i=1..1000000):`
	
	`time = 42.98,`
	`bytes = 64865766`
		`sum(i^2,i=1..1000000):`
		
		`time = 0.03,`
		`bytes = 53002`

`s:=0:`
`for i from 1`
`to 100000 do`
`s:=s+sqrt(1.0*i)`
`od:`
`time = 37.86,`
`bytes = 61130498`
	`add(sqrt(1.0*i),`
	`i=1..100000):`
	
	`time = 34.81,`
	`bytes = 60973990`
		`sum(sqrt(1.0*i),`
		`i=1..100000):`
		
		`Error, object too large`
		`time = 8.89,`
		`bytes = 10909746`

Figure 1. Three different ways of computing $\sum_{i=1}^{1,000,000} i^2$ and $\sum_{i=1}^{100,000} \sqrt{i}$ using Maple Vr5.1 on a Sun SPARCstation 20 with a 125 MHz hyperSPARC processor and 384 MB of memory.

For the first series of sums, a loop is indeed slow, add is much faster, but sum in this special case (because it can compute a symbolic sum and then evaluate it) is even faster. However, if sum cannot find a simple symbolic solution quickly, then this is the wrong choice and one should use add as we see in the second series of sums. Interestingly, add is still faster than using an explicit loop, but not by a great deal this time. To confuse the situation even more, we note that the fastest way to compute the second sum in Maple is none of the above methods but rather

```
evalhf(add(sqrt(i), i = 1..100000));
```

evalhf evaluates its argument using hardware floats. For our example, the time required was 0.36 second with 24534 bytes used.

As a second example, in Macsyma, if one works with polynomials, CRE (canonical rational expression) form is much more efficient than the standard tree-structured general representation for manipulating polynomials. Figure 2 presents three ways of computing Fibonacci polynomials in Macsyma: no intermediate simplification, intermediate simplification, using CRE representation (i.e., rat). The latter is the most efficient method. Note also the use of arrays (indicated by [ ]) to remember intermediate results so that they need not be recomputed.

```
(c1) f[n](x):= if n <= 1 then 1 else x*f[n-1](x) + f[n-2](x)$
(c2) g[n](x):= if n <= 1 then 1 else ratsimp(x*g[n-1](x) + g[n-2](x))$
(c3) h[n](x):= if n <= 1 then 1 else rat(x*h[n-1](x) + h[n-2](x))$
(c4) f[20](x)$
Time= 26120 msecs
(c5) g[100](x)$
Time= 26680 msecs
(c6) h[100](x)$
Time= 2360 msecs
```

Figure 2. Three different ways of computing Fibonacci polynomials using Macsyma 422 on a Sun SPARCstation 10 with a 61 MHz superSPARC processor and 128 MB of memory.

One might write a Mathematica program that is twice as fast as a Maple program, but perhaps another user can play with the Maple program by using various tricks and make it twice as fast as the Mathematica program. Thus, speed comparisons can lead to quite different results for a complex program which can be written in many different ways.

CPU time is nevertheless very important for many practitioners working in the various areas of scientific computing. With the advent of high performance desktop computers, however, the focus has shifted from the price per hour of CPU time to how many man-hours should the mathematician spend to solve his specific task. This is especially important when parametric runs are to be undertaken and the context of each run depends strongly on the results of previous runs. Under these conditions, it is extremely important that each run take as little CPU time as possible.

Despite an apparent chaos in the CPU timings that may be obtained within a chosen CAS by using different built-in functions and programming styles, it is nevertheless possible to formulate a number of general recommendations for speeding up computations:

1. Use built-in functions whenever possible as they will likely have been programmed by the CAS developers to be fairly efficient. Of course, there may be several possible options, so some study will often be needed to evaluate what is most appropriate for the problem at hand.
2. Some problems can take great advantage of a particularly efficient representation of the quantities being manipulated. For example, Macsyma's CRE form is good for working with sparse, univariate and multivariate polynomials.[1] A similar feature in MuPAD is invoked by converting polynomial expressions to `DOM_POLY`.
3. In most CASs, it is possible to remember values so that they need not be recomputed. This is especially useful when working with recurrences or other recursive procedures. The trade off is that additional memory (perhaps a great deal) is consumed to hold these remembered values. Traditionally, one can define arrays to hold saved values, although Maple also employs "remember tables" to preserve function values.
4. If a CAS provides a compilation facility, take advantage of it. For example, Macsyma can translate Macsyma code into Lisp and then compile it for great efficiency. In Mathematica, compilation of numerical (but not symbolic) procedures can be performed by using the built-in function `Compile`.
5. Finally, be sure to study the programming guide/reference manual/online help/discussion group of the CAS for any helpful tips. The trouble here is that useful suggestions will likely be scattered, so more than one iteration of careful reading will probably be needed.

3 Contour Plots

The plots of the isolines (contour plots) are very often used in computational fluid dynamics and computational physics to present results. With the aid of contour plots, one can, for example, easily watch the propagation of shock waves and interfaces between different fluids [1–4]. For the numerical modeling of incompressible fluids, the contour lines $\psi(x, y)$ = constant show the streamlines, where x and y are the spatial coordinates and $\psi(x, y)$ is the stream function [5].

In such CASs as Maple and Mathematica, there are the built-in functions for the construction of contour lines on a rectangular uniform grid in the (x, y) plane. There are no built-in functions for making the contour plots on a curvilinear grid of quadrilateral cells, however.

[1] Also rational functions.

On the other hand, many practically important problems of computational fluid dynamics are now solved numerically on curvilinear grids [2, 3, 6, 7]. It is therefore of practical interest to develop programs in a CAS environment which would enable users to make contour plots on arbitrary curvilinear grids.

We will assume that the fluid flow problem is solved in a bounded region D of the (x, y) plane. The boundary of D is generally a curved line. In addition, the region D may be multiply connected. We will assume further that there is a nonsingular transformation

$$x = x(\xi, \eta), \quad y = y(\xi, \eta) \tag{1}$$

which maps the region D onto a rectangle D_ξ in the (ξ, η) plane of the curvilinear coordinates ξ and η. The inequality $J \neq 0$ ensures that the transformation (1) is nonsingular [6, 7], where J is the determinant of the Jacobian matrix of the transformation (1),

$$J = \frac{\partial x}{\partial \xi}\frac{\partial y}{\partial \eta} - \frac{\partial y}{\partial \xi}\frac{\partial x}{\partial \eta}\,.$$

We will assume in the following that the inequality $J \neq 0$ is satisfied everywhere in the region D_ξ.

We now describe two algorithms for the construction of contour lines on a curvilinear mesh of quadrilateral cells.

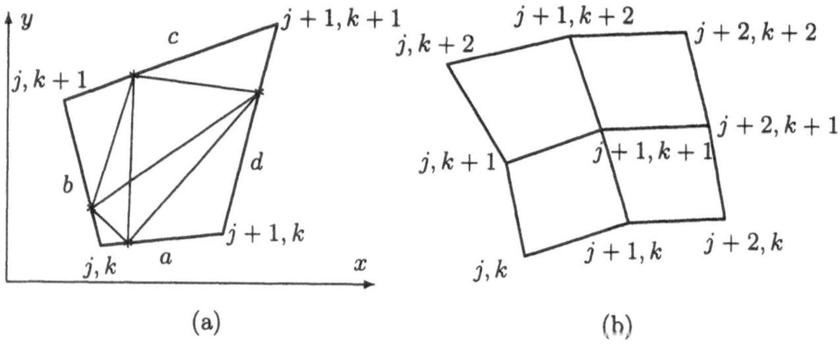

Figure 3. The construction of contour lines: (a) drawing the contour segments in an individual cell; (b) illustration for the implementation of Algorithm 2.

Algorithm 1. Let us assume that the contour lines $F(x, y) = C_i$, $i = 1, \ldots, M$ $(M \geq 1)$ are to be determined, where C_i are user-specified constants. Let $C_{i_0} = 0$ be one of the specific values of C_i $(1 \leq i_0 \leq M)$. In the process of a search for the contour line $F(x, y) = 0$, the overall curvilinear mesh is scanned by focusing attention on each cell formed by four neighboring grid nodes $(j, k), (j + 1, k), (j, k + 1), (j + 1, k + 1)$; see Figure 3a [8]. Let a, b, c, d be the cell sides. Suppose we are plotting an $F = 0$ line. If the extreme values of F on the side a (i.e., $F_{j,k}$ and $F_{j+1,k}$) differ in sign, then $F = 0$ passes through some point P between. One can find this point P by linear interpolation. Likewise for sides b, c, d. Then all points found on the sides where F vanishes (see the starlets in

Figure 3a) are connected by straight lines as shown in Figure 3a and the next cell is scanned.

The following eventuality is taken into account in this simple procedure. If the contour line should pass through a corner point, say $(j+1,k)$, then two "points" are located, one on side a and one on side d. These points have the same location and are connected, in the logic of the program, by a zero-length line.

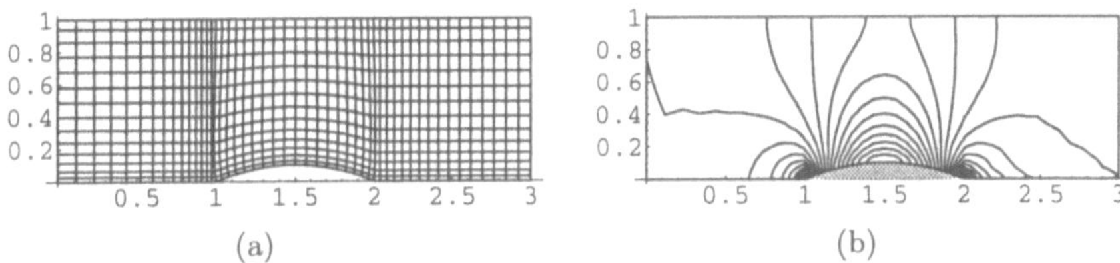

Figure 4. Inviscid fluid flow through a channel with a circular arc bump in the lower wall where the subsonic flow proceeds at a free stream Mach number value of $M_\infty = 0.5$: (a) curvilinear grid of 50×15 nodes; (b) computed Mach number contours.

Using this definition of the number of contour value points, one can easily verify that the number of points per cell must be 0, 2 or 4. Following the circuit (a, b, c, d) starting from $(j+1,k)$ with $F_{j+1,k}$, there must be an even number of sign changes to return to the $(j+1,k)$ node; hence, there must be an even number of $F=0$ values.

If four locations are found, they are connected in all six possible ways as is shown in Figure 3a [8].

Algorithm 2. One can implement the above algorithm on a computer in different ways. A faster algorithm may be programmed as follows. Let us take the following two neighboring cells (see Figure 3b):

Cell 1: the vertices $(j,k),(j+1,k),(j+1,k+1),(j,k+1)$.
Cell 2: the vertices $(j+1,k),(j+2,k),(j+2,k+1),(j+1,k+1)$.

Let us assume that we have already scanned cell 1. Cell 2 obviously has the side $\{(j+1,k),(j+1,k+1)\}$, which belongs also to cell 1. Then in the process of scanning cell 2, there is already no need to scan this side, and one can simply take the information about the availability or absence of a point of the line $F=0$ on the side $\{(j+1,k),(j+1,k+1)\}$ from the result of scanning cell 1.

We can implement the same idea when considering the following neighboring cells:

Cell 1: the vertices $(j,k),(j+1,k),(j+1,k+1),(j,k+1)$.
Cell 3: the vertices $(j,k+1),(j+1,k+1),(j+1,k+2),(j,k+2)$.

The implementation of these observations naturally makes the program logic more complex.

As an example of application of the above Algorithm 1, we present in Figure 4b the contour plot of the Mach number contours obtained in [9] as a result

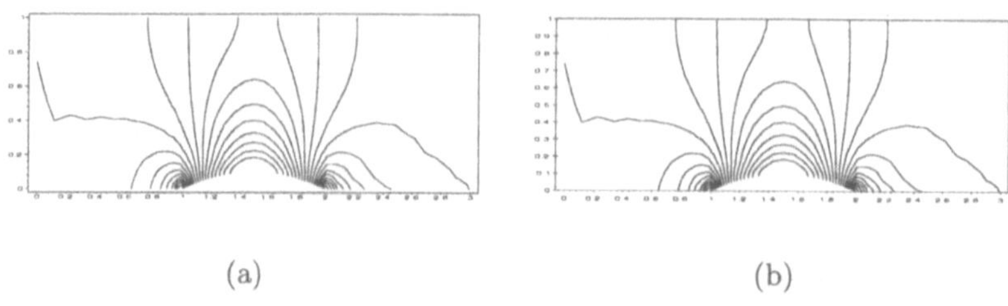

(a) (b)

Figure 5. The Mach number contours obtained by Algorithm 2 on the mesh of Figure 4a obtained by a (a) Maple program, (b) MuPAD program.

of the numerical solution of a fluid dynamics problem in a Mathematica 4.0 environment. In Figure 4a, we show the curvilinear grid of the nodes of which the Mach number values $M_{j,k}$ are specified.

The same picture of the Mach number contours was obtained by a Mathematica program implementing Algorithm 2. We show in Figure 5 the Mach number contours obtained by the Maple and MuPAD programs. In Table 1, we summarize the computer times that were needed by different CASs to make the contour plots on the curvilinear grid of Figure 4a for both algorithms. For all the runs whose results are presented in this table, we have chosen the same 20 contour levels of the Mach number.

Table 1. Computer times (in seconds) needed to make the contour plot for different CA systems.

CAS	Maple Vr5	Mathematica 4.0	MuPAD 1.4
Algorithm 1	11.225	26.048	22.132
Algorithm 2	10.136	27.666	22.402

All the computations presented throughout this article have been performed on the same Silicon Graphics, Inc. desktop computer with the following technical characteristics:

2 180 MHz IP27 Processors
CPU: MIPS R10000 Processor Chip Revision: 2.6
FPU: MIPS R10010 Floating Point Chip Revision: 0.0
Main memory size: 2048 Mbytes
Instruction cache size: 32 Kbytes
Data cache size: 32 Kbytes
Secondary unified instruction/data cache size: 2 Mbytes
Integral SCSI controller 0: Version QL1040B (rev. 2), single ended.

It should be noted that the timings in Table 1 are relatively small. As a result, errors in the determination of CPU times can occur whose magnitude are of the order of the timing granularity of the above computer. In order to increase the reliability of the CPU times in Table 1, we have computed the arithmetic mean of five runs of the same program.

It can be seen from Table 1 that Maple Vr5 is the fastest among the three considered CASs on these problems. For Mathematica 4.0 and MuPAD 1.4, the use of Algorithm 2 has not reduced the CPU time in comparison with Algorithm 1. A possible reason for this is that the overhead imposed by the two systems for additional algorithm complexity here exceeds any savings that the more sophisticated algorithm provides.

4 Numerical Quadrature after Archimedes

In this section, we want to consider one of the classical problems of numerical analysis: numerical quadrature. There are many algorithms for the solution of this task; see, for example, [10, 11]. The algorithm of Archimedes is an interesting example because it is implemented by a recursive algorithm. The efficiency in the implementation of such algorithms differs for different CASs as we shall see in the following.

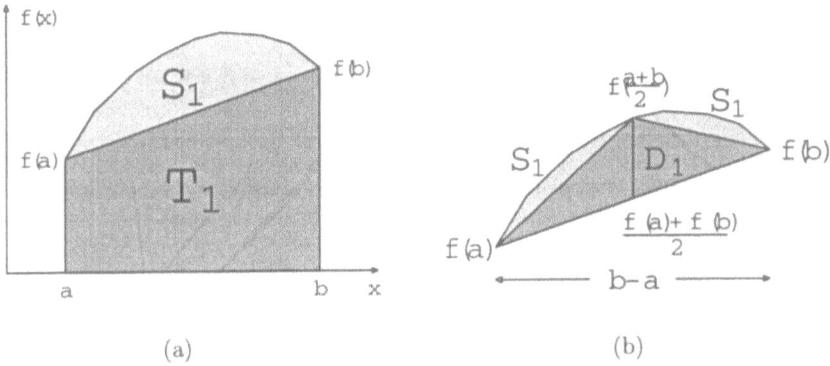

Figure 6. Archimedes' algorithm: (a) the representation of (2) as a sum of the trapezoid area T_1 and the remaining segment area S_1; (b) the approximation of S_1 by the triangular area D_1.

Numerical quadrature solves the task of approximating the definite integral

$$F(f(x), a, b) = \int_a^b f(x)\, dx \,. \tag{2}$$

In Archimedes' algorithm, F is presented as a sum of the trapezoid T_1 and the remaining segment S_1 (see Figure 6a):

$$F(f(x), a, b) = \underbrace{\frac{b-a}{2} \cdot (f(a) + f(b))}_{T_1} + S_1(f(x), a, b) \,.$$

S_1 is then approximated by the inscribed triangle D_1 as shown in Figure 6b. The process is then continued recursively. Let

$$\bar{S}_1(f(x), a, b) = \left(f\left(\frac{a+b}{2}\right) - \frac{f(a) + f(b)}{2} \right) \cdot \frac{b-a}{2} \,.$$

Then

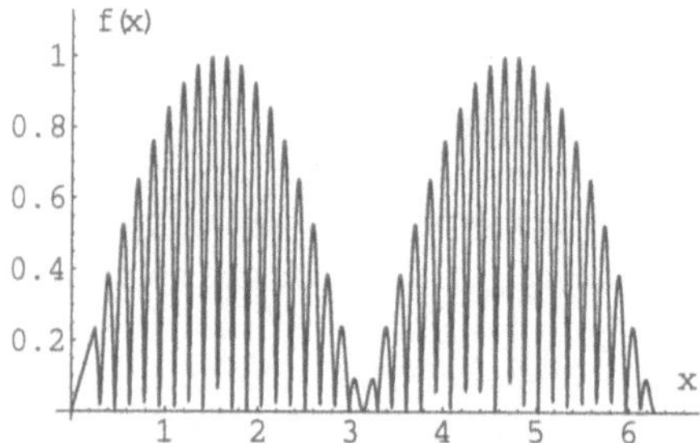

Figure 7. The plot of $f(x) = |\sin x|\,|\sin nx|$ at $n = 20$: $f(\frac{a+b}{2}) - \frac{f(a)+f(b)}{2} = 0$, but the integral is not small.

$$S_1(f(x),a,b) \approx \bar{S}_1(f(x),a,b) + \bar{S}_1\Big(f(x),a,\frac{a+b}{2}\Big) + \bar{S}_1\Big(f(x),\frac{a+b}{2},b\Big)\,. \quad (3)$$

The recursion needs a convergence criterion. There are several possibilities here:

(i) Fixed recursion depth t. For $t = 1$, the numerical quadrature process is terminated as soon as the sum

$$F(f(x),a,b) \approx T_1 + \bar{S}_1(f(x),a,b)$$

has been computed. At $t = 2$, the numerical quadrature rule (3) coincides with the well-known trapezoid rule with three equidistant nodes $a, (a+b)/2$ and b and the step size $h = (b-a)/2$. In the case of a general depth t, one obtains the trapezoid sum with $2^{t-1}+1$ equidistant nodes and the step size $h = (b-a)/2^{t-1}$.

(ii) An adaptive criterion. There are two possibilities here. One can terminate the process as soon as the area of a local triangle D_1 falls below a user-specified small positive number ε, so that

$$S_1 := D_1 \quad \text{if} \quad \left|f\Big(\frac{a+b}{2}\Big) - \frac{f(a)+f(b)}{2}\right| \cdot \frac{b-a}{2} < \varepsilon\,.$$

Or one can interrupt the recursion with $S_1 := D_1$ if the inequality

$$\left|f\Big(\frac{a+b}{2}\Big) - \frac{f(a)+f(b)}{2}\right| < \varepsilon \quad (4)$$

is satisfied. In this case, the area of the local triangle D_1 is smaller than $\varepsilon \cdot h$.

The problem, which arises in the both cases, is obvious: the triangle area may vanish although the integral does not vanish; see Figure 7. The following criterion for the interruption of the recursive process can be useful here: one interrupts the process, if the convergence criterion is satisfied at a number of the last recursion stages.

As an example of the application of Archimedes' quadrature rule, let us take the following integrand:

$$f(x,n) = |\sin x|\,|\sin nx|\,, \tag{5}$$

where n is a natural number. The factor $|\sin x|$ represents the amplitude of the modulated signal. For large n, the function (5) rapidly oscillates. Let us plot $f(x,n)$ using Mathematica 4.0 for $n = 20$ (see also Figure 7):

```
f[x_, n_]:= Abs[Sin[x]]*Abs[Sin[n*x]]; Plot[f[x, 20], {x, 0, 2 Pi}];
```

The integral

$$F(n) = \int_0^{2\pi} f(x,n)\,dx \tag{6}$$

may be found in closed form for a given numerical value of n; for example,

$$F(10) = \frac{40}{99}(\sqrt{5}+1) + \frac{20}{99}\sqrt{2}\left(\sqrt{5-\sqrt{5}} + \sqrt{5+\sqrt{5}}\right) \approx 2.551010712\,.$$

This gives the possibility of estimating the accuracy of Archimedes' rule. Let us calculate the absolute error $\mathit{Error} = |F(n) - F_A(n)|$, where $F_A(n)$ is the numerical value of the integral (6) which has been computed with the aid of Archimedes' algorithm.

The numerical quadrature after Archimedes for the case of a fixed recursion depth t may be implemented in Maple as follows:

```
depth := 2:        # the recursion depth
n := 10:           # the numerical value of n in the integrand

f := x -> abs(sin(x))*abs(sin(n*x)):

minsearch:= proc(f::procedure, a::numeric, b::numeric, m::numeric)
   local i, mincand, raster;
   raster := evalf((b - a)/m);
   mincand := [seq(a + raster*i, i = 1..m-1)];
   [a, op(select(x -> f(x) < f(x-raster) and f(x) < f(x+raster),
                 mincand)), b]
   end:

# This function determines adaptively the number of minima
# of the integrand.
minsearchfinal:= proc(f::procedure, a::numeric, b::numeric)
   local min0, min1, minima, mx;
   mx := 40;        min0 := nops(minsearch(f, evalf(a), evalf(b), mx));
   mx := mx + 2;    minima := minsearch(f, evalf(a), evalf(b), mx);
   min1 := nops(minima);
   while min1 <> min0 do
      min0 := min1;
      mx := mx + 2;    minima := minsearch(f, evalf(a), evalf(b), mx);
      min1 := nops(minima);
```

```
   od;
   RETURN(minima)
   end:

Archimedes1:= proc(f::procedure, a::numeric, b::numeric)
   local T1, S1, sum1, min_list, i;
   # Convergence criterion: fixed recursion depth

   T1:= proc(f::procedure, a::numeric, b::numeric)
      RETURN((b - a)/2*(f(a) + f(b)))
      end;

   S1:= proc(f::procedure, a::numeric, b::numeric, t::integer)
      local D1;
      if t = 0 then RETURN(0) fi;
      D1 := (f((a + b)/2) - (f(a) + f(b))/2)*(b - a)/2;
      RETURN(D1 + S1(f,a,(a+b)/2,t-1) + S1(f,(a+b)/2,b,t-1)
      end;

   sum1 := 0;
   min_list := minsearchfinal(f, a, b);
   for i from 1 to nops(min_list) - 1 do
      sum1 := sum1 + evalf(T1(f, min_list[i], min_list[i+1]) +
              S1(f, min_list[i], min_list[i+1], depth));
   od;
   RETURN(sum1)
   end:
```

The oscillatory character of the integrand (5) may give rise to an increase in the numerical quadrature error if one uses a uniform grid in the interval $[0, 2\pi]$ which does not take into account the positions of local minima of the integrand. This becomes clear from geometric considerations when one approximates the integral over each elementary subinterval by the trapezoid T_1 in accordance with the initial stage of Archimedes' quadrature rule.

In this connection, we have implemented Archimedes' quadrature rule as a two-stage process. In the first stage, we find with the aid of the procedure `minsearchfinal` the abscissas x_i^* of the local minima of the integrand $f(x)$. Since for a crude uniform mesh in the interval $[0, 2\pi]$, some of the local minima of the integrand may be missed, we have implemented the search for these local minima in `minsearchfinal` as follows. Let $\mathcal{N}(m)$ denote the number of local minima of the integrand found at a given value m of the grid nodes in $[0, 2\pi]$ with the aid of the procedure `minsearch`. Now, compute $\mathcal{N}(m+2)$. If these two values do not coincide, we increase m by 2 and continue the search for the value of m at which the equality $\mathcal{N}(m) = \mathcal{N}(m+2)$ is satisfied.

In the second stage, Archimedes' quadrature rule is applied to each interval $[x_i^*, x_{i+1}^*]$. For this purpose, the procedure `S1` calls itself recursively. The desired integral over $[0, 2\pi]$ is then computed as the sum of integrals over the intervals $[x_i^*, x_{i+1}^*]$.

For $n = 10$ and at the recursion depth $t = 2$, the above procedure `Archimedes1` produces the following result:

```
Fa1 := Archimedes1(f, 0, evalf(2*Pi));
```

$$Fa1 := 2.626135433$$

```
Fn1 := evalf(int(f(x), x = 0..2*Pi));
```

$$Fn1 := 2.551010713$$

```
Error1 := abs(Fn1 - Fa1);
```

The exact determination of CPU times for calculations which last 1–5 seconds presents certain difficulties, because the actual CPU time will vary from run to run with typical differences amounting to 0.05–0.5 second. It is tempting to define the final CPU time for such calculations as the statistical mean of the values obtained from 100 runs of the same numerical task. However, we note that both Maple and Mathematica use "remember tables". Thus, on the second run through of a given task, Maple and Mathematica can use previous results that have been computed and saved by the system in memory without explicit user intervention, and so avoid unnecessarily re-evaluating expressions [12]. As a result, the CPU time computed for 100 consecutive runs of the same task will be much smaller than the actual CPU time for 100 independent runs. A variety of other factors can also influence timings performed in the middle of a session.

In addition, there is a certain "granularity" in CPU time measurements. For Mathematica 4.0, this granularity is given as the value of the global variable `$TimeUnit`. For our runs, `$TimeUnit` $= \frac{1}{1000}$ second, but the actual differences in the CPU times obtained from independent runs of the same task are much larger, so this effect can be ignored.

In connection with the foregoing, we have determined the CPU times for cases in which these times are less than 10 seconds as the arithmetic mean of five independent runs of the same numerical task. This means that we terminated the CAS and started a new session after each calculation.

In Tables 2 and 3, we present some computational results obtained on a Silicon Graphics desktop computer. An analysis of these tables enables us to draw the following conclusions:

– The CPU times are about the same for Maple Vr5 and Mathematica 4.0 for moderate values of the frequency [$n \leq 10$ in (5)] at a fixed recursion depth. However, for larger frequencies, the Mathematica 4.0 program ran faster than the one in Maple Vr5.

– In the case of using the convergence criterion (4), the Maple Vr5 implementation runs faster than the one for Mathematica 4.0 for all frequencies considered. Overall, Maple Vr5 provides the best performance among the three CASs considered.

– MuPAD 1.4 is considerably slower than both Maple Vr5 and Mathematica 4.0. In addition, for high frequencies in the integrand (5) ($n = 15$ and $n = 20$), MuPAD produces much larger errors than either Maple Vr5 or Mathematica 4.0.

Table 2. The error *Error* in the case of using a stopping criterion based on a fixed recursion depth of $t = 2$.

n	5	10	15	20
Error	.054385164	.075124720	.06000103	.046251743
CPU time (sec)	0.148	0.170	1.248	2.609

(a) Maple Vr5

n	5	10	15	20
Error	.05438516	.07512472	.06000107	.0588083
CPU time (sec)	0.15	0.23	0.74	1.72

(b) Mathematica 4.0

n	5	10	15	20
Error	.05438516	.07512471	.89400060	1.27107944
CPU time (sec)	0.545	0.774	2.995	5.971

(c) MuPAD 1.4

Table 3. The error *Error* in the case of using the convergence criterion (4).

n	5	10	15	20
ε	8^{-4}	8^{-4}	8^{-4}	8^{-4}
Error	.00002312	.00002358	.00003027	.00003066
CPU time (sec)	8.122	15.263	23.567	32.790

(a) Maple Vr5

n	5	10	15	20
ε	8^{-4}	8^{-4}	8^{-4}	8^{-4}
Error	.00002310	.00002359	.00003027	.00003145
CPU time (sec)	11.28	21.27	31.73	40.57

(b) Mathematica 4.0

n	5	10	15	20
ε	8^{-4}	8^{-4}	8^{-4}	8^{-4}
Error	.00002312	.00002358	.83397474	1.22479702
CPU time (sec)	33.868	64.480	95.596	129.776

(c) MuPAD 1.4

– In the case of Mathematica 4.0, the CPU times presented in Tables 2 and 3 were obtained by using a Do-loop in the function `minsearch`. We note in passing that a Do-loop in Mathematica 4.0 cannot directly increment a real number as can be done in Maple. Since it is recommended in the Mathematica programming guides to avoid using Do-loops whenever possible in order to obtain better timings, we have also made another version of the function `minsearch` in which we used the built-in construct `While` instead of `Do`. In the case of the variant $n = 10$ in (5), we have obtained a CPU time of 0.26 seconds, whereas the use of the Do-loop resulted in 0.25 seconds (see Table 2b at $n = 10$). Thus, the construct `While` proved in this numerical problem to be slightly less efficient than the Do-loop. The same is true for all other Mathematica 4.0 runs presented in Tables 2 and 3.

5 Efficiency of Symbolic Computations by Different CASs

We have considered above the CPU time efficiency of the three specific CASs in the case where these systems are used for the numerical floating-point computation.

The overall efficiency of any CAS should generally be estimated from both numerical and symbolic computations. Fast symbolic computations are generally desirable for those problems of scientific computing in which analytic expressions of many coefficients or integrals are needed. One example is the solution of linear problems in mathematical physics in which the solution vector components are expanded into Fourier series [13]. The analytic expressions for the Fourier coefficients help the researcher to understand how the solution depends on the problem parameters.

In our opinion, for detailed future comparative studies of the capabilities and efficiencies of different CASs on intermediate size computations, it will be necessary to

(i) formulate general requirements and criteria for choosing test problems, and then

(ii) select a comprehensive suite of about 15–30 sufficiently nontrivial examples that involve symbolic and numeric computation, including some in which the results are displayed graphically. Some possible tasks are
   - decomposition of matrices with complex entries involving the imaginary unit $i = \sqrt{-1}$,
   - computation of definite integrals,
   - the solution of initial value and boundary value problems for ordinary differential equations,
   - the same but for partial differential equations.

In the special case of computing definite integrals, we suggest examples in which

- the integrand contains a free parameter as, for example, the parameter n in (5) [as we saw, higher frequencies made the CASs work harder],
- integration is over one, two and three or more dimensions,
- the integration curve/surface/volume is defined simply as well as complexly (e.g., a complicated volume, which is not untypical in scientific applications),
- the integrand involves a removable singularity.

Table 4. Computer times (in seconds) needed to find analytic expressions for integral (6) with integrand (5) at different n by different CASs.

n	5	10	15	20
Maple Vr5	0.4	1.412	2.368	2.878
Mathematica 4.0	49.41	77.33	99.95	101.92
MuPAD 1.4	Failed	Failed	Failed	Failed

As a first step in the investigation of the efficiency of the three CASs under consideration in performing symbolic computation tasks, we consider the symbolic computation of integral (6) with integrand (5) at different values of n. The results of these symbolic integrations are summarized in Table 4. Maple Vr5 performed the symbolic computation of the integral faster, by at least an order of magnitude, than Mathematica 4.0. MuPAD 1.4 failed to find the analytic expression for the integral (6), although it can produce the numerical value of it for the given finite values of n considered in Table 4.

6 Hierarchical Bases

In this section, we consider the task of the representation of functions in hierarchical bases. This task can be conveniently implemented by means of both object-oriented programming and parallel programming.

Let a finite set of grid points x_i be given together with the values $u_i = u(x_i)$ $(1 \leq i \leq N)$, $N \in \mathbb{N}$, of a given function $u(x)$. The interpolation condition for an interpolant $u^I(x)$ is $u^I(x_i) = u_i$. Let $\{\psi_1, \ldots, \psi_N\}$ be a basis for the interpolation functions. We can then present the interpolant $u^I(x)$ in the form

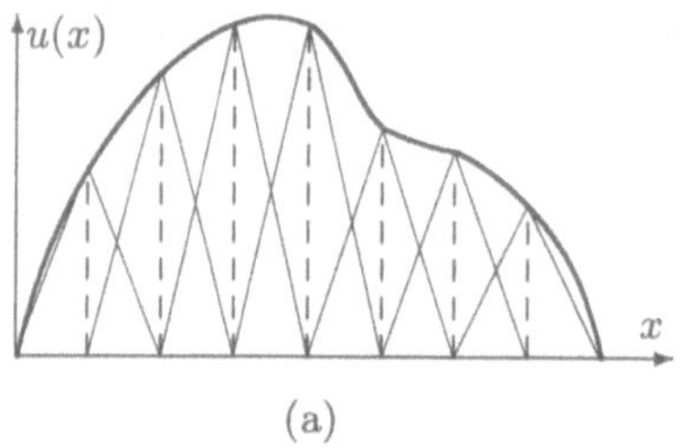

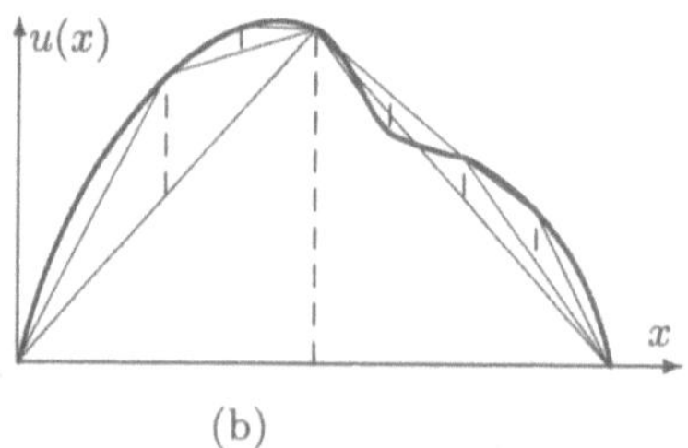

Figure 8. Function representation in different bases: (a) piecewise linear basis; (b) piecewise linear hierarchical basis.

$$u^I(x) = \sum_{j=1}^{N} a_j \psi_j(x). \tag{7}$$

In the case of the functions of a single variable, the piecewise linear basis functions $\psi_j(x)$ (see Figure 8a) represent a simple form of a basis for the interpolation on an uniform grid having the grid stepsize h. The individual basis functions satisfy the following conditions:

$$\psi_i(x_j) = \delta_{ij}, \quad \mathrm{supp}\,(\psi_i) = [x_i - h, x_i + h].$$

Here, δ_{ij} is the Kronecker symbol: $\delta_{ij} = 1$ if $i = j$ and $\delta_{ij} = 0$ if $i \neq j$. The interval $\mathrm{supp}\,(\psi_i)$ in which the basis function ψ_i is defined is the support of ψ_i, that is, $\psi_i(x) = 0$ outside $\mathrm{supp}\,(\psi_i)$.

There is one more type of bases, the hierarchical bases (Figure 8b). Each grid node is treated here in a different way, so that a hierarchy is defined on the nodes [14]. As a result, the function representation is refined successively in such a basis. At each refinement, the step size is halved. After n refinement stages, the procedure of the construction of a hierarchical basis leads to the number $N = 2^n + 1$ of grid nodes ($n \in \mathbb{N}$).

The conditions for a hierarchical basis are

$$\varphi_i(x) = 1; \quad \varphi_i(x_i \pm h_k) = 0; \quad \mathrm{supp}\,(\varphi_i) = [x_i - h_k, x_i + h_k].$$

Here, k denotes the kth of n refinement steps. The grid point x_i appears for the first time in the grid of the kth stage. The basis function and its support depend on its stage k. The coarsest grid in this hierarchy corresponds to the highest stage. The grid of stage k is denoted to lie hierarchically higher than the grid $k+1$ and hence is a refinement of this grid.

The simple functions which satisfy the above conditions are the piecewise linear functions. These functions form a space which is decomposed into the partial spaces of piecewise linear functions with the aid of the above refinement strategy.

The coefficients c_i of the hierarchical basis functions in the interpolant

$$u^I(x) = \sum_{j=1}^{N} c_j \varphi_j(x)$$

are called the hierarchical overshoots. The basis functions $\varphi_j(x)$ are linear shape functions and are defined as follows:

$$\varphi_j(x) = \begin{cases} 1 - |x - x_j|/h_k, & |x - x_j| \leq h_k; \\ 0, & |x - x_j| > h_k. \end{cases} \tag{8}$$

The hierarchical overshoot is computed in the case of a single independent variable as the difference between the actual value at a given grid node and the value obtained by interpolation from the end points of the support into the given node:

$$c_j = u_j - \frac{u_a + u_b}{2} \quad \text{with} \quad a = x_j - h_k,\ b = x_j + h_k.$$

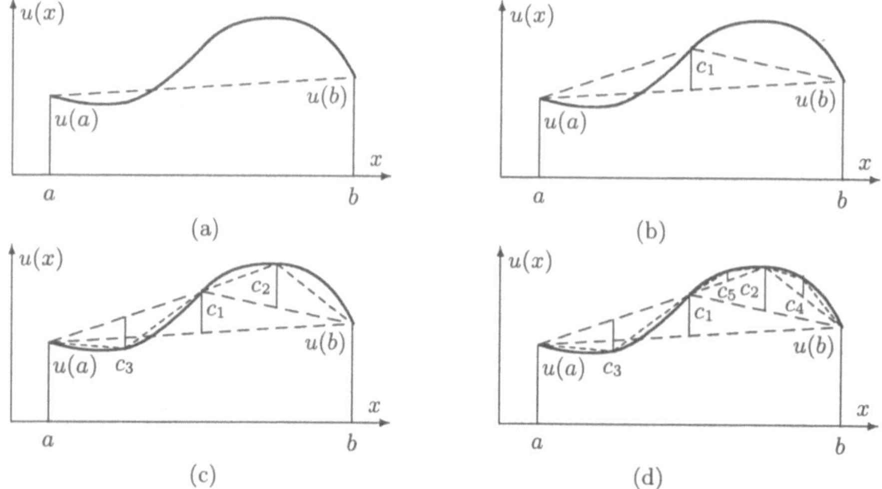

Figure 9. The adaptive representation of a function using hierarchical bases: (a) the first approximation of the function; (b) the second approximation by using the interpolation error from the middle of the interval $[a, b]$; (c) the 3rd approximation; (d) the 4th approximation.

The construction of an adaptive hierarchical basis proceeds as follows. As the first approximation for the function $u(x)$ in the interval $[a, b]$, one can take a linear interpolation of this function in which the end points a and b are taken as two grid nodes of a grid with step $h_1 = b - a$ (Figure 9a). The evaluation of the interpolation error in the middle of the interval $[a, b]$ enables us to compute a correction to the first approximation as a product of a basis function in the form of a linear shape function (8) and the coefficient

$$c_2 = u\left(\frac{a+b}{2}\right) - \frac{u(a) + u(b)}{2} .$$

This correction is added to the previous approximation (Figure 9b). This process is continued recursively on the partial intervals (Figure 9c). The process is interrupted as soon as the height of the corrective linear shape function, c_j, becomes smaller than a user-specified error ε : $|c_j| < \varepsilon$; see Figure 9d.

As the data structure for the hierarchical corrections, we can construct a binary tree. For the representation of the total interpolant, we need the top node of the graph in which the original trapezoid is stored (Figure 10).

As an example, we take the function

$$u(x) = \frac{1}{(1 + 2.4x)^3}, \quad 0 \leq x \leq 1. \tag{9}$$

The representation (7) can be obtained with the aid of the following Mathematica program:

```
u[x_]:= 1./((1. + 2.4 x)^3);
```

```
ψ[x_,j_]:= If[Abs[x - (i-1)*h] ≤h, 1. - Abs[x/h - i+1.], 0];
uI[x_]:= Sum[a[[i]]*ψ[x, i],{i, np}];

np = 7; h = 1./(np - 1); a = Table[u[(i - 1)*h], {i, 1, np}];
gr1 = Plot[u[x], {x, 0, 1}, DisplayFunction -> Identity];
gr2 = Plot[uI[x], {x,0, 1.}, PlotStyle -> {Dashing[{0.02, 0.015}]},
                              DisplayFunction -> Identity];
x = Graphics[Text[FontForm["x", {"Times-Oblique", 12}],{1.0, 0.1}]];
Show[gr1, gr2, x, DisplayFunction -> $DisplayFunction];
```

We show in Figure 11 the graphical result of the work of this program for the case $N = 7$ in (7).

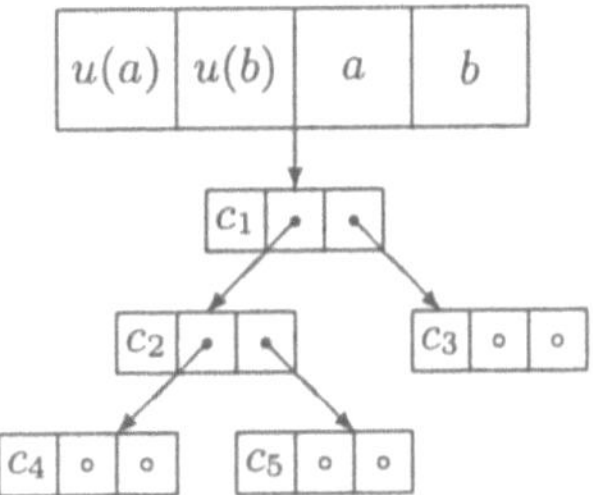

Figure 10. The binary tree for the hierarchical corrections.

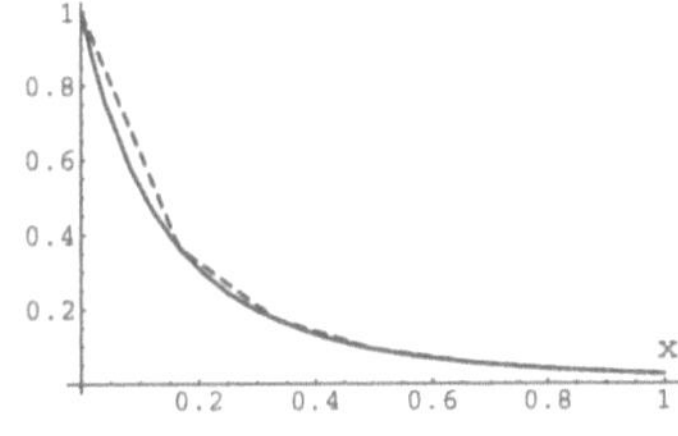

Figure 11. The interpolation of function (9) by the conventional basis (7) of linear shape functions: $u(x)$ (solid line), $u^I(x)$ (dashed line).

Let

$$\varepsilon_N(x) = \max_{x\in[0,1]} |u(x) - u_N^I(x)|$$

be the interpolation error, where $u_N^I(x)$ denotes the interpolant (7) for a given value of N.

It can be seen from Tables 5–7 that Maple Vr5 is the most efficient in terms of the CPU time needed while computing function approximations using different basis representations. At a fixed tree depth t, the construction of a hierarchical basis with Mathematica 4.0 proved to be slower than in the case of the conventional basis representation (7) (cf., Tables 5b and 6b). The CPU time needed by MuPAD 1.4 to construct the hierarchical basis at a fixed tree depth t is smaller than for Mathematica 4.0 (cf., Tables 6b and c), however, in the case of the conventional interpolant (7), MuPAD 1.4 was slower.

Tables 7a and b show that in the case of an adaptive hierarchical basis, a much smaller number N of grid nodes is needed to achieve the same accuracy ε than is the case with using the conventional interpolant (7). As a result, a much smaller CPU time is needed to achieve the same accuracy. This difference in CPU times is very striking at $\varepsilon = 10^{-6}$: in this case, the interpolant (7) requires a CPU time three orders of magnitude larger than in the case of the adaptive hierarchical basis.

Table 5. The interpolation error $\varepsilon_N(x)$ for different values of N in the case of using the interpolant (7).

N	17	33	65	129
$\varepsilon_N(x)$	.02379754	.00704194	.00192390	.00050341
CPU time (sec)	0.013	0.024	0.046	0.077

(a) Maple Vr5

N	17	33	65	129
$\varepsilon_N(x)$	.02379754	.00704194	.00192390	.00050341
CPU time (sec)	0.02	0.03	0.06	0.130

(b) Mathematica 4.0

N	17	33	65	129
$\varepsilon_N(x)$	.02379754	.00704194	.00192390	.00050341
CPU time (sec)	0.023	0.043	0.084	0.173

(c) MuPAD 1.4

Table 6. The interpolation error $\varepsilon_N(x)$ for a fixed tree depth t in the case of using a hierarchical basis.

t	4 (N=17)	5 (N=33)	6 (N=65)	7 (N=129)
$\varepsilon_N(x)$	.02379754	.00704194	.00192390	.00050341
CPU time (sec)	0.013	0.023	0.046	0.077

(a) Maple Vr5

t	4 (N−17)	5 (N−33)	6 (N−65)	7 (N−129)
$\varepsilon_N(x)$	.02379754	.00704194	.00192390	.00050341
CPU time (sec)	0.03	0.06	0.12	0.25

(b) Mathematica 4.0

t	4 (N=17)	5 (N=33)	6 (N=65)	7 (N=129)
$\varepsilon_N(x)$	.02379754	.00704194	.00192390	.00050341
CPU time (sec)	0.022	0.045	0.092	0.184

(c) MuPAD 1.4

Table 7. (Maple Vr5) The grid node number N in the case of using ...

ε	10^{-3}	10^{-4}	10^{-5}	10^{-6}
N	33	100	310	1006
CPU time (sec)	0.035	0.164	0.393	1.388

(a) an adaptive hierarchical basis.

ε	10^{-3}	10^{-4}	10^{-5}	10^{-6}
N	88	288	924	2932
CPU time (sec)	1.281	14.722	153.364	1562.397

(b) the interpolant (7).

7 Conclusions

From the viewpoint of applications in scientific computing, the comparative analysis of the efficiency of the existing general purpose CASs is of considerable interest. Our studies presented above show that Maple Vr5 generally ran the fastest among the three considered CASs for our numerical and symbolic computations. One possible reason is that the two other CASs, Mathematica and MuPAD, are younger than Maple and so their languages are not as highly optimized.

These results provide a stimulus for more detailed future comparative studies of the capabilities and efficiencies of different CASs on intermediate size computations.

References

1. Vorozhtsov, E. V., Yanenko, N. N.: *Methods for the Localization of Singularities in Numerical Solutions of Gas Dynamics Problems*, Springer-Verlag, New York, Berlin, Heidelberg (1990)
2. Fletcher, C. A. J.: *Computational Techniques for Fluid Dynamics,* 3rd edition, *Volumes I and II.* Springer-Verlag, Berlin (1996)
3. Ferziger, J. H., Perić, M.: *Computational Methods for Fluid Dynamics.* Springer, Berlin, Heidelberg, New York (1996)
4. Ganzha, V. G., Vorozhtsov, E. V.: *Numerical Solutions for Partial Differential Equations. Problem Solving Using Mathematica.* CRC Press, Boca Raton, New York, London (1996)
5. Kiselev, S. P., Vorozhtsov, E. V., Fomin, V. M.: *Foundations of Fluid Mechanics with Applications: Problem Solving Using Mathematica.* Birkhäuser, Boston (1999)
6. Thompson, J. F., Warsi, Z. U. A., Mastin, C. W.: *Numerical Grid Generation. Foundations and Applications.* North-Holland, New York (1985)
7. Knupp, P., Steinberg, S.: *Fundamentals of Grid Generation.* CRC Press, Boca Raton (1993)
8. Roache, P. J.: *Computational Fluid Dynamics*, Hermosa, New Mexico (1976)

9. Ganzha, V. G., Vorozhtsov, E. V.: Implementation of aerodynamic computations with Mathematica. In: *Computer Algebra in Scientific Computing*, Ganzha, V. G., Mayr, E. W. and Vorozhtsov, E. V. (eds.), Springer-Verlag, Berlin (1999) 101–114
10. Strampp, W., Ganzha, V. G., Vorozhtsov, E. V.: *Höhere Mathematik mit Mathematica. Band 3: Differentialgleichungen und Numerik*. Vieweg, Braunschweig (1997)
11. Bungartz, H., Zenger, Chr.: Error Control for Adaptive Sparse Grids. In: *Error Control and Adaptivity in Scientific Computing*, Bulgak, H. and Zenger, C. (eds.), Kluwer (1999) 125–157
12. Wolfram, S., *The Mathematica Book, Fourth Edition*, Wolfram Media and Cambridge University Press, Champaign, Illinois, USA (1999)
13. Tikhonov, A. N., Samarskii, A. A.: *Equations of Mathematical Physics* (in Russian), 5th Edition. Nauka, Moscow (1977)
14. Yserentant, H.: On the multi-level splitting of finite element spaces. *Numerische Mathematik* **49** (1986) 379–412

On the Relation between Pommaret and Janet Bases*

Vladimir P.Gerdt

Laboratory of Computing Techniques and Automation, Joint Institute for Nuclear Research, 141980 Dubna, Russia

Abstract. In this paper the relation between Pommaret and Janet bases of polynomial ideals is studied. It is proved that if an ideal has a finite Pommaret basis then the latter is a minimal Janet basis. An improved version of the related algorithm for computation of Janet bases, initially designed by Zharkov, is described. For an ideal with a finite Pommaret basis, the algorithm computes this basis. Otherwise, the algorithm computes a Janet basis which need not be minimal. The obtained results are generalized to linear differential ideals.

1 Introduction

Pommaret and Janet bases may be cited as typical representatives of involutive bases [1–4]. Involutive bases are Gröbner bases, though, generally redundant, and involutive methods provide an alternative approach to computation of Gröbner bases. In so doing, polynomial Pommaret bases which were first introduced in [5] have become a research subject in commutative algebra. They can be considered as generalized left Gröbner bases in the commutative ring with respect to non-commutative grading [6]. Pommaret bases of homogeneous ideals in generic position coincide with their reduced Gröbner bases [7]. The use of these bases makes more accessible the structural information of zero-dimensional ideals [8]. Pommaret bases provide an algorithmic tool for determining combinatorial decompositions of polynomial modules and for computations in the syzygy modules [9].

Linear differential Pommaret bases form the main tool in formal theory of linear partial differential equations [10, 11] whereas linear differential Janet bases form an algorithmic tool in Lie symmetry analysis of nonlinear differential equations [12, 13]. Unlike reduced Gröbner bases, Pommaret and Janet bases along with any other involutive bases lead to explicit formulae for Hilbert functions and Hilbert polynomials [3, 4, 11, 14].

Basic properties of Pommaret and Janet bases are determined by the underlying involutive divisions [1–3]. Non-noetherity of Pommaret division [1] is responsible for non-existence of finite Pommaret bases for some polynomial (linear differential) ideals of positive (differential) dimension. On the other hand, any polynomial ideal as well as any linear differential ideal has a finite Janet

* Dedicated to the memory of Alyosha Zharkov.

basis due to the noetherity of Janet division. The two divisions differ greatly by their definition: unlike Janet divisibility, Pommaret divisibility do not depend on the leading terms of generators. Given an ideal and an admissible monomial ordering, or ranking in the differential case, its monic Pommaret basis is unique whereas there are infinitely many different monic Janet bases and among them only the minimal Janet basis is uniquely defined [2].

However, in spite of the above differences, Pommaret and Janet bases are closely related, and Zharkov was the first to observe this fact in the last paper [15] of his life. He argued that if a polynomial ideal has a finite Pommaret basis and a Janet basis which is Pommaret autoreduced, then they have identical monic forms (c.f. [9]). Zharkov put also forward an algorithm for construction of Janet bases by sequential treatment of Janet nonmultiplicative prolongations followed by Pommaret autoreduction.

The goal of this paper is to study the relation between polynomial Pommaret and Janet bases in more details and to improve the Zharkov algorithm. Our analysis is based on the properties of Janet and Pommaret divisions and involutive algorithms studied in [1, 2, 4, 16].

This paper is organized as follows. The next section sketches some definitions, notations and conventions which are used in the sequel. Section 3 deals with analysis of the relationships between polynomial Pommaret and Janet bases. In particular, we prove that if an ideal has a finite Pommaret basis, then it is a minimal Janet basis. We describe here an algorithm of the combined Pommaret and Janet autoreduction which, given a Janet basis, converts it into another Janet basis. Since a minimal Janet basis is both Janet and Pommaret autoreduced, the existence of a finite Pommaret basis is equivalent to Pommaret-Janet autoreducibility of any non-minimal Janet basis into a minimal one. In Section 4 we describe an algorithm for computation of polynomial Janet bases which is an improved version of the Zharkov algorithm [15]. One of the improvements is the use of Pommaret-Janet autoreduction rather than the pure Pommaret autoreduction. Another improvement is incorporation of the involutive analogue [1] of Buchberger's chain criterion [17]. Section 5 contains generalization of the results of Sections 3 and 4 to linear differential ideals. The generalization is based on paper [4] where general involutive methods and algorithms of papers [1, 2, 16] are extended from commutative to differential algebra.

2 Basic Definitions and Notations

Let $\mathbb{N}$ be the set of nonnegative integers, and $\mathbb{M} = \{x_1^{d_1} \cdots x_n^{d_n} \mid d_i \in \mathbb{N}\}$ be a set of monomials in the polynomial ring $\mathbb{R} = K[x_1, \ldots, x_n]$ over a field K of characteristic zero.

By $deg(u)$ and $deg_i(u)$ we denote the total degree of $u \in \mathbb{M}$ and the degree of variable x_i in u, respectively. If monomial u divides monomial v we shall write $u|v$. Throughout the paper we restrict ourselves to admissible monomial orderings [18] $\succ$ which are compatible with

$$x_1 \succ x_2 \succ \cdots \succ x_n \,. \tag{1}$$

The leading monomial of the polynomial $f \in \mathbb{R}$ with respect to $\succ$ will be denoted by $lm(f)$. If $F \subset \mathbb{R}$ is a polynomial set, then by $lm(F)$ we denote the leading monomial set for F, and $Id(F)$ will denote the ideal in R generated by F. The initial ideal of an ideal $I \in \mathbb{R}$ with respect to the monomial ordering $\succ$ will be denoted by $in_{\succ}(I)$. The support of a polynomial f, that is, the set of monomials occurring in f with nonzero coefficients will be denoted by $supp(f)$. For the least common multiple of two monomials $u, v \in \mathbb{M}$ we shall use the conventional notation $lcm(u, v)$.

Definition 1. *[10, 14] For a monomial $u = x_1^{d_1} \cdots x_k^{d_k}$ with $d_k > 0$ the variables $x_j, j \geq k$ are considered as* Pommaret multiplicative *or $P-$*multiplicative *and the other variables as $P-$*nonmultiplicative. *For $u = 1$ all the variables are $P-$multiplicative.*

Definition 2. *[14] Let $U \subset \mathbb{M}$ be a finite monomial set. For each $1 \leq i \leq n$ divide U into groups labeled by non-negative integers $d_1, \ldots, d_i$:*

$$[d_1, \ldots, d_i] = \{\ u\ \in U \mid d_j = deg_j(u),\ 1 \leq j \leq i\ \}.$$

A variable x_i is called Janet multiplicative *or $J-$*multiplicative *for $u \in U$ if $i = 1$ and $deg_1(u) = \max\{deg_1(v) \mid v \in U\}$, or if $i > 1$, $u \in [d_1, \ldots, d_{i-1}]$ and $deg_i(u) = \max\{deg_i(v) \mid v \in [d_1, \ldots, d_{i-1}]\}$. If a variable is not $J-$multiplicative for $u \in U$, it is called $J-$*nonmultiplicative *for u.*

For a polynomial $f \in \mathbb{R}$ the Pommaret separation of variables into multiplicative and nonmultiplicative is done in accordance with Definition 1 where $u = lm(f)$. Analogously, for an element $f \in F$ in a finite polynomial set $F \subset \mathbb{R}$ the Janet multiplicative and nonmultiplicative variables are determined by Definition 2 with $u = lm(f) \in U = lm(F)$.

We denote by $M_P(f)$, $NM_P(f)$ and by $M_J(f, F)$, $NM_J(f, F)$, respectively, the sets of $P-$multiplicative, $P-$nonmultiplicative and $J-$multiplicative, $J-$nonmultiplicative variables for f. A set of monomials in $P-$multiplicative variables for u and $J-$multiplicative variables for $u \in U$ will be denoted by $P(u)$ and $J(u, U)$, respectively.

Remark 1. The monomial sets $P(u)$ and $J(u, U)$ for any u, U such as $u \in U$ satisfy the following axioms

(a) If $w \in L(u, U)$ and $v | w$, then $v \in L(u, U)$.
(b) If $u, v \in U$ and $uL(u, U) \cap vL(v, U) \neq \emptyset$, then $u \in vL(v, U)$ or $v \in uL(u, U)$.
(c) If $v \in U$ and $v \in uL(u, U)$, then $L(v, U) \subseteq L(u, U)$.
(d) If $V \subseteq U$, then $L(u, U) \subseteq L(u, V)$ for all $u \in V$

if one takes either $P(u)$ or $J(u, U)$ as $L(u, U)$. The axioms characterize an involutive monomial division, a concept invented in [1]. Every monomial set $L(u, U)$ satisfying the axioms generates an appropriate separation of variables into ($L-$)multiplicative and ($L-$)nonmultiplicative. As this takes place, an element $u \in U$ is an $L-$divisor of a monomial $w \in \mathbb{M}$ if $w/u \in L(u, U)$. In this case w is $L-$multiple of u. Using the axioms, a number of new divisions was constructed [2, 16] which may be also used for algorithmic computation of involutive bases.

All the next definitions in this section are those in [1, 2] specified to Pommaret and Janet divisions.

Definition 3. *Given a finite monomial set U, its* cone $C(U)$, $P-$cone $C_P(U)$ *and* $J-$cone $C_J(U)$ *are the following monomial sets*

$$C(U) = \cup_{u \in U} u\mathbb{M}, \quad C_P(U) = \cup_{u \in U} uP(u), \quad C_J(U) = \cup_{u \in U} uJ(u,U).$$

Definition 4. *Given an admissible ordering $\succ$, a polynomial set $F \subset \mathbb{R}$ is called* Pommaret autoreduced *or* $P-$autoreduced *if every $f \in F$ has no terms $P-$multiple of an element in $lm(F) \setminus lm(f)$. Similarly, a finite polynomial set F is* Janet autoreduced *or* $J-$autoreduced *if each term in every $f \in F$ has no $J-$divisors among $lm(F) \setminus lm(f)$. A finite set F will be called* Pommaret-Janet autoreduced *or* $PJ-$autoreduced *if it is both $P-$autoreduced and $J-$autoreduced.*

Remark 2. From Definition 2 it follows that any finite set U of distinct monomials is $J-$autoreduced

$$(\forall u, v \in U)\ (u \neq v)\ \ [\ uJ(u,U) \cap vJ(v,U) = \emptyset\ ].$$

Definition 5. *Given an admissible ordering $\succ$ and a polynomial set $F \subset \mathbb{R}$, a polynomial $h \in \mathbb{R}$ is said to be in $P-$*normal form modulo F *if every term in h has no $P-$ divisors in $lm(F)$. Similarly, if all the terms in h have no $J-$divisors among the leading terms of a finite polynomial set F, then h is in $J-$*normal form modulo F.

A general involutive normal form algorithm is described in [1], and an involutive normal form of a polynomial modulo any involutively autoreduced set is uniquely defined. We denote by $NF_P(f,F)$ and $NF_J(f,F)$, respectively, $P-$normal and $J-$normal form of polynomial f modulo F.

Pommaret or Janet autoreduction of a finite polynomial set F may be performed [1] similarly to the conventional autoreduction [17, 18]. If H is obtained from F by the conventional, or $J-$autoreduction we shall write $H = Autoreduce(F)$ or $H = Autoreduce_J(F)$, respectively.

Definition 6. *A $P-$ autoreduced set F is called a* Pommaret basis *($P-$basis) of the ideal $Id(F)$ generated by F if*

$$(\forall f \in F)\ (\forall x \in NM_P(f))\ \ [\ NF_P(f \cdot x, F) = 0\ ]. \tag{2}$$

Similarly, a $J-$autoreduced set F is called a Janet basis *($J-$basis) of $Id(F)$ if*

$$(\forall f \in F)\ (\forall x \in NM_J(f,F))\ \ [\ NF_J(f \cdot x, F) = 0\ ]. \tag{3}$$

In accordance with Definition 5 the nonmultiplicative prolongation $f \cdot x$ with the vanishing $P-$ or $J-$normal form modulo polynomial set $F = \{f_1, \cdots, f_m\}$ can be rewritten as

$$f \cdot x = \sum_{i=1}^{m} f_i \, h_i, \quad h_i \in \mathbb{R}, \quad lm(f) \cdot x \succeq lm(f_i \, h_i) \quad (1 \leq i \leq m)$$

where $supp(h_i) \subset P(f_i)$ or $supp(h_i) \subset J(f_i, F)$, respectively, for every polynomial product $f_i \, h_i$.

Let G_P and G_J be Pommaret and Janet bases of an ideal I, respectively. Then, from Definition 6 it follows [1] that

$$h \in I \quad \text{iff} \quad NF_P(h, G_P) = 0 \quad \text{and} \quad h \in I \quad \text{iff} \quad NF_J(h, G_J) = 0 \,. \tag{4}$$

This implies, in particular, the equalities

$$C_P(lm(G_P)) = C(lm(G_P)) \,, \quad C_J(lm(G_J)) = C(lm(G_J)) \,. \tag{5}$$

It is immediate from (5) that $lm(G_P)$ and $lm(G_J)$ are $P-$ and $J-$bases of the initial ideal $in_{\succ}(I)$.

Corollary 1. *If for a $P-$autoreduced set G_P the equality (5) of its cone and $P-$cone holds and $lm(G_P)$ is a basis of the initial ideal $in\,(Id(G_P))\}$, then G_P is a $P-$ basis of $Id(G_P)$. Analogous statement holds for a $J-$basis.*

Whereas monic Pommaret bases much like to reduced Gröbner bases are unique, every ideal, by property (2), has infinitely many monic Janet bases. Among them there is the unique $J-$basis defined as follows.

Definition 7. *A Janet basis G of ideal $Id(G)$ is called* minimal *if for any other $J-$basis F of the ideal the inclusion $lm(G) \subseteq lm(F)$ holds.*

Remark 3. Every zero-dimensional polynomial ideal has a finite Pommaret basis, and for a positive dimensional ideal the existence of finite Pommaret basis can be always achieved by means of an appropriate linear transformation of variables [5, 6, 10, 11].

3 Relation Between Polynomial Pommaret and Janet Bases

Given a finite monomial set U, Definitions 1 and 2 generally give different separations of variables for elements in U.

Example 1. $U = \{x_1x_2, x_2x_3, x_3^2\}$.

monomial	Pommaret		Janet	
	M_P	NM_P	M_J	NM_J
x_1x_2	x_2, x_3	x_1	x_1, x_2, x_3	$-$
x_2x_3	x_3	x_1, x_2	x_2, x_3	x_1
x_3^2	x_3	x_1, x_2	x_3	x_1, x_2

Here is, however, an important relation between Pommaret and Janet separations:

Proposition 1. [15](see also [1]). *If a finite monomial set U is $P-$autoreduced, then for any $u \in U$ the following inclusions hold*

$$M_P(u,U) \subseteq M_J(u,U), \quad NM_J(u,U) \subseteq NM_P(u,U)\,.$$

For U in Example 1 the minimal Janet basis U_J and the Pommaret basis U_P of the monomial ideal $Id(U)$ are

$$U_J = U \cup \{x_1x_3^2\}, \quad U_P = U_J \cup_{i=2}^{\infty} \{x_1^i x_2\} \cup_{j=2}^{\infty} \{x_1^j x_3^2\} \cup_{k=2}^{\infty} \{x_2^k x_3\}\,. \qquad (6)$$

Note that U_P is infinite and $U_J \subset U_P$. Below we show that the inclusion $G_J \subseteq G_P$ holds for any minimal Janet basis G_J and Pommaret basis G_P if both of them are monic and generate the same ideal. Furthermore, the proper inclusion $G_J \subset G_P$ holds iff P is infinite.

The following algorithm, given a finite polynomial set $F \in \mathbb{R}$ and an admissible ordering $\succ$, performs $PJ-$autoreduction of F and outputs a PJ-autoreduced set H. In this case we shall write $H = Autoreduce_{PJ}F$.

Since the involutive $P-$ and $J-$reductions which are performed in the course of the algorithm form subsets of the conventional reductions [1], the algorithm terminates. Furthermore, the **while**-loop generates the $P-$autoreduced monomial set $lm(H)$. In accordance with Remark 2 the Janet autoreduction in line 12 does not affect the leading terms, and, hence, the output polynomial set is both Pommaret and Janet autoreduced.

Algorithm: **Pommaret-JanetAutoreduction**(F, $\succ$)

Input: $F \in \mathbb{R}$, a finite set; $\succ$, an admissible ordering
Output: $H = Autoreduce_{PJ}(F)$
begin
1 $G := \emptyset;\ H := F$
2 **repeat**
3 $\tilde{H} := H$
4 **while** exist $h \in H$ such that
5 $lm(h) \in C_P\,(lm(H \setminus \{h\}))$ **do**
6 $H := H \setminus \{h\};\ G := G \cup \{h\}$
7 **for each** $g \in G$ **do**
8 $G \setminus \{g\};\ f := NF_J(g, H)$
9 **if** $f \neq 0$ **then**
10 $H := H \cup \{f\}$
11 **until** $H = \tilde{H}$
12 $H := Autoreduce_J(H)$
end

Proposition 2. *If algorithm* **Pommaret-JanetAutoreduction** *takes a Janet basis F as an input its output H is also a Janet basis of the same ideal, and $H \subseteq F$.*

Proof. Let F be a Janet basis of the ideal $Id(F)$ and H be a polynomial set which is computed by the algorithm. Apparently, $Id(H) = Id(F)$. Consider $U = lm(F)$. If U is $P-$autoreduced, then H initiated as F in line 1 does not change in the process of the algorithm, and, hence, $H = F = Autoreduce_{PJ}(F)$. Otherwise, consider the output polynomial sets H. Denote $lm(H)$ by V. Then, by construction, $V \subset U$.

Consider a monomial $t \in C_J(U)$ and show that $t \in C_J(V)$. Let $u \in U$ be a $J-$divisor of t, that is, $t \in uJ(u,U)$, and $v \in V$ be such that $v|u$. If $u = v$, by property (d) of Janet division in Remark 1, we are done. Let now $u \in U \setminus V$. The **while**-loop provides that $u \in vP(v)$. If $v = x_1^{d_1} \cdots x_k^{d_k}$ with $d_k > 0$ $(1 \le k \le n)$, then from Definition 1 it follows that $u = x_1^{d_1} \cdots x_k^{d_k+e_k} \cdots x_n^{e_n}$.

We have to prove that any variable $x_i \in J(u,U)$ $(1 \le i \le n)$ satisfy $x_i \in J(v,V)$. Consider two alternative cases: $i < k$ and $i \ge k$. In the first case, by Definition 2, both u, v belong to the same group $[d_1, \cdots, d_{i-1}]$ of monomials in U. It follows that $x_i \in J(v,V)$. In the second case, by Definition 1, $x_i \in P(v)$ and Proposition 1 we find again that x_i is $J-$multiplicative for v as an element in V.

Therefore, V is a Janet monomial basis of $Id(U)$. Thus, by Corollary 1, H is a $J-$basis. In so doing, every $J-$normal form computed in line 8 of the algorithm apparently vanishes. This implies the inclusion $H \subseteq F$. □

Corollary 2. *A minimal Janet basis is Pommaret autoreduced.*

Proof. Let G be a minimal Janet basis. From Proposition 2 and Definition 7 it follows that $lm(G)$ is $P-$autoreduced. Thus, G is $P-$autoreduced. □

It is clear that, given a Janet basis, its $PJ-$autoreduction yields, generally, more compact basis than the pure $P-$autoreduction.

Example 2. Consider polynomial set

$$F = \{xyzt - xz, xyz + z^2, xzt + x^2, xy + z, zt + x\}$$

which is a Janet basis of the ideal $Id(F)$ with respect to the degree-reverse-lexicographical ordering $\succ$ such that $x \succ y \succ z \succ t$. Given F and $\succ$ as input, the algorithm **Pommaret-JanetAutoreduction**$(F, \succ)$ outputs the minimal $J-$basis

$$G_1 = \{xzt + x^2, xy + z, zt + x\}.$$

If one uses the Pommaret normal form computation in line 7 instead of that of Janet, then the algorithm leads to the output

$$G_2 = \{xzt + x^2, z^2t + xz, xy + z, zt + x\} = G_1 \cup \{z^2t + xz\}.$$

The following theorem is the main theoretical result of the present paper and forms a basis of an algorithm for construction of Janet and Pommaret bases described in the next section.

Theorem 1. *Given an ideal $I \subseteq R$ and an admissible monomial ordering $\succ$ compatible with (1), the following are equivalent:*

(*i*) *I has a finite Pommaret basis.*
(*ii*) *A minimal Janet basis of I is its Pommaret basis.*
(*iii*) *If F is a Janet basis of I, then $G = Autoreduce_{JP}(F)$ is its Pommaret basis.*

Proof. (*i*) $\Longrightarrow$ (*iii*): Suppose $G = \{g_1, \ldots, g_m\}$ which, by Proposition 2, is also a $J-$basis of I, is not its $P-$basis. Our goal is to prove that a Pommaret basis of I is an infinite polynomial set. From Corollary 1 it follows that

$$(\exists g \in G)\ (\exists x_k \in NM_P(g))\ \ [\ lm(g) \cdot x_k \notin C_P\,(lm(G))\ ]\,. \tag{7}$$

Among nonmultiplicative prolongations $g \cdot x_k$ satisfying (7) choose one with the lowest $lm(g) \cdot x_k$ with respect to the pure lexicographical ordering $\succ_{Lex}$ generated by (1). If there are several such prolongations choose that with the lexicographically lowest x_k, that is, with the lexicographically highest g. This choice is unique since G is $J-$autoreduced.

We claim that $x_k \in M_J(g, G)$. Assume for a contradiction that $x_k \in NM_J(g, G)$. Then from Janet involutivity conditions (3) we obtain

$$(\exists f \in G)\ \ [\ lm(g) \cdot x_k = lm(f) \cdot w \mid w \in J\,(lm(f), lm(G))\ ]\,. \tag{8}$$

In accordance with condition (7) w contains $P-$nonmultiplicative variables for f and from (8) it follows [1] that $f \succ_{Lex} g$. If $w = x_j$, then $x_j \succ x_k$, and both $lm(f) \cdot x_j$ and $lm(g)$ belong to the same monomial group $[\,deg_1(g), \ldots, deg_{k-1}(g)]$ appearing in Definition 2. Hence, $x_j \in NM_J(f, G)$ in contradiction to (8). Therefore, $deg(w) \geq 2$ and (8) can be rewritten as

$$lm(g) \cdot x_k = (lm(f) \cdot v) \cdot x_j$$

with $w = v \cdot x_j$. Show that $lm(f) \cdot v \in P\,(lm(f))$. If $v \in C_P\,(lm(G))$ we are done. Otherwise there is $x_m | v$ such that $x_m \in NM_P\,(lm(f))$. Since $lm(f) \cdot x_m \prec_{Lex} lm(g) \cdot x_k$, our choice of g and x_k implies the existence $g_1 \in G$ such that $f \cdot x_m = g_1 \cdot v_1$ and $v/x_m \in P\,(lm(g_1))$. If $v_1 = v/x_m \in P\,(lm(g_1))$ we are done. Otherwise we select again a $P-$nonmultiplicative variable for g_1 occurring in v_1 and rewrite the corresponding prolongation in terms of its $P-$divisor $lm(g_2) \in lm(G)$. Continuity of Pommaret division [1] provides termination of the rewriting process with an element in $\tilde{g} \in G \setminus \{g\}$ such that $lm(\tilde{g})$ is a $P-$divisor of $lm(f) \cdot v$. Because G is $P-$autoreduced, by Proposition 1 $lm(\tilde{g})$ is also a $J-$divisor of $lm(f) \cdot v$. By this means there are two different Janet divisors $lm(f)$ and $lm(\tilde{g})$ of the same monomial that contradicts Remark 2 and proves the claim.

Let now H be a $P-$basis of $Id(G)$. Denote $lm(g)\cdot x_k$ by u and show that $u\in lm(H)$. Suppose there is an element $h\in H\setminus G$, such that $u=lm(h)\cdot v$ with $v\in P(lm(h))$. Then $h\prec_{Lex} u$ and there is $q\in G$ satisfying $lm(h)=lm(q)\cdot w$ where $w\notin P(lm(q))$. Thus, there is a $P-$nonmultiplicative prolongation $q\cdot x_j$ with $x_j|w$ such that $lm(q)\cdot x_j \prec_{Lex} u$ and $lm(q)\cdot x_j\notin C_P(lm(G))$ that contradicts the above choice of g and x_k.

Now consider monomial set $U=lm(G)\cup\{u\}\subseteq lm(H)$. By Definition 2, $x_k\in NM_P(u)$ and $u\cdot x_k$ is obviously the lexicographically lowest $P-$nonmultiplicative prolongation of elements in U. It is easy to see that $u\cdot x_k\notin C_P(U)$. Indeed, since $x_k\in M_J(g,G)$, it follows that $u\cdot x_k\notin U$, and a $P-$divisor of $u\cdot x_k$ would also $P-$divide $lm(g)$ that is impossible as G is $P-$autoreduced. Thus, we find that $u\cdot x_k\in lm(H)$. By sequential repetition of this reasoning for $u\cdot x_k^i$ $(i=2,3\ldots)$ we deduce that every such monomial is an element in $lm(H)$, and, therefore, H is infinite.

$(iii)\Longrightarrow(ii)$: If G is a minimal Janet basis of $Id(G)$, then Corollary 2 implies $G=Autoreduce_{JP}(G)$, and the above arguments show that either G is also a $P-$basis or the latter is infinite.

$(ii)\Longrightarrow(i)$: This implication is obvious. □

Corollary 3. *Let G_J be a monic minimal Janet basis and G_P be a monic Pommaret basis for the same polynomial ideal and monomial ordering. Then $G_J\subseteq G_P$ and $G_J\subset G_P$ iff G_P is infinite.*

Proof. This follows from the above proof of Theorem 1. □

Corollary 4. *If a polynomial ideal I is in generic position [6] with respect to the given admissible monomial ordering, then a minimal Janet basis of I is also its Pommaret basis. Additionally, if I is homogeneous, then every of these bases is the reduced Gröbner basis.*

Proof. I has a finite Pommaret basis [6, 7]. If I is homogeneous, then its Pommaret basis is the reduced Gröbner basis [7]. □

4 Algorithm for Construction of Janet Bases

In this section we present an algorithm for constructing $J-$bases of polynomial ideals which is based on Theorem 1 and will be called **JanetBasis**. Whenever the ideal generated by an input polynomial set has a finite $P-$basis for a given admissible ordering, the algorithm outputs just this basis which is also a minimal $J-$basis. Otherwise, the $J-$basis computed by the algorithm is not necessarily minimal as we demonstrate below by the explicit example.

Algorithm: **JanetBasis**(F, $\succ$)

```
Input:  F ∈ ℝ, a finite set; ≻, an admissible ordering
Output:  G, an involutive basis of the ideal Id(F)
begin
1     G := Autoreduce(F)
2     T := ∅
3     for each g ∈ G do
4        T := T ∪ {(g, lm(g), ∅)}
5     while exist (g, u, D) ∈ T and x ∈ NM_J(g, G) \ D do
6        choose such (g, u, D), x with the lowest lm(g) · x w.r.t. ≻
7        T := T \ {(g, u, D)} ∪ {(g, u, D ∪ {x})}
8        if Criterion(g · x, u, T) is false then
9           h := NF_J(g · x, G)
10          if h ≠ 0 then
11             if lm(h) = lm(g · x) then
12                T := T ∪ {(h, u, ∅)}
13             else
14                T := T ∪ {(h, lm(h), ∅)}
15                G := Autoreduce_PJ(G ∪ {h})
16       Q := T
17       T := ∅
18       for each g ∈ G do
19          if exist (f, u, D) ∈ Q s.t. lm(f) = lm(g) then
20             choose g̃ ∈ G s.t. u ∈ lm(g̃)J(lm(g̃), lm(G))
21             T := T ∪ {(g, lm(g̃), D ∩ NM_J(g, G))}
22          else
23             T := T ∪ {(g, lm(g), ∅)}
end
```

$Criterion(g,u,T)$ is true provided that if there is $(f,v,D) \in T$ such that $lcm(u,v) \prec lm(g)$ and $lm(g) \in lm(f)J\,(lm(f), lm(G))$.

The structure of algorithm **JanetBasis** is very close to that of the algorithm **InvolutiveBasis** devised in [1] for construction of involutive bases for arbitrary constructive involutive divisions, and, hence, applicable to Janet division. The main difference between the algorithms is the form of their intermediate autoreduction. Whereas the previous algorithm, when specified for Janet division, uses the pure Janet autoreduction which do not affect the leading terms, the below one uses the above algorithm **Pommaret-JanetAutoreduction**. By this reason, the algorithm **InvolutiveBasis**, unlike the below one, almost never outputs a minimal Janet basis or a Pommaret basis if the latter is finite. In paper [2] we designed the algorithm **MinimalInvolutiveBasis** which always outputs a minimal involutive basis whenever the latter exists. As we now see from Theorem 1, this algorithm in the case of Janet division outputs also a Pommaret basis if it is finite.

Besides, in the algorithm **InvolutiveBasis** the involutive autoreduction is performed whenever nonzero normal form is obtained in the **while**-loop. In the algorithm **JanetBasis** the subalgorithm **Pommaret-JanetAutoreduction** is caused by a nonzero $J-$normal form h in line 15 only if the leading term of the related prolongation is $J-$reducible. Otherwise, since G is always $P-$autoreduced before its enlargement with h, $lm(h)$ cannot $P-$divide, in accordance with Definition 1, any other element in $lm(G)$.

Note that we indicated the intersection in line 21 to emphasize that elements in D must be nonmultiplicative variables for the corresponding polynomial that is always understood in algorithms of papers ([1, 2, 4, 16]).

Noetherity of Janet division provides termination of the algorithm **JanetBasis** [1]. To show this consider the intermediate bases $G_0 = Autoreduce(F)$ and G_i $(i = 1, 2, \ldots)$ generated after the $i-th$ iteration of the **while**-loop. It is clear that

$$Id\,(lm(G_0)) \subseteq Id\,(lm(G_1)) \subseteq Id\,(lm(G_2)) \subseteq \cdots \tag{9}$$

and this chain is stabilized after finitely many steps. Namely, the stabilization starts when the intermediate polynomial set G becomes a (non-necessarily reduced) Gröbner basis of $Id(F)$. By partial involutivity of G [1], the proper inclusion in chain (9) holds only when G is enlarged by h in line 15. In between of such proper inclusions and after the chain stabilization $lm(G)$ is completed with $lm(h) = lm(g \cdot x)$ as stands in line 11, and this completion cannot be infinite by noetherity of Janet division [1]. Once algorithm terminates, it produces, by Proposition 2, a $PJ - autoreduced$ Janet basis of F because the involutivity conditions (3) are satisfied as is checked in lines 6 and 9 where $h = 0$. In so doing, correctness of the criterion which is verified in line 8 is proved exactly as done in [1] for the algorithm **InvolutiveBasis**.

Remark 4. In line 6 of the algorithm **JanetBasis** one can use any admissible ordering for selection of the current $J-$nonmultiplicative prolongation $g \cdot x$ to be treated in the following lines. This selection ordering may not only be different from the main ordering $\succ$ but also may vary at every step when the selection is done. Correctness of this arbitrariness in the choice of selection ordering follows from the correctness of this arbitrariness for the monomial completion procedure [16].

As mentioned above, the algorithm **JanetBasis** for an ideal of positive dimension may not output its minimal $J-$basis. We demonstrate this by the following example.

Example 3. Let F be a set $\{x^2y - z, x\,y^2 - y\}$ generating one-dimensional ideal, and $\succ$ be the degree-reverse-lexicographical ordering with $x \succ y \succ z$. Then the algorithm **JanetBasis**$(F, \succ)$ outputs the following Janet basis

$$G = \{x^2y - z, x^2z - z^3, x\,y - y\,z, x\,z - z^2, y^2z - y, y\,z^2 - z\}$$

whereas the minimal Janet basis coinciding with the reduced Gröbner basis is

$$\{x\,y - y\,z, x\,z - z^2, y^2z - y, y\,z^2 - z\}\,.$$

Remark 5. The algorithm **JanetBasis** is an improved version of the algorithm designed by Zharkov [15]. The first improvement is the use of the mixed Pommaret-Janet autoreduction instead of the pure Pommaret autoreduction as Zharkov proposed. Let G_1 and G_2 be output Janet bases computed with the use of $PJ-$ and $P-$autoreduction, respectively. Then, Proposition 1 implies $G_1 \subseteq G_2$ and below we give an example when the proper inclusion holds. The second improvement is the criterion used in line 7. This criterion is an involutive analogue [1] of the Buchberger's chain criterion [17] and is superior to the criterion used in [15].

Example 4. [20] The following polynomial set generates three-dimensional ideal

$$F = \left\{ \begin{array}{l} 4x_5(x_6 + 2a_1 - 8x_1)(a_2 - a_3) - x_2x_3x_4 + x_2 + x_4 \\ 4x_5(x_6 + 2a_1 - 8x_2)(a_2 - a_3) - x_1x_3x_4 + x_1 + x_3 \\ 4x_5(x_6 + 2a_1 - 8x_3)(a_2 - a_3) - x_1x_2x_4 + x_2 + x_4 \\ 4x_5(x_6 + 2a_1 - 8x_4)(a_2 - a_3) - x_1x_2x_3 + x_1 + x_3 \end{array} \right\} \tag{10}$$

where

$$a_1 = x_1 + x_2 + x_3 + x_4, \ a_2 = x_1x_2x_3x_4, \ a_3 = x_1x_2 + x_2x_3 + x_3x_4 + x_4x_1 .$$

For the polynomial set (10) and the degree-reverse-lexicographical ordering compatible with (1) the algorithm **JanetBasis** outputs set G_1 with 71 polynomials. The pure Pommaret autoreduction in line 15 generates set $G_2 \supset G_1$ of 75 elements. Note that a minimal Janet basis contains 49 polynomials whereas the reduced Gröbner basis contains 44 polynomials.

5 Linear Differential Bases

Let $\mathbb{K}$ be a zero characteristic differential field with a finite number of mutually commuting derivation operators $\partial/\partial x_1, \ldots, \partial/\partial x_n$. Consider the differential polynomial ring $\mathbb{DR} = \mathbb{K}\{y_1, \ldots, y_m\}$ with the set of differential indeterminates $\{y_1, \ldots, y_m\}$. Elements in $\mathbb{DR}$ are differential polynomials in $\{y_1, \ldots, y_m\}$. An ideal in $\mathbb{DR}$ generated by linear differential polynomials is called linear differential ideal [23].

In [4], by exploiting the well-known algorithmic similarities between polynomial and linear differential systems and the association between monomials and the derivatives, we extended the general involutive methods and algorithms designed in [1, 2, 16] for polynomial ideals to linear differential ideals. The statements and algorithms of Sect. 3 and 4 admit similar extension. As this takes place, all the statements proved above for the polynomial case are proved by parallel arguments for the differential case. In the following table we give a short correspondence between these two cases. In particular, this correspondence allows one to rewrite the algorithms **Pommaret-JanetAutoreduction** and **JanetBasis** for linear differential bases.

Commutative Algebra	Differential Algebra
$\mathbb{R} = K[x_1, \ldots, x_n]$	$\mathbb{DR} = \mathbb{K}\{y_1, \ldots, y_m\}$
$f, g, h \in \mathbb{R}$	$f, g, h \in \mathbb{DR}$
$F, G, H \subset \mathbb{R}$	$F, G, H \subset \mathbb{DR}$
$x^\alpha \equiv x_1^{\alpha_1} \cdot x_n^{\alpha_n}$	$\partial_\alpha y_j \equiv \frac{\partial^{\alpha_1+\cdots+\alpha_n} y_j}{\partial x_1^{\alpha_1} \cdots \partial x_n^{\alpha_n}} \Longleftrightarrow [x^\alpha]_j$
$g \cdot x$	$\partial_x g$
monomial ordering	ranking
$lm(g)$	leading derivative $ld(g)$
$lm(G)$	$ld(G) = \cup_{g \in G}\{ld(g)\}$

6 Conclusion

As we have seen finite Pommaret bases of polynomial and linear differential ideals are minimal Janet bases. The above proof of Theorem 1 shows that, given a $P-$autoreduced Janet basis G, it is easy to verify the existence of a finite Pommaret basis. One suffices to check the condition (7). Another check which may be even easier in practice is to look at the Pommaret and Janet separation of variables for elements in G. As shown in [9], the existence of a finite Pommaret basis implies coincidence of both separations for every element in G. Moreover, the leading monomial structure of an infinite Pommaret basis can be read off the structure of $lm(G)$ (cf. (6) for Example 1).

Therefore, minimal Janet bases can be used in commutative and differential algebra as well as Pommaret bases with the advantage of finiteness. For example, in the formal theory of differential equations [9–11] infinity of a Pommaret basis signals on δ-singularity of the coordinate system chosen, and the condition (7) gives the same signal for Janet bases.

On the face of it, Pommaret division looks like more attractive than Janet one since its separation is easier to compute than the Janet separation. Besides, Pommaret division, unlike that of Janet, is globally defined [2]. Hence, after enlargement of the intermediate polynomial set with an irreducible $P-$nonmultiplicative prolongation there are no needs to recompute the Pommaret separation for other elements in the set. However, the careful implementation of both divisions do not reveal any notable advantage of Pommaret division over Janet division in construction of polynomial bases. This rather surprising fact was firstly observed by Zharkov [15]. One of the explanations of this experimental phenomenon is given by Proposition 1: one must generally treat more $P-$nonmultiplicative prolongations than $J-$nonmultiplicative ones. In so doing, the search for an involutive divisor in the process of involutive reduction can be done similarly for both divisions as we show in our forthcoming paper [21].

The algorithm **JanetBasis** presented above is now under implementation in C in parallel with the algorithm **MinimalInvolutiveBasis** [2] specified for

Janet division. Our first experimentation with the codes shows that sometimes the former algorithm runs faster than the latter one and needs less computer memory. For example, modular computation of the degree-reverse-lexicographic Janet basis for the Cyclic 7 example [22] is about twice faster with the algorithm **JanetBasis** than with the algorithm **MinimalInvolutiveBasis**. Currently, the timings for computation modulo 31013 are about 5 and 10 minutes, respectively, on a Pentium Pro 333 Mhz computer.

7 Acknowledgements

I am grateful to W.M.Seiler for useful discussions and for providing me with his recent paper [9]. I would also like to thank D.A.Yanovich for his assistance in computing the above examples. This work was partially supported by grant INTAS-99-1222 and grants No. 98-01-00101 and 00-15-96691 from the Russian Foundation for Basic Research.

References

1. Gerdt, V.P., Blinkov, Yu.A.: Involutive Bases of Polynomial Ideals. *Math. Comp. Sim.* **45** (1998) 519–542.
2. Gerdt, V.P., Blinkov, Yu.A.: Minimal Involutive Bases. *Math. Comp. Simul.* **45** (1998) 543–560.
3. Apel, J.: Theory of Involutive Divisions and an Application to Hilbert Function. *J. Symb. Comp.* **25** (1998) 683–704.
4. Gerdt, V.P.: Completion of Linear Differential Systems to Involution. In: *Computer Algebra in Scientific Computing / CASC'99*, V.G.Ganzha, E.W.Mayr and E.V.Vorozhtsov (Eds.), Springer-Verlag, Berlin, 1999, pp. 115–137.
5. Zharkov, A.Yu., Blinkov, Yu.A.: Involutive Approach to Investigating Polynomial Systems. In: Proceedings of "SC 93", International IMACS Symposium on Symbolic Computation: New Trends and Developments (Lille, June 14-17, 1993). *Math. Comp. Simul.* **42** (1996), 323–332.
6. Apel, J.: A Gröbner Approach to Involutive Bases. *J. Symb. Comp.* **19** (1995) 441–458.
7. Mall, D.: On the Relation Between Gröbner and Pommaret Bases. *AAECC* **9** (1998) 117–123.
8. Zharkov, A.Yu.: Solving Zero-Dimensional Involutive Systems. In: *Algorithms in Algebraic Geometry and Applications*, Gonzales-Vega, L., Recio, T. (eds.). Progress in Mathematics, Vol. 143, Birkhäuser, Basel, 1996, pp. 389–399.
9. Seiler, W.M.: A Combinatorial Approach to Involution and δ-Regularity. *Preprint*, University of Mannheim, 2000. See **www.ira.uka.de/iaks-calmet/werner/publ.html**.
10. Pommaret, J.F.: *Systems of Partial Differential Equations and Lie Pseudogroups*, Gordon & Breach, New York, 1978.
11. Pommaret J.-F.: *Partial Differential Equations and Group Theory. New Perspectives for Applications*, Kluwer, Dordrecht, 1994.
12. Schwarz, F.: An Algorithm for Determining the Size of Symmetry Groups. *Computing* **49** (1992) 95–115.

13. Schwarz, F.: Janet Bases of 2nd Order Ordinary Differential Equations. In: *Proceedings of the ISSAC'96*, Lakshman Y.N. (ed.), ACM Press, 1996, pp.179–188.
14. Janet, M.: *Leçons sur les Systèmes d'Equations aux Dérivées Partielles*, Cahiers Scientifiques, IV, Gauthier-Villars, Paris, 1929.
15. Zharkov, A.Yu.: Involutive Polynomial Bases: General Case. *Preprint JINR E5-94-224*, Dubna, 1994. Submitted to ISSAC'94.
16. Gerdt, V.P.: Involutive Division Technique: Some Generalizations and Optimizations, *Zapiski Nauchnykh Seminarov POMI (St.Petersburg)* **258** (1999) 185–207. To appear in *J. Math. Sci.* **258** (2000).
17. Buchberger, B.: Gröbner Bases: an Algorithmic Method in Polynomial Ideal Theory. In: *Recent Trends in Multidimensional System Theory*, Bose, N.K. (ed.), Reidel, Dordrecht, 1985, pp. 184–232.
18. Becker, T., Weispfenning, V., Kredel, H.: *Gröbner Bases. A Computational Approach to Commutative Algebra.* Graduate Texts in Mathematics **141**, Springer-Verlag, New York, 1993.
19. Gerdt, V.P., Kornyak, V.V., Berth, M., Czichowsky G.: Construction of Involutive Monomial Sets for Different Involutive Divisions. In: *Computer Algebra in Scientific Computing / CASC'99*, V.G.Ganzha, E.W.Mayr and E.V.Vorozhtsov (Eds.), Springer-Verlag, Berlin, 1999, pp. 147–157.
20. Verschelde, J., Gaterman, K.: Symmetric Newton Polytopes for Solving Polynomial Systems, *Preprint SC 94-3*, Konrad-Zuse-Zentrum für Informationstechnik, Berlin, 1994.
21. Gerdt, V.P., Blinkov, Yu.A., Yanovich D.A. *Computation of Monomial Janet Bases.* In preparation.
22. Bini, D., Mourrain, B.: *Polynomial Test Suite*, 1996. See **www-sop.inria.fr/saga/POL**.
23. Kolchin, E.R.: *Differential Algebra and Algebraic Groups*, Academic Press, New York, 1973.

An (Asymptotic) Error Bound for the Evaluation of Polynomials at Roots

Manfred Göbel

Department of Electronic Systems Engineering
University of Essex, Wivenhoe Park
Colchester CO4 3SQ, United Kingdom
mkgoebel@essex.ac.uk

Abstract. Given is a univariate polynomial

$$F(X) = X^n - \sigma_1 X^{n-1} + \ldots + (-1)^{n-1}\sigma_{n-1}X + (-1)^n \sigma_n \in \mathbb{C}[X]$$

and a "poor" numerical solution $\alpha_1, \ldots, \alpha_n \in \mathbb{C}$ of $F(X) = 0$ such that $1 \leq |\alpha_i - x_i| \leq \delta \in \mathbb{R}$ and $F(x_i) = 0$ for $1 \leq i \leq n$. We show that $O(2^{n-1}n! \cdot \delta^{\frac{n(n-1)}{2}})$ is an (asymptotic) error bound for the evaluation of an arbitrary but fixed multivariate polynomial $f \in \mathbb{C}[X_1, \ldots, X_n]$ at the n-tuple $(\alpha_1, \ldots, \alpha_n)$ instead of the n-tuple of the roots $(x_1, \ldots, x_n)$, and that for some polynomials the (asymptotic) error bound is $\Omega(n! \cdot \delta^{\frac{n(n-1)}{2}})$.
Keywords. Analysis of algorithms, data structures, rewriting techniques, evaluation of polynomials, roots, error bounds.

1 Introduction

Many problems arising in nature have a symmetric structure, and a recent trend in symbolic-numerical computation and algorithm development, which contribute to solutions of such problems, is to explicitly utilize these symmetries (cf. [1, 4]). This makes solutions for difficult problems feasible, or leads to significant complexity improvements compared to conventional methods, which neglect the symmetric structure.

In [5] a rewriting technique to compute a representation of any permutation-invariant polynomial $f \in \mathbb{C}[X_1, \ldots, X_n]$ as a finite linear combination of special permutation-invariant orbits with symmetric polynomials as coefficients was described. This method has solved an open problem in invariant theory concerning the degree bounds of generators of permutation-invariant polynomials over arbitrary ground rings R. The classical algorithm [11, §33], which was developed by Gauß, and which rewrites any symmetric polynomial as a polynomial in the elementary symmetric polynomials, can be considered as a special case of our method w.r.t. complexity and reduction technique [7]. Gauß' algorithm plays an important role in invariant theory and has many other significant applications, e.g. in Galois theory [2, 3], or the computation of resultants [11] and Sturm sequences [8]. We want to show here that the general rewriting technique [5] has also (theoretical and practical) applications outside the area of classical invariant theory.

The solution of an equation

$$F(X) = X^n - \sigma_1 X^{n-1} + \ldots + (-1)^{n-1}\sigma_{n-1}X + (-1)^n\sigma_n = 0$$

is — as first shown by Galois at the beginning of the 19th century — in general not possible by means of radical expressions [2]. The equation $F(X) = 0$ can be solved by radical expressions, if the corresponding Galois group of $F(X)$ is solvable [3, section 4.6]. In most cases the Galois group of $F(X)$ is non-solvable, and one has to deal with numerical solutions $\alpha_1, \ldots, \alpha_n$ of $F(X) = 0$ over the real or complex numbers (cf. [9, chapter 10]). The fact that many mathematical applications demand to evaluate a given polynomial $f \in \mathbb{C}[X_1, \ldots, X_n]$ at the roots $x_1, \ldots, x_n$ of $F(X) = 0$, i.e. to compute $f(x_1, \ldots, x_n)$, is our starting point for an application of the general rewriting technique. Three typical examples are as follows:

- In order to obtain statistics about the roots, one can consider to compute various moments like average, variance, skewness and so forth w.r.t. an assumed distribution (cf. [9, chapter 14]). These statistical quantities are polynomials which have to be evaluated at the roots of F.
- The evaluation of a univariate polynomial $H(X)$ at a root x_i of $F(X) = 0$ for some $1 \leq i \leq n$ is — by considering $H(X)$ as $H(X_i) \in \mathbb{C}[X_1, \ldots, X_n]$ — a special case of the above situation. The computation of local extrema, stationary points, etc. are typical examples, which require the evaluation of $H(X)$ at the roots $F(X) = \frac{d^k H(X)}{d^k x}$ for some $k \in \mathbb{N}$.
- A similar example is the (fast) Fourier transformation of a polynomial $p(X) = \sum_{i=0}^n a_i X^n$ with $a_i \in \mathbb{C}$, which computes a sequence of numbers $b_0, b_1, \ldots, b_n$, where $b_i = p(r_i)$ and r_i is a root of $F(X) = X^n - 1$ for $1 \leq i \leq n$.[1]

We show in the following that an *almost* symmetric representation of $f \in \mathbb{C}[X_1, \ldots, X_n]$ can be used to maintain the quality of a numerical solution $\alpha_1, \ldots, \alpha_n$ of $F(X) = 0$ during the computation of $f(\alpha_1, \ldots, \alpha_n)$. Our *almost* symmetric representation can be used to control the (asymptotic) propagation of rounding errors in this specific algebraic process [12]. The algorithmic approach behind our (asymptotic) error bound can be applied to polynomials of any degree, even to the polynomial $F(X)$ itself. Theoretical and practical improvements can be expected, whenever the degree of f is sufficiently large compared to the number roots of $F(X) = 0$.

We proceed now as follows: After briefly recalling our notation, we state and prove our (asymptotic) error bound for "poor" numerical solutions in Section 2. Section 3 presents some further analysis of our (asymptotic) error bound, and eventually, Section 4 concludes with some remarks about "good" numerical solutions.

[1] Thanks to Prof. Fateman (Berkeley) for this example.

Notation. $\mathbb{N}$, $\mathbb{R}$ and $\mathbb{C}$ denote the natural, real and complex numbers. The commutative polynomial ring over $\mathbb{C}$ in the indeterminates $X_1, \ldots, X_n$ is denoted by $\mathbb{C}[X_1, \ldots, X_n]$. The coefficients $\sigma_1, \ldots, \sigma_n$ of the univariate polynomial $F(X) = X^n - \sigma_1 X^{n-1} + \ldots + (-1)^{n-1}\sigma_{n-1}X + (-1)^n \sigma_n \in \mathbb{C}[X]$ can be viewed as the elementary symmetric polynomials [11, §33]

$$\begin{aligned}
\sigma_1 &= X_1 + \ldots + X_n, \\
\sigma_2 &= X_1X_2 + \ldots + X_{n-1}X_n, \\
\ldots &\quad \ldots \\
\sigma_n &= X_1 \ldots X_n,
\end{aligned}$$

which are evaluated at the roots $x_1, \ldots, x_n$ of $F(X)$.

2 The (Asymptotic) Error Bound

Theorem 1. *Let $F(X) = X^n - \sigma_1 X^{n-1} + \ldots + (-1)^{n-1}\sigma_{n-1}X + (-1)^n \sigma_n \in \mathbb{C}[X]$, let $x_1, \ldots, x_n \in \mathbb{C}$ be the exact solutions of $F(X) = 0$, and let $f \in \mathbb{C}[X_1, \ldots, X_n]$. Then the following holds for all "poor" numerical solution $\alpha_1, \ldots, \alpha_n \in \mathbb{C}$ of $F(X) = 0$ with $1 \leq |\alpha_i - x_i| \leq \delta \in \mathbb{R}$ and $1 \leq i \leq n$:*

$$|f(\alpha_1, \ldots, \alpha_n) - f(x_1, \ldots, x_n)| \in O(2^{n-1} n! \cdot \delta^{\frac{n(n-1)}{2}})$$

Note that $O(2^{n-1} n! \cdot \delta^{\frac{n(n-1)}{2}})$ depends only on δ and n. The bound is in particular independent of the degree of f.[2]

Proof. According to [5], $f(X_1, \ldots, X_n)$ is a polynomial invariant of the trivial permutation group and has therefore the following formal representation as a $\mathbb{C}[\sigma_1, \ldots, \sigma_n]$-linear combination of special terms, where $\sigma_1, \ldots, \sigma_n$ denote the unevaluated elementary symmetric polynomials:

$$f(X_1, \ldots, X_n) = \sum_{t \text{ spec.}} p_t(\sigma_1, \ldots, \sigma_n) \cdot t \tag{1}$$

Any special term $t = X_1^{e_1} \ldots X_n^{e_n}$ has by construction a total degree of at most $\frac{n(n-1)}{2}$. The number of special terms is finite, and there are at most $2^{n-1}n!$ of them [6]. By evaluating the elementary symmetric polynomials $\sigma_1, \ldots, \sigma_n$ at the given roots $x_1, \ldots, x_n$, it follows that

$$\begin{aligned}
|f(\alpha_1, \ldots, \alpha_n) - f(x_1, \ldots, x_n)| &= |\sum_{t \text{ spec.}} p_t(\sigma_1, \ldots, \sigma_n) \cdot (t_\alpha - t_x)| \\
&\leq \sum_{t \text{ spec.}} |p_t(\sigma_1, \ldots, \sigma_n)| \cdot |t_\alpha - t_x|
\end{aligned}$$

[2] We cannot establish the "same" bound involving the "same" constants for all $f \in \mathbb{C}[X_1, \ldots, X_n]$, because the estimations in the proof depend on f. We only claim in Theorem 1 that the "same" bound holds for all δ w.r.t. a fixed f.

with $t_\alpha = \alpha_1^{e_1} \dots \alpha_n^{e_n}$ and $t_x = x_1^{e_1} \dots x_n^{e_n}$. Now let $\alpha_i - x_i = \delta_i$ and $1 \le |\delta_i| \le \delta$ for $1 \le i \le n$. Then $\alpha_i = x_i + \delta_i$ and

$$\begin{aligned} |t_\alpha - t_x| &= |\alpha_1^{e_1} \dots \alpha_n^{e_n} - x_1^{e_1} \dots x_n^{e_n}| \\ &= |(x_1 + \delta_1)^{e_1} \dots (x_n + \delta_n)^{e_n} - x_1^{e_1} \dots x_n^{e_n}| \\ &\le g(|x_1|, \dots, |x_n|) \cdot \delta^{\frac{n(n-1)}{2}} \end{aligned} \tag{2}$$

with $g \in \mathbb{R}[X_1, \dots, X_n]$. The separation of $x_1, \dots, x_n$ and δ in the last step of equation (2) was obtained by expanding the product, applying the triangle inequality, estimating the δ_i for $1 \le i \le n$ by δ, and subsequently collecting and further estimating all products containing δ's by $\delta^{\frac{n(n-1)}{2}}$. We obtain $|t_\alpha - t_x| \in O(\delta^{\frac{n(n-1)}{2}})$, because the polynomial $g(|x_1|, \dots, |x_n|)$ is independent of f and can be viewed as constant.

Furthermore, we have $|p_t(\sigma_1, \dots, \sigma_n)| \in \mathbb{R}$ for at most $2^{n-1}n!$ special terms t; each coefficient $p_t(.)$ of the linear combination in equation (1) is independent of the error δ and can also be viewed as constant w.r.t. the given polynomial f. Hence, $|f(\alpha_1, \dots, \alpha_n) - f(x_1, \dots, x_n)| \in O(2^{n-1}n! \cdot \delta^{\frac{n(n-1)}{2}})$. □

The bound $O(2^{n-1}n! \cdot \delta^{\frac{n(n-1)}{2}})$ [3] is constructive, because the representation can be effectively computed (cf. [5, Algorithm 3.12 with $G = \{id\}$]). The proof of Theorem 1 estimates all special terms by $\delta^{\frac{n(n-1)}{2}}$, but not all special terms have a total degree of $\frac{n(n-1)}{2}$, and further, there is no need to consider descending special terms. The bound is therefore a rough estimation leading to the correct power of δ, but the term in front of δ can be reduced.[4] Similar rough estimations are used in the following to obtain straightforward error bounds, which have the same properties w.r.t. δ. As the following example shows, the representation (1) is a useful data structure to control the propagation of errors during the evaluation of a polynomial at roots.

Example 1. Let $F(X) = X^2 - 7X + 10 \in \mathbb{C}[X]$ with $\sigma_1 = 7$ and $\sigma_2 = 10$. We assume that we do not know the roots of $F(X) = 0$, which are $x_1 = 2$ and $x_2 = 5$, and that a "poor" numerical solution of $F(X) = 0$ with $\alpha_1 = 1$ and $\alpha_2 = 6$ is given.

Our goal is to evaluate $f(X_1, X_2) = X_1^5 + X_2^5 + 3X_1^2 \in \mathbb{C}[X_1, X_2]$ at the roots of $F(X)$. The exact value is $f(x_1 = 2, x_2 = 5) = 3169$. The straightforward evaluation of f leads to $f(\alpha_1 = 1, \alpha_2 = 6) = 7780$. Our symbolic-numerical approach via the proof of Theorem 1 requires the computation of a representation of f, which is $f(X_1, X_2) = \sigma_1^5 - 5\sigma_1^3\sigma_2 + 3\sigma_1^2 + 5\sigma_1\sigma_2^2 - 3\sigma_2 - 3\sigma_1 X_2$. The evaluation

[3] The bound can be improved w.r.t. δ, whenever the total degree of f is less than $\frac{n(n-1)}{2}$.

[4] The exact number of non-descending special terms in $n = 1, 2, \dots$ variables is $\frac{d^k}{x^k} \frac{1}{2-e^x} |_{x=0} - 2^{k-1}$, which is 0, 1, 9, 67, 525, 4651, 47229, 545707, … [6]. This is obviously much better than $2^{n-1}n!$, which has an integer sequence commencing with 1, 4, 24, 192, 1920, 23040, 322560, 5160960, …. The exact number of non-descending special terms in $n = 1, 2, \dots$ variables with a total degree $\frac{n(n-1)}{2}$ is $n! - 1$.

of this representation gives $f(\alpha_1 = 1, \alpha_2 = 6) = 3148$, which is much closer to the exact value.

The (asymptotic) error bound for the evaluation of $f \in \mathbb{C}[X_1, \ldots, X_n]$ with maximal variable degree d at a "poor" numerical solution without making use of representation (1) is $O(d^n \delta^{nd})$, because f can have d^n terms, which depend on δ and the total degree of a term is at most nd. Figure 1 compares the boundary functions $h_1(n, d, \delta) := 2^{n-1} n! \cdot \delta^{\frac{n(n-1)}{2}}$ and $h_2(n, d, \delta) := d^n \delta^{nd}$ for $\delta \in \{1.1, 1.5, 2.0\}$ by showing the breakeven points w.r.t. the maximal variable degree d and the number of roots n.

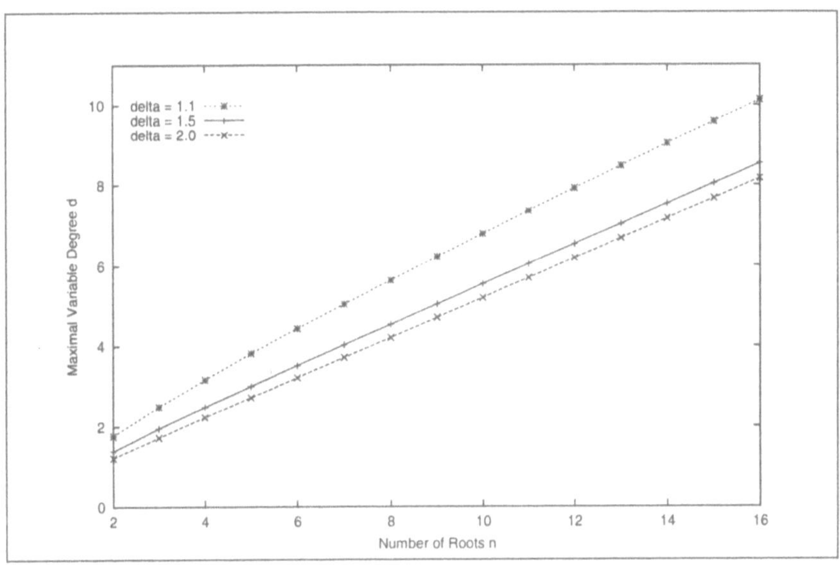

Fig. 1. Breakeven points for multivariate polynomials w.r.t. maximal variable degree and number of roots.

The (asymptotic) error bound for the evaluation of a univariate $H \in \mathbb{C}[X]$ with degree d at a "poor" numerical solution without making use of representation (1) is $O(d\delta^d)$, because H can have d terms, which depend on δ, and the degree of a term is at most d. Figure 2 compares the boundary functions $h_1(n, d, \delta) := 2^{n-1} n! \cdot \delta^{\frac{n(n-1)}{2}}$ and $h_2(d, \delta) := d\delta^d$ for $\delta \in \{1.1, 1.5, 2.0\}$ by showing the breakeven points w.r.t. the degree d and the number of roots n.

3 Some Further Analysis

We cannot improve our bound very much from the point of view of worst case complexity, if we do not allow radical expressions [2].

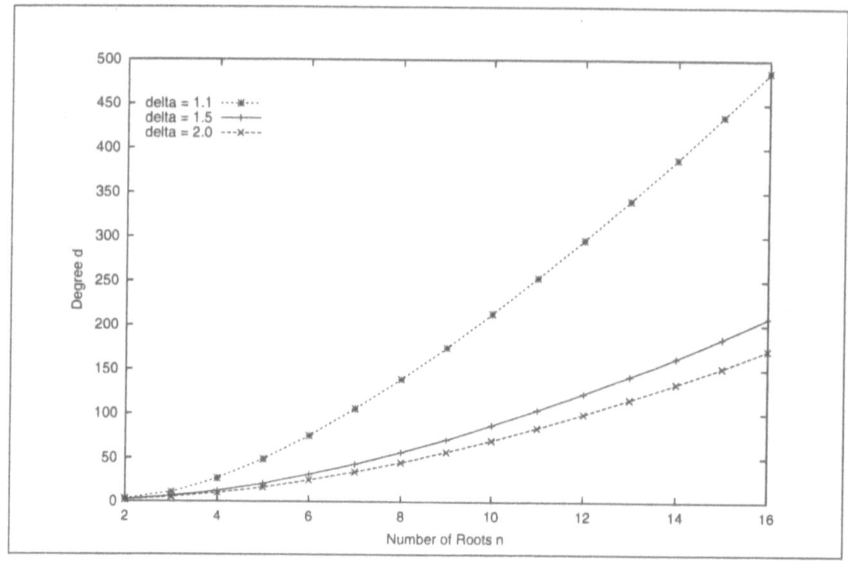

Fig. 2. Breakeven points for univariate polynomials w.r.t. degree and number of roots.

Lemma 1. *Let $F(X)$, $x_1, \ldots, x_n$, $\alpha_1, \ldots, \alpha_n$ and δ be as in Theorem 1. Then there exists a $f \in \mathbb{C}[X_1, \ldots, X_n]$ such that $|f(\alpha_1, \ldots, \alpha_n) - f(x_1, \ldots, x_n)| \in \Omega(n! \cdot \delta^{\frac{n(n-1)}{2}})$.*

Proof. The representation of polynomial invariants of the alternating group A_n requires $\frac{n!}{2}$ special terms with a total degree of $\frac{n(n-1)}{2}$ [5]. □

Note that we can derive the bound $O(n! \cdot \delta^{\frac{n(n-1)}{2}})$ already within the proof of Theorem 1, if we — according to our remarks above — only consider the $n!$ special terms with a total degree $\frac{n(n-1)}{2}$. From this point of view $\Theta(n! \cdot \delta^{\frac{n(n-1)}{2}})$ is the *optimal bound* which can be established for every given $f \in \mathbb{C}[X_1, \ldots, X_n]$.

The proof of Lemma 1 does not hold if we allow radical expressions.

Lemma 2. *Let $F(X)$, $x_1, \ldots, x_n$, $\alpha_1, \ldots, \alpha_n$ and δ be as in Theorem 1. Then any invariant $f \in \mathbb{C}[X_1, \ldots, X_n]$ of the symmetric group S_n and the alternating group A_n can be evaluated at $f(x_1, \ldots, x_n)$ by radical expressions.*

Proof. See [2, §10] for the S_n case. For A_n the polynomial f has a unique representation as a $\mathbb{C}[\sigma_1, \ldots, \sigma_n]$-linear combination of the form

$$f = p_1(\sigma_1, \ldots, \sigma_n) + p_2(\sigma_1, \ldots, \sigma_n) \cdot \Delta,$$

where $\Delta = \prod_{1 \leq i < j \leq n}(x_i - x_j)$ denotes the Schur polynomial [10]. Δ^2 is a symmetric function evaluated at the roots of $F(X)$, i.e. $\Delta^2 = p(\sigma_1, \ldots, \sigma_n)$ for some unique $p \in \mathbb{C}[X_1, \ldots, X_n]$. Hence, $\Delta = \sqrt{\Delta^2} = \sqrt{p(\sigma_1, \ldots, \sigma_n)}$ can be evaluated at the roots of $F(X)$ by radical expressions. □

Theorem 1 can be improved for $n < 5$ variables if we allow radical expressions.

Lemma 3. *Let $F(X)$, $x_1, \ldots, x_n$, $\alpha_1, \ldots, \alpha_n$ and δ be as in Theorem 1, and let $n < 5$. Then any $f \in \mathbb{C}[X_1, \ldots, X_n]$ can be evaluated at $f(x_1, \ldots, x_n)$ by radical expressions.*

Proof. Any root $x_1, \ldots, x_n$ of an equation $F(X) = 0$ with $n < 5$ can be expressed in terms of known quantities. Thus any $f \in \mathbb{C}[X_1, \ldots, X_n]$ can be evaluated at $f(x_1, \ldots, x_n)$ by radical expressions [2]. □

4 "Good" Numerical Solutions

The proof of Theorem 1 does not hold for "good" numerical solutions, i.e. in the case $\delta < 1$, because we have $O(\delta^{n_1}) \subset O(\delta^{n_2})$ for $n_1 > n_2$. Nevertheless, Example 2 shows that our symbolic-numerical evaluation technique seems to work still quite well.

Example 2. Let $F(X)$ and $f(X_1, X_2)$ be as in Example 1, and let $\alpha_1 = 1.9$ and $\alpha_2 = 5.1$ be a "good" numerical solution of $F(X) = 0$.

The straightforward evaluation of f at the given approximation of the roots of $F(X)$ leads to $f(\alpha_1 = 1.9, \alpha_2 = 5.1) = 3485.8435$; our symbolic-numerical approach results in $f(\alpha_1 = 1.9, \alpha_2 = 5.1) = 3166.9$, which is again much closer to the exact value.

We can use a similar argument as in Theorem 1 to derive a lower bound for the error in the case of "good" numerical solutions.

Theorem 2. *Let $F(X) = X^n - \sigma_1 X^{n-1} + \ldots + (-1)^{n-1}\sigma_{n-1}X + (-1)^n\sigma_n \in \mathbb{C}[X]$, let $x_1, \ldots, x_n \in \mathbb{C}$ be the exact solutions of $F(X) = 0$, and let $f \in \mathbb{C}[X_1, \ldots, X_n]$. Then the following holds for all "good" numerical solution $\alpha_1, \ldots, \alpha_n \in \mathbb{C}$ of $F(X) = 0$ with $|\alpha_i - x_i| \le \delta \in \mathbb{R}$, $\delta < 1$ and $1 \le i \le n$:*

$$|f(\alpha_1, \ldots, \alpha_n) - f(x_1, \ldots, x_n)| \in \Omega(\delta^{\frac{n(n-1)}{2}})$$

Note that $\Omega(\delta^{\frac{n(n-1)}{2}})$ depends only on δ and n. The bound is in particular independent of the degree of f.

Proof. Consider $f = X_1^{n-1}X_2^{n-2}\ldots X_{n-1} \in \mathbb{C}[X_1, \ldots, X_n]$. Then f has a total degree of $\frac{n(n-1)}{2}$ and consists only of one special term. f cannot be further rewritten as a linear combination of special terms with smaller degree and elementary symmetric polynomials as coefficients. It follows that the error for evaluating f at the "good" approximation $(\alpha_1, \ldots, \alpha_n)$ is at least $\Omega(\delta^{\frac{n(n-1)}{2}})$.

The polynomial f reflects the best case situation; any other polynomial has a representation involving special terms with the same or lower degrees, but this does not lead to a further reduction of the lower bound. □

Example 1 and Example 2 are showing the good symbolic-numerical behavior of our *almost* symmetric representation even if the degree of $f \in \mathbb{C}[X_1, \ldots, X_n]$ is small compared to the number of roots, and even if we deal with "good" numerical solutions. Our approach is based on a finite linear combination of special terms evaluated at approximated roots with exact coefficients. The straightforward evaluation can be seen — via representation (1) — as a linear combination of special terms evaluated at approximated roots with approximated coefficients. The approximated coefficients are computed by using the elementary symmetric polynomials evaluated at the approximated roots. This point of view explains roughly, why our *almost* symmetric representation is in most cases superior to the conventional approach and how more exact information flows into the evaluation process.

Acknowledgment. The author would like to thank Volker Heun (Munich), Heinz Kredel (Mannheim), Patrick Maier (Saarbrücken), Ursula Walter (Colchester), Andreas Weber (Darmstadt) and the anonymous referees for their comments and remarks.

References

1. Allgower, E. L., Georg, K., Miranda, R. (1992). Exploiting Symmetry in Applied and Numerical Analysis. Proc. of the 22th AMS-SIAM Summer Seminar on Applied Mathematics, Colorado State University, Fort Collins, Colorado. Lectures in Applied Mathematics, 29, American Mathematical Society, Providence, RI.
2. Edwards, H. M. (1993). Galois Theory. Springer
3. Gaal, L. (1988). Classical Galois Theory. Chelsea Publishing Company, New York, 4th edition
4. Gatermann, K. (1990). Symbolic Solution of Polynomial Equation Systems with Symmetry. In Watanabe, Sh., Nagata, M., (eds.), Proc. International Symposium on Symbolic and Algebraic Computation ISSAC'90, 112-119, Tokyo, Japan, August, ACM Press
5. Göbel, M. (1995). Computing Bases for Permutation-Invariant Polynomials. Journal of Symbolic Computation 19, 285-291
6. Göbel, M. (1997). On the Number of Special Permutation-Invariant Orbits and Terms. Applicable Algebra in Engineering, Communication, and Computation 8(6), 505–509
7. Göbel, M. (1998). On the Reduction of G-invariant Polynomials for Arbitrary Permutation Groups G. In: Manuel Bronstein, Johannes Grabmeier, Volker Weispfenning, (eds.), Symbolic Rewriting Techniques, Progress in Computer Science and Applied Logic (PCS 15), 71–92. Birkhäuser
8. Netto, E. (1896). Vorlesungen über Algebra. Leipzig, Teubner
9. Press, W. H., Flannery, B. P., Teukolsky, S. A., Vetterling, W. T. (1988). Numerical Recipes in C. The Art of Scientific Computing. Cambridge University Press.
10. Sturmfels, B. (1993). Algorithms in Invariant Theory. Springer
11. van der Waerden, B. L. (1971). Algebra I. Springer, Berlin
12. Wilkinson, J. H. (1963). Rounding Errors in Algebraic Processes. Prentice-Hall, Inc., Englewood Cliffs, NJ

Three Remarks on Comprehensive Gröbner and SAGBI Bases

Manfred Göbel[1] and Patrick Maier[2]

[1] Department of Electronic Systems Engineering
University of Essex, Wivenhoe Park
Colchester CO4 3SQ, United Kingdom
mkgoebel@essex.ac.uk

[2] Max-Planck-Institut für Informatik
Im Stadtwald
66123 Saarbrücken, Germany
maier@mpi-sb.mpg.de

Abstract. This note presents new complexity results for the comprehensive Gröbner bases (CGB) algorithm in the special case of one main variable and two polynomials, a general remark about CGB for parameterized binomial ideals, and it introduces the concept of comprehensive SAGBI bases together with a first application in invariant theory.
Keywords. Comprehensive Gröbner bases, parameterized binomial ideals, comprehensive SAGBI bases, algorithmic invariant theory, permutation groups.

1 Introduction

The investigation of Gröbner bases is one of the major topics in computational algebra [7]. The various aspects of the Buchberger algorithm and the theory of Gröbner bases have been studied in the past years in several directions like generalizations, extensions, special cases, implementation issues, complexity analysis and applications. Typical examples for generalizations and extensions are universal [25], comprehensive [26] and non-commutative [18] Gröbner bases computations. Special cases are dealing with Gröbner bases for binomial ideals [8] and their relationship to integer programming [4] and convex polytopes [22], or with Hilbert-driven Buchberger algorithms [23] and their applications to invariant theory [21]. Many criteria and swift heuristics have been developed to deal with S-polynomials [10], to walk around in the Gröbner fan [3] or to exploit parallelism [1], in order to speed up implementations of the Buchberger algorithm, and to solve algebraic equations [5, 9]. For most of these approaches, detailed complexity studies, degree bounds and empirical results have been presented [15, 17, 24].

Our goal in the paper is to present some new results and aspects of Weispfenning's comprehensive Gröbner bases (CGB) algorithm [26]. In Section 2, this rather complex algorithm is analyzed w.r.t. the important case of one main variable and two polynomials. The study of this *comprehensive greatest common divisor algorithm* enables us to present a couple of new complexity results. Section 3 briefly addresses the computation of CGB for parameterized binomial ideals, and eventually, Section 4 introduces the concept of *comprehensive SAGBI bases* together with a first application in invariant theory.

We follow the notation and setting used in [26] for all aspects of CGB. The notation and setting for comprehensive SAGBI bases refers to [12, 14].

2 Comprehensive GCDB

We are dealing in this section with two parameterized polynomials which have only one main variable X.

Definition 1. *Let $F \subset K[U_1, \ldots, U_m, X]$ finite. Then $G \subset K[U_1, \ldots, U_m, X]$ finite is a comprehensive greatest common divisor basis (GCDB) for F, if*

1. *$\sigma(G)$ contains a greatest common divisor g of $\sigma(F)$ in $K'[X]$, and*
2. *$g \mid p$ for all $0 \neq p \in \sigma(G)$*

for all fields K' and for all specializations $\sigma\colon K[U_1, \ldots, U_m] \longrightarrow K'$ of $K[U_1, \ldots, U_m, X]$.

A CGB contains after specialization only one — not necessarily reduced — Gröbner basis. Our first example shows that a comprehensive GCDB contains usually more than one element after specialization.

Example 1. Let $F = \{U, X + 1\}$. Then F is a comprehensive GCDB of F, and we need both elements of F to cover all greatest common divisors for all possible specializations. We have $\sigma(F) = \{1, X + 1\}$, if $\sigma(U) = 1$.

Our second example shows that the greatest common divisors in a comprehensive GCDB are not necessarily unique.

Example 2. Let $F = \{U_1X + 1, U_2X + 2\}$. Then $G = \{U_1X + 1, U_2 - 2U_1\}$ is a comprehensive GCDB of F, and we need both elements of G to cover all greatest common divisors for all possible specializations. We have $\sigma(G) = \{1, 2\}$, if $\sigma(U_1) = 0$ and $\sigma(U_2) = 2$.

The criterion in our first lemma tells us how we can find the greatest common divisor in a comprehensive GCDB after specialization.

Lemma 1. *Let F, G and σ be as in Definition 1. Then any $0 \neq f \in \sigma(G)$ with minimal degree is a greatest common divisor of $\sigma(F)$, if $\sigma(G) \neq \{0\}$. (The greatest common divisor is 0, if $\sigma(G) = \{0\}$.)*

Furthermore, if $f_1, f_2 \in \sigma(G)$ are of minimal degree, then there exists $c \in K$ such that $f_1 = cf_2$.

Proof. If $0 \neq f$ is not of minimal degree, it implies that there exists $0 \neq g \in \sigma(G)$ of minimal degree. Then we must have $f \xrightarrow[g]{*} 0$, because otherwise g is not of minimal degree.

Assume that f_1, f_2 are of minimal degree and $f_1 \neq cf_2$ for some $c \in K$. Then $f_1 \xrightarrow[f_2]{} g$ and $0 \neq g \in \sigma(G)$ implies that f_1 and f_2 was not of minimal degree (contradiction). □

Theorem 1. *Let $F = \{f_1, f_2\} \subset K[U_1, \ldots, U_m, X]$. Then there exists an algorithm to construct a finite comprehensive GCDB.*

Proof. This is a consequence of Weispfenning's CGB algorithm [26]. □

In order to prove a couple of complexity results, we present a simplified and slightly restructured version of Weispfenning's CGB algorithm[1] (GROEBNERSYSTEM) for the special case of one main variable. Our modifications are based on the fact that we don't have to deal with S-polynomials. See [26] for the definitions of $HT_\gamma(.)$ and $T_{green,\gamma}(.)$ and the algorithms DET(.) and NORMALFORM(.).[2]

We need one more definition for the algorithm and the subsequent complexity analysis.

Definition 2. *Let $f = \sum_{i=0}^{n} g_i(U_1, \ldots, U_m)X^i \in K[U_1, \ldots, U_m, X]$, let*

$$mdeg(g_i) = \max\{\max\{e_1, \ldots, e_m\} \,|U_1^{e_1} \ldots U_m^{e_m} \in T(g_i)\},$$

*and let $tdeg(g_i) = \max\{\sum_{j=1}^{m} e_j \,|U_1^{e_1} \ldots U_m^{e_m} \in T(g_i)\}$ with $g_i \in K[U_1, \ldots, U_m]$ for $0 \leq i \leq n$. Then we say f has **X**-degree n, total **U**-degree*

$$\max\{tdeg(g_i(U_1, \ldots, U_m)) \,|\, 0 \leq i \leq n\}$$

*and maximal **U**-degree*

$$\max\{mdeg(g_i(U_1, \ldots, U_m)) \,|\, 0 \leq i \leq n\}.$$

*The $\mathbf{X}_\gamma$-degree, total $\mathbf{U}_\gamma$-degree and maximal $\mathbf{U}_\gamma$-degree of f are the degrees computed w.r.t. the condition γ, i.e. these degrees are the **X**-degree, total **U**-degree and maximal **U**-degree of f without considering the terms in $T_{green,\gamma}(f)$.*

Algorithm 1. COMPREHENSIVE GCDB(f_1, f_2)

```
INPUT {f1, f2} ⊂ K[U1, ..., Um, X];
IF (X-degree of f2) > (X-degree of f1) THEN
    h := f1; f1 := f2; f2 := h;
    ENDIF;
B := ∅;
```

[1] Weispfenning's CGB algorithm is implemented in MAS [16], SIMATH [27] and REDUCE [6]. The first implementation was done by Schönfeld [20] in ALDES/SAC-2.

[2] We use a slightly different version of NORMALFORM(.): We always reduce the head term of the polynomial in a reduction step.

```
6. P := {(γ, f1, f2) | γ ∈ DET({}, {f2})};
7. WHILE P ≠ ∅ DO
8.      select (γ, f, g) ∈ P; P := P \ {(γ, f, g)};
9.      IF  T(g) = T_green,γ(g)  THEN B := B ∪ {f};
10.     ELSE
11.         (h, c) := NORMALFORM(γ, f, {g});
12.         P := P ∪ {(δ, g, h) | δ ∈ DET(γ, {h})};
13.         ENDIF;
14.     ENDWHILE;
15. OUTPUT Comp. GCDB B ⊂ K[U1, ..., Um, X];
```

Termination is based on the finiteness of P at the end of step 6, and the fact that any 3-tuple added to P in step 12 is constructed by using the result of a normalform operation. The algorithm is correct, because the $GCD(\sigma(f), \sigma(g))$ is equal to the $GCD(\sigma(g), \sigma(h))$ in step 11 for all specializations σ which belong to condition γ.

Note that Algorithm 1 coincides with the classical Euclidian algorithm to compute a greatest common divisor of two polynomials, if there are no parameters at all.

Lemma 2. *Let $f_1, f_2 \in K[U_1, \dots, U_m, X]$ such that the $\mathbf{X}$-degree of f_2 is at most the $\mathbf{X}$-degree of f_1. Let x be the $\mathbf{X}$-degree of f_2. Then the cardinality of P at the end of step 6 of Algorithm 1 is at most $x + 2$; at most $x + 1$ of these elements induce a normalform operation in the algorithm.*

Proof. The polynomial $f_2 = \sum_{i=0}^{x} g_i(U_1, \dots, U_m)X^i$ has at most $x+1$ terms, and we need the following $x + 2$ conditions $\{g_x \neq 0\}$, $\{g_x = 0, g_{x-1} \neq 0\}$, ..., $\{g_x = 0, \dots, g_1 = 0, g_0 \neq 0\}$, $\{g_x = 0, \dots, g_1 = 0, g_0 = 0\}$ to determine f_2. Hence, P has at most $x + 2$ elements at the end of step 6. The last condition is trivial and induces later no normalform operation in the while loop of Algorithm 1. □

Lemma 3. *Let $f, g, h \in K[U_1, \dots, U_m, X]$, let γ be a condition, and let x_f, x_g, x_h and u_f, u_g, u_h be the $\mathbf{X}_\gamma$-degree and total $\mathbf{U}_\gamma$-degree of f, g, h, respectively. Then $f \xrightarrow[g]{*} h\ [\gamma]$ can be computed via* NORMALFORM$(\gamma, f, \{g\})$ *with at most $\max\{0, x_f - x_g + 1\}$ reduction steps, x_h is at most $x_g - 1$, and u_h is at most $u_f + u_g \max\{0, x_f - x_g + 1\}$.*

The cardinality of P is increased by at most $x_g + 1$ elements at the end of step 12 of Algorithm 1.

Proof. The polynomial f has at most $\max\{0, x_f - x_g + 1\}$ terms with a $\mathbf{X}_\gamma$-degree greater or equal to the $\mathbf{X}_\gamma$-degree of g. Each reduction step reduces the degree of f modulo γ at least by one. At the end of this process x_h can be at most $x_g - 1$, and u_h can be at most $u_f + u_g \max\{0, x_f - x_g + 1\}$.

The cardinality of P is increased by only x_g elements at the end of step 12, because g is already determined by γ. Any successor of γ has no effect on $HT_\gamma(g)$ and we need at most $x_g + 1$ successors of γ to determine h. □

Note that Lemma 3 is also true if we replace "total **U**-degree" by "maximal **U**-degree".

Lemma 4. *Let $f, g \in K[U_1, \ldots, U_m, X]$ and let x be the maximal* **X***-degree of f and g. Then the Algorithm 1 needs at most $2^{x+1} - 1$ normalform operations and at most $(x+1)(2^{x+1} - 1)$ reduction steps in order to compute a comprehensive GCDB.*

Proof. Let (γ, h_1, h_2) be an element of P occurring sometime in the algorithm, and assume that the $\mathbf{X}_\gamma$-degree of h_2 is at most r. Then the number of normalform operations $\omega(r)$ initiated by this 3-tuple and including all its successors added in step 12 to P can be obtained recursively as follows:

$$\omega(r) = 1 + \sum_{i=0}^{r-1} \omega(i)$$

It is easy to verify that a generating function for this recursion is $\omega(r) = 2^r$.

Because of Lemma 2, P has at the beginning at most $x+1$ non-trivial elements (γ, h_1, h_2) with a $\mathbf{X}_\gamma$-degree of h_2 equal to $0, \ldots, x$. Hence, the total number of normalform operations is at most

$$\sum_{i=0}^{x} \omega(i) = \sum_{i=0}^{x} 2^i = 2^{x+1} - 1.$$

Because each normalform operation needs at most $x + 1$ reductions, we have $(x+1)(2^{x+1} - 1)$ as upper bound for the number of reduction steps. □

Close inspection of the proof of Lemma 4 shows that the number of normalform operations does not depend on the maximal **X**-degree of f and g. Instead, Algorithm 1 takes at most $2^{\tilde{x}+1} - 1$ normalform operations where $\tilde{x}$ is the minimal **X**-degree of f and g.

We can prove a better bound for the number of reduction steps as follows:

Lemma 5. *Let $f, g \in K[U_1, \ldots, U_m, X]$ and let x be the maximal* **X***-degree of f and g. Then the Algorithm 1 needs at most $3(2^{x+1} - 1) - 2(x+1)$ reduction steps in order to compute a comprehensive GCDB.*

Proof. Let (γ, h_1, h_2) be an element of P occurring sometime in the algorithm, and assume that the $\mathbf{X}_\gamma$-degree of h_1 and h_2 is at most r_1 and at most r_2, respectively. Since no reduction step is performed, if $r_1 < r_2$, we can assume w.l.o.g. that $r_1 \geq r_2$. Then the number of reduction steps $\kappa(r_1, r_2)$ initiated by this 3-tuple and including all its successors added to P in step 12 can be obtained recursively as follows:

$$\kappa(r_1, r_2) = r_1 - r_2 + 1 + \sum_{i=0}^{r_2-1} \kappa(r_2, i)$$

It is easy to verify that a generating function for this recursion is $\kappa(r_1, r_2) = r_1 - 2(r_2 + 1) + 3 \cdot 2^{r_2}$.

Because of Lemma 2, P has at the beginning at most $x+1$ non-trivial elements (γ, h_1, h_2) with a $\mathbf{X}_\gamma$-degree of h_2 equal to $0, \ldots, x$. Hence, the total number of reduction steps is at most

$$\sum_{i=0}^{x} \kappa(x, i) = \sum_{i=0}^{x} (x - 2(i+1) + 3 \cdot 2^i) = 3(2^{x+1} - 1) - 2(x+1).$$

□

A comparison of Lemma 4 and Lemma 5 shows that the bound for the number of reduction steps is roughly three times the bound for the number of normalform operations. It is interesting that this factor is independent of the **X**-degree of the given f and g.

Lemma 6. *Let $f, g \in K[U_1, \ldots, U_m, X]$ and let x be the maximal* **X***-degree of f and g. Then we can construct a comprehensive GCDB B with at most 2^{x+1} elements.*

Proof. According to Lemma 4, we need at most $2^{x+1} - 1$ normalform operations in Algorithm 1 to compute B. This takes all elements of P into account occurring sometime in the algorithm except the trivial one at the end of step 6, which is not involved in a normalform operation and directly put into B. Hence, the number of elements in B is at most 2^{x+1}. □

Lemma 7. *Let $f = \sum_{i=0}^{l_1} U_{f,i} X^i$ and $g = \sum_{i=0}^{l_2} U_{g,i} X^i$ be "generic" polynomials with $l_1, l_2 \geq 0$ and assume that $\max\{l_1, l_2\} \geq 1$. Then any comprehensive GCDB B of f and g contains more than one element.*

Proof. First, we assume that $l_1 = 1$ and $l_2 = 0$, i.e. $f = U_{f,1}X + U_{f,0}$ and $g = U_{g,0}$. Then we must have both $f \in B$ and $g \in B$, because otherwise B does not cover all greatest common divisors of f and g for all possible specializations.

Any other case with $\max\{l_1, l_2\} \geq 1$ contains $l_1 = 1$ and $l_2 = 0$ as subcase.

□

3 CGB for Parameterized Binomial Ideals

Binomial ideals have received considerable attention recently. The reason for this is that the computation of Gröbner bases for such ideals is related to many fundamental mathematical questions arising in integer programming, statistics and combinatorics [2]. We would like to point out that Weispfenning's CGB algorithm maintains the main property of binomials. Therefore, progress in algorithms and degree bounds for computing Gröbner bases for binomial ideals [2, 15] may be applicable in the comprehensive case.

Definition 3. *Let $F \subset K[U_1, \dots, U_m, X_1, \dots, X_n]$ finite such that for any $f \in F$ the set $T(f)$ contains at most two elements. Then F is a set of parameterized binomials, and $Id(F)$ is a parameterized binomial ideal.*

Lemma 8. *Let $F \subset K[U_1, \dots, U_m, X_1, \dots, X_n]$ finite be a set of parameterized binomials. Then $Id(\sigma(F))$ is a binomial ideal for any specialization σ.*

Furthermore, any $f \in K[U_1, \dots, U_m, X_1, \dots, X_n]$ occurring in some step of Weispfenning's CGB algorithm has the property that $\sigma(f)$ is a binomial for all specializations σ.

Proof. Coefficients of polynomials are labeled either green, red or white in the CGB algorithm w.r.t. to the given condition. Green marks vanishing, red invertible, and white either vanishing or invertible coefficients. The algorithm commences with at most two non-green labeled terms for each polynomial, and maintains this property during any normalization and building of S-polynomials. □

4 Comprehensive SAGBI Bases

SAGBI bases play an important role in computational algebra. They were introduced in [19] and — from the term rewriter's point of view — in [13]. Comprehensive SAGBI bases are similarly defined as CGB.

Definition 4. *A (finite) $F \subset S \subseteq K[U_1, \dots, U_m, X_1, \dots, X_n]$ is called a (finite) comprehensive SAGBI basis of the subalgebra S, if $\sigma(F)$ is a SAGBI basis for all fields K' and all specializations $\sigma\colon K[U_1, \dots, U_m] \longrightarrow K'$ of S.*

We are not going to present an algorithm to construct a (finite) comprehensive SAGBI bases out of a given finite set of polynomials. This can (most likely) be achieved by combining Weispfenning's CGB algorithm [26] and Robbiano's and Sweedler's SAGBI bases algorithm [19], but the details have to be worked out. In contrary to the Gröbner bases case, the construction of a finite comprehensive SAGBI basis is not guaranteed, because a SAGBI basis can be already infinite by itself.

Non-trivial and interesting parameterized subalgebras are, for example, rings of polynomial invariants of conjugates of permutation groups [14]. In Lemma 9 we present the "smallest" non-trivial ring of polynomial invariants of conjugates of a permutation group, which has a finite SAGBI basis for all its conjugates. "Smallest" refers to the number of variables (= 2), the number of generators for each individual invariant ring (= 2), the degree bound of the generators (= 2), and the group order (= 2). Our setting is the same as in [14]; $GL(n, \mathbb{C})$ denotes the general linear group, S_n the symmetric group, and $S_n^\delta = \{\delta\pi\delta^{-1} | \pi \in S_n\}$ a conjugate of S_n with $\delta \in GL(n, \mathbb{C})$.

Lemma 9. *Let* $\delta = \begin{pmatrix} U_{11} & U_{12} \\ U_{21} & U_{22} \end{pmatrix} \in GL(2,\mathbb{C})$. *Then* $\mathbb{C}[X_1,X_2]^{S_2^\delta}$ *has a finite comprehensive SAGBI basis B w.r.t.* $<_{lex}$ *which is*

$$\left\{\begin{array}{l} (U_{11}+U_{21})X_1+(U_{12}+U_{22})X_2, \\ U_{11}U_{21}X_1^2+(U_{12}U_{21}+U_{11}U_{22})X_1X_2+U_{12}U_{22}X_2^2, \\ (U_{11}^2U_{21}U_{12}+U_{11}U_{21}^2U_{22}-U_{11}^3U_{22}-U_{21}^3U_{12})X_1X_2+ \\ (U_{11}U_{21}U_{12}^2+U_{11}U_{21}U_{22}^2-U_{11}^2U_{12}U_{22}-U_{21}^2U_{12}U_{22})X_2^2 \end{array}\right\}.$$

Proof. We know that $\det(\delta) = U_{11}U_{22} - U_{21}U_{12} \neq 0$, and further, $w_1 = \delta(\sigma_1) = \delta(X_1+X_2) = (U_{11}+U_{21})X_1 + (U_{12}+U_{22})X_2$ and $w_2 = \delta(\sigma_2) = \delta(X_1X_2) = U_{11}U_{21}X_1^2 + (U_{12}U_{21}+U_{11}U_{22})X_1X_2 + U_{12}U_{22}X_2^2$. We consider all cases for the coefficients:

$U_{11}+U_{21} = 0$: Then we have $U_{11} = -U_{21}$ and $w_1 = (U_{12}+U_{22})X_2$ and $w_2 := -U_{21}^2X_1^2 + U_{21}(U_{12}-U_{22})X_1X_2 + U_{12}U_{22}X_2^2$. We have to look at the following subcases:

- $U_{12}+U_{22} = 0$: This is impossible, because $U_{11} = -U_{21}$ and $U_{12} = -U_{22}$ implies $U_{11}U_{22} - U_{21}U_{12} = 0$.
- $U_{12}+U_{22} \neq 0$: This leads to a SAGBI basis B_1, because we have $U_{21} \neq 0$. The basis polynomials (including vanishing coefficients) are $w_1 = (U_{11}+U_{21})X_1 + (U_{12}+U_{22})X_2$ and $w_2 = U_{11}U_{21}X_1^2 + (U_{12}U_{21}+U_{11}U_{22})X_1X_2 + U_{12}U_{22}X_2^2$.

$U_{11}+U_{21} \neq 0$: We have to look at the following subcases:

- $U_{11}U_{21} = 0$:
  - $U_{11} = 0$: This is a SAGBI basis B_2, because we have $U_{12} \neq 0$ and $U_{21} \neq 0$. The basis polynomials are $w_1 = (U_{11}+U_{21})X_1 + (U_{12}+U_{22})X_2$ and $w_2 = U_{11}U_{21}X_1^2 + (U_{12}U_{21}+U_{11}U_{22})X_1X_2 + U_{12}U_{22}X_2^2$.
  - $U_{21} = 0$: This is a SAGBI basis B_3, because we have $U_{11} \neq 0$ and $U_{22} \neq 0$. The basis polynomials are $w_1 = (U_{11}+U_{21})X_1 + (U_{12}+U_{22})X_2$ and $w_2 = U_{11}U_{21}X_1^2 + (U_{12}U_{21}+U_{11}U_{22})X_1X_2 + U_{12}U_{22}X_2^2$.
- $U_{11}U_{21} \neq 0$: We know that $U_{11} \neq U_{21} \neq 0$, and further, $w_1 = (U_{11}+U_{21})X_1 + (U_{12}+U_{22})X_2$ and $w_2 = U_{11}U_{21}X_1^2 + (U_{12}U_{21}+U_{11}U_{22})X_1X_2 + U_{12}U_{22}X_2^2$. Hence, we can reduce w_2 with w_1, which leads to $\hat{w}_2 = U_{11}U_{21}w_1^2 - (U_{11}+U_{21})^2w_2 = (U_{11}^2U_{21}U_{12} + U_{11}U_{21}^2U_{22} - U_{11}^3U_{22} - U_{21}^3U_{12})X_1X_2 + (U_{11}U_{21}U_{12}^2 + U_{11}U_{21}U_{22}^2 - U_{11}^2U_{12}U_{22} - U_{21}^2U_{12}U_{22})X_2^2$, and we obtain a SAGBI basis $B_4 = \{w_1, \hat{w}_2\}$.

This finishes the proof, because we have considered all possible cases. There were finitely many of them and we obtained finite SAGBI basis in each allowed case, i.e. we have a finite comprehensive SAGBI basis $B = B_1 \cup B_2 \cup B_3 \cup B_4$. □

It is well known that the elementary symmetric functions are a finite SAGBI basis for $\mathbb{C}[X_1,\ldots,X_n]^{S_n}$ w.r.t. any admissible order [19, theorem 1.14]. We suppose that a similar result holds w.r.t. comprehensive SAGBI bases for the ring of symmetric functions.

Conjecture 1. Let $\delta \in GL(n,\mathbb{C})$ and $n \geq 3$. Then $\mathbb{C}[X_1,\ldots,X_n]^{S_n^\delta}$ has a finite comprehensive SAGBI basis.

The rest of this section illustrates the finiteness problem of (comprehensive) SAGBI bases in this type of application a little more.

Lemma 10. *Let G be not a direct product of symmetric groups. Then the invariant ring $\mathbb{C}[X_1,\ldots,X_n]^{G^\delta}$ has no finite comprehensive SAGBI basis w.r.t. $<_{lex}$ and all $\delta \in GL(n,\mathbb{C})$.*

Proof. This is a consequence of [12]. The case $\delta = id$ leads already to an infinite SAGBI basis. □

Lemma 11. *Let $\delta \in GL(n,\mathbb{C})$ denote an arbitrary upper triangular matrix. Then the invariant ring $\mathbb{C}[X_1,\ldots,X_n]^{G^\delta}$ has a finite comprehensive SAGBI basis w.r.t. $<_{lex}$ and all δ iff G is a direct product of symmetric groups.*

Proof. According to [12] $\mathbb{C}[X_1,\ldots,X_n]^G$ has a finite SAGBI basis w.r.t. $<_{lex}$ iff G is a direct product of symmetric groups. Any upper triangular matrix δ implies that the set of head terms of polynomials in $\mathbb{C}[X_1,\ldots,X_n]^G$ is equivalent to the set of head terms in $\mathbb{C}[X_1,\ldots,X_n]^{G^\delta}$. Hence, $\mathbb{C}[X_1,\ldots,X_n]^{G^\delta}$ has a finite SAGBI iff $\mathbb{C}[X_1,\ldots,X_n]^G$ has a finite SAGBI basis, which is the case whenever G is a direct product of symmetric groups. □

It is well-known that $\mathbb{C}[X_1,\ldots,X_n]^{A_n}$ is generated by the elementary symmetric polynomials $\sigma_1 = X_1 + \ldots + X_n$, $\ldots$, $\sigma_n = X_1 \ldots X_n$, and

$$\sigma_{n+1} = orbit_{A_n}(X_1^{n-1}X_2^{n-2}\ldots X_{n-2}^2 X_n).$$

Furthermore, $\mathbb{C}[X_1,\ldots,X_n]^{A_n}$ has no finite SAGBI bases w.r.t. $<_{lex}$, and especially, $B = \{\sigma_1,\ldots,\sigma_{n+1}\}$ is not such a finite SAGBI basis [12]. Now let $\delta \in GL(n,\mathbb{C})$ be defined as lower triangular matrix with $\delta_{ii} = 1$ for $1 \le i \le n$, $\delta_{n,n-1} = -1$, and $\delta_{ij} = 0$, otherwise.

Lemma 12. *Let σ_1, $\ldots$, σ_{n+1} and δ be as above. Then the following holds:*

1. $\det(\delta) = 1$
2. $HC(\delta(\sigma_i)) = 1$, $HT(\delta(\sigma_i)) = X_1X_2\ldots X_i$ *for* $1 \le i \le n-2$
3. $HC(\delta(\sigma_{n-1})) = 1$, $HT(\delta(\sigma_{n-1})) = X_1X_2\ldots X_{n-2}X_n$
4. $HC(\delta(\sigma_n)) = -1$, $HT(\delta(\sigma_n)) = X_1X_2\ldots X_{n-2}X_{n-1}^2$
5. $HC(\delta(\sigma_{n+1})) = -1$, $HT(\delta(\sigma_{n+1})) = X_1^{n-1}X_2^{n-2}\ldots X_{n-2}^2X_{n-1}$

Furthermore, the set $\hat{B} = \{\delta(\sigma_1),\ldots,\delta(\sigma_{n+1})\}$ is a basis for $\mathbb{C}[X_1,\ldots,X_n]^{A_n^\delta}$.

Proof. These are consequences of the definition of δ and the properties of invariant rings of a conjugate of A_n. □

Theorem 2. *Let $\hat{B}$ be as in Lemma 12. Then $\hat{B}$ is a finite SAGBI basis for $\mathbb{C}[X_1,\ldots,X_n]^{A_n^\delta}$ w.r.t. $<_{lex}$.*

Proof. We know from Lemma 12 that $\hat{B}$ is a basis for $\mathbb{C}[X_1, \ldots, X_n]^{A_n^\delta}$. Once we know this, we only have to check, according to [19], if $\hat{B}$ is in addition a SAGBI basis. An analysis of the set of head terms $\{HT(\delta(\sigma_1)), \ldots, HT(\delta(\sigma_{n+1}))\}$ reveals that we only have to verify that the polynomial[3]

$$f = (\delta(\sigma_{n+1}))^2 + \delta(\sigma_n) \cdot (\delta(\sigma_{n-2}))^3 \cdot (\delta(\sigma_{n-3}) \ldots \delta(\sigma_1))^2 \tag{1}$$

can be reduced to zero by means of $\hat{B}$ in order to ensure that $\hat{B}$ is a SAGBI basis. From the reduction point of view, f is the only relevant critical polynomial, which can be made up from the initial set $\hat{B}$.

Let $g = \delta^{-1}(f)$. Of course, then we know that $g \in \mathbb{C}[X_1, \ldots, X_n]^{A_n}$. Thus, g can be reduced with [11, algorithm 3.12] and represented as

$$g = p_1(\sigma_1, \ldots, \sigma_n) + p_2(\sigma_1, \ldots, \sigma_n) \cdot \sigma_{n+1} \tag{2}$$

with $p_1, p_2 \in \mathbb{C}[X_1, \ldots, X_n]$. The representation (2) is different from

$$g = \sigma_{n+1}^2 + \sigma_n \sigma_{n-2}^3 (\sigma_{n-3} \ldots \sigma_1)^2$$

and obtained by computing $\delta^{-1}(f)$ via equation (1). The algorithm terminates after a finite number $l \in \mathbb{N}$ of reduction steps. At any reduction step we have to subtract a certain product of the basis polynomials $\sigma_1, \ldots, \sigma_{n+1}$, say

$$h_i = c_i \sigma_1^{e_{i_1}} \ldots \sigma_n^{e_{i_n}} \sigma_{n+1}^{e_{i_{n+1}}}$$

with $(e_{i_1}, \ldots, e_{i_n}, e_{i_{n+1}}) \in \mathbb{N}^n \times \{0, 1\}$ and $0 \neq c_i \in \mathbb{C}$ in step $1 \leq i \leq l$. Eventually, after termination of the algorithm, the representation (2) of g has been computed as $g = h_1 + \ldots + h_l$.

The algorithm ensures that $h_i \neq h_j$ for $1 \leq i \neq j \leq l$. This implies that

$$\delta(h_i) = c_i \delta(\sigma_1)^{e_{i_1}} \ldots \delta(\sigma_n)^{e_{i_n}} \delta(\sigma_{n+1})^{e_{i_{n+1}}} \neq \delta(h_j),$$

and, because of our choice of δ, that $HT(\delta(h_i)) \neq HT(\delta(h_j))$ for $1 \leq i \neq j \leq l$. We can rearrange now the sequence $\delta(h_1), \ldots, \delta(h_l)$ in such a way that

$$HT(\delta(h_i)) >_{lex} HT(\delta(h_j)) \tag{3}$$

for $1 \leq i < j \leq l$. This new sequence is then, due to our construction, able to reduce our polynomial f as follows

$$f \xrightarrow{\delta(h_1)} f_1 \xrightarrow{\delta(h_2)} \ldots \xrightarrow{\delta(h_{l-1})} f_{l-1} \xrightarrow{\delta(h_l)} f_l$$

with $f_1, \ldots, f_{l-1} \in \mathbb{C}[X_1, \ldots, X_n]$ and $f_l = 0$. This reduction sequence has to be a SAGBI basis reduction w.r.t. $\hat{B}$, because the relations (3) imply $HT(\delta(h_1)) \in T(f)$, $HT(\delta(h_2)) \in T(f_1)$, ..., and $HT(\delta(h_l)) \in T(f_{l-1})$. □

[3] Note that we define $(\delta(\sigma_{n-3}) \ldots \delta(\sigma_1))^2 = 1$, if $n = 3$.

We have learned so far that at least some lower triangular matrices lead to invariant rings of conjugates of some permutation groups with a finite SAGBI basis. Generally speaking, the classification of conjugates of permutation groups and their corresponding rings of polynomial invariants, which have a finite SAGBI basis w.r.t. $<_{lex}$, is an open problem.

Problem 1. Characterize subsets $\Gamma \subseteq GL(n, \mathbb{C})$ such that the invariant ring $\mathbb{C}[X_1, \ldots, X_n]^{G^\delta}$ has a finite SAGBI basis w.r.t. $<_{lex}$ for any $\delta \in \Gamma$; compute the corresponding comprehensive SAGBI basis.

Our last theorem shows that such subsets are probably extremely difficult to characterize.

Theorem 3. *Let G be not a direct product of symmetric groups, and let $\Gamma \subseteq GL(n, \mathbb{C})$ be such that $\mathbb{C}[X_1, \ldots, X_n]^{G^\delta}$ has a finite SAGBI basis w.r.t. $<_{lex}$ for any $\delta \in \Gamma$. Then Γ is not a subgroup of $GL(n, \mathbb{C})$.*

Proof. Assume Γ is a subgroup of $GL(n, \mathbb{C})$. Then we have $id \in \Gamma$, but $\mathbb{C}[X_1, \ldots, X_n]^G$ has — according to [12] — no finite SAGBI basis (contradiction). □

Acknowledgment. The authors would like to thank Heinz Kredel (Mannheim), Ursula Walter (Colchester), Andreas Weber (Darmstadt) and the anonymous referees for their comments and remarks.

References

1. Attardi, G., Traverso, C. (1996). Strategy-Accurate Parallel Buchberger Algorithms. Journal of Symbolic Computation 21(4–6), 411–426
2. Bigatti, A. M., La Scala, R., Robbiano, L. (1999). Computing Toric Ideals. Journal of Symbolic Computation 27(4), 351–365
3. Collart, S., Kalkbrener, M., Mall, D. (1997). Converting Bases with the Gröbner Walk. Journal of Symbolic Computation 24(3–4), 465–470
4. Conti, P., Traverso, C. (1991). Buchberger Algorithm and Integer Programming. In: Mattson, H. F., Mora, T., Rao, T. R. N., (eds.), Applied Algebra, Algebraic Algorithms and Error-Correcting Codes, 6th Intl. Conf., AAECC-91, volume 539 of *LNCS*, Springer, 130–139
5. Czapor, S. R. (1989). Solving Algebraic Equations: Combining Buchberger's Algorithm with Multivariate Factorization. Journal of Symbolic Computation 7, 49–53
6. Dolzmann, A., Sturm, T. (1999). Redlog User Manual. MIP-9905, Fakultät für Mathematik und Informatik, Universität Passau.
7. Eisenbud, D. (1994). Commutative Algebra with a View Towards Algebraic Geometry. Springer
8. Eisenbud, D., Sturmfels, B. (1996). Binomial Ideals. Duke Mathematical Journal 84, 1–45.
9. Gatermann, K. (1990). Symbolic Solution of Polynomial Equation Systems with Symmetry. In: Watanabe, Sh., Nagata, M., (eds.), International Symposium on Symbolic and Algebraic Computation ISSAC'90, 112–119, Tokyo, Japan, August, ACM Press

10. Giovanni, A., Mora, T., Niesi, G., Robbiano, L., Traverso, C. (1991). One Sugar Cube, Please or Selection Strategies in the Buchberger Algorithm. In: Watt, S. M., (ed.), International Symposium on Symbolic and Algebraic Computation, ISSAC-91, ACM, 49–54
11. Göbel, M. (1995). Computing Bases for Permutation-Invariant Polynomials. Journal of Symbolic Computation 19, 285-291
12. Göbel, M. (1998). A Constructive Description of SAGBI Bases for Polynomial Invariants of Permutation Groups. Journal of Symbolic Computation 26, 261–272
13. Kapur, D., Madlener, K. (1989). A Completion Procedure for Computing a Canonical Basis of a k-Subalgebra. In: Kaltofen, E., Watt, S., (eds.), Proceedings of Computers and Mathematics 89. MIT, Cambridge, 1–11
14. Khuller, S., Göbel, M., Walter, J. (1999). Bases for Polynomial Invariants of Conjugates of Permutation Groups. Journal of Algorithms 32(1), 58–61
15. Koppenhagen, U., Mayr, E. W. (1999). An Optimal Algorithm for Constructing the Reduced Gröbner Basis of Binomial Ideals. Journal of Symbolic Computation 28(3), 317–338
16. Kredel, H. (1990). MAS: Modula-2 Algebra System. In: Gerdt, V. P., Rostovtsev, V. A., and Shirkov, D. V. (eds.), IV International Conference on Computer Algebra in Physical Research. World Scientific Publishing Co., Singapore, 31-34
17. Möller, H. M., Mora, F. (1984). Upper and Lower Bounds for the Degree of Gröbner Bases. In: Fitch, J., (ed.), International Symposium on Symbolic and Algebraic Computation, EUROSAM-84, volume 174 of *LNCS*, Springer, 172–183
18. Mora, T. (1994). An Introduction to Commutative and Non-Commutative Gröbner Bases. Theoretical Computer Science 134(1), 131–173
19. Robbiano, L., Sweedler, M. (1990). Subalgebra Bases. In: Bruns, W., Simis, A., (eds.), Commutative Algebra (Lect. Notes Math. 1430). Springer, 61-87
20. Schönfeld, E. (1991). Parametrische Gröbner Basen im Computeragebra System ALDES/SAC-2, Diplomarbeit, Universität Passau.
21. Sturmfels, B. (1993). Algorithms in Invariant Theory. Springer
22. Sturmfels, B. (1995). Gröbner Bases and Convex Polytopes. AMS University Lecture Series, Vol. 8, Providence RI
23. Traverso, C. (1996). Hilbert Functions and the Buchberger Algorithm. Journal of Symbolic Computation 22(4), 355–376
24. Weispfenning, V. (1987). Some Bounds for the Construction of Gröbner Bases. In: Beth, T., Clausen, M., (eds.), Applicable Algebra, Error-Correcting Codes, Combinatorics and Computer Algebra, 4th Intl. Conf., AAECC-87, volume 307 of *LNCS*, Springer, 195–201
25. Weispfenning, V. (1989). Constructing Universal Gröbner Bases. In: Huguet, L., Poli, A., (eds.), Applied Algebra, Algebraic Algorithms and Error-Correcting Codes, 5th Intl. Conf., AAECC-89, volume 356 of *LNCS*, Springer, 195–201
26. Weispfenning, V. (1992). Comprehensive Gröbner Bases. Journal of Symbolic Computation 14(1), 1–30
27. Zimmer, H.-G. (1998). SIMATH-Manual. Fachbereich Mathematik, Universtät Saarbrücken / Siemens AG München. (*http://emmy.math.uni-sb.de/~simath/*)

Computing the Cylindrical Algebraic Decomposition Adapted to a Set of Equalities

Neila Gonzalez–Campos[1] and Laureano Gonzalez–Vega*[2]

[1] Departamento de Matemática Aplicada y Ciencias de la Computación, Universidad de Cantabria, Santander (Spain)
gcampos@matesco.unican.es

[2] Departamento de Matemáticas, Estadística y Computación, Universidad de Cantabria, Santander (Spain)
gvega@matesco.unican.es

Abstract. The Cylindrical Algebraic Decomposition algorithm, in its projection phase, proceeds by eliminating one variable from a given set of polynomials $\mathcal{P}$ by means of the computation of the principal subresultant coefficients of a certain set of pairs of polynomials in $\mathcal{P}$ (including their derivatives and reducta). Since this method produces usually a big number of polynomials, and since the process must be iterated several times, any improvement in the projection phase would convey to dramatically speed up the efficiency of the Cylindrical Algebraic Decomposition algorithm.
The purpose of this paper is to present two approaches allowing, in some cases, to simplify the projection phase in the Cylindrical Algebraic Decomposition algorithm when some of the involved polynomials are prescribed to have a particular sign behaviour.

1 Introduction

One of the main problems in Computational Real Algebraic Geometry is the development of efficient Quantifier Elimination algorithms. As this problem is well–known to be not solvable, for the general case, in polynomial time (see [11] and [28]) the only way to attack it is the isolation of specific and particular cases where efficient algorithms can be applied. The word efficient does not mean in our context the searching of a polynomial time algorithm: we look for algorithms, methods and criteria allowing to perform Quantifier Elimination on formulae with a particular structure and low degrees of the involved polynomials. This strategy has been already considered in [15] and [29] where the particular case of degree two polynomial constraints is regarded, or in [22], which studies the cases in which the general Quantifier Elimination algorithm presented in [21] works in a fast way.

One of the general ways of performing Quantifier Elimination is by means of the Cylindrical Algebraic Decomposition algorithm (CAD algorithm in the

* Partially supported by the grant DGESIC PB 98-0713-C02-02 (Ministerio de Educación y Cultura)

sequel) introduced by G. Collins in [8]. Given a finite family of polynomials $\mathcal{F}$ in $\mathbb{Q}[x_1, \ldots, x_n]$, this algorithm produces a decomposition of $\mathbb{R}^n$ in a finite family of subsets $\{C_i\}_i$ with the following properties:

- every C_i is homeomorphic to some $(0,1)^{s_i}$ $(0 \leq s_i \leq n)$,
- every C_i can be defined by a boolean combination of sign conditions on polynomials in $\mathbb{Q}[x_1, \ldots, x_n]$, and
- the signs of the polynomials in $\mathcal{F}$ are constant on every C_i.

Thus any question regarding the existence of real solutions for any boolean combination of sign conditions on polynomials in $\mathcal{F}$ can be easily answered by merely checking the sign conditions verified by the polynomials in $\mathcal{F}$ on every C_i (the algorithm produces also in an explicit way, involving algebraic numbers, one point belonging to every C_i). The CAD algorithm solves it too even if the question is more complicated, in the sense that the problem to solve is the determination of the conditions to be verified by $x_{m+1}, \ldots, x_n$ $(m < n)$ in order that a fixed boolean combination of sign conditions on polynomials in $\mathcal{F}$ is verified. In the way the decomposition of $\mathbb{R}^n$ is produced, the CAD algorithm computes also similar decompositions of $\mathbb{R}$, $\mathbb{R}^2$, …, $\mathbb{R}^{n-1}$ with the additional property that the projection (forgetting the last coordinate) of any set in the decomposition of $\mathbb{R}^{k+1}$ is a set of the decomposition of $\mathbb{R}^k$. Thus the searched condition is obtained by collecting the boolean combinations defining the subsets of the decomposition of $\mathbb{R}^m$ arising from the projections $(\mathbb{R}^n \to \mathbb{R}^m)$ of the subsets in the decomposition of $\mathbb{R}^n$ verifying the original boolean combination of sign conditions on polynomials in $\mathcal{F}$.

The theoretical existence of this kind of decomposition (also called in French "sauccisonage") with respect to a family of polynomials can be traced back to S. Łojasiewicz. It has been widely used in Real Algebraic Geometry to prove important facts about semialgebraic sets: for example, that a semialgebraic set in $\mathbb{R}^n$ has a finite number of connected components, and that each of them is also semialgebraic (see [6]).

Collins algorithm and improvements, introduced by his collaborators, were implemented into the SAC–2 system and currently are being applied to solve Quantifier Elimination problems arising, for example, in the derivation of stable difference schemes for partial differential equations (see [20] and [19]).

From our point of view the main drawback of Collins algorithm is that it solves a problem which is usually more general than the one considered. If the problem to solve is the quantifier elimination on the formula:

$$\exists x_3\, \exists x_4\, (F_1(x_1,x_2,x_3,x_4) = 0 \wedge F_2(x_1,x_2,x_3,x_4) = 0 \wedge F_3(x_1,x_2,x_3,x_4) > 0)$$

then the first thing to do is to compute the CAD of $\mathbb{R}^4$ corresponding to the family $\{F_1, F_2, F_3\}$. But the result of this computation will not only produce the solution of the considered problem but also the solution to any quantifier elimination problem involving the polynomials F_1, F_2 and F_3.

The main purpose of this paper is to introduce new algorithms computing the CAD adapted to a set of equalities which can simplify the CAD construction when some sign conditions (in this case equality constraints) on some polynomials in the considered initial family are known in advance.

This paper is divided into three parts. In the first one the Cylindrical Algebraic Decomposition definition is reviewed together with Collins algorithm. In the second part the definition of CAD adapted to a set of equalities is introduced and a first way of producing this CAD, different from Collins original algorithm is presented by using Barnett's Method which allows to easily parametrize the greatest common divisor of a finite family of univariate polynomials with parametric coefficients. The final part is devoted to show how to compute a CAD adapted to a set of equalities by eliminating a whole set of variables when the algebraic structure of the initial polynomials is of a particular type (a "Pham System").

2 The Cylindrical Algebraic Decomposition

We will present here the definition of the Cylindrical Algebraic Decomposition and the way it can be constructed algorithmically. For that, we will firstly review the notion of semialgebraic set.

Definition 1. *Let $\mathbb{K}$ be a field, $\mathbb{K} \subset \mathbb{R}$.*

- *We will call quantifier-free formula (or simply formula) over $\mathbb{K}$ to a boolean combination $\Phi(\underline{x})$ of sign conditions involving a finite number of polynomials $P_1, \ldots, P_r \in \mathbb{K}[\underline{x}] = \mathbb{K}[x_1, \ldots, x_n]$:*

$$\Phi(\underline{x}) = \{P_1(\underline{x}) \lhd_1 0 \;\; ?_1 \;\; P_2(\underline{x}) \lhd_2 0 \;\; ?_2 \;\; \ldots \;\; ?_{r-1} \;\; P_r(\underline{x}) \lhd_r 0\}$$

  where $\lhd_i \in \{<, >, =\}$ and $?_i \in \{\vee, \wedge\}$.
- *$\mathbf{A} \subset \mathbb{R}^n$ is said to be a $\mathbb{K}$-semialgebraic set if there exists a formula $\Phi(\underline{x})$ over $\mathbb{K}$ such that $\mathbf{A}$ is the set of those points which verify the formula, i.e.*

$$\mathbf{A} = \{\underline{\alpha} \in \mathbb{R}^n \;:\; \Phi(\underline{\alpha})\} \;.$$

Here we will confine ourselves to $\mathbb{K} = \mathbb{Q}$ and $\mathbb{Q}$-semialgebraic sets, which we will call simply semialgebraic sets. The Cylindrical Algebraic Decomposition, or CAD, is a set of partitions of the spaces $\mathbb{R} \subset \mathbb{R}^2 \subset \cdots \subset \mathbb{R}^n$. More precisely, a partition C_k of $\mathbb{R}^k \subset \mathbb{R}^n$ for every $k \in \{1, \ldots, n\}$, each of which must satisfy the following conditions:

- C_k is *finite* (i.e. consists of a finite number of subsets of $\mathbb{R}^k$, which we will call cells).
- C_k is *semialgebraic* (i.e. every cell is a semialgebraic set).
- Every cell of C_k is *semialgebraically homeomorphic* to $(0,1)^d$ for some $d \in \{1, \ldots, k\}$ (where $(0,1)^0$ is a single point), and therefore connected.
- C_k is *"cylindrically arranged"*, that is, for every cell $S \in C_k$ the cylinder $S \times \mathbb{R} \in \mathbb{R}^{k+1}$ is a finite union of some cells of C_{k+1} $(k < n)$; if $C_1, \ldots, C_t$ are those cells, then the canonical projection $\pi : \mathbb{R}^{k+1} \longrightarrow \mathbb{R}^k$ satisfies $\pi(C_i) = S$ $(i = 1, \ldots, t)$.

In other words, we should say that, for every $S \in \mathcal{C}_k$, there exists a finite collection of continuous semialgebraic functions $S \longrightarrow \mathbb{R}$, whose graphs are pairwise disjoint, and such that the cylinder $S \times \mathbb{R}$ is a (disjoint) union of cells of $\mathcal{C}_{k+1}$ which are either one of these graphs, or the piece of cylinder comprised between two of these graphs or between one graph and $\pm\infty$.

Now let us consider a finite family of polynomials $\mathcal{P} = (P_1, \dots, P_r)$, $P_i = P_i(x_1, \dots, x_n) \in \mathbb{Q}[\underline{x}]$. If a CAD of $\mathbb{R}^n$ fulfills the condition that every cell $S \in \mathcal{C}_k$ is $\mathcal{P}$-invariant (i.e. that the sign of P_i does not change in S, $i = 1, \dots, r$) then we say that the CAD is adapted to the family $\mathcal{P}$.

The CAD construction allows quantifier elimination in the following way: assume that we have a CAD adapted to $(P_1, \dots, P_r)$, and a semialgebraic set $\mathbf{A} \subset \mathbb{R}^k$ given by a formula with quantifiers

$$\mathbf{A} = \{(x_1, \dots, x_k) : \exists x_{k+1}, \dots, \exists x_n \Phi(x_1, \dots, x_n)\}$$

where Φ is a conjunction of sign conditions over $P_1, \dots, P_r$:

$$\Phi(\underline{x}) = \{P_{i_0}(\underline{x}) \lhd_1 0 \ \wedge \ P_{i_1}(\underline{x}) \lhd_2 0 \ \wedge \ \dots \ \wedge \ P_{i_s}(\underline{x}) \lhd_s 0\},$$

with $\underline{x} = (x_1, \dots, x_n)$ and $\lhd_i \in \{<, >, =\}$. This set is regarded as the projection onto $\mathbb{R}^k$ of $\mathbf{B} = \{\underline{x} : \Phi(\underline{x})\} \subset \mathbb{R}^n$ which is also a semialgebraic set. In view of the CAD adapted to $(P_1, \dots, P_r)$ definition, it is clear that $\mathbf{B}$ is the (disjoint) union of certain cells of $\mathcal{C}_n$ (and we can find them exactly, as $\mathcal{C}_n$ consists of a finite number of cells). Since the projection of every cell of $\mathcal{C}_n$ is a cell of $\mathcal{C}_k$, $\mathbf{A}$ is the union of some cells of $\mathcal{C}_k$: exactly the projection of those cells that form $\mathbf{B}$. Thus we can find a description of $\mathbf{A}$ without quantifiers.

To eliminate the universal quantifier $\forall$, let us note that

$$\{(x_1, \dots, x_k) : \forall x_{k+1}, \dots, \forall x_n \Phi(\underline{x})\} = \mathbb{R}^k \setminus \{(x_1, \dots, x_k) : \exists x_{k+1}, \dots, \exists x_n \neg\Phi(\underline{x})\}$$

which is also a union of cells of $\mathcal{C}_k$, as long as $\mathcal{C}_k$ is a partition of $\mathbb{R}^k$. Of course, if we can eliminate one quantifier, we can eliminate a finite number of them.

2.1 Constructing the CAD

The CAD (or one CAD, as it is not unique) of $\mathbb{R}^n$ adapted to $\mathcal{P} = (P_1, \dots, P_r)$ can be constructed, in the sense of determining its structure, in the following inductive way: it is enough to see how it can be constructed for $n = 1$, and how to construct it for $n > 1$ if we have done it for $n - 1$.

- $n = 1$: Suppose we have $P_1, \dots, P_r \in \mathbb{Q}[x]$. The cells of $\mathcal{C}_1$ are just points and open intervals, as they must be homeomorphic to $(0,1)^1$ or $(0,1)^0$; so the real roots of $P_1, \dots, P_r$ and the intervals between then provide a CAD adapted to $(P_1, \dots, P_r)$.
- $n > 1$: Suppose we have already constructed $\mathcal{C}_{k-1}$. The goal is to decompose the cylinder $S \times \mathbb{R}$ for each $S \in \mathcal{C}_{k-1}$. It is clear (we will justify it next)

that if we have only one polynomial $P_1 = P \quad (r = 1)$, we can consider the semialgebraic functions

$$\gamma_i: \alpha = (\alpha_1, \ldots, \alpha_{n-1}) \in S \rightsquigarrow i^{-th} \text{ real root of } P(\alpha; x_n)$$

for $1 \leq i \leq \#\{\text{different real roots of } P \text{ over } S\}$. The graphs of these functions and the "pieces" of cylinder between them will be cells of $\mathcal{C}_k$ semialgebraically homeomorphic to $(0,1)^d$ and $(0,1)^{d+1}$respectively, if S is semialgebraically homeomorphic to $(0,1)^d$.

Next two propositions justify this, and its extension to $r > 1$.

Proposition 1. *Let $S \subseteq \mathbb{R}^{n-1}$ be a semialgebraic connected set, and $P(\underline{x}) \in \mathbb{Q}[x_1, \ldots, x_n]$ verifying, for every $\alpha = (\alpha_1, \ldots, \alpha_{n-1}) \in S$, that $\deg(P(\alpha; x_n))$ and the number of different complex roots of $P(\alpha; x_n)$ are constant. Then, the number of different real roots of $P(\alpha; x_n)$ is also constant in S. Therefore, there exist $\gamma_1, \ldots, \gamma_l : S \longrightarrow \mathbb{R}$ continuous semialgebraic functions such that $\gamma_1 < \cdots < \gamma_l$ in S, and, for all $\alpha \in S$, $\gamma_i(\alpha)$ is the ith real root of $P(\alpha; x_n)$.*

The proof of this result can be found in [8] or [10]. It is based on the fact that, in S connected, the roots of $P(\alpha; x_n)$ change continuously as we move $\alpha \in S$; besides, complex roots always appear in conjugate pairs, so, as $\deg_{x_n}(P)$ is constant, the number of real roots cannot vary without changing the total number of roots.

Now we will see how to decompose the cylinder $S \times \mathbb{R}$ for $r = 2$, say $\mathcal{P} = (P, Q)$; it works in a similar way for $r > 2$. We define the functions

$$\gamma_i : \alpha \in S \rightsquigarrow i^{-th} \text{ real root of } P(\alpha; x_n)$$

$$\eta_j : \alpha \in S \rightsquigarrow j^{-th} \text{ real root of } Q(\alpha; x_n)$$

for $1 \leq i \leq \#\{\text{real roots of } P(\alpha; x_n)\}$ and $1 \leq j \leq \#\{\text{real roots of } Q(\alpha; x_n)\}$. These functions are continuous and semialgebraic, and their graphs decompose the cylinder in the same way we said before. But for that we have to see that the graphs are pairwise disjoint:

Proposition 2. *Let $P, Q \in \mathbb{Q}[x_1, \ldots, x_n]$, $S \in \mathbb{R}^{n-1}$ a semialgebraic connected set and suppose the numbers*

a) $\#\{$*complex roots of* $P(\alpha; x_n)$ *over* $S\}$, $\deg(P(\alpha; x_n))$
b) $\#\{$*complex roots of* $Q(\alpha; x_n)$ *over* $S\}$, $\deg(Q(\alpha; x_n))$
c) $\deg[\gcd(P(\alpha; x_n), Q(\alpha; x_n))]$

are constant for all $\alpha \in S$. Then, the functions γ_i and η_j satisfy, for each pair (i, j) and any $\alpha \in S$: $\gamma_i(\alpha) < \eta_j(\alpha)$, or $\gamma_i(\alpha) > \eta_j(\alpha)$, or $\gamma_i(\alpha) = \eta_j(\alpha)$.

The hypotheses (a) and (b) are needed as in the previous proposition, and (c) is needed since the number of common complex roots of $P(\alpha; x_n)$ and $Q(\alpha; x_n)$ is represented by $\deg[\gcd(P(\alpha; x_n), Q(\alpha; x_n))]$. The proof is similar to the previous Proposition and can be found in [8] or [10].

So, if we can reduce the CAD of $\mathbb{R}^n$ to the CAD of $\mathbb{R}^{n-1}$, we can iterate $n-1$ times until we reach $\mathbb{R}$ and perform the CAD of $\mathbb{R}$. This is achieved by means of the projection operators: given $\mathcal{P} = (P_1, \ldots, P_r)$ polynomials in n variables, the projection of $\mathcal{P}$, $\mathbf{PROJ}(\mathcal{P})$, must be a set of polynomials in $n-1$ variables verifying the following condition:

> Let $\mathcal{C}_{n-1}$ be the partition of $\mathbb{R}^{n-1}$ given by the CAD adapted to the set of polynomials $\mathbf{PROJ}(\mathcal{P})$, and for each cell $S \in \mathcal{C}_{n-1}$ let $\gamma_1, \ldots, \gamma_t$ be the functions that describe the real roots of the product
>
> $$\prod_{i=1}^{r} \{P_i : P_i \not\equiv 0\}$$
>
> (i.e. the real roots of all polynomials not identically zero in $\mathcal{P}$) with respect to x_n. The projection operator must satisfy the condition that, if we decompose the cylinder $S \times \mathbb{R} \subset \mathbb{R}^n$ according to the functions $\gamma_1, \ldots, \gamma_t$, and this for every $S \in \mathcal{C}_{n-1}$, we obtain a CAD of $\mathbb{R}^n$ adapted to $\mathcal{P}$.

The classical projection operator $\mathbf{PROJ}(\mathcal{P})$ which fulfills this condition can be found in [10]. It is based on the computing of the reducta (the reducta of a polynomial is defined as the same polynomial without its leading term) of the given polynomials $\mathcal{P}$ and their *principal subresultant coefficients* (**psc**), in such a way that Proposition 2 can be applied to guarantee the possibility of constructing a CAD of $\mathbb{R}^n$ adapted to $\mathcal{P}$. The way the algorithm works is, consequently, the following:

Input: $\mathcal{P} = (P_1, \ldots, P_r) \subset \mathbb{Q}[x_1, \ldots, x_n]$
Output: $\{\mathcal{C}_1, \ldots, \mathcal{C}_n\}$ a CAD of $\mathbb{R}^n$ adapted to $\mathcal{P}$.

- Find the sets of polynomials (**Projection phase**)
  - $\mathbf{PROJ}^1(\mathcal{P}) := \mathbf{PROJ}(\mathcal{P})$
  - $\mathbf{PROJ}^i(\mathcal{P}) := \mathbf{PROJ}\ (\mathbf{PROJ}^{i-1}(\mathcal{P}))$ for $i = 2, \ldots, n-1$.

  $\mathbf{PROJ}^i(\mathcal{P})$ is a set of polynomials in $n-i$ variables.
- With $\mathbf{PROJ}^{n-1}(\mathcal{P})$ decompose $\mathbb{R}$ in cells (points and intervals) according to the real roots of these polynomials. This gives $\mathcal{C}_1$.
- Once $\mathcal{C}_k$ has been constructed, consider the polynomials, in $k+1$ variables, of $\mathbf{PROJ}^{n-k-1}(\mathcal{P})$ and find their real roots with respect to x_{k+1} on each cell of $\mathcal{C}_k$. This gives $\mathcal{C}_{k+1}$ (**Extension phase**).
- Finish when we reach $\mathcal{C}_n$.

In order to give the output, roots can be isolated by disjoint intervals (isolating intervals), and each cell of the CAD can be characterized by a *"test point"* (whose coordinates should be rational or real algebraic numbers) belonging to the cell, which allows to check the sign of $P_1, \ldots, P_r$ in that cell.

The usual **PROJ** operator can be improved, as some of its elements can be removed to get a smaller **PROJ**. Different ways to do it are explained in [8]. On the other hand, different projection operators can also be defined, such as the one using principal subdiscriminants for 2 or 3 variables (see [24] or [4]), or the cluster-based CAD presented in [1], [2] and [3], which intends to get a smaller number of cells.

3 CAD Adapted to a Set of Equalities

We are going to consider now a slight variation in the notion of CAD. As we said in the first section, a CAD adapted to the set of polynomials $\mathcal{P} = (P_1, \dots, P_r)$ allows to solve any quantifier elimination problem expressed by a formula $\Phi(x_1, \dots, x_n)$, which is a conjunction of equalities and inequalities on the polynomials $P_1, \dots, P_r$ (see Section 2). But, if we are interested in eliminating quantifiers for one particular formula $\Phi(x_1, \dots, x_n)$, we do not need to find the whole decomposition of $\mathbb{R}^n$. Suppose that our formula is

$$\Phi(x_1, \dots, x_n) := \begin{cases} P_i(x_1, \dots, x_n) = 0 & i = 1, \dots, s \\ P_j(x_1, \dots, x_n) \lhd_j 0 & j = s+1, \dots, r \end{cases}$$

with $\lhd_j \in \{<, >\}$. Then we can decompose $\mathcal{V} := \{P_1 = 0, \dots, P_s = 0\} \subset \mathbb{R}^n$ instead of $\mathbb{R}^n$ (this approach was already introduced in [25] and [9]).

Definition 2. *A CAD of $\mathbb{R}^n$ adapted to $(P_1 = 0, \dots, P_s = 0;\ P_{s+1}, \dots, P_r)$ is a set of partitions $\mathcal{C}_1, \dots, \mathcal{C}_{n-1}, \mathcal{C}_\mathcal{V}$, where $\mathcal{C}_k$ is a partition of $\mathbb{R}^k$ for $k = 1, \dots, n-1$ and $\mathcal{C}_\mathcal{V}$ is a partition of the algebraic set $\mathcal{V} = \mathcal{V}(P_1, \dots, P_s)$, such that:*

- *The partitions $\mathcal{C}_1, \dots, \mathcal{C}_{n-1}$ form a CAD of $\mathbb{R}^{n-1}$.*
- *The partition $\mathcal{C}_\mathcal{V}$ is also finite, and its cells are semialgebraic sets, semialgebraically homeomorphic to $(0,1)^d$ for some $d \leq n$.*
- *For every cell $S \in \mathcal{C}_{n-1}$, $(S \times \mathbb{R}) \cap \mathcal{V}$ is a finite union of cells of $\mathcal{C}_\mathcal{V}$.*
- *Every cell $\mathcal{C}_\mathcal{V}$ is $(P_{s+1}, \dots, P_r)$-invariant.*

3.1 Barnett's Method

Barnett's method is a Linear Algebra-based method to compute the degree of the gcd of several univariate polynomials over an integral domain. With this, in particular, we can parametrize $\deg(\gcd_{x_n}(P_1, \dots, P_r))$ for $P_i \in \mathbb{Q}[x_1, \dots, x_{n-1}][x_n]$, considering the variation of $x_1, \dots, x_{n-1}$ as parameters. The method allows also to compute explicitily the gcd.

We will consider $\mathbb{D}$ an integral domain, $P, Q_1, \dots, Q_r \in \mathbb{D}[x]$ with $\deg(P) = p$. Let us write $P = x^p + \dots + a_1 x + a_0$, and let us consider the companion matrix of P:

$$\Delta_P := \begin{pmatrix} 0 & \dots & \dots & 0 & -a_0 \\ 1 & \dots & \dots & 0 & -a_1 \\ & \ddots & & \vdots & \vdots \\ & & \ddots & \vdots & \vdots \\ 0 & & & 1 & -a_{p-1} \end{pmatrix}.$$

Let us consider the matrix $Q(\Delta_P)$ for any polynomial $Q(x) \in \mathbb{Q}[x]$, in particular the matrices $Q_i(\Delta_P)$ for $i = 1, \dots, r$, and the $rp \times p$ matrix

$$\mathcal{Q}_P := \begin{pmatrix} Q_1(\Delta_P) \\ \vdots \\ Q_r(\Delta_P) \end{pmatrix}.$$

Then we have the following result whose proof can be found in [13].

Theorem 1 (Barnett's Theorem).

$$\deg(\gcd(P, Q_1, \ldots, Q_r)) = p - \operatorname{rank}(\mathcal{Q}_P) \ .$$

The computation of the degree of $\gcd(P, Q_1, \ldots, Q_r)$ is made easier by two facts (see also [13]):

1. The computation of the entries of each matrix $Q_i(\Delta_P)$ requires just one euclidean remainder: the k-th column of $Q_i(\Delta_P)$ is the vector of coefficients of $\operatorname{rem}(Q_i x^k, P)$.
2. The fact that, if $\operatorname{rank}(\mathcal{Q}_P) = d$, then the first d columns are linearly independent (and therefore, the rest are linearly dependent of them). Therefore $\deg[\gcd(P, Q_1, \ldots, Q_r)] = h$ if and only if the first $p-h$ columns are linearly independent and the first $p - h + 1$ are not.

In order to check the linear independence of the columns, the next theorem, that can be found in [23], will be useful.

Theorem 2 (Gram's Criterion). *Let V be a vector space provided with a scalar product $\langle\ ,\ \rangle$. Then, for any $v_1, \ldots, v_s \in V$, $v_1, \ldots, v_s$ are linearly independent if and only if their Gram determinant*

$$\mathbf{Gram}(v_1, \ldots, v_s) = \begin{vmatrix} \langle v_1, v_1 \rangle & \cdots & \langle v_1, v_s \rangle \\ \vdots & & \vdots \\ \langle v_s, v_1 \rangle & \cdots & \langle v_s, v_s \rangle \end{vmatrix}$$

is nonzero (Note that $\mathbf{Gram}(v_1, \ldots, v_s) \geq 0$ *always).*

Thus we can compute Gram determinants for the columns of $\mathcal{Q}_P$, with respect to the usual scalar product in $\mathbb{R}^p$, to characterize the rank of $\mathcal{Q}_P$ when this matrix depends on some parameters. An explicit expression of $\gcd(P, Q_1, \ldots, Q_t)$ can also be obtained in terms of the matrix $\mathcal{Q}_P$ (see [13]).

3.2 CAD Adapted to a Set of Equalities via Barnett's Method

Note that Barnett's method is especially suitable for the problem of finding a CAD adapted to $(P_1 = 0, \ldots, P_s = 0)$, as we only have to consider the common zeros of the polynomials $P_1, \ldots, P_s$ all together, not pairwise: that is, the zeros of $\gcd(P_1, \ldots, P_s)$. For this, we will define a new projection operator that considers the matrix

$$\mathcal{Q}_{P_1} := \begin{pmatrix} P_2(\Delta_{P_1}) \\ \vdots \\ P_s(\Delta_{P_1}) \end{pmatrix}$$

whose rank can be characterized by using Gram determinants. Let us denote by $\rho_1, \ldots, \rho_p$ the columns of this matrix ($p = \deg_{x_n}(P_1)$). Then we define the projection operator $\mathbf{PROJ}_{\text{Gram}}(P_1 = 0, \ldots, P_s = 0)$ as the set consisting of all the $G_k = \mathbf{Gram}(\rho_1, \ldots, \rho_k)$ for $k = 2, \ldots, p$ replacing the principal subresultants of every pair P_i and P_j. Note that, if the leading coefficient of P_i is a constant for some i then we can assume $i = 1$ by reordering the polynomials; if this is not the case, we have to add to the projection operator the coefficients of P_1 with respect to x_n.

Next theorem shows how these polynomials can replace the computation of the different subresultants for every pair of polynomials in the initial family. Its proof is similar to the proof of proposition 2.

Theorem 3. *Let $S \in \mathbb{R}^{n-1}$ be a semialgebraic connected set, $P_1, \ldots, P_s \in \mathbb{Q}[x_1, \ldots, x_n]$, and suppose the numbers*

a) $\deg(P_1(\alpha; x_n))$,
b) $\#\{$*complex roots of* $P_i(\alpha; x_n)$ *over* $S\}$ $(i \in \{1, \ldots, s\})$, *and*
c) $\deg[\gcd(P_1(\alpha; x_n), \ldots, P_s(\alpha; x_n))]$

are constant for all $\alpha \in S$. Then, there exits a finite number, m, of functions $\gamma_i : S \longrightarrow \mathbb{R}$ such that for every $\alpha \in S$ $\gamma_1(\alpha) < \ldots < \gamma_m(\alpha)$ are the different real roots of $P_1(\alpha; x_n) = 0, \ldots, P_s(\alpha; x_n) = 0$.

Note that, due to special properties of the polynomials in $\mathbf{PROJ}_{\mathrm{Gram}}(P_1 = 0, \ldots, P_s = 0)$ (i.e. the G_i's), only the following sign conditions are feasible over the reals:

$$[G_1 \neq 0, G_2 = 0], [G_2 \neq 0, G_3 = 0], \ldots\ldots, [G_{p-1} \neq 0, G_p = 0], [G_p = 0]$$

and this can be very helpful in the construction of the next projection operator. This construction can be extended to a CAD adapted to $(P_1 = 0, \ldots, P_s = 0;\ P_{s+1}, \ldots, P_r)$.

This operator can be used only in the first step of the projection phase, that is, $\mathcal{V} \subset \mathbb{R}^n \longrightarrow \mathbb{R}^{n-1}$. In the rest of the process, the usual operator **PROJ** is used. As far as the extension phase is concerned, the test points are located in the algebraic set $\mathcal{V}$.

Example 1. The following polynomials in $\mathbb{Q}[x, y, z]$ are considered:

$$P_1 = x^3 + yx^2 + x + z, \quad P_2 = x^2 y + 3x + 2, \quad P_3 = x^2 - yx + z$$

and we intend to compute a CAD of $\mathbb{R}^3$ adapted to $(P_1 = 0, P_2 = 0, P_3 = 0)$. Let $\mathcal{V} = \mathcal{V}(P_1, P_2, P_3) \subset \mathbb{R}^3$. The projection phase will be $\mathcal{V} \longrightarrow \mathbb{R}^2 \longrightarrow \mathbb{R}$. In the step $\mathcal{V} \longrightarrow \mathbb{R}^2$ the operator $\mathbf{PROJ}_{\mathrm{Gram}}(P_1 = 0, P_2 = 0, P_3 = 0)$ can be used: let us compute this operator, eliminating the variable x. As P_1 is monic in x, we consider the matrix

$$\begin{pmatrix} P_2(\Delta_{P_1}) \\ \\ P_3(\Delta_{P_1}) \end{pmatrix} = \begin{pmatrix} 2 & -yz & zy^2 - 3z \\ 3 & 2 - y & y^2 - 3 - yz \\ y & -y^2 + 3 & 2 - 4y + y^3 \\ z & -z & 2yz \\ -y & z - 1 & 2y - z \\ 1 & -2y & z - 1 + 2y^2 \end{pmatrix}.$$

We have $\deg_x(P_1) = 3$, then we must compute $G_2 = \mathbf{Gram}(\rho_1, \rho_2)$ and $G_3 = \mathbf{Gram}(\rho_1, \rho_2, \rho_3)$ where ρ_1, ρ_2 and ρ_3 are the columns of the matrix. Then,

$$\begin{aligned} G_2 = {} & 8y^2z^2 + 160 + 54z^2 + 10y^4 - 6y^4z - 6yz^3 + z^4 + y^6 - 2z^3 - 2y^3z^2 - \\ & -44y - 28z + 3y^4z^2 + z^4y^2 - 6z^2y + 4y^3 - 10zy^2 + 13y^2 + 36yz \end{aligned}$$

$$\begin{aligned} G_3 = {} & 4y^8 - 200y^3z^3 + 7y^6z^4 - 42y^5z^2 - 8y^4z^3 + 2y^8z^2 + 150y^2z^3 - 312y^3z - \\ & -2y^7z^2 + 103y^2z^2 - 106z^4y + 20y^5z - 44y^6z + 8y^7z + 50y^3z^4 - 28z^5y + 10z^6 + \\ & +2z^6y^4 - 32z^5 - 2z^6y^3 + 6z^6y + 2z^5y^4 + 62z^5y^2 - 20z^5y^3 + 2268z^2 - 16y^5 - \\ & -4y^8z - 18y^5z^4 - 20y^6z^3 - 44y^4 + 212y^4z + 412yz^3 + 537z^4 + 16y^6 - 486z^3 + \\ & +50y^5z^3 + 25y^4z^4 + 45y^6z^2 - 1136y - 2032z + 2y^7z^3 + 2y^5z^5 - 3z^6y^2 - 114y^4z^2 + \\ & +15z^4y^2 - 1434z^2y + 598y^3z^2 + 64y^3 - 852zy^2 + 560y^2 + 2476yz + 840\ . \end{aligned}$$

With this, $\mathbf{PROJ}_{\text{Gram}}(P_1 = 0, P_2 = 0, P_3 = 0)$ contains $\{[G_2 \neq 0, G_3 = 0], [G_3 \neq 0]\}$ instead of the principal subresultants coefficients of P_i and P_j and also some of their reductums (note that P_2 is not x-monic).

4 Global Projection Operator for a CAD Adapted to a Set of Equalities

In this section, we intend to eliminate several variables in one step of the CAD by using Hermite's method. Given a zero-dimensional ideal in $\mathbb{Q}[\underline{x}] = \mathbb{Q}[x_1, \ldots, x_n]$, the number of real roots of the ideal with respect to $x_1, \ldots, x_s$ is, under certain conditions, the signature of a matrix whose entries are polynomials in $x_{s+1}, \ldots, x_n$ considered as parameters: thus we can eliminate the variables $x_1, \ldots, x_s$. The method will prove especially useful for a certain kind of zero-dimensional systems, called Pham systems. These systems have already been considered in [14] where Quantifier Elimination over them was used to analyze discretization schemes or in [27] where they appeared to be, after a deformation procedure, the kind of systems to solve when checking the emptyness of a real hypersurface. First we review Hermite's method and see how it can be applied to our purposes.

4.1 Hermite's Method

If J is a zero-dimensional ideal in $\mathbb{Q}[\underline{x}]$, then $\mathbb{B} = \mathbb{Q}[\underline{x}]/J$ is a finite-dimensional $\mathbb{Q}$-vector space. This allows one to define, for every $f \in \mathbb{B}$, the $\mathbb{Q}$-linear mapping:

$$\begin{array}{rccc} P_f\colon & \mathbb{B} & \longrightarrow & \mathbb{B} \\ & g + J & \longmapsto & fg + J \end{array}$$

whose geometric meaning is provided by Stickelberger Theorem (see [5] or [30]). The trace of the linear mapping P_f verifies:

$$\mathrm{Trace}(P_f) = \sum_{i=1}^{s} e_i f(\alpha_i)$$

where $\alpha_1, \ldots, \alpha_s$ are the zeroes in $\mathbb{C}^n$ of J and $e_1, \ldots, e_s$ their multiplicities. Next we define the trace symmetric bilinear mapping as:

$$\begin{array}{rccc} B\colon & \mathbb{B} \times \mathbb{B} & \longrightarrow & \mathbb{Q} \\ & (f,g) & \longmapsto & \mathrm{Trace}(P_{fg}) \ . \end{array}$$

Next theorem characterizes the number of different complex or real zeroes of J by means of the computation of the rank or the signature of a matrix with entries in $\mathbb{Q}$ (see [5] or [26] for a proof). If J is an ideal in $\mathbb{Q}[\underline{x}]$, then $\mathcal{V}_{\mathbb{R}}(J)$ (resp. $\mathcal{V}_{\mathbb{C}}(J)$) will denote the algebraic set in $\mathbb{R}^n$ (resp. $\mathbb{C}^n$) defined by J.

Theorem 4. *If J is a zero-dimensional ideal in $\mathbb{Q}[\underline{x}]$ then*

$$\mathrm{rank}(B) = \#\mathcal{V}_{\mathbb{C}}(J) \quad \textit{and} \quad \mathrm{signature}(B) = \#\mathcal{V}_{\mathbb{R}}(J) \ .$$

So, if $\omega_1, \ldots, \omega_t$ is a basis for the $\mathbb{Q}$-vector space $\mathbb{Q}[x_1, \ldots, x_n]/J$ then the number of zeroes of J in $\mathbb{C}^n$ (resp. in $\mathbb{R}^n$) is equal to the rank (resp. signature) of the associated matrix of the quadratic form B:

$$\begin{pmatrix} \mathrm{Trace}(P_{\omega_1\omega_1}) & \ldots & \mathrm{Trace}(P_{\omega_1\omega_t}) \\ \vdots & & \vdots \\ \mathrm{Trace}(P_{\omega_t\omega_1}) & \ldots & \mathrm{Trace}(P_{\omega_t\omega_t}) \end{pmatrix} .$$

4.2 Algebraic Structure of a Pham System

We will study a particular type of zero-dimensional system. Let J be the ideal generated by the polynomials

$$F_1 = x_1^{d_1} + U_1(x_1, \ldots, x_n), \ldots\ldots, F_n = x_n^{d_n} + U_n(x_1, \ldots, x_n)$$

where every $U_i(x_1, \ldots, x_n)$ is a polynomial in $\mathbb{Z}[x_1, \ldots, x_n]$ with total degree smaller than d_i. The restriction on the structure (and number) of the polynomials in the system is due to the need to control the algebraic structure of the quotient ring $\mathbb{B} = \mathbb{Q}[x_1, \ldots, x_n]/J$ with $J = \langle F_1, \ldots, F_n \rangle$. In a Pham system, for any choice of polynomials U_j, $\mathbb{B}$ is a $\mathbb{Q}$-vector space of dimension $D = d_1 \cdots d_n$, and

$$\mathcal{B} = \{x_1^{\gamma_1} \cdots x_n^{\gamma_n} + J \ : \ 0 \le \gamma_i < d_i \ \ \forall i\}$$

is a basis of this vector space. In this case it is very easy to prove that the entries of the matrix associated to trace symmetric bilinear mapping B (with respect to the previous basis) are polynomials in the coefficients of the polynomials U_j. Next, we assume that polynomials U_j depend on several parameters $\underline{t} = (t_1, \ldots, t_m)$, $U_j = U_j(\underline{t}, x_1, \ldots, x_n)$, and thus the entries of the matrix of B (with respect to the natural monomial basis) are elements in $\mathbb{Q}[\underline{t}]$. Besides, the structure and the basis $\mathcal{B}$ of $\mathbb{B}$ as $\mathbb{Q}$-vector space does not depend on the parameters. Next proposition gives the way of parametrizing the number of real solutions of the ideal J.

Proposition 3. *Let S be a connected subset of $\mathbb{R}^n$. If the number of different complex solutions for J is constant for any $\underline{t} \in S$, then the number of different real solutions of J is also constant.*

Proof. Due to Theorem 4, this proposition can be reformulated as: "if rank(B) is constant for any $\underline{t} \in S$ then signature(B) is also constant". As the matrix B is real symmetric, its rank and signature are given by the sign of its eigenvalues:

$$\begin{aligned}\text{rank}(B) &= \#\{\text{nonzero eigenvalues}\},\\ \text{signature}(B) &= \#\{\text{positive eigenvalues}\} - \#\{\text{negative eigenvalues}\}\ .\end{aligned}$$

The eigenvalues are the real roots of the characteristic polynomial of B, so they depend continuously on the coefficients of this polynomial; and these, in turn, depend continuously on the parameters. For this reason, and the fact of S being connected, easy continuity arguments show that an eigenvalue cannot change its sign without being zero. As the number of nonzero eigenvalues is constant, it follows that the sign of each eigenvalue is constant in S and therefore signature(B) is constant. □

4.3 Simultaneous Elimination of Several Variables

Let $P_1, \ldots, P_s$ be polynomials in $\mathbb{Z}[x_1, \ldots, x_n]$. It is assumed that these polynomials verify $s \leq n$ and, for $i \in \{1, \ldots, s\}$, P_i has the following structure:

$$P_i(x_1, \ldots, x_s; x_{s+1}, \ldots, x_n) = x_i^{d_i} + U_i(x_1, \ldots, x_n)$$

with U_i a polynomial with total degree smaller than d_i in the variables $x_1, \ldots, x_s$. Under these assumptions, the trace matrix B is a polynomial matrix whose entries are elements in $\mathbb{Q}[x_{s+1}, \ldots, x_n]$. It suffices to describe the rank of this matrix as a function of the parameters, to assure the constancy for the number of real roots of the system with respect to the variables $x_1, \ldots, x_s$.

For that purpose, we can use the characteristic polynomial: if $\chi(\lambda) = k_0 + k_1\lambda + k_2\lambda^2 + \cdots + k_D\lambda^D$ is the characteristic polynomial of B, then its coefficients k_i are polynomials in $\mathbb{Q}[x_{s+1}, \ldots, x_n]$. Then we consider as a projection operator the set $\{k_0, k_1, \ldots, k_D\}$ as long as, for every subset C of $\mathbb{R}^{n-s}$ such that the sign of these polynomials is constant for all $(\alpha_{s+1}, \ldots, \alpha_n) \in C$, the rank of B is constant, and so is the number of real roots of the system with respect to $x_1, \ldots, x_s$. As in the application of Barnett's Theorem to define $\mathbf{PROJ}_{\text{Gram}}$, we have also some extra information about the signs of the polynomials inside the projection operator: $[k_0 \neq 0], [k_0 = 0, k_1 \neq 0], [k_0 = 0, k_1 = 0, k_2 \neq 0]$, etc .

This projection operator performs the projection phase $\mathbb{R}^n \longrightarrow \mathbb{R}^{n-s}$ and therefore this step avoids the one by one **PROJ** computations related to the variables $x_1, \ldots, x_s$ in the usual algorithm. The rest of the projection phase is performed as usual, so the whole phase will be

$$\mathbb{R}^n \longrightarrow \mathbb{R}^{n-s} \longrightarrow \mathbb{R}^{n-s-1} \longrightarrow \cdots \longrightarrow \mathbb{R}^2 \longrightarrow \mathbb{R}\ .$$

For the extension phase, the steps $\mathbb{R} \longrightarrow \mathbb{R}^2 \longrightarrow \cdots \longrightarrow \mathbb{R}^{n-s}$ work also as usual; for the step $\mathbb{R}^{n-s} \longrightarrow \mathbb{R}^n$ we remember that the CAD is adapted to $(P_1 = 0, \ldots, P_s = 0)$, so given a test point $\underline{\alpha} = (\alpha_{s+1}, \ldots, \alpha_n)$ in a cell of $\mathbb{R}^{n-s}$, we only have to consider the real roots of the system

$$P_1(x_1, \ldots, x_s, \underline{\alpha}) = 0, \ldots\ldots, P_n(x_1, \ldots, x_s, \underline{\alpha}) = 0$$

and we will get a decomposition of the algebraic set

$$\mathcal{V} = \{(\alpha_1, \ldots, \alpha_n) \in \mathbb{R} : P_i(\alpha_1, \ldots, \alpha_n) = 0, \quad i = 1, \ldots, s\}$$

that is, a CAD of $\mathbb{R}^n$ adapted to $(P_1 = 0, \ldots, P_s = 0)$.

Next we study a special case in which it is not necessary even to compute the characteristic polynomial of the matrix B.

The Case of a Single Parameter. Let us assume that $s = n-1$: the variables that will be eliminated are $x_1, \ldots, x_{n-1}$, and the only parameter is x_n. Then, the projection phase is reduced to one step, $\mathbb{R}^n \longrightarrow \mathbb{R}$. Besides, let us note that:

- The entries of B are polynomials in one variable, x_n, so if the determinant of B is not identically zero, it is also a polynomial in x_n.
- The intervals where this polynomial does not vanish, are cells in which $\operatorname{rank}(B) = D$.
- In a similar way, the zeros of $\det(B)$ are cells where $\operatorname{rank}(B)$ is strictly smaller than D; these cells cannot be further decomposed.

For these reasons, $\det(B)$ gives a decomposition of $\mathbb{R}$ in cells where $\operatorname{rank}(B)$ is constant, and therefore, the number of real roots of the system is constant. In view of this, the projection operator consists only of the polynomial $\det(B)$. The projection phase is thus performed in one step and with just one polynomial.

In case $\det(B)$ is identically zero, the projection operator can still be formed with a single polynomial: let $\chi(\lambda) = k_0 + k_1\lambda + k_2\lambda^2 + \cdots + k_D\lambda^D$ be the characteristic polynomial of B (the coefficients k_i depend on t); $k_0 = \det(B)$, so if $k_0 = 0$ then $\chi(\lambda) = \lambda^m(k_m + k_{m+1}\lambda + \cdots + k_D\lambda^{D-m})$ with k_m the first non identically zero coefficient. If $k_m \neq 0$ then $\operatorname{rank}(B) = D - m$, so for the same reasons as above, we form the projection operator with just the polynomial k_m.

Example 2. Consider these polynomials in $\mathbb{Q}[x, y, z, t]$:

$$P_1 = x^2 + tx + y - z, \quad P_2 = y^2 + x + ty - tz + 1, \quad P_3 = z^2 - x + ty + z + t$$

which form a Pham system with respect to the variables x, y, z, with t as a parameter. For all values $t \in \mathbb{R}$, $\{1,\ x,\ y,\ z,\ xy,\ xz,\ yz,\ xyz\}$ is a basis of $\mathbb{Q}[x,y,z]/\langle P_1, P_2, P_3\rangle$ as a $\mathbb{Q}$-vector space. Next some of the columns of the matrix B, whose entries are polynomials in t, are displayed (we write its columns $B^{(1)}, \ldots, B^{(8)}$):

$$B^{(1)} = \begin{bmatrix} 8 \\ 4t \\ 4t \\ 4 \\ 2(t-1)(1+t) \\ 2t-2 \\ 2t(1+t) \\ (t-1)(1+t)^2 \end{bmatrix} \quad B^{(2)} = \begin{bmatrix} 4t \\ -4t+4-4t^2 \\ 2(t-1)(1+t) \\ 2t-2 \\ 8+4t+6t^2-2t^3 \\ -4-4t-8t^2 \\ (t-1)(1+t)^2 \\ -t(-13-3t+t^3-9t^2) \end{bmatrix} \quad B^{(3)} = \begin{bmatrix} 4t \\ 2(t-1)(1+t) \\ -8-4t^2 \\ 2t(1+t) \\ \vdots \\ \vdots \end{bmatrix}.$$

Its determinant is:

$$\det(B) = 50625t^{24} + 1404000t^{23} + 24232500t^{22} + 112893720t^{21} + \dots\dots\dots$$
$$\dots\dots\dots\dots\dots\dots\dots\dots\dots\dots$$
$$\dots\dots + 7814332368t^3 + 4206515028t^2 + 1446358008t + 288191657$$

which has only two real roots, say a and b with $a < b$, so the decomposition of $\mathbb{R}$ is: $(-\infty, a) \cup \{a\} \cup (a, b) \cup \{b\} \cup (b, +\infty)$. Now, for t belonging to each of these cells of $\mathbb{R}$, we must compute the number of real roots of the system with respect to x, y, z, obtaining

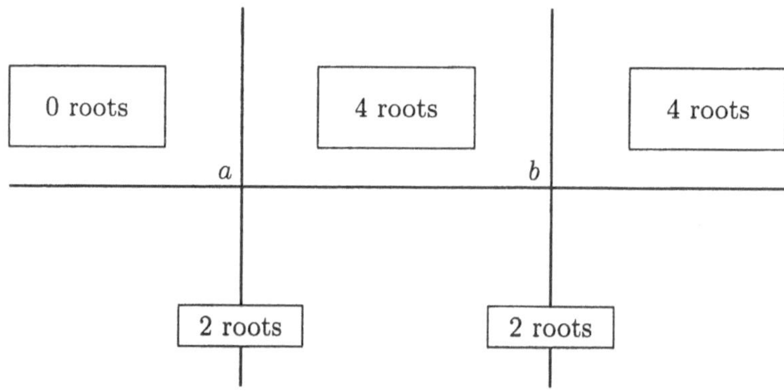

which represents the structure of the CAD adapted to $(P_1 = 0, P_2 = 0, P_3 = 0)$.

References

1. Arnon D. S.: A cluster-based Cylindrical Algebraic Decomposition algorithm. J. of Symb. Comp. **5** (1988) 189–212
2. Arnon D. S., Collins G. E., McCallum S.: Cylindrical algebraic decomposition. II. An adjacency algorithm for the plane. SIAM J. of Computing **13** (1984) 878–889
3. Arnon D. S., Collins G. E., McCallum S.: An adjacency algorithm for Cylindrical Algebraic Decompositions of three–dimensional space. J. of Symb. Comp. **5** (1988) 163–187
4. Brown C.: Improved Projection for CAD's of $\mathbb{R}^3$. ISSAC-00 (Saint Andrews) Proceedings (2000), ACM-Press
5. Becker E., Wörmann T.: Radical computation of a zero–dimensional ideal and real root counting. Mathematics and Computers in Simulation **42** (1996) 561–569
6. Bochnak J., Coste M., Roy M.–F.: Géométrie Algébrique Réelle. Ergebnisse der Mathematik und ihrer Grenzgebiete **12** (1987) Springer–Verlag
7. Collins G. E., Hong H.: Partial cylindrical algebraic decomposition for quantifier elimination. J. of Symb. Comp. **12** (1991) 299–328
8. Collins G. E.: Quantifier elimination for real closed fields by Cylindrical Algebraic Decomposition. Lecture Notes in Computer Science **33** (1975) 134–183
9. Collins G. E.: Quantifier elimination by Cylindrical Algebraic Decomposition - 20 years of progress. Quantifier elimination and cylindrical algebraic decomposition, Texts and Monographs in Symbolic Computation (1998) 8–23, Springer–Verlag

10. Coste M.: An Introduction to Semialgebraic Geometry. Available at `http://www.maths.univ-rennes1.fr/~coste/Enseignement.html` (2000)
11. Davenport J. H., Heintz J.: Real Quantifier Elimination is Doubly Exponential. J. of Symb. Comp. **5** (1988) 29–36 .
12. Gonzalez–Lopez M. J., Gonzalez–Vega L.: Newton identities in the multivariate case: Pham systems. Grobner Bases and Applications. London Mathematical Society Lecture Notes Series **251** (1998) 351–366, Cambridge University Press
13. Gonzalez-Vega L.: An elementary proof of Barnett's Theorem about the greatest common divisor of several univariate polynomials. Linear Algebra and Applications **247** (1996) 185–202
14. Gonzalez-Vega L.: A special quantifier elimination algorithm for Pham systems. Real Algebraic Geometry and Ordered Structures (Charles N. Delzell and James J. Madden eds.). Comtemporary Mathematics **253** (2000) 115–134, AMS
15. Hong H.: Quantifier elimination for formulas constrained by quadratic equations via slope resultants. The Computer Journal **36** (1993) 439–449.
16. Hong H.: An improvement of the projection operator in cylindrical algebraic decomposition. Quantifier elimination and cylindrical algebraic decomposition, Texts and Monographs in Symbolic Computation (1998) 166–173, Springer–Verlag
17. Hong H.: Quantifier Elimination for Formulas Constrained by Quadratic Equations. ISSAC-93 (Kiev) Proceedings (1993) 264–274, ACM–Press
18. Hong H.: Simple Solution Formula Construction in Cylindrical Algebraic Decomposition based Quantifier Elimination. ISSAC-92 (Tokyo) Proc. (1992) 177–188, ACM–Press
19. Hong H., Liska R. (eds): Applications of quantifier elimination. Journal of Symbolic Computation **24** (1997) 123–231
20. Hong H., Liska R., Steinberg S.: Testing stability by quantifier elimination. J. of Symb. Comp. **24** (1997) 161–187
21. Heintz J., Roy M.–F., Solerno P.: Sur la complexité du Principe de Tarski-Seidenberg. Bulletin de la Societe Mathematique de France **118** (1990), 101–126
22. Heintz J., Roy M.–F., Solerno P.: On the theoretical and practical complexity of the existential theory of the reals. The Computer Journal **36** (1993) 427–431
23. Lancaster P., Tismenetsky M.: The theory of matrices. Computer Science and Applied Mathematics (1985) Academic Press
24. McCallum S.: An improved projection operator for Cylindrical Algebraic Decomposition of three-dimensional space. J. of Symb. Comp. **5** (1988) 141–161
25. McCallum S.: On projection in cad–based quantifier elimination with equational constraint. ISSAC-99 (Vancouver) Proceedings (1999) 145–199, ACM–Press
26. Pedersen P., Roy M.–F., Szpirglas A.: Counting real zeros in the multivariate case. Progress in Mathematics **109** (1993) 203-224, Birkhauser
27. Roy M.–F.: Basic algorithms in real algebraic geometry: from Sturm theorem to the existential theory of reals. Expositions in Mathematics **23** (1996) 1-67, de Gruyter
28. Weispfenning V.: The complexity of linear problems in fields. Journal of Symbolic Computation **5** (1988) 3–27
29. Weispfenning V.: Quantifier elimination for real algebra—the quadratic case and beyond. Appl. Algebra Engrg. Comm. Comput. **8** (1997) 85–101
30. Yokoyama K., Noro M., Takeshima T.: Solutions of Systems of Algebraic Equations and Linear Maps on Residue Class Rings. J. of Sym. Comp. **14** (1992) 399–417

Symbolic Algorithms of Algebraic Perturbation Theory for a Hydrogen Atom: the Stark Effect

Alexander Gusev[1], Valentin Samoilov[1], Vitaly Rostovtsev[2], and Sergue Vinitsky[3]

[1] Science Center of Applied Investigations, Joint Institute for Nuclear Research, Dubna, Moscow Region 141980, Russia
[2] Laboratory of Computer Techniques and Automatics, Joint Institute for Nuclear Research, Dubna, Moscow Region 141980, Russia
[3] Bogoliubov Laboratory of Theoretical Physics, Joint Institute for Nuclear Research, Dubna, Moscow Region 141980, Russia

Abstract. We present symbolic algorithms realized in REDUCE 3.6 for evaluation of eigenvalues and eigenfunctions of the 3-D and 2-D hydrogen atoms in weak uniform electric fields. Algebraic perturbation theory schemes are built up using the irreducible representations of the dynamical symmetry algebras $so(4,2)$ and $so(3,2)$, which are connected by the tilting transformations with wave functions of the 3-D and 2-D hydrogen atoms. Such a construction is based on a representation of the unperturbed Hamiltonian and polynomial perturbation operator via generators of the algebra. It was done without an assumption on the separation of independent variables of the perturbation operator and without using fractional powers of the parabolic quantum numbers in recurrence relations determining the effects of generators of the algebra on the corresponding basis. The efficiency of the proposed schemes and algorithms is demonstrated by calculations of coefficients of the Stark effect perturbations series for the hydrogen atoms with arbitrary parabolic quantum numbers.

1 Introduction

As is known, the group $SO(4)$ is the symmetry group of the 3D hydrogen atom, while the group $SO(4,2)$ in the 6D pseudo-Euclidean space with metric $g_{\alpha\alpha} = (1,1,1,1,-1,-1)$ is the dynamical symmetry group[1, 2]. We can consider in a similar way formally the group $SO(3,2)$ in the 5D pseudo-Euclidean space with metric $g_{\alpha\alpha} = (1,1,1,-1,-1)$ as the dynamical symmetry group of the 2D hydrogen atom using the fact that the group $SO(3)$ is the symmetry group of the 2D hydrogen atom [3]. The irreducible representation of the algebra $so(4,2)$ or $so(3,2)$ is indeed related by the tilting transformation to eigenfunctions of a discrete spectrum of the 3D or 2D hydrogen atom according to [4]. This circumstance allows us to construct an algebraic perturbation scheme in terms of the representation of the algebra $so(4,2)$ or $so(3,2)$ via substitution of a polynomial perturbation for an appropriate combination of generators of the algebra. For the construction of such perturbation schemes one usually applies the representation of the algebra $so(2,1)$ starting from the fact of separation of independent variables, which takes place in the Stark effect [5]. In the case of polynomial perturbations of a general type, for example, for a hydrogen atom in uniform electric and magnetic fields or nonuniform electric fields, i.e. when the independent variables are not separated, such a representation is, however, inconvenient, and the development of a general perturbation scheme is needed [6]. Moreover, a special normalized factor of the irreducible representation, which is conventionally used following [7], leads to a necessity of explicit calculations of matrix elements that are also inconvenient for the implementation of computer algebra algorithms via known recurrence relations for generators of the algebra only. To solve a linear eigenvalue problem without troubles, we can obviously choose an appropriate normalized factor without fractional powers of parabolic quantum numbers involved in an evaluation process and can renormalize only final eigenvectors.

The aim of the present talk consists in a formulation of effective algebraic perturbation schemes and algorithms of an arbitrary high order by using computer algebra facilities. The advantage of the proposed approach consists in avoiding the restrictions of separation of independent variables. This is more efficient for numerical calculations of hydrogen atom in electric and magnetic fields [8] or asymptotics of effective potentials in the Coulomb three-body problem [9]. To simplify presentation, we implement the scheme to calculate the Stark effect for a hydrogen atom in the 3D and 2D cases. It provides a closed form of the zero-order wave functions in diagonal representation of the parabolic quantum numbers for the perturbation operator under consideration. To organize evaluation of perturbation corrections to an eigenvalue in a more effective way within the framework of the conventional algorithms, we choose an appropriate minimal set of unknown coefficients in the expansions of eigenvector corrections over the irreducible representation.

In Sections 2 and 3, the description of the 3D hydrogen atom wave functions with help of the irreducible representation of the algebra $so(4,2)$ is given for free and Stark effect cases. In Section 4, the new efficient recursive algorithm within the framework of algebraic version of the conventional perturbation theory without separation of the independent variables is presented. In Section 5, we consider in a similar way the Stark effect for the 2D hydrogen atom. In Section 6, we present the program **Stark** and present the results for the energy spectrum up to seventh order. The proposed algorithm[1] was realized with help of REDUCE 3.6 and the evaluation of seventh order of eigenvalues and the fifth order of eigenfunctions of the Stark effect of the 3D and 2D hydrogen atoms was produced.

[1] The authors are trying to realize the same procedure in Maple V

2 The Dynamical Symmetry Group of Hydrogen Atom

The Lie algebra $so(4,2)$ is composed of 15 generators

$$L_{\alpha\beta} = -L_{\beta\alpha}, \quad [L_{\alpha\beta}, L_{\alpha\gamma}] = \imath g_{\alpha\alpha} L_{\beta\gamma}, \quad \alpha, \beta, \gamma = 1, ..., 6, \tag{1}$$

where $g_{\alpha\alpha} = (1,1,1,1,-1,-1)$. In the configuration space the generators $L_{\alpha\beta}$ are defined by the relations

$$L_{ij} = x_i p_j - x_j p_i = \varepsilon_{ijk} L_k, \quad i,j,k = 1,2,3,$$

$$L_{i4} = \frac{1}{2}(x_i \boldsymbol{p}^2 + \chi \imath p_i - 2\boldsymbol{x}\boldsymbol{p} p_i - x_i) = A_i, \quad L_{46} = \frac{1}{2}(r\boldsymbol{p}^2 - r),$$

$$L_{i5} = \frac{1}{2}(x_i \boldsymbol{p}^2 + \chi \imath p_i - 2\boldsymbol{x}\boldsymbol{p} p_i + x_i), \quad L_{56} = \frac{1}{2}(r\boldsymbol{p}^2 + r),$$

$$L_{45} = -\imath(\frac{\chi}{2} + \imath \boldsymbol{x}\boldsymbol{p}), \qquad L_{i6} = -r p_i, \tag{2}$$

where, for the 3D hydrogen atom

$$\chi = 2, \quad p_k = -\imath \frac{\partial}{\partial x_k}, \quad r = \sqrt{x_1^2 + x_2^2 + x_3^2}.$$

Note that the coordinates may be expressed in terms of the generators

$$x_i = L_{i5} - L_{i4}, \quad r = L_{56} - L_{46}. \tag{3}$$

The above operators act in the Hilbert space of functions with the inner scalar product

$$\langle f | g \rangle = \int f^*(\boldsymbol{x}) r^{-1} g(\boldsymbol{x}) d\boldsymbol{x} \tag{4}$$

with respect to which they are Hermitian. The basis vectors in the representation space with given n will be classified over eigenvalues of the operators L_{56}, L_{34}, L_{12}, i.e. we use the parabolic basis $|n_1, n_2, m\rangle$

$$L_{56}|n_1 n_2 m\rangle = n|n_1 n_2 m\rangle, \quad L_{34}|n_1 n_2 m\rangle = (-n_1 + n_2)|n_1 n_2 m\rangle,$$

$$L_{12}|n_1 n_2 m\rangle = m|n_1 n_2 m\rangle.$$

As the basis that realizes the infinite-dimensional irreducible representation of the algebra $so(4,2)$, we take eigenfunctions of these operators, which in the parabolic coordinates $\{y_1, y_2, y_3\}$

$$y_1 = r + x_3, \quad y_2 = r - x_3, \quad y_3 = arctg(x_2/x_1), \quad d^3y = \frac{1}{4}(y_1 + y_2) dy_1 dy_2 dy_3,$$

have the form:

$$\Phi^{(0)}_{n_1 n_2 m}(\boldsymbol{y}) = C_{n_1 n_2 |m|} \phi_{n_1 |m|}(y_1) \phi_{n_2 |m|}(y_2) \frac{e^{\imath m y_3}}{\sqrt{2\pi}} \left\{ \begin{array}{l} -1,\, m > 0 \\ 1,\ \ m \le 0 \end{array} \right. , \tag{5}$$

$$C_{n_1 n_2 |m|} = \left[2 \frac{n_1!}{(n_1+|m|)!} \frac{n_2!}{(n_2+|m|)!} \right]^{\frac{1}{2}},$$

$$\phi_{n_j |m|}(y_j) = [(n_1+|m|)!]^{-1} y_j^{\frac{|m|}{2}} e^{-\frac{y_j}{2}} L^{|m|}_{n_j+|m|}(y_j), \quad j=1,2.$$

Here $L^{|m|}_{n_j+|m|}$ are the Laguerre polynomials [10], and the basis functions $\Phi^{(0)}_{n_1 n_2 m}(\boldsymbol{y})$ are orthonormalised according to (4) with $d\boldsymbol{x} = d^3 y$.

It is known that the discrete spectrum of a hydrogen atom with charge Z (atomic units are used), $E^{(0)} = -\frac{Z^2}{2n^2}$ is numbered by eigenvalues $n = n_1 + n_2 + |m| + 1$ of the operator L_{56}, and the wave functions $\langle \boldsymbol{y} | \overline{n_1, n_2, m} \rangle \equiv \overline{\Phi}^{(0)}(\boldsymbol{y})$ normalized by a standard condition

$$\langle \overline{n_1, n_2, m} | \overline{n'_1, n'_2, m'} \rangle = \int \overline{\Phi}^{*(0)}_{n_1 n_2 m}(\boldsymbol{y}) \overline{\Phi}^{(0)}_{n'_1 n'_2 m'}(\boldsymbol{y}) d^3 y = \delta_{n_1 n'_1} \delta_{n_2 n'_2} \delta_{mm'}.$$

They are related to the eigenfunctions of the operator L_{56} by the tilting transformation

$$\begin{aligned} \langle \boldsymbol{y} | \overline{n_1, n_2, m} \rangle &= \frac{(-2E^{(0)})^{1/4}}{n^{1/2}} U^{-1} \langle \boldsymbol{y} | n_1, n_2, m \rangle \\ &= \frac{(-2E^{(0)})^{3/4}}{n^{1/2}} \langle \sqrt{(-2E^{(0)})} \boldsymbol{y} | n_1, n_2, m \rangle, \end{aligned} \tag{6}$$

where U is the unitary operator for $E^{(0)} < 0$

$$U = \exp(\imath \theta^{(0)} L_{45}), \quad \operatorname{th} \theta^{(0)} = \frac{1/2 + E^{(0)}}{1/2 - E^{(0)}}. \tag{6a}$$

To confirm these relations for the Shroedinger equation of a hydrogen-like atom

$$\left(\frac{p^2}{2} - \frac{Z}{r} - E^{(0)} \right) | \bar{\Phi}^{(0)} \rangle = 0, \tag{7}$$

it will be multiplied by r, rewritten in the notations of the algebra $so(4,2)$,

$$((1/2 + E^{(0)}) L_{46} + (1/2 - E^{(0)}) L_{56} - Z) | \bar{\Phi}^{(0)} \rangle = 0,$$

and reduced by means of the tilting transformation to the equation for $|\Phi^{(0)}\rangle$,

$$U \{ (1/2 + E^{(0)}) L_{46} + (1/2 - E^{(0)}) L_{56} - Z \} U^{-1} | \Phi^{(0)} \rangle = 0. \tag{8}$$

The latter is equivalent to the passage to the new variables in equation (8), because

$$U^{-1} x_k U = \sqrt{-2E^{(0)}} x_k. \tag{9}$$

Using the relations

$$\begin{aligned} U L_{56} U^{-1} &= L_{56} ch\theta^{(0)} - L_{46} sh\theta^{(0)}, \\ U L_{46} U^{-1} &= -L_{56} sh\theta^{(0)} + L_{46} ch\theta^{(0)}, \end{aligned} \tag{9a}$$

instead of (8), we obtain

$$\left[L_{56} - \frac{Z}{\sqrt{-2E^{(0)}}}\right]|\Phi^{(0)}\rangle = 0, \quad E^{(0)} = -\frac{Z^2}{2n^2}.$$

If we take into account (6),(6a) and scalar product (4), the normalization condition for $|\overline{\Phi}^{(0)}\rangle$ assumes the form

$$\overline{\langle n_1, , n_2, m|}\overline{n_1', n_2', m'\rangle} = \int \overline{\Phi}^{(0)*}_{n_1 n_2 m}(y)\overline{\Phi}^{(0)}_{n_1' n_2' m}(y) d^3y$$

$$= \frac{1}{n}\int \Phi^{(0)*}_{n_1 n_2 m}(\sqrt{-2E^{(0)}}y)\Phi^{(0)}_{n_1' n_2' m'}(\sqrt{-2E^{(0)}}y)(\sqrt{-2E^{(0)}})^3 d^3y \tag{10}$$

$$= \frac{1}{n}\int \Phi^{(0)*}_{n_1 n_2 m}(y) r \Phi^{(0)}_{n_1' n_2' m'}(y) r^{-1} d^3y$$

$$= \frac{1}{n}\langle n_1, n_2, m|L_{56} - L_{46}|n_1', n_2', m'\rangle = \delta_{n_1 n_1'}\delta_{n_2 n_2'}\delta_{mm'}.$$

Here, we have used the relation

$$\langle n_1, n_2, m|L_{56} - L_{46}|n_1', n_2', m'\rangle = n\delta_{n_1 n_1'}\delta_{n_2 n_2'}\delta_{mm'}. \tag{10a}$$

So, using the representation of the algebra $so(4,2)$, we can readily find the wave functions normalized by the standard condition and the energy of the discrete spectrum of a hydrogen-like atom.

3 Problem Statement

The Shroedinger equation of a hydrogen-like atom with the charge Z in uniform electric field $\boldsymbol{F}(0,0,F)$ in atomic units takes the form

$$\left(\frac{1}{2}p^2 - \frac{Z}{r} + Fx_3 - E\right)|\overline{\Phi}\rangle = 0. \tag{11}$$

Under condition $F \ll 1$, the wave functions $|\overline{\Phi}\rangle$ and the energy $E = E(F)$ are sought for in the form

$$|\overline{\Phi}\rangle = \sum_{k=0} F^k |\overline{\Phi}^{(k)}\rangle, \quad E = \sum_{k=0} F^k E^{(k)}. \tag{12}$$

Let us now multiply equation (11) by r and rewrite it using definitions (12)

$$\left(\frac{1}{2}rp^2 - Z - \sum_{k=1} F^k V^{(k)}(x_3) - rE^{(0)}\right)|\overline{\Phi}\rangle = 0, \tag{13}$$

where

$$V^{(1)}(x_3) = r(E^{(1)} - x_3), \quad V^{(k)}(x_3) = rE^{(k)}, \quad k > 1. \tag{13a}$$

Then we can write (13) in the notations (2) and (3). Making use of the tilting transformation (6), we go over from the hydrogen states $|\overline{\Phi}\rangle$ to the basis states $|\Phi\rangle$, and taking into account (6a), (9), (9a), we arrive at the following equation:

$$\left[L_{56} - \frac{Z}{\sqrt{-2E^{(0)}}} - \frac{1}{\sqrt{-2E^{(0)}}} \sum_{k=1} F^k V^{(k)} \left(\frac{L_{35} - L_{34}}{\sqrt{-2E^{(0)}}}\right)\right] |\Phi\rangle = 0. \qquad (14)$$

The normalization condition for basis states $|\overline{\Phi}\rangle$ in this equation in terms of scalar product (4) for states $|\Phi\rangle$ is analogous by the condition of eq. (10)

$$\langle\overline{\Phi}|\overline{\Phi}\rangle = \frac{\sqrt{-2E^{(0)}}}{n} \langle\Phi|U(L_{56} - L_{46})U^{-1}|\Phi\rangle$$

$$= \frac{\sqrt{-2E^{(0)}}}{n} \langle\Phi| \frac{L_{56} - L_{46}}{\sqrt{-2E^{(0)}}} |\Phi\rangle = \frac{1}{n} \langle\Phi|L_{56} - L_{46}|\Phi\rangle = 1. \qquad (10b)$$

Hence it follows that in the course of the passage from the wave functions $|\Phi\rangle$ to the basis states $|\overline{\Phi}\rangle$, the first one should be normalized by the condition

$$\langle\Phi|L_{56} - L_{46}|\Phi\rangle = n, \qquad (15)$$

which becomes identity for states $|\Phi^{(0)}\rangle$.

4 The Scheme and Algorithm of Perturbation Theory

We look for the solution of equation (14) in the form of the perturbation series

$$|\Phi\rangle = \sum_{k=0} F^k |\Phi^{(k)}\rangle. \qquad (16)$$

The unknown coefficients $|\Phi^{(k)}\rangle$ satisfy the system of inhomogeneous differential equations

$$L(n)|\Phi^{(0)}\rangle - (L_{56} - n)|\Phi^{(0)}\rangle = 0 \equiv f^{(0)}, \qquad (17)$$

$$L(n)|\Phi^{(1)}\rangle = \frac{n^2}{Z^2}(L_{56} - L_{46})[E^{(1)} - \frac{n}{Z}(L_{35} - L_{34})]|\Phi^{(0)}\rangle \equiv f^{(1)}, \qquad (17a)$$

$$L(n)|\Phi^{(k)}\rangle = \frac{n^2}{Z^2}(L_{56} - L_{46}) \left[\sum_{p=0}^{k-1} E^{(k-p)}|\Phi^{(p)}\rangle - \frac{n}{Z}(L_{35} - L_{34})|\Phi^{(k-1)}\rangle\right] \equiv f^{(k)}. \qquad (17b)$$

As the basis that realizes the infinite-dimensional irreducible representation of the algebra $so(4,2)$ in x-space, we take the eigenfunctions $\langle y|s,t\rangle$ of commuting operators L_{56}, A_3 and L_3, which differ from the above introduced basis functions $\langle y|n_1 + s, n_2 + t, m\rangle$ only by the normalization factor and coincide with them at $s = t = 0$

$$\langle y|s,t\rangle = \frac{C_{n_1 n_2 |m|}}{C_{n_1+s, n_2+t, |m|}} \langle y|n_1 + s, n_2 + t, m\rangle. \qquad (18)$$

The operators $L_{56}, L_{34}, L_{46}, L_{35}$ and L_{12} on the functions $\langle \boldsymbol{y}|s,t\rangle$ are defined by the relations without fractional powers of parabolic quantum numbers

$$L_{56}|s,t\rangle = (n_1+n_2+|m|+1+s+t)|s,t\rangle = (n+s+t)|s,t\rangle, \tag{19}$$

$$L_{34}|s,t\rangle = (-(n_1+s)+(n_2+t))|s,t\rangle, \tag{19a}$$

$$L_{12}|s,t\rangle = |m||s,t\rangle, \tag{19b}$$

$$\begin{aligned} L_{45}|s,t\rangle = \tfrac{1}{2}(\quad & (n_1+s+|m|)|s-1,t\rangle + (n_1+s+1)|s+1,t\rangle \\ & +(n_2+t+|m|)|s,t-1\rangle + (n_2+t+1)|s,t+1\rangle), \end{aligned} \tag{19c}$$

$$\begin{aligned} L_{35}|s,t\rangle = \tfrac{1}{2} \quad & (-(n_1+s+|m|)|s-1,t\rangle - (n_1+s+1)|s+1t\rangle \\ & +(n_2+t+|m|)|s,t-1\rangle + (n_2+t+1)|s,t+1\rangle). \end{aligned} \tag{19d}$$

We assume that the azimuthal quantum number m will be positive $m = |m|$. Applying relations (19)-(19d), we expand the right-hand side $f^{(k)}$ and solutions $|\Phi^{(k)}\rangle$ of the system (17) over basis states $|s,t\rangle$ (18):

$$f^{(k)} = \sum_{s=-2k}^{2k} \sum_{\substack{t=\\-2k+|s|}}^{2k-|s|} f_{st}^{(k)}|s,t\rangle, \quad |\phi^{(k)}\rangle = \sum_{s=-2k}^{2k} \sum_{\substack{t=\\-2k+|s|}}^{2k-|s|} b_{st}^{(k)}|s,t\rangle. \tag{20}$$

Substituting (20) into (17)–(17b) and taking into account the relation

$$L(n)|s,t\rangle = (s+t)|s,t\rangle,$$

and orthogonality condition (4) of basis (18) for each $k \geq 1$, we obtain the system of linear algebraic equations for unknown coefficients $b_{st}^{(k)}$ and perturbation corrections $E^{(k)}$

$$(s+t)b_{st}^{(k)} - f_{st}^{(k)} = 0, \quad |s|+|t| \leq 2k. \tag{21}$$

It enables us to find the coefficients $b_{st}^{(k)}$ using known coefficients $f_{st}^{(k)}$ from the above definitions. In the first order $k=1$, we obtain at $s=t=0$ the energy correction $E^{(1)} = 3nd/2Z$ and at $s+t \neq 0$ eight coefficients $b_{st}^{(1)}$:

$$b_{0,-1}^{(1)} = (-(d-m-n+1)(d+2n-2)n^3)/(8Z^3),$$

$$b_{0,1}^{(1)} = ((d+m-n-1)(d+2n+2)n^3)/(8Z^3),$$

$$b_{0,2}^{(1)} = ((d+m-n-1)(d+m-n-3)n^3)/(32Z^3),$$

$$b_{0,-2}^{(1)} = (-(d-m-n+3)(d-m-n+1)n^3)/(32Z^3).$$

Note that for coefficients $b_{st}^{(k)} = b_{st}^{(k)}(n,d,m)$ as functions of $n, d = n_2 - n_1, m$, the following relations of symmetry with respect to sign of d take place: $b_{st}^{(k)}(d) = (-1)^k b_{ts}^{(k)}(-d)$. The other coefficients $b_{s,-s}^{(1)}$ are equal to zero , as it follows from equations $f_{s-s}^{(2)} = 0$ of second order $k=2$ with $s+t=0$. In the general case of $k \geq 2$, the coefficients $b_{s,-s}^{(k)}$ are non-zero and calculated from

$(k+1)$th order equation (22b) at $s+t=0$. Indeed, at each k the functions $f_{st}^{(k)}$ depend on the unknown correction $E^{(k)}$ and known coefficients $E^{(p)}$ and $b_{st}^{(p)}$ for $p=0,1,\cdots,k-1$, which are evaluated from previous $k-1$ equations. The coefficients $b_{st}^{(p)}$ also depend on the corrections $E^{(q)}$ and $b_{st}^{(q)}$ with $q=0,1,\cdots,p-1$, which are evaluated from previous equations (21) by recurrence. Thus, in each order ($k \geq 1$) we calculate step-by-step the needed corrections $E^{(k)}, b_{s,-s}^{(k-1)}, b_{st}^{(k)}$ by solving the following algebraic equations:

$$f_{00}^{(k)}\left(E^{(k)}, E^{(p)}, b_{i0}^{(p)}, b_{0i}^{(p)}, 0 \leq p \leq k-1, -2 \leq i \leq 2\right) = 0 \to E^{(k)}, \tag{22}$$

$$f_{s-s}^{(k)}\left(E^{(p)}, b_{i-s,s}^{(p)}, b_{s,i-s}^{(p)}, 1 \leq p \leq k-1, -2 \leq i \leq 2\right) = 0 \to b_{s-s}^{(k-1)}, \tag{22a}$$

$$b_{st}^{(k)} = (s+t)^{-1} f_{st}^{(k)}\left(E^{(p)}, b_{s+i,t}^{(p)}, b_{s,t+j}^{(p)}, 0 \leq p \leq k-1, -2 \leq i,j \leq 2\right). \tag{22b}$$

The initial conditions for the recurrence procedure are given by

$$E^{(0)} = -\frac{Z^2}{2n^2}, \quad b_{00}^{(0)} = 1, \quad b_{st}^{(0)} = 0 \quad \text{for} \quad s,t \neq 0. \tag{23}$$

Note that $b_{00}^{(k)} = 0$ and if $|s|+|t| > 2k$, then $b_{st}^{(k)} = f_{st}^{(k)} = 0$. If we wish to apply the above algorithm for the evaluation of only the energy spectrum corrections up to $k_{max} = 2k+1$, we can restrict the range of summation for s,t in equations (20), (21),(22a) and (22b) for each kth order to the region of $0 \leq |s|+|t| \leq 2\min(k, k_{max}-k)$. It means that in this case, we use only the minimal set of coefficients $b_{st}^{(k)}$ needed for calculation of $E^{(k_{max})}$ that gives significant savings in computer resources.

To obtain the normalized wave function Φ up to the kth order we must redefine the coefficient $b_{00}^{(k)}$ by the following relation:

$$b_{00}^{(k)} = -\frac{1}{2n} \sum_{p=0}^{k} \sum_{\substack{s'= \\ -2p}}^{2p} \sum_{\substack{t'= \\ -2p+|s'|}}^{2p-|s'|} \sum_{\substack{s= \\ -2(k-p)}}^{2(k-p)} \sum_{\substack{t= \\ -2(k-p)+|s|}}^{2(k-p)-|s|} b_{st}^{(k-p)} \langle s,t|r|s',t'\rangle b_{s't'}^{(p)}. \tag{24}$$

The passage from the kth order approximate solution Φ in the form (16), (20) and (24) to the approximate solution $\overline{\Phi}$ of equation (13) normalized by condition (10) with the same accuracy is made by means of tilting transformation (6), (6a).

5 Two-Dimensional Hydrogen Atom

The Lie algebra $so(3,2)$ is composed of 10 generators

$$L_{\alpha\beta} = -L_{\beta\alpha}, \quad [L_{\alpha\beta}, L_{\alpha\gamma}] = \imath g_{\alpha\alpha} L_{\beta\gamma}, \quad \alpha,\beta,\gamma = 2,...,6,$$

where $g_{\alpha\alpha} = (1,1,1,-1,-1)$. In the x-representation these generators are given by the same relations as in equations (2), taking into account the following definitions:

$$\chi = 1, \quad p_k = -\imath\frac{\partial}{\partial x_k}, \quad r = \sqrt{x_2^2 + x_3^2}, \quad k = 2,3.$$

The basis vectors in representation space with given n will be classified over eigenvalues of the operators L_{56}, L_{34}, i.e. we use the parabolic basis $|n_1, n_2\rangle$

$$L_{56}|n_1 n_2 m\rangle = n|n_1 n_2\rangle, \quad L_{34}|n_1 n_2 m\rangle = (-n_1 + n_2)|n_1 n_2\rangle.$$

As the basis, that realizes the infinite-dimensional irreducible representation of the algebra $so(3,2)$, we take eigenfunctions of these operators, which in the parabolic coordinates $\{y_1, y_2\}$

$$y_1 = r + x_3, \quad y_2 = r - x_3, \quad d^2y = \frac{1}{4}\frac{(y_1+y_2)}{\sqrt{y_1 y_2}} dy_1 dy_2,$$

have the form

$$\langle \boldsymbol{y}|n_1, n_2\rangle = \Phi^{(0)}_{n_1 n_2}(\boldsymbol{y}) = C_{n_1 n_2}\phi_{n_1}(y_1)\phi_{n_2}(y_2), \tag{5'}$$

$$C_{n_1 n_2} = \left[2\frac{n_1!}{\Gamma(n_1 - \frac{1}{2} + 1)}\frac{n_2!}{\Gamma(n_2 - \frac{1}{2} + 1)}\right]^{\frac{1}{2}},$$

$$\phi_{n_j}(y_j) = e^{-\frac{y_j}{2}} L_{n_j}^{-\frac{1}{2}}(y_j), \quad j = 1, 2.$$

Here, $L_{n_j}^{-\frac{1}{2}}(y_j)$ are the Laguerre polynomials [11], and the $\Phi^{(0)}_{n_1 n_2}(\boldsymbol{y})$ are orthonormalised according to (4) with $d\boldsymbol{x} = d^2y$. It is known that the discrete spectrum of the 2D hydrogen atom $E^{(0)} = -\frac{Z^2}{2n^2}$ is numbered by eigenvalues $n = n_1 + n_2 + \frac{1}{2}$ of the operator L_{56}, and the wave functions normalized by standard condition are related to the eigenfunctions of the operator L_{56} by the tilting transformation (6)–(9a) in the 2D notation. Analogously to the explanation of eq. (10), we have:

$$\overline{\langle n_1, n_2}\,\overline{|n_1', n_2'\rangle} = \int \overline{\Phi}^{(0)*}_{n_1 n_2}(y)\overline{\Phi}^{(0)}_{n_1' n_2'}(y) d^2y \tag{10'}$$

$$= \frac{1}{n}\langle n_1, n_2|L_{56} - L_{46}|n_1', n_2'\rangle = \delta_{n_1 n_1'}\delta_{n_2 n_2'}.$$

After substitution of eqs. (12), (13a), (16) into Shroedinger's equation (11) of a hydrogen-like atom in the 2D notation, we obtain a system of equations (17)–(17b). As the basis that realizes the infinite-dimensional irreducible representation of the algebra $so(3,2)$, we take the eigenfunctions $\langle \boldsymbol{y}|st\rangle$ of commuting operators L_{56} and L_{34}, which differ from the above introduced functions $\langle \boldsymbol{y}|n_1 + s, n_2 + t\rangle$ only by the normalization factor and coincide with them at $s = t = 0$

$$\langle \boldsymbol{y}|s, t\rangle = \frac{C_{n_1 n_2}}{C_{n_1+s, n_2+t}}\langle \boldsymbol{y}|n_1 + s, n_2 + t\rangle. \tag{18'}$$

The operators L_{56}, L_{34}, L_{46} and L_{35} on functions $\langle \boldsymbol{y}|s, t\rangle$ are defined by the relations similar to (19)–(19d)

$$L_{56}|s, t\rangle = (n_1 + n_2 + \frac{1}{2} + s + t)|s, t\rangle = (n + s + t)|s, t\rangle, \tag{19'}$$

$$L_{34}|s,t\rangle = (-(n_1+s)+(n_2+t))|s,t\rangle, \tag{19'a}$$

$$\begin{aligned} L_{45}|s,t\rangle = \tfrac{1}{2}(\quad &(n_1+s-\tfrac{1}{2})|s-1,t\rangle + (n_1+s+1)|s+1,t\rangle \\ &+(n_2+t-\tfrac{1}{2})|s,t-1\rangle + (n_2+t+1)|s,t+1\rangle), \end{aligned} \tag{19'c}$$

$$\begin{aligned} L_{35}|s,t\rangle = \tfrac{1}{2}(\quad &-(n_1+s-\tfrac{1}{2})|s-1,t\rangle - (n_1+s+1)|s+1,t\rangle \\ &+(n_2+t-\tfrac{1}{2})|s,t-1\rangle + (n_2+t+1)|s,t+1\rangle). \end{aligned} \tag{19'd}$$

Applying relations (19′)–(19′d), we expand the right-hand side $f^{(k)}$ and solutions $|\Phi^{(k)}\rangle$ of the system (17) in the 2D notation over basis states $|s,t\rangle$ (18′). As a result we arrive at the algorithm (22)–(24) described in Section 4.

6 Program Stark and Results

In this Section we show the program **Stark** realized in REDUCE 3.6 as well as the test run output file for the energy corrections **ee(i)**$= E^{(i)}$ of the 2D and 3D hydrogen-like atoms with charge Z in atomic units.

PROGRAM STARK

```
% PROGRAM STARK
kmax:=7;
operator x3,r,f,b,ee,ket; array o(kmax);
for i:=0:kmax do o(I):=min(I,kmax-I);

procedure v(k,x,y);
  if k<1 then 0 else
   if k=1
     then sub(x1=x,y1=y,ee(1)*y1-y1*x1)
     else sub(y1=y,ee(k)*y1);


for all k let f(k)=n/za*(v(k,x3,r)*ket(0,0)+
 for p1:=1:k-1 sum v(k-p1,x3,r)*(for s:=-2*o(p1):2*o(p1) sum
   for tt:=-2*o(p1)+abs(s):2*o(p1)-abs(s) sum b(p1,s,tt)*ket(s,tt)));


let x3=x3(), r=r();
for all x,y let x3()*ket(x,y)=x3(x,y),r()*ket(x,y)=r(x,y);
for all x,y let r(x,y)=((n+x+y)*ket(x,y)
 -1/2*((x+n1+m)*ket(x-1,y)+(x+n1+1)*ket(x+1,y)
 +(y+n2+m)*ket(x,y-1)+(y+n2+1)*ket(x,y+1)))*n/za;
for all x,y let x3(x,y)=(1/2*(-(y-x-(n1-n2))*ket(x,y)
 -(x+n1+m)*ket(x-1,y)-(x+n1+1)*ket(x+1,y)
 +(y+n2+m)*ket(x,y-1)+(y+n2+1)*ket(x,y+1)))*n/za;

procedure fk(k);
   begin scalar u,u1,u2;
     u:=f(k);u1:=den u; u:=num u;
     u2:=coeffn(u,ket(0,0),1); u2:=-u2/coeffn(u2,ee(k),1)+ee(k);
```

```
    ee(k):=u2;
      for i:=1:o(k-1) do
      for each i2 in {-1,1} do
      <<
        i1:=i*i2; u2:=coeffn(u,ket(i1,-i1),1)/u1;
        b(k-1,i1,-i1):=sub(solve(u2,b(k-1,i1,-i1)),b(k-1,i1,-i1));
        u:=sub(ket(i1,-i1)=0,u);
      >>;
    b(k,0,0):=0;

    for i1:=1/2 step 1/2 until o(k) do
      for j1:=-o(k)+(if fixp(i1) then 0 else 1/2)
              :o(k)-(if fixp(i1) then 0 else 1/2) do
       for each i2 in {-1,1} do
        <<
          ii:=i1*i2+j1;jj:=i1*i2-j1;
          u2:=coeffn(u,ket(ii,jj),1)/u1;
          b(k,ii,jj):=u2/(ii+jj);
          u:=sub(ket(ii,jj)=0,u);
        >>
  end;

let n1=(n+d-m-1)/2, n2=(n-d-m-1)/2;
 for all x,y,z such that x<=0 let b(x,y,z)=0;
off nat, echo; out starkout;  order n,d,m;
 for i:=1:kmax do fk(i);
 for i:=1:kmax do write "e(",i,"):=",ee(i);
write showtime; shut starkout;
bye;% END OF PROGRAM STARK
```

Comments to the program The array o is the range of summation for s and t (see the description after eq. (23)). Procedure v realizes the substitution of the difference between energy and perturbation corrections multiplied by the factor r in the kth order. The variables **f(k)**$=f^{(k)}$ are the right-hand sides in equations (17)–(17b). The procedure fk(k) solves the equations (17)–(17b) with substitution (20), and in the k-th order, after calculation of the **f(k)**, finds sequentially the **ee(k)**$= E^{(k)}$, **b(k-1,s,-s)**$= b^{(k-1)}_{s,-s}$ and **b(k,s,tt)**$= b^{(k)}_{st}$ in accordance with eqs. (22).

Other variables are auxiliary. So far we consider a class of the polynomial perturbations in the x-representation. It takes the form of the multiply operators, i.e. $[x_k, r] = 0 => [L_{56} - L_{46}, L_{35} - L_{34}] = 0$. In this case, we do not need to use the declaration **noncom** x_k, r. In the case of a nonlocal perturbation operator, for an example, depending on L_{45}, etc., it is necessary to apply the declaration **noncom** and commutator algebraic relations (1) for the corresponding generators, which will compose the perturbation operator under consideration.

TEST RUN OUTPUT

```
e(1):=(3*n*d)/(2*za)$

e(2):=(n**4*( - 17*n**2 + 3*d**2 + 9*m**2 - 19))/(16*za**4)$
```

```
e(3):=(3*n**7*d*(23*n**2 - d**2 + 11*m**2 + 39))/(32*za**7)$

e(4):=(n**10*( - 5487*n**4 - 1806*n**2*d**2 + 3402*n**2*m**2 -
35182*n**2 - 147*d**4 + 1134*d**2* m **2 - 5754*d**2 + 549*m**4 +
8622*m**2 - 16211))/(1024*za**10)$

e(5):=(3*n**13*d*(10563*n**4 + 98*n**2*d**2 + 772*n**2*m**2 +
90708*n**2 - 21*d**4 + 220*d**2*m**2 + 780*d**2 + 725*m**4 +
830*m**2 + 59293))/(1024*za**13)$

e(6):=(n**16*( - 547262*n**6 - 685152*n**4*d**2 + 429903*n**4*m**2
- 9630693*n**4 - 390*n**2*d**4 + 25470*n**2*d**2*m**2 -
7787370*n**2*d**2 + 16200*n**2*m**4 + 4786200*n**2*m**2 -
22691096*n**2 + 372*d**6 - 765*d**4*m**2 - 1185*d**4 +
36450*d**2*m**4 + 62100*d**2*m **2 - 7028718*d**2 + 6951*m**6 +
16845*m**4 + 4591617*m**2 - 7335413))/(8192*za**16)$

e(7):=(3*n**19*d*(7071885*n**6 + 1502283*n**4*d**2 -
1530561*n**4*m**2 + 152685291*n**4 + 1947* n**2 *d**4 +
21410*n**2*d**2*m**2 + 22015690*n**2*d**2 + 94915*n**2*m**4 -
22686230*n**2*m**2 + 458448411*n**2 + 957*d**6 - 6321*d**4*m**2 +
5691*d**4 + 66115*d**2*m**4 + 64330*d**2*m **2 + 25667355*d**2 +
55937*m**6 + 155435*m**4 - 26168905*m**2 + 196899293))/(32768*za**19)$
Time: 2650000 ms
```

This test was calculated on computer PC-2 350MHz 64MB memory. In the above formulas, we use the following notations: **n**$\equiv n = n_1 + n_2 + |m| + 1$, **m**$\equiv |m|$, **d**$\equiv n_1 - n_2$, **za**$\equiv Z$. These results also apply in the 2D case, if we will make the formal identification $n = n_1 + n_2 + \frac{1}{2}$ and **m**$\equiv m^2 = \frac{1}{4}$. Note that our results coincide up to the sixth order with [5].

7 Conclusion

We have demonstrated the efficiency of the proposed recursive symbolic algorithm within the framework of an algebraic version of the conventional perturbation theory without assumption on separation of independent variables. The generalities of this algorithm in the case of hydrogen atom in nonuniform fields will be presented elsewhere.

8 Acknowledgements

We are grateful to Prof. V.P. Gerdt for his support of our work. The authors (AG, VR and SV) thank the Russian Foundation for Basic Research for the support by grants No. 00-02-16337 and No. 00-02-81023.

References

1. Malkin, I.A., Man'ko, V.I.: *Dynamical symmetry and coherent satates of quantum systems*, Nauka, Moscow, 1979 (in Russian).
2. Adams, B.G., Cizek, J., and Paldus, J.: Lie algebraic methods and their applications to simple quantum systems. In: *Advances in Quantum Chemistry*, Per-Olov Lowdin (Ed.), Academic Press, New York, **18** (1988), pp. 1–85.
3. Cisneros, A., and McIntosh, H.V.: Symmetry or Two-Dimension Hydrogen Atom. *J. Math. Phys.* **10** (1969) 277–286.
4. Engerfield, M.J.: *Group Theory and the Coulomb Problem*, Monash university, Victoria (1972).
5. Adams, B.G.: Unified treatment of high-order perturbation theory for Stark effect in a two- and three-dimensional hydrogen atom. *Phys. Rev. A.* **46** (1992) 4060–4064.
6. Kadomtsev, M.B. and Vinitsky, S.I.: Perturbation theory within the O(4,2) group for hydrogen atom in the field of distant charge. *J. Phys. A: Math. Gen.* **18** (1985) L689–L695.
7. Silverstone, H.J.: Perturbation theory of the Stark effect in hydrogen to arbitrarily high order. *Phys. Rev. A* **18** (1978) 1853–1864.
8. Silverstone, H.J., Moats, R.K.: Practical recursive solution of degenerate Rayleigh-Shroedinger perturbation theory and application to high-order calculation of the Zeeman-effect in hydrogen. *Phys. Rev. A* **23** (1981) 1645–1654.
9. Abrashkevich, A.G., Puzynin, I.V., Vinitsky, S.I.: ASYMPT: a program for calculating asymptotics of hyperspherical potential curves and adiabatic potentials. *Computer Physics Communications* **125** (2000) 259–281.
10. Courant, R. and Hilbert, D.: *Methods of Mathematical Physics*, Interscience publishers, New-York, London, 1953.
11. Lebedev, N.N.: *Special Functions and Their Application*, GITTL, Moscow, 1953 (in Russian).

CADECOM: Computer Algebra software for functional DECOMposition

Jaime Gutierrez and Rosario Rubio

Departamento de Matemáticas, Estadística y Computación
Facultad de Ciencias, Universidad de Cantabria
Santander E–39071, SPAIN
e-mail:{jaime, sarito}@matesco.unican.es

Abstract. In this paper we present the Maple package Cadecom which is designed for performing computations in rational function fields. The main objects that Cadecom deals with are multivariate rational functions over any computable field, and the key tool are the functional decomposition algorithms. The functional decomposition problem has many applications in computer science, engineering (CAGD), pure mathematics or robotics. We motivate the interest of this program package by presenting applications on computing roots, simplifying sine–cosine equations, integrating rational functions, computing subfields, computing Gröbner bases and reparametrizing parametric curves. We also include a short overview of the package from the Maple system point of view.

1 Introduction

The general functional decomposition problem can be stated as follows: given f in a class of functions, we want to represent f as a composition of two "simpler" functions g and h in the same class, i.e. $f = g \circ h = g(h)$. Although not every function can be decomposed in this manner, when such a decomposition does exist many problems become significantly simpler. Nowadays, this problem has become more important for the simplification of some algebraic objects/structures: polynomials, rational functions, sine–cosine equations or more generally multivariate rational functions module a polynomial ideal. Over the last ten years, there have been several new results in the area of polynomial decomposition, see the references in [8].

Univariate polynomial decomposition has applications in computer science, computational algebra, and robotics. In fact, computer algebra system such as AXIOM, MAPLE, MATHEMATICA and REDUCE support polynomial decomposition for univariate polynomials. However we do not know a program package dedicated to functional decomposition algorithms. In this paper we present the program package CADECOM (Computer Algebra software for functional DECOMposition) which is built on the computer algebra system MAPLE and designed for performing computations over rational function fields. The germ of Cadecom was a collection of procedures (c.f. [3]), which were designed as a help for the manipulation of univariate rational function fields.

The main objects that Cadecom deals with are multivariate rational functions over any computable field K, in the sense of the underlying computer algebra system Maple, that is, all the arithmetic operations have to be available in the system. Typically, a finite field $K = F_q$, or the rational numbers $K = Q$, or a finite algebraic extension $K(\alpha)$ of K, or a multivariate rational function field $K(x_1, \ldots, x_n)$.
The key tool of Cadecom are the functional decomposition algorithms. The main implementations are based on algorithms presented in [4, 9, 22] for decomposing univariate polynomial and rational functions; in [8] for decomposing multivariate polynomials in several ways; in [19] for reparametrizing parametric curves and in [10] for decomposing bivariate polynomials module the unit circle.
We motivate the importance of this program package by presenting some of the most interesting applications: computing roots, evaluating functions, simplifying sine–cosine equations, integrating rational functions, computing subfields, computing Gröbner bases and reparametrizing parametric curves. We describe the functionalities by examples, including the computation time on a personal computer Macintosh Centris 650.

The structure of the paper is as follows: In Section 2 we present some of the problems that can be simplified using the package and we illustrate them by some examples. In Section 3 we outline some aspects on Cadecom's procedures.

2 Motivation

In this section we present some applications of the package Cadecom and we illustrate them by some examples.

2.1 Evaluating Functions

Evaluation is a common calculation in mathematics. The evaluation of a function f in a point requires, in the general case, $O(n)$ multiplications; but if f is decomposable, it can be computed with $O(\sqrt{n})$ multiplications.
A rational function $f(x) \in K(x)$ is called *decomposable* if there exist two rational functions $g(x), h(x) \in K(x)$ with degree greater than one and such that $f(x) = g(h(x))$. The degree of a rational function is the maximum of the degree of numerator and the degree of the denominator.
For instance, suppose we want to evaluate the rational function

$$f = \frac{x^{35} + 5\,x^{28} + 10\,x^{21} - 3\,x^{19} + 10\,x^{14} - 3\,x^{12} + 5\,x^7 + 1}{x^{19} + x^{15} + x^{12}}$$

Using the `decomp` function in Cadecom package we get:

> `decomp`(f, x);

$$\left[\frac{3\,x^4 - 1}{x^4\,(x-1)}, -\frac{x^3}{x^7+1}\right]$$

time = 60.35 bytes = 7783850.

Then, $f = g(h)$ where $h = -\dfrac{x^3}{x^7+1}$ and $g = \dfrac{3\,x^4-1}{x^4\,(x-1)}$.

Evaluation of multivariate polynomials can also be simplified via functional decomposition. A multivariate polynomial $f(x_1,\ldots,x_n) \in K[x_1,\ldots,x_n]$ is *uni–multivariate decomposable* if there exist a univariate polynomial $g(y) \in K[y]$ and a multivariate polynomial $h(x_1,\ldots,x_n) \in K[x_1,\ldots,x_n]$ with total degree greater than one and such that $f(x_1,\ldots,x_n) = g(h(x_1,\ldots,x_n))$.

Given the multivariate polynomial $f \in K[x,y,z]$:

$$f = 3x^4y^2 + 6x^2y^3z + 18x^3y + 3z^2y^4 + 18zy^2x + 27x^2 + 6x^2y + 6zy^2 + 18x + 2$$

Using the umdecompoly function, we get:

> umdecompoly(f, u);

$$\left[3u^2 + 6u + 2, u = x^2y + zy^2 + 3x\right]$$

time = 45.21 bytes = 5727569.

2.2 Computing Roots

In general, the equation $f(x) = 0$ can be numerically solved more efficiently if f is decomposable. Moreover, in the particular case when f is a polynomial, it is easier to determine if the zeroes of f can be expressed in terms of radicals. Suppose we want to find out all the roots of a polynomial f; if $f = g(h)$, in general, it would be more effective to compute the roots of g, say α and then solve the equations $h - \alpha$. For example, let

$$f = x^6 - 6x^5 - 17x^4 + 112x^3 + 67x^2 - 442x + 290$$

f is an irreducible polynomial over $Q[x]$. But it is decomposable, this can be seen using the decompoly code:

> decompoly(f, x);

$$\left[x^2 + 34x + 290, x^3 - 3x^2 - 13x\right]$$

time = 1.47, bytes = 133434.

Now we can compute all the roots of f: first solve the equation $x^2+34x+290 = 0$: we have two roots $\alpha = -17 + \sqrt{-1}$ and its conjugate $\overline{\alpha}$; then solve the equation $x^3 - 3x^2 - 13x - \alpha = 0$, we get three roots of f, β_1, β_2 and β_3. The other roots of f are $\overline{\beta}_1, \overline{\beta}_2$ and $\overline{\beta}_3$. Summarizing, we can compute the roots of the polynomial f, with degree 6, solving a second order equation and a third order equation.

Sometimes you come across an irreducible and indecomposable univariate polynomial, but you can still simplify the polynomial equation. We say that a univariate polynomial $f(x) \in K[x]$ is *ideal decomposable* if there exist $g(x)$ and $h(x)$

in $K[x]$ forming a non–trivial decomposition, $g(h(x))$, i.e. $\deg h, \deg g < \deg f$, and so that $f(x)$ divides $g(h(x))$. The *ideal decomposition* problem is to decide if f is ideal decomposable; and in the affirmative case compute $\bar{f}$, g and h such that $f\bar{f} = g(h)$ (c.f. [6]).

Suppose we want to compute the zeroes of the polynomial

$$f = x^6 + 9x^4 - 10^3 + 27x^2 + 90x + 52$$

with rational coefficients. First of all, we check that f is irreducible and indecomposable. In fact, Maple was unable to compute symbolically the roots of this polynomial. The `idealdecompoly` function in Cadecom package give us:

> `idealdecompoly`(f, x);

$$\left[x^3 + 9x^2 + 27x + 22, \left[x^3 + 12x^2 + 1128x + 1144, x^3 + 3x^2 + 3x\right]\right]$$

time = 567.22, bytes = 67838253.

We have that $f\bar{f} = g(h)$ where

$$\bar{f} = x^3 + 9x^2 + 27x + 22, g = x^3 + 12x^2 + 1128x + 1144, \text{ and } h = x^3 + 3x^2 + 3x.$$

So, we have reduced the problem to compute the zeroes of two polynomial of degree 3.

2.3 Reparametrizing Parametric Curves

A parametrization $(g(t), h(t))$ of a parametric curve $f(x, y) = 0$ is called *faithful* if every point (x_0, y_0) of the curve (except a finite number of them) corresponds to a unique value of the parameter t_0, that is, given a point (x_0, y_0) of the curve, i.e. $f(x_0, y_0) = 0$, there exists a unique t_0 such that

$$x_0 = g(t_0), \; y_0 = h(t_0).$$

Not every algebraic curve is parametric, but if it is, there exists a faithful parametrization. The computation of a faithful parametrization is an important topic in Computer Aided Geometric Design (CAGD); see for example [7, 19].

In algebraic terms, a parametrization $(g(t), h(t))$ of a curve f is faithful if and only if $K(g(t), h(t)) = K(t)$. In Cadecom there are implemented several procedures to test the faithfulness: for example, `TRfaithful` which is based on the concept of Taylor resultant introduce by Abhyancar around 1972 (see [1]). We can also find a faithful parametrization from a non-faithful one: for instance, using the procedure `netto` based on the constructive proof of the classical result of Lüroth's theorem or the procedure `sedeberg` based on the paper [19].

Suppose we are given the parametrization

$$(g,h) = \left(\frac{t^6+3t^4+675t^2+2745-64t^3-2352t}{(2t-7)^3}, \frac{(2t-7)(t^2+1)}{t^4+26t^2+295-168t}\right).$$

of the plane algebraic curve $f(x,y) = y^3x^2+16y^3x+280y^3+144y^2+18y^2x-x-8$. In Cadecom we get:

> k:=`sedeberg`$([g,h],t)$;

$$k = \frac{31557-8980t+127t^2}{13t^2-202t+720}$$

time = 0.40, bytes = 83230.

In order to find a faithful parametrization of f from (g,h), we call `lcomp` procedure:

> g':=`lcomp`(k,g,t);

$$g' = 9\frac{-10055829104+678164228t-15210262t^2+112525t^3}{-2048383+629031t-64389t^2+2197t^3}$$

time = 1.85, bytes = 226074.

> h':=`lcomp`(k,h,t);

$$h' = \frac{570230-71197t+1313t^2}{20256874-926792t+11215t^2}$$

time = 1.12, bytes = 142782.

Then (g',h') is a faithful parametrization of the algebraic curve defined by the polynomial f.

Another interesting test over parametric curves is to decide if a parametrization is quasi–polynomial. A parametrization $(g(t),h(t))$ is *quasi–polynomial* if the union field $K(g(t),h(t))$ contains a non–constant polynomial. To simplify this kind of parametrization we can use decomposition polynomial algorithms which are very fast. We can test it using cadecom function `quasipol`.

Given the plane algebraic curve, $-31+140y^3x+296x-99y-92yx^2-4x^2+196y^2x^2-100y^3x^2-117y^2+672y^2x-49y^3+764xy=0$, by the parametrization

$$(g,h) = \left(\frac{t^6+3t^4+3t^2+2}{2t^2-4}, \frac{t^4-5}{t^4+6t^2+9}\right).$$

> `quasipol`$([g,h],t)$;

`true`

time = 2.72, bytes = 196062.

2.4 Computing Gröbner Bases

Gröbner bases computation can also be reduced via multivariate decomposition. For multivariate decomposition there exist several definitions of decomposable, see [8]. For Gröbner bases we are interesting in multi–univariate decomposable polynomial, since this decomposition is compatible with Gröbner bases computation:

Given G be a reduced Gröbner basis —under some term ordering— of the ideal generated by H, where H is a finite set of polynomials in the variables $x_1, \dots, x_n$; let Θ be a polynomial map, that is, $\Theta = (\theta_1, \dots, \theta_n)$ is a list of n polynomials in the variables $x_1, \dots, x_n$. Now, we consider two new polynomial sets: H^* and G^*, obtained from H and G, respectively, by replacing x_i with θ_i. A natural question that arises is: *Under which circumstances is G^* the reduced Gröbner basis of the ideal generated by H^* under the same term ordering?* In [11] the authors give a complete answer: this happens if and only if the composition by Θ is "compatible" with the term ordering and Θ is a list of permuted univariate and monic polynomials. Similar results were obtained in [12] for Gröbner bases. This problem has two natural applications. One of them is in the computation of reduced Gröbner bases of the ideal generated by composed polynomials: so, in order to compute a reduced Gröbner basis of H^*, we first compute a reduced Gröbner basis G of H and carry out the composition on G, obtaining a reduced Gröbner basis of H^*. This appears to be more efficient than computing a reduced Gröbner basis of H^* directly. On the other hand, the opposite application is decomposing the input polynomials $f \in H$ as $f = g(\theta_1(x_1), \dots, \theta_n(x_n))$, where $\theta_i(x_i)$ is an univariate polynomial in the variable x_i, and then check if the composition by $\Theta = (\theta_1, \dots, \theta_n)$ is "compatible" with the term ordering.

We say that a multivariate polynomial $f(x_1, \dots, x_n) \in K[x_1, \dots, x_n]$ is *multi–univariate* if there exist a multivariate polynomial $g \in K[x_1, \dots, x_n]$ and univariate polynomials $h_i(x_i) \in K[x_i]$ with degree greater than 1 such that $f = g(h_1, \dots, h_n)$.

Suppose we want to find the Gröbner basis of the ideal generated by

$$
\begin{aligned}
(g,h) = (&-94038\,y + 1707\,y^6 + 320\,x^9 - 720\,x^7 + 540\,x^5 + 61452\,y^2 - 4800\,x^6 \\
&+11082\,x - 8808\,y^3x + 864\,x^8 + 54\,x^2y^6 + 648\,x^2y^4 + 20484\,y^4 - 4428\,y^5 \\
&-738\,y^7 - 270\,y^7x - 15\,y^9x + 20\,y^9x^3 - 1020\,x^3y^6 - 36720\,x^3y^2 4128\,x^6y \\
&-12240\,x^3y^4 + 1152\,x^6y^4 + 3456\,x^6y^2 + 96\,x^6y^6 + 27540\,xy^2 + 765\,xy^6 \\
&+9180\,xy^4 + 256\,x^{12} + 1944\,x^2y^2 - 24529\,y^3 - 33408\,xy - 5184\,x^4y^2 \\
&+45354\,x^3y - 144\,x^4y^6 - 768\,x^{10} + 42684 - 1728\,x^4y^4 + 6633\,x^4 \\
&-2457\,x^2 - 1920\,x^9y + 2322\,x^2y - 14911\,x^3 - 1032\,x^4y^3 - 6192\,x^4y \\
&+4320\,x^7y - 3240\,x^5y + 387\,x^2y^3 - 320\,x^9y^3 + 11879\,x^3y^3 + 720\,x^7y^3 \\
&-1620\,y^5x - 540\,x^5y^3 - 41\,y^9 + 688\,x^6y^3 + 360\,y^7x^3 + 2160\,y^5x^3, \\
&-2898\,y - 20\,y^6 - 720\,y^2 + 736\,x^6 + 3030\,x - 330\,y^3x - 240\,y^4 + 108\,y^5 \\
&+18\,y^7 - 267\,y^3 - 1980\,xy + 2640\,x^3y - 1104\,x^4 + 414\,x^2 - 270\,x^2y \\
&-4040\,x^3 + 120\,x^4y^3 + 720\,x^4y - 45\,x^2y^3 + 440\,x^3y^3 + 5363 + y^9 \\
&-80\,x^6y^3 - 480\,x^6y),
\end{aligned}
$$

the Maple V (release 5) function **gbasis** was not able to compute it -after two days or work- but if we use **mudecompoly** function we get

> `mudecompoly`(g);

$$[-15673\,y + 7424\,xy + 688\,x^2y - 320\,x^3y + 42684 - 14776\,x - 4368\,x^2 + 320\,x^3 \\ +256\,x^4 + 1707\,y^2 - 1020\,xy^2 + 96\,x^2y^2 - 41\,y^3 + 20\,y^3x, [y^3 + 6\,y, -3/4\,x + x^3]]$$

time = 7.60, bytes = 492534.

> `lmdecompoly`$([y^3 + 6\,y, -3/4\,x + x^3], h)$;

$$-483\,y + 440\,xy - 80\,x^2y + 5363 - 4040\,x + 736\,x^2 - 20\,y^2 + y^3$$

time = 3.45, bytes = 257258.

Now, we can compute the Gröbner basis of the ideal generated by

$$\begin{aligned}(g', h') = (&-15673\,y + 7424\,xy + 688\,x^2y - 320\,x^3y + 42684 - 14776\,x - 4368\,x^2 \\ &+320\,x^3 + 256\,x^4 + 1707\,y^2 - 1020\,xy^2 + 96\,x^2y^2 - 41\,y^3 + 20\,y^3x, \\ &-483\,y + 440\,xy - 80\,x^2y + 5363 - 4040\,x + 736\,x^2 - 20\,y^2 + y^3)\end{aligned}$$

> G:=`gbasis`$([g', h']$,`plex`$(x, y))$;

$$\begin{aligned}G = {}&17928427558923\,y + 69652831043084\,y^2 - 50521292884496\,y^3 \\ &+6589152\,y^{10} - 124688\,y^{11} - 203979830\,y^9 + 1039\,y^{12} + 17619477962188\,y^4 \\ &-3821455949733\,y^5 - 57443632170\,y^7 + 560626353463\,y^6 + 4127863857\,y^8, \\ &833974769726125719632772957356034283 20\,x - 130766609289177 \\ &+36191299508720060447403063118197599\,y^{11} \\ &-404237972597957167470875071652599 5067\,y^{10} \\ &+19587394609627373384105613215746659 9879\,y^9 \\ &-54727831961748548946516185426468879 50369\,y^8 \\ &+98113406517428783182772567336042491 854966\,y^7 \\ &-11810323585918841697531281186278360 46029776\,y^6 \\ &+96455180188792866321694705321784400 50271479\,y^5 \\ &-52292039510395436170685577583223902 062229272\,y^4 \\ &+17499194706943631318837497005013946 0701264940\,y^3 \\ &-28951858247902984329434755929497531 8913272396\,y^2 \\ &-10527356026123798669705159258688493 019930760\,y \\ &+54111194887877251263513180464902424 5435417683]\end{aligned}$$

time = 12.20, bytes = 6460538.

The Gröbner basis of (g, h) is $\{f(-3/4\,x + x^3, y^3 + 6\,y) \mid f \in G\}$.

2.5 Computing Subfields

A classical issue in Algebra is to describe the lattice of fields related to subfields $K(g)$ and $K(h)$, where g, h are two univariate rational functions. In other words, to determine the union field, the intersection field or compute the intermediate subfields F: $K(g) \subset F \subset K(x)$. We know that there exists only a finite number of them, since by Lüroth's theorem every field F is generated by one rational function $f(x) \in K(x)$, i.e. $F = K(f)$. The following diagram illustrates this:

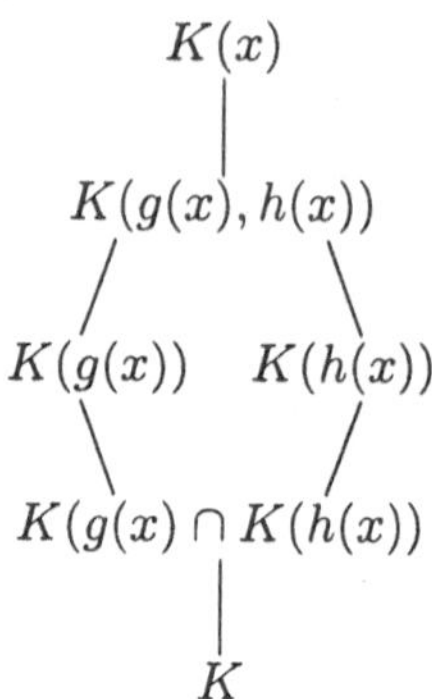

In order to construct symbolically this lattice, we use the Cadecom procedures `maxcomponent` or `netto` (to compute the union field) and `inters` (to compute the intersection field) for rational functions and `maxcompoly1`, `maxcompoly2` and `interspol` for polynomials. Again we have distinguished between rational functions and polynomials since for polynomials we have faster algorithms. Each procedure for rational functions calls the respective polynomial one when the input is a polynomial.

Suppose we want to compute the lattice of the fields $K(g)$ and $K(h)$ for $g = x^4$ and $h = \frac{x}{1-x^2}$. The computation in Cadecom:

> `maxcomponent`$([g, h], x)$;

$$x$$

time = 0.75, bytes = 67650,

> `inters`(g, h, x);

$$\frac{x^4}{1 - 2x^4 + x^8}$$

time = 5.52, bytes = 295662.

Then the subfield lattice is:

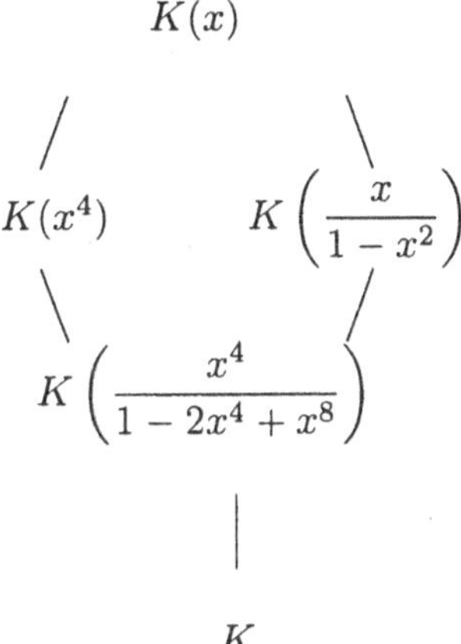

Another subfield computation problem is the calculation of all the fields F which contains $K(f)$. For this computation we use the function `intermediate`. Moreover, we can order the subfields under inclusion relation using the procedure `intermtree`. In the particular case when the field $K(f)$ contains a non-constant polynomial, we can use the procedure `intermpol` (c.f. [2]). Of course, this code is much faster than the previous one.
In this example, we want to compute all intermediate subfields ordered with respect to inclusion. Suppose

$$f = \frac{2x^4 - 2x^3 - 8x - 1}{4x^4 + 2x^3 - 16x + 1},$$

if we compute F such that $Q(f) \subset F \subset Q(x)$ there is no proper intermediate field, but if we compute F such that $Q(\alpha)(f) \subset F \subset Q(\alpha)(x)$ where $\alpha^3 = 2$ we get an intermediate field.

> `intermtree`(f, x);

$$[\,]$$

time = 0.77, bytes = 89630.

> `alias(alpha=RootOf(`$Z^3 - 2$`))` :
`intermtree`(f, x, α);

$$\left[\frac{\alpha x(\alpha x - 2)}{\alpha x + 1}\right]$$

time = 29.88, bytes = 2285522.

2.6 Simplifying Sine–Cosine Equations

By a *sine-cosine* equation we will understand a polynomial equality $f(s, c) = 0$, with f in the quotient ring $K[s, c]/(s^2 + c^2 - 1)$, and where K is a field of characteristic zero (typically, a numerical field such as the rational numbers field Q or a field of parameters $Q(x_1, \ldots, x_m)$).

Therefore, when we write $f(s,c)$ we regard this sine–cosine polynomial expression $f(s,c) = 0$ as implicitly univariate in some unknown angle θ such that:

$$s = sin(\theta), \; c = cos(\theta).$$

We are interested in simplifying or solving equations of the sort $f(s,c) = 0$; and thus, equivalently, for solving or simplifying systems

$$f(s,c) = 0$$
$$s^2 + c^2 - 1 = 0.$$

Polynomial systems, where the variables are interpreted as trigonometric functions of unknown angles, are quite ubiquitous, arising, for instance, in electrical networking, in molecular kinematics and in concrete situations, like in tilting effects on a double pendulum. Here, our applications will be taken from the field of robot kinematics. Besides refering to the many situations described in the recent book of [13], we will sketch, for the sake of being self-contained, an example of the role of sine–cosine systems in robotics:

Given a robot arm with six revolute joints, i.e. a $6R$ robot, the *inverse kinematics* problem is to find the values of the different joint angles (with respect to some standard way of measuring them) that place the tip (or hand) of the robot at some desired position and orientation.
So, the inverse kinematics problem amounts to solving a non-linear polynomial system where the unknowns are the sines and cosines :

$$\{s_i = sin(\theta_i), c_i = cos(\theta_i), i = 1, \ldots, 6\}$$

of the six joint angles:

$$\{\theta_i, i = 1, \ldots, 6\}.$$

The solution of such systems, in general, is quite involved, and depends on the particular geometry of the robot. After decades of research, a symbolic solution (though not in closed form) for the general $6R$ manipulator inverse kinematics system has been found (see [14, 17]). By a clever elimination method it turns out that in this system θ_3 can be determined as the solution of a 16–degree polynomial in the tangent of $\theta_3/2$; then θ_1 and θ_2 are found by solving a system of sine–cosine polynomials, linear in these trigonometric functions, with coefficients in θ_3. Of course, the determining 16–degree polynomial can be also expressed as a 8–degree polynomial in the sine and cosine of θ_3. However, the degree of this solution, together with the complexity of its coefficients, which may contain thousand of terms (c.f. [20]), limits the practical use of this approach.

In practice, the control of a robot requires the solution of the kinematic problem to be of low degree, so that the joint angles can be quickly found. Thus, it is of primordial interest to simplify, when possible, such a univariate sine–cosine equation. The recent paper [10] contains several methods for solving or simplifying sine–cosine equations.

Although $K[s,c]/(s^2 + c^2 - 1)$ is not a unique factorization domain, we can still look for lower degree factors of f. More precisely, factoring f over the domain $K[s,c]/(s^2+c^2-1)$ essentially means: finding sine–cosine polynomials $g(s,c), h(s,c)$, verifying $f = gh$ modulo s^2+c^2-1, plus the conditions: $deg(f) > deg(g)$ and $deg(f) > deg(h)$, in order to avoid trivial factorizations.

The following example is an irreducible sine–cosine polynomial with coefficients over the rational function field $Q(a,b)$.

$f = 70 + 21060\,b - 2\,b^5cs + 2\,ac^2 - 10530\,b^2s + 256\,c^5a^2 - 35\,bs + 8960\,ac^3 + 2947200\,c^3a^2 - 4912\,sa + 2456\,ab - 627536\,c^3a^2s - 256\,c^5ba + 12636000\,bsac - 256\,b^5c^4sa - 1200\,c^3bsa - ac^2bs - c^3b^6 + b^6c + 2695680\,bac^3 + c^2b^2s - 2\,c^2b + 42000\,sac - 2947200\,a^2c + 1200\,c^4b^5a - 1200\,b^5c^2a - 2456\,c^2ab.$

We are looking for a factorization of the sine–cosine polynomial f over the field $Q(a,b)$. Then we use the Cadecom procedure

> `scfacpol` (f, s, c);

$$\left[1,\ \frac{ac^2}{2} - \frac{c^2b}{2} - \frac{b^5cs}{2} + 5265\,b + \frac{35}{2} - 1228\,sa,\ 512\,ac^3 + 2400\,sac + 4 - 2\,bs\right]$$

time : 0.88, bytes 94331.

The natural notion of decomposability for sine–cosine polynomials $f(s,c)$ states, therefore, the existence of a standard polynomial $g(x)$ and of a sine–cosine polynomial $h(s,c)$, such that

$$f(s,c) = g(h(s,c)) \text{ modulo } s^2+c^2-1.$$

As in the case of factorization, we look for composition factors which are simpler than the given polynomial.

The following example is uni-multivariate indecomposable sine–cosine polynomial with coefficients in the rational function field $Q(a,b,m,n)$.

$$\begin{aligned}
f = {} & -21662\,c + 3938\,cs + 114874\,c^2s - 260864\,c^2 - 121438\,c^3 + 22022\,c^3bs \\
& -547515\,c^5s + 934\,c^2sb + 71415\,c^6b + 12420\,c^4sb - 6210\,c^5a^2 - 8078\,c^3b \\
& +135\,c^6a^2b^2 + 540\,c^5a^2b + 556830\,c^3s - 12420\,c^4a^2s + 142830\,c^5 + 8078\,c^3a^2 \\
& -934\,c^2a^2s - 22022\,c^3a^2s + 3105\,c^5a^4s + 135\,c^4a^4s + 5001\,n^7 - 71415\,c^4s \\
& -71415\,c^6a^2 - 270\,c^4a^2sb + 135\,c^4b^2s - 90\,c^2b - 467 - 467\,c^4a^4 + 934\,c^4a^2b \\
& +90\,ca^5b - 934\,c^3a^7b - 934\,c^3a^2m^3 + 934\,c^3b^2a^5 + 934\,c^3bm^3 - 22022\,c^2sm^3 \\
& +8078\,c^2a^5b + 90\,c^2a^2 - 71415\,c^5a^5b + 259463\,c^4 + b^2 - 6210\,c^4m^3 \\
& -6210\,c^4a^5b - 467\,c^2a^{10}b^2 - 934\,cm^3s - 6210\,c^5a^2bs - 270\,c^5a^2bm^3 \\
& +270\,c^3a^2m^3s + 270\,c^4a^7bm^3 + 270\,c^3a^7bs + 6210\,c^4a^7sb + 135\,c^5b^3a^5 \\
& -934\,c^2a^5bm^3 - 934\,ca^5bs + 135\,c^5a^9b + 135\,c^5a^4m^3 - 270\,c^5a^7b^2 - 540\,c^4a^7b \\
& -540\,c^4a^2m^3 + 135\,c^4a^2m^6 - 22022\,c^2sa^5b + 8078\,c^2m^3 - 467\,c^2m^6 \\
& -270\,c^3m^6 + 45\,c^3m^9 + 71820\,c^4a^2 - 6210\,c^4b^2sa^5 + 45\,c^3a^{15}b^3 + 540\,c^4bm^3 \\
& -270\,c^5a^4 + 45\,c^6a^6 - 71820\,c^4b - 270\,c^5b^2 + 135\,c^4a^{12}b^2 - 45\,c^6b^3 \\
& +6210\,c^3sa^5bm^3 + 71820\,c^3a^5b - 270\,c^3bm^3s - 270\,c^4b^2a^5m^3 - 270\,c^3a^{10}b^2 \\
& -12420\,c^3sm^3 + 3105\,c^3sm^6 - 6210\,c^4bsm^3 + 540\,c^4b^2a^5 - 135\,c^4b^3a^{10} \\
& -135\,c^4bm^6 + 135\,c^2m^6s + 270\,c^2a^5bm^3s + 135\,c^3a^{10}b^2m^3 + 135\,c^2a^{10}b^2s \\
& +135\,c^3a^5bm^6 + 90\,cm^3 + 6210\,c^5b - 540\,c^3a^5bm^3 - 12420\,c^3sa^5b \\
& -467\,c^4b^2 - 135\,c^6a^4b + 3105\,c^5b^2s + 135\,c^5b^2m^3 + 6210\,c^4a^2sm^3 \\
& -71415\,c^5m^3 + 71820\,c^3m^3 - 270\,c^3b^2a^5s + 3105\,c^3sa^{10}b^2.
\end{aligned}$$

We are interested in a decomposition of f modulo the unit circle over the field $Q(a, b, n, m)$. We use the Cadecom primitive `scdecpol`.

> `scdecpol` (f, s, c);

$$\Big[\big[b^2 + 5001\, n^7 + \left(45\, b - 45\, a^2\right) x + \left(-467\, b^2 + 934\, a^2 b - 467\, a^4\right) x^2 + \left(-45\, b^3 + 135\, a^2 b^2 - 135\, a^4 b + 45\, a^6\right) x^3, \frac{1}{b - a^2}(c^2 b - c^2 a^2 - 23\, cs - ca^5 b + 2\, c - cm^3 - s)\big]\Big]$$

time: 6.68, bytes: 557710.

2.7 Integrating

Assume we want to integrate an indefinite integral of the form:

$$\int f(x)h(x)dx$$

where f, h are rational functions. If the $\int h$ is a suitable leftcomponent of f, we can simplify the integral; i.e. if the there exists a rational function g such that $f = g\left(\int h\right)$. If we call $y = \int h$, then $dy = d(\int h) = h(x)dx$. Therefore,

$$\int f(x)h(x)dx = \int g(y)dy.$$

If such g exists satisfying some additional conditions, then the previous integral is reduced to the integral of a rational function, which is simpler.
The computation of the left component can be made with the function `lcomp` for rational functions and `lcompoly` for polynomials. Suppose we want to integrate

$$\int \frac{x^2}{x^6 + x^3 + 1} dx = \frac{1}{3} \int \frac{(x^3)'}{x^6 + x^3 + 1} dx.$$

$h = 3x^2$ and $\int h = x^3$ is a left component of $f = \dfrac{1}{x^6 + x^3 + 1}$:

> `lcomp` $\left(x^3, \dfrac{1}{x^6 + x^3 + 1}, x\right)$;

$$\frac{1}{1 + x + x^2}$$

time = 0.97, bytes = 104210.

Thus

$$\int \frac{x^2}{x^6 + x^3 + 1} dx = \frac{1}{3} \int \frac{1}{x^2 + x + 1} dx$$

$$= \frac{2}{9}\sqrt{3} \arctan\left(\frac{1}{3}(2x + 1)\sqrt{3}\right).$$

2.8 Other Applications

The computation of a decomposition was conjectured to be computationally hard: the security of a cryptographic protocol was based on its hardness (c.f. [5]), but it was broken by Berkovits and Lidl & Niederreiter.

We have seen that decomposition can be applied in many other topics and the main aim is simplification. In the following we highlight others: the n–partition problem (see [15]), the problem of characterizing the class of automorphisms of $K[x_1, \ldots, x_n]$ and computing their inverses (see [21, 9]).

3 Cadecom Package

Cadecom is an ordinary Maple package of about 3 megabytes, it has been developed at Departamento de Matemáticas, Estadística y Computación in Universidad de Cantabria over the last years, starting in 1992, by a research group under the direction of the first author. Several grants by Spanish Ministerio de Educación have been instrumental for reaching this stage of the system. Many people have contributed to Cadecom in different ways. The first version of this package was called FRAC(=Funciones RACionales) [3] and it was mainly implemented by Dr. Alonso.

The package is loaded via the `with` function. Each function is put into a separate file to be loaded via `readlib` into Maple session. Generally, functions are loaded at the time of their first invocation, in order to save memory. There are more than 80 auxilary functions and 42 of them are principal. The library also contains a Maple help; you can load it with `?function;` or `help(function);` commands. We also have included in the library the sypnosis of the procedures and some other extra information. The procedures in Cadecom work over the ground field, that is, the field generated by the coefficients of the input. In some of the procedures you can also work in other fields; you just need to add an argument K to work in such field. For instance, if you type

```
> ?decomp;
```

decomp - decompose a rational function

Calling sequence:
`decomp`(f, x)
`decomp`(f, x, K)

Parameters:
f – multivariate rational function
x – a variable
K – a field extension over which to decompose

Description:

- The procedure `decomp` computes a complete decomposition of f with respect to the variable x, following the algorithm in [4].
- If the input is a polynomial calls `decompoly` function.
- If a third argument is given the decomposition is made over K, otherwise computes the decomposition in the ground field.

Examples: (We omit them)

See Also:
rcomp,Brcomp,lcomp,decompoly

The examples have been tested on a personal computer Macintosh Centris 650 in the system MapleV Release 5. We wrote down low degree functions in order to illustrate what the procedures may be used for. To give an idea of about the performance of our implementation, it is important to highlight that you can decompose instantaneously polynomials of degree 50 in a SUN machine. Moreover, the authors were able to decomposing sine–cosine equations of eight-degree with hundred of digits in the coefficients, which were highly complex terms, within 20 seconds of CPU time on an SUN machine. Therefore, we think that our package can now be a useful tool for solving sine–cosine equations.

The package Cadecom is available by anonymous ftp from `ftp.hall.matesco.unican.es` or by e-mail from the authors.

Acknowledgments

This research is partially supported by the National Spanish project PB97-0346 and "Acción Integrada Alemana-Española" HA1997-0124.

References

1. S. Abhyankar, C. Bajaj: *Computations with algebraic curves.* ISSAC–89 L.N.C.S., Springer–Verlag no. 358, 1989, pp. 274 284.
2. C. Alonso: *Desarrollo análisis e implementación de algoritmos para la manipulación de variedades paramétricas.* Ph. dissertation, Dep. Math. and Computing, Universidad de Cantabria, 1994.
3. C. Alonso, J. Gutierrez, T. Recio: *Frac: A Maple package for computing in the rational function field K(X).* Proc. of Maple Summer Workshop and Symposium'94. Birkhäuser, 1994, pp.107–115.
4. C. Alonso, J. Gutierrez, T. Recio: *A rational function decomposition algorithm by near–separated polynomials.* J. Symbolic Comput. **19**, 1995, pp.527–544.
5. J.J. Cade: *A new public–key which allows signatures.* Proc. 2nd SIAM Conf. on Appl. Linear Algebra, raleigh NC, 1985.
6. D. Casperson, D.Ford, J. McKay: *Ideal decompositions and subfields.* J. Symbolic Comput, **21**, 1996, 133–137.
7. G. Farin: *Curves and surfaces for computer aided geometric design.* Academic Press, Boston, 1998.

8. J. von zur Gathen, J. Gutierrez, R. Rubio: *On multivariate polynomial decomposition.* Computer Algebra in Scientific Computing, CASC'99, 1999, pp.463–478.
9. J. Gutierrez: *A polynomial decomposition algorithm over factorial domains.* Compt. Rendues Math. Acad. **XIII–2**, 1991, pp 437–452.
10. J. Gutierrez, T. Recio: *Advances on the simplification of sine–cosine equations.* J. Symbolic Comput. **26**, 1998, pp.31–70.
11. J. Gutierrez, R. Rubio: *Reduced Gröbner Basis Under Composition.* J. Symbolic Comput, **26**, 1998, 433–444.
12. H. Hong: *Gröbner Basis Under Composition I.* J. Symbolic Comput. **25**, 1998, pp.643–663.
13. P. Kovács: *Rechnergestützte symbolische Roboterkinematik.* Vieweg Verlag. 1993.
14. H-Y. Lee, C.-G. Liang: *A new vector theory for the Analysis of Spatial Mechanisms.* Mechanisms and Machine Theory, **23-3**, 1988, pp.209-217.
15. A.K. Lenstra, H.W. Lenstra, L. Lovasz: *Factoring polynomials with rational coefficients.* Math. Ann. 261, 1982, pp.515–534.
16. F. Ollivier: *Inversibility of rational mappings and structural identifiability in automatics.* Proc. ISSAC'89, ACM, 1989, pp. 43-53.
17. M. Raghavan, B. Roth: *Kinematic Analysis of the 6R Manipulator of General Geometry.* Proc. Intl. Symposium on Robotic Research, Tokyo. 1989, pp. 314-320.
18. A. Schinzel: *Selected Topics on Polynomials.* Ann Arbor, University Michigan Press, 1982.
19. T.W. Sederberg: *Improperly parametrized rational curves.* Computed Aided Geometric Design, **3**, 1986, pp.67–75.
20. R. Selfridge: *Analysis of 6 Link Revolute Arms.* Mechanism and Machine Theory. **24-1**, 1989, pp.1-8.
21. D. Shannon, M. Sweedler: *Using Gröbner basis to determine algebra membership, split surjective algebra homomorphisms, determine birational equivalence.* J. Symbolic Comput, **6**, 1988, pp. 267-273.
22. R. Zippel: *Rational Function Decomposition* Proc. of ISSAC-91. ACM press, 1991, 1–6.

Computeralgebra and the Systematic Construction of Finite Unlabeled Structures

Adalbert Kerber

Dep. of Mathematics
University of Bayreuth
D-95447 Bayreuth, Germany
kerber@uni-bayreuth.de

Abstract. This review is concerned with mathematical structures that can be defined as equivalence classes on finite sets. The method used is to *replace the equivalence relation by a finite group action* and then to apply all what is known about such actions, i.e. to apply a mixture of quite general methods, taken from combinatorics as well as from algebra.
For this purpose group actions will be introduced, enumerative methods will be reported, but the main emphasize is put on the *constructive aspects,* the generation of orbits representatives, and several *applications* of these methods, in particular to graph theory, design theory, coding theory and to mathematical chemistry.
These methods have been successfully implemented in various computer-algebra packages like MOLGEN (for the generation of molecular graphs and applications to molecular structure elucidation) as well as in DISC-RETA (for the evaluation of combinatorial designs and linear codes as well as other finite discrete structures).
Finally we shall discuss actions on posets, semigroups, lattices, where the action is compatible with the order of the lattice or the composition of the semigroup.

1 Structures defined as equivalence classes

Let me begin with briefly describing a few interesting mathematical structures that are defined by equivalence relations:

– Here is, to begin with, a **labeled graph** with 4 vertices:

and here are the **unlabeled graphs** on 4 vertices:

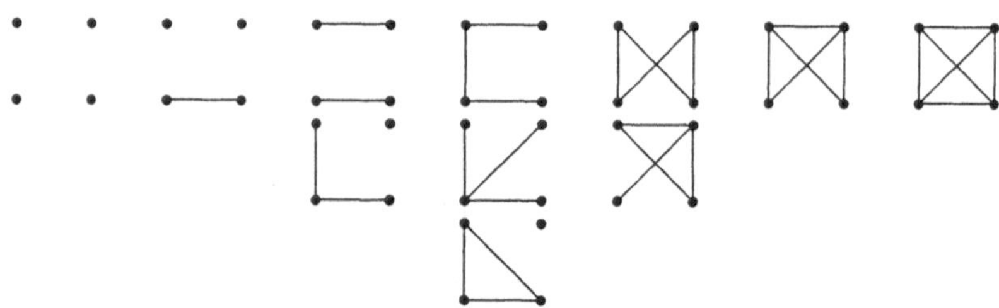

Unlabeled graphs, on v vertices, say, are important since they can be considered as *interaction models,* which describe the interactions between objects, represented by points, nodes, vertices, atoms, and where the names of the objects do not matter at all. The pairs of objects that interact are joint by edges (sometimes the strength of the interaction is also emphasized, by a weight of the edge or by a multiple edge, but, for sake of simplicity, we do not consider this refinement).

Unlabeled graphs are obviously defined as equivalence classes on *the set of labeled graphs on v vertices,* and the equivalence relation is the following one: An equivalence class consists of the labeled graphs that arise from one of them by *renumbering* the vertices. In other words, unlabeled graphs with v vertices can be defined by an *action of the symmetric group S_v.*

- A second interesting example is formed by the error correcting **linear codes,** i.e. by finite vector spaces. Here is a concrete example, a basis of a linear code, forming the rows of the following *generator matrix:*

$$\begin{pmatrix} 1\,0\,0\,0\,1\,0\,1 \\ 0\,1\,0\,0\,1\,1\,1 \\ 0\,0\,1\,0\,1\,1\,0 \\ 0\,0\,0\,1\,0\,1\,1 \end{pmatrix}$$

Now we recall that the *Hamming metric* and the *minimal distance* decide about error correcting quality, and hence the equivalence relation is *isometry* with respect to Hamming metric. Moreover, isometries are simultaneous permutations of coordinates, and simultaneous multiplications of specific coordinates by nonzero elements of the ground field $GF(q)$, and therefore we are faced in this case with an action of the wreath product $GF(q)^* \wr S_n$ (more details will follow down below, see also [3], [7] or [8]).

- Here comes a third example, the **permutational isomers of dioxine.** These molecules have the following skeleton, consisting of two benzene rings that are joint via two oxygen atoms:

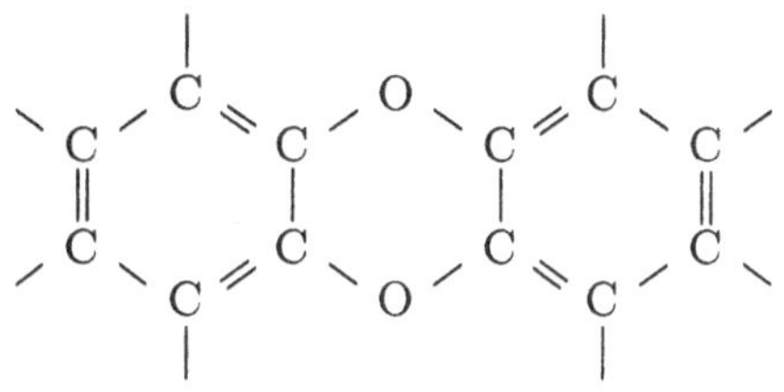

```
          |                       |
          C           O           C
   \    /   \\     /     \     //   \    /
     C         C             C          C
     ||        |             |          ||
     C         C             C          C
   /    \    //   \       /    \\     /   \
          C           O           C
          |                       |
```

Over the 8 free valences of this molekular skeleton we have to distribute 4 hydrogen atoms and 4 chlorine atoms in all the essentially different ways. Hence the equivalence relation is defined by the symmetry of skeleton, which is, as we (better say: the chemists) suppose the skeleton to be planar in space, the symmetry of the Kleinian 4-group V_4.

Here is an isomer:

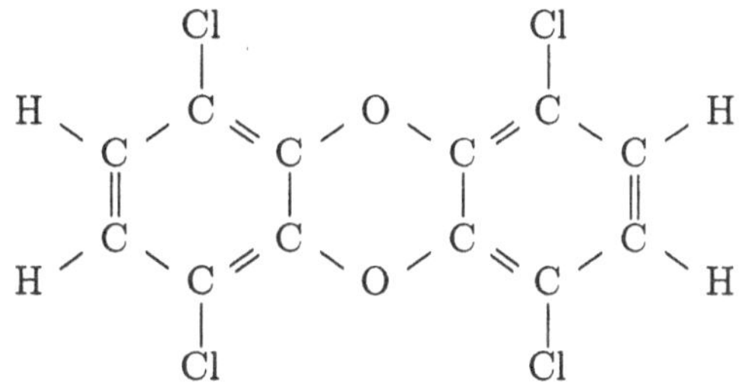

```
          Cl                      Cl
          |                       |
  H       C           O           C       H
   \    /   \\     /     \     //   \    /
     C         C             C          C
     ||        |             |          ||
     C         C             C          C
   /    \    //   \       /    \\     /   \
  H       C           O           C       H
          |                       |
          Cl                      Cl
```

and here is another one:

```
          H                       Cl
          |                       |
  Cl      C           O           C       H
   \    /   \\     /     \     //   \    /
     C         C             C          C
     ||        |             |          ||
     C         C             C          C
   /    \    //   \       /    \\     /   \
  H       C           O           C       H
          |                       |
          Cl                      Cl
```

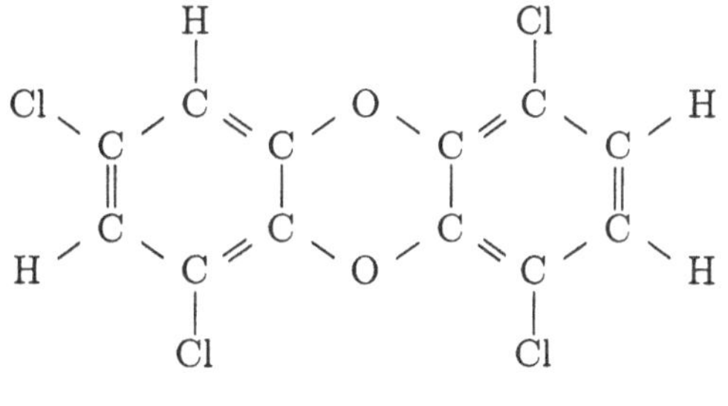

There are altogether 22 of them, and the interested reader might like to see how this number 22 can be calculated and how the corresponding molecular graphs can be obtained!

- A fourth example stems from **combinatorial chemistry.** In order to describe it we use a central molecule, e.g. triacid chloride

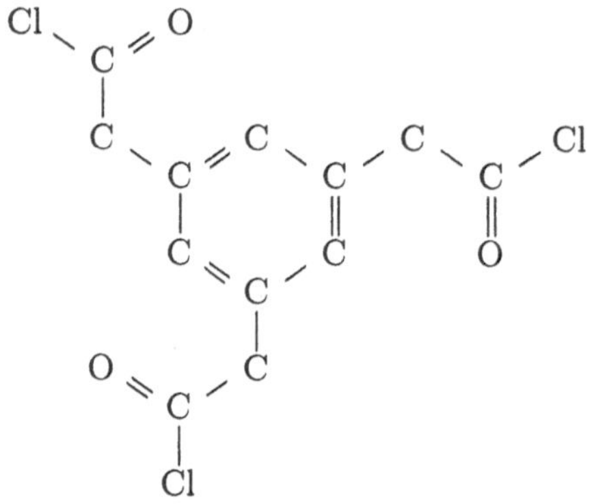

  to which we attach building blocks, e.g. amino acids via a well defined chemical reaction. The problem is to evaluate the *different* attachments, the attachments that result in pairwise different molecules. Also here the equivalence relation is induced by the *symmetry group* of the central molecule, which consists of the proper roations of space that leave the central molecule invariant. And the chemists are very interested in the corresponding **combinatorial library,** i.e. in a system of representatives of the equivalence classes.

2 Finite group actions

Having introduced a few structures that are defined as equivalence classes we are now going to define the basic notions of the theory of finite group actions which we later on will apply to the examples. Let G denote a finite group and X a finite set. An *action* of G on X (from the left) is defined to be a mapping

$$G \times X \to X \colon (g, x) \mapsto gx,$$

such that

$$g(g'x) = (gg')x, \; 1x = x.$$

We will abbreviate this by writing

$$_GX.$$

The most important basic aspects of finite group actions are the following ones:

- The action defines *substructures* on G and on X :
  - $G(x) := \{gx \mid g \in G\}$, the *orbit* of x,
  - $G\backslash\backslash X := \{G(x) \mid x \in X\}$, the set of *all* orbits,
  - $G_x := \{g \in G \mid gx = x\}$, the *stabilizer* of x,
  - $X_g := \{x \in X \mid gx = x\}$, the *fixed points* of g.

- The *fundamental facts* are:
  - Two orbits are either equal or disjoint,
  - the orbits form the *set partition* $G\backslash\backslash X$ of X,
  - the stabliziers G_x are *subgroups* of G.

□

And it is very helpful to note that *many mathematical structures, and also various structures in sciences, are orbits (or stabilizers),* so that it is immediately clear that, being orbits, they are either equal or disjoint (for example, conjugacy classes of elements in groups, conjugacy classes of subgroups, double cosets,...) and that structures that can be recognized as stabilizers, are groups.

But it is even more important that, conversely,

- each set partition $\{X_\iota \mid i \in I\}$ of a set X is an orbit set:

$$(\oplus_i S_{X_\iota})\backslash\backslash X = \{X_1, X_2, \ldots\},$$

  where S_{X_ι} is the symmetric group on X_i.

Hence each equivalence relation is an orbit set, and so its definition can be replaced by choosing an appropriate set, a suitable group and its action on that set!

- **Paradigmatic examples of orbit sets** which cover the above mentioned important applications are the *symmetry classes of mappings:* Given actions ${}_GX$, ${}_HY$ yield canonic actions on the set of mappings

$$Y^X := \{f: X \to Y\},$$

  namely:

$$\begin{aligned} G \times Y^X &\to Y^X: (g, f) \mapsto f \circ g^{-1}, \\ H \times Y^X &\to Y^X: (h, f) \mapsto h \circ f, \\ (H \times G) \times Y^X &\to Y^X: ((g, h), f) \mapsto h \circ f \circ g^{-1}, \end{aligned}$$

  The most general action is the following action of the *wreath product*

$$H \wr G := H^X \times G := \{(\psi; g) \mid \psi: X \to H, g \in G\}$$

  on the set of mappings:

$$(H \wr G) \times Y^X \to Y^X: ((\psi; g), f) \mapsto \tilde{f},$$

  where

$$\tilde{f}(x) := \psi(x) \circ f(g^{-1}x).$$

  It contains the actions of G, H, and $H \times G$, since these groups can be considered as subgroups of the wreath product. The corresponding sets of orbits are

$$G\backslash\backslash Y^X,\ H\backslash\backslash Y^X,\ H \times G\backslash\backslash Y^X,\ H \wr G\backslash\backslash Y^X.$$

Many structures can be described as elements of such sets, for suitably chosen $G, H, {}_GX, {}_HY$ and the corresponding actions of $G, H, H \times G, G \wr H$.

- **Graphs on v vertices:**

$$Y := 2 := \{0,1\},\ X := \binom{v}{2},\ G := S_v.$$

  The set of all the labeled graphs is

$$2^{\binom{v}{2}},$$

  while the set of *unlabeled* graphs on v vertices is

$$S_v \backslash\backslash 2^{\binom{v}{2}}.$$

- Similarly we can define **multigraphs on v vertices** with edge multiplicity restricted by m :

$$S_v \backslash\backslash (m+1)^{\binom{v}{2}},$$

  as $m+1 = \{0,1,\ldots,m\}$.
- A slightly more complicated and very important application is the definition of **connectivity isomers** of a given chemical formula that comprises v atoms, say. It is the set of orbits of the symmetric group S_v on the set of multigraphs the vertices of which are colored by atom names and the vertex degrees of which agree with the (prescribed) valences of the atoms. Generators of connectivity isomers play a central role in *molecular modeling* and in the *molecular structure elucidation,* since they allow to generate the *candidates for the structure elucidation* which are compatible with the given spectroscopic data. In Bayreuth we have developed several such generators, the most recent one is **MOLGEN 4.0,** a demo version can be downloaded from

  http://www.mathe2.uni-bayreuth.de/molgen4/

  and the mathematical background was published recently ([6]).

- **Isometry classes of linear (n, k)−codes** are the orbits of the group of isometries of the vector space $GF(q)^n$. In order to describe this group in detail, we recall that the isometries (with respect to the Hamming metric!) are just the linear mappings that map a unit vector onto a multiple of a unit vector, and hence they are simultaneous permutations of coordinates or simultaneous multiplications of a particular coordinate by a nonzero element of the ground field, or compositions of such permutations/multiplications. Hence the isometry group of the space $GF(q)^n$ is the wreath product

$$GF(q)^* \wr S_n.$$

  It acts on the set $\left[{n \atop k}\right]$ of subspaces of dimension k, and the corresponding orbits are the isometry classes.

We can rephrase this in terms of symmetry classes of mappings: The isometry group acts on the space $X := GF(q)^n$, and therefore also on its power set

$$Y^X := 2^{(GF(q)^n)}.$$

Some of the orbits are unions of vector spaces, and these are just the isometry classes. Hence the set of isometry classes of (n, k)-codes is a subset of the orbit set

$$GF(q)^* \wr S_n \backslash\backslash 2^{(GF(q)^n)}.$$

(The other orbits are also of interest, they are classes of (nonlinear) *block codes*.)

Now we are going to describe the basic *algebraic tools* that we can apply to finite group actions:

2.1 The Basic Lemma: *If* ${}_GX$ *is an action, then, for each element* $x \in X$ *we have:*

- *The mapping*

$$G(x) \to G/G_x : gx \mapsto gG_x$$

between the orbit of x *and the set of left cosets of its stabilizer, is a bijection that commutes with the action:* $\varphi(gx) = g\varphi(x)$.
- *In other words:* G *acts on* $G(x)$ *in more or less the same way as it acts on the set* G/G_x *of left cosets, i.e. we have equivalent actions*

$${}_G(G(x)) \approx {}_G(G/G_x).$$

- *In particular, the length of the orbit is equal to the index of the stabilizer, and in the finite case, we can express this as a quotient of orders,*

$$|G(x)| = \frac{|G|}{|G_x|}.$$

- *Moreover, if* U *is a subgroup of* G, *then*

$$U\backslash\backslash G(x) \to U\backslash G/G_x : U(gx) \mapsto UgG_x$$

is a bijection, where $U\backslash G/G_x := \{UgG_x \mid g \in G\}$, *the set of* double cosets.

□

Summary of this section: Mathematical structures as well as structures occurring in the sciences which can be defined as equivalence classes on finite sets X, say, for example unlabeled graphs, isometry classes of linear codes, permutational isomers of molecules, combinatorial libraries in chemistry etc., can be redefined as orbits of finite groups on finite sets.

- Such orbits can be considered as sets of left cosets of particular subgroups, the stabilizers of elements. Moreover,

- we can map the set of orbits of a subgroup (and therefore a set of finite structures) bijectively onto a set of double cosets. Hence we can move from an unstructured set X to a set which has a well known algebraic structure (covered by group theory).

This opens a vast field of algebraic results for applicational purposes, some of them we are going to describe now.

3 The enumeration of finite structures

We now assume that our favourite finite structure is already defined as an orbit of a finite group G on a finite set X. In order to examine that structure we first want to get an idea how many of them are there, for example, how many unlabeled graphs on v vertices are there, or how many isometry classes of linear (n, k)-codes do exist. Thus we are going to do *enumeration* first. And we shall see that also enumeration can be done in different ways:

3.1 Plain enumeration:

Here we can apply the following consequence of the Basic Lemma:

3.1 The Lemma of Cauchy-Frobenius *In the finite case, the number of orbits of G on X is the average number of fixed points:*

$$|G\backslash\backslash X| = \frac{1}{|G|} \sum_{g \in G} |X_g|,$$

□

An easy application to symmetry classes of mappings gives

Corollary 1. *The number of symmetry classes of mappings, with respect to an action of G on Y^X, induced by an action ${}_GX$ is*

$$|G\backslash\backslash Y^X| = \frac{1}{|G|} \sum_{g \in G} |Y|^{c(g)},$$

where $c(g) = |\langle g\rangle\backslash\backslash X|$, the number of cyclic factors of g on X which is the number of orbits of $\langle g\rangle$ on X. □

Applications:

- The number of graphs on 4 vertices turns out to be

$$|S_4\backslash\backslash 2^{\binom{4}{2}}| = 11,$$

in accordance with the above drawing.
- More generally the number of graphs with given number v of vertices is

$$|S_v\backslash\backslash 2^{\binom{v}{2}}|.$$

- The number of isometry classes of linear codes of length n and of dimension k can be evaluated (cf. [4],[5]) analogously.
- The number 22 of permutational isomers of dioxin is obtained in a similar way.
- The size of the combinatorial library triacid+amino acids can be calculated, too ([10]).

3.2 Enumeration by weight:

There is a refinement of plain enumeration, where the number of orbits with given properties can be evaluated under certain conditions. The main result is

3.2 Lemma of Cauchy-Frobenius, weighted form *Let again $_GX$ denote a finite action, and consider now a mapping $w\colon X \to R$, where R is a commutative ring containing $\mathbb{Q}$ as a subring, and where w is constant on the orbits. Then, if T is a transversal of the orbits, we have*

$$\sum_{t\in T} w(t) = \frac{1}{|G|}\sum_{g\in G}\sum_{x\in X_g} w(x).$$

□

In order to apply this to actions of the form $_G(Y^X)$, we take the weight function

$$w\colon Y^X \to \mathbb{Q}[Y]\colon f \mapsto \prod_x f(x),$$

the *multiplicative weight,* obtaining

$$\sum_{t\in T} w(t) = \frac{1}{|G|}\sum_{g\in G}\prod_{i=1}^{|X|}\left(\sum_{y\in Y} y^i\right)^{a_i(g)},$$

where $a_i(g)$ is the number of i-cycles of g on X, which is the number of orbits of length i of $\langle g\rangle$ on X. For example, in the case of graphs, it gives that the number of graphs with v vertices and e edges is the coefficient of y^e in

$$\frac{1}{v!}\sum_{\pi\in S_v}\prod_{i=1}^{\binom{v}{2}}(1+y^i)^{a_i(\pi)}.$$

This function is obtained from the *cycle index* polynomial

$$C(G,X) := \frac{1}{v!}\sum_{\pi\in S_v}\prod_{i=1}^{\binom{v}{2}} z_i^{a_i(\pi)}$$

by replacing z_i by the i-th power sum symmetric function $\sum_{y\in Y} y^i$. Here is an example:

$$C(S_4,\binom{4}{2}) = \frac{1}{24}(z_1^6 + 9z_1^2z_2^2 + 8z_3^2 + 6z_2z_4),$$

yields for the enumeration of graphs on 4 vertices by number of edges (replace z_1 by $1+y$, z_2 by $1+y^2$, ...):

$$1+y+2y^2+3y^3+2y^4+y^5+y^6.$$

More generally the generating function

$$\frac{1}{|G|}\sum_{g\in G}\prod_{i=1}^{|X|}(\sum_{y\in Y} y^i)^{a_i(\pi)}$$

can be expressed using representation theory of symmetric groups, in terms of Schur polynomials:

3.3 Foulkes' Lemma

$$\frac{1}{|G|}\sum_{g\in G}\prod_{i=1}^{|X|}(\sum_{y\in Y} y^i)^{a_i(\pi)} = \sum_{\alpha\vdash|X|}(IG\uparrow S_X,[\alpha])\{\alpha\}.$$

Using this result, unimodality questions can be answered. The most general result obtained so far in this context is the following theorem:

Theorem 1. (Kerber,Stanley) *If we replace the indeterminate z_i of the cycle index by $u(x^i)$, where u is a unimodal polynomial with natural coefficients, then we obtain a unimodal polynomial.*

This implies, for example, the unimodality of the Gaussian polynomials, of the numbers of graphs, multigraphs, directed graphs, and so on.

3.3 Enumeration by stabilizer class

To begin with we note that

$$\{G_{x'} \mid x'\in G(x)\} = \widetilde{G_x} = \{gG_xg^{-1} \mid g\in G\},$$

i.e. the set of stabilizers of the elements in an orbit is a full conjugacy class of subgroups! We can therefore speak of *the orbits of type* $\widetilde{U}$, which means the orbits the elements of which form the conjugacy class $\widetilde{U}$ of subgroups conjugate to U. Now we introduce the following notation for the set of orbits of stabilizer type $\widetilde{U}$, $U\leq G$:

$$G\backslash\!\backslash_{\widetilde{U}}X = \{G(x)\in G\backslash\!\backslash X \mid G_x\in\widetilde{U}\}$$

The order of this set is related to the no. of fixed points $|X_U|$ via

$$|X_U| = \sum_V \zeta(U,V)\frac{|G/V|}{|\widetilde{V}|}|G\backslash\!\backslash_{\widetilde{V}}X|,$$

where ζ denotes the zeta function of the poset of subgroups:

$$\zeta(U,V) := \begin{cases} 1, & \text{if } U\leq V,\\ 0, & \text{otherwise.}\end{cases}$$

And so, by Moebius inversion,

$$|G\backslash\backslash_{\widetilde{U}} X| = \frac{|\widetilde{U}|}{|G/U|}\sum_V \mu(U,V)|X_V|.$$

We can simplify this equation (since these numbers do only depend on the conjugacy class and not on the chosen representative of it) by introducing the poset of conjugacy classes of subgroups

$$\widetilde{L}(G) := \{\widetilde{U}_1,\ldots,\widetilde{U}_d\}, \text{ with representatives } U_i \in \widetilde{U}_i,$$

and putting

$$b_{ik} := \frac{|\widetilde{U}_i|}{|G/U_i|}\sum_{V\in\widetilde{U}_k} \mu(U_i,V),\ B(G) := (b_{ik}),$$

the *Burnside matrix* of G.

3.4 Burnside's Lemma *The vector of the lengths* $|G\backslash\backslash_{\widetilde{U}_i} X|$ *of the* strata $G\backslash\backslash_{\widetilde{U}_i} X$ *of* G *on* X *satisfies the equation*

$$\begin{pmatrix} \vdots \\ |G\backslash\backslash_{\widetilde{U}_i} X| \\ \vdots \end{pmatrix} = B(G)\cdot\begin{pmatrix} \vdots \\ |X_{U_i}| \\ \vdots \end{pmatrix}.$$

□

Application to $_G(Y^X)$: The number of symmetry classes of G on Y^X of type $\widetilde{U}_i$ is the i-th entry of the one column matrix

$$\begin{pmatrix} \vdots \\ |G\backslash\backslash_{\widetilde{U}_i} Y^X| \\ \vdots \end{pmatrix} = B(G)\cdot\begin{pmatrix} \vdots \\ |Y|^{|U_i\backslash\backslash X|} \\ \vdots \end{pmatrix}.$$

◇

The inverse $M(G) = B(G)^{-1}$ is Burnside's *table of marks* $M(G) := (m_{ik})$,

$$m_{ik} = \frac{|G/U_k|}{|\widetilde{U}_k|}\zeta(U_i,\widetilde{U}_k) = \frac{|G/U_k|}{|\widetilde{U}_i|}\zeta(\widetilde{U}_i,U_k)\ \in \mathbb{N},$$

$\zeta(\widetilde{U},V) := \sum_{W\in\widetilde{U}} \zeta(W,V)$, $\zeta(U,\widetilde{V})$ correspondingly.

Example: Let us consider, for example, the symmetric group S_4. A transversal of the conjugacy classes of subgroups is

$$U_0 = \langle 1\rangle, U_1 = \langle(1,3)\rangle, U_2 = \langle(0,2)(1,3)\rangle, U_3 = \langle(0,2,1)\rangle,$$

$$U_4 = \langle(0,2),(1,3)\rangle, U_5 = \langle(0,1,2,3)\rangle,$$

$$U_6 = \langle (0,1)(2,3), (0,3)(1,2) \rangle, U_7 = \langle (0,2,1), (0,2) \rangle,$$
$$U_8 = \langle (0,1,2,3), (1,3) \rangle, U_9 = \langle (0,2,1), (0,3,1) \rangle,$$
$$U_{10} = \langle (0,2,1,3), (0,2,3,1) \rangle = S_4.$$

The table of marks is

$$\begin{pmatrix}
24 & 12 & 12 & 8 & 6 & 6 & 6 & 4 & 3 & 2 & 1 \\
 & 2 & . & . & 2 & . & . & 2 & 1 & . & 1 \\
 & & 4 & . & 2 & 2 & 6 & . & 3 & 2 & 1 \\
 & & & 2 & . & . & . & 1 & . & 2 & 1 \\
 & & & & 2 & . & . & . & 1 & . & 1 \\
 & & & & & 2 & . & . & 1 & . & 1 \\
 & & & & & & 6 & . & 3 & 2 & 1 \\
 & & & & & & & 1 & . & . & 1 \\
 & & & & & & & & 1 & . & 1 \\
 & & & & & & & & & 2 & 1 \\
 & & & & & & & & & & 1
\end{pmatrix}.$$

The Burnside matrix of S_4 is

$$\begin{pmatrix}
1/24 & -1/4 & -1/8 & -1/6 & 1/4 & . & 1/12 & 1/2 & . & 1/6 & -1/2 \\
 & 1/2 & . & . & -1/2 & . & . & -1 & . & . & 1 \\
 & & 1/4 & . & -1/4 & -1/4 & -1/4 & . & 1/2 & . & . \\
 & & & 1/2 & . & . & . & -1/2 & . & -1/2 & 1/2 \\
 & & & & 1/2 & . & . & . & -1/2 & . & . \\
 & & & & & 1/2 & . & . & -1/2 & . & . \\
 & & & & & & 1/6 & . & -1/2 & -1/6 & 1/2 \\
 & & & & & & & 1 & . & . & -1 \\
 & & & & & & & & 1 & . & -1 \\
 & & & & & & & & & 1/2 & -1/2 \\
 & & & & & & & & & & 1
\end{pmatrix}$$

The numbers of k–graphs on 4 vertices by stabilizer type are

type\\k	0	1	2	3	4
$\widetilde{U}_0$	0	0	11	100	465
$\widetilde{U}_1$	0	2	21	84	230
$\widetilde{U}_2$	0	1	9	36	100
$\widetilde{U}_3$	0	0	0	0	0
$\widetilde{U}_4$	0	2	9	24	50
$\widetilde{U}_5$	0	0	0	0	0
$\widetilde{U}_6$	0	0	1	4	10
$\widetilde{U}_7$	0	2	6	12	20
$\widetilde{U}_8$	0	2	6	12	20
$\widetilde{U}_9$	0	0	0	0	0
$\widetilde{U}_{10}$	1	2	3	4	5
$\sum$	1	11	66	276	900

The rows of this matrix which consist of zeros only motivate a **pattern recognition:** To begin with, we note that

$$U_5 \simeq C_4, U_9 \simeq A_4.$$

This raises, for example, the following question:

Which subgroups of G occur as stabilizers of mappings $f \in Y^X$?

In order to solve this problem we recall that the stabilizers of the $f \in Y^X$ are the $U \leq G$ with

$$\bar{U} = \left(\oplus_{\omega \in U \backslash\backslash X} S_\omega\right) \cap \bar{G}.$$

Corollary 2. *The groups A_v, C_v do not occur as stabilizers, if $v \geq 3$.*

3.4 Enumeration by weight and stabilizer class

There is, of course, also a weighted form of Burnside's Lemma:

3.5 Burnside's Lemma, weighted form *If w is constant on the orbits of X, T a transversal of the orbits, then*

$$\begin{pmatrix} \vdots \\ \sum_{t: G_t \in \tilde{U}_i} w(t) \\ \vdots \end{pmatrix} = B(G) \cdot \begin{pmatrix} \vdots \\ \sum_{x: U_i \leq G_x} w(x) \\ \vdots \end{pmatrix}.$$

□

An application to symmetry classes of mappings is:

Corollary 3. *For the enumeration of G-classes on Y^X by and by weight $w: f \mapsto \prod f(x) \in \mathbb{Q}[Y]$, we obtain*

$$B(G) \cdot \begin{pmatrix} \vdots \\ \prod_{\nu \in |U_i \backslash\backslash X|} \sum_y y^{l_\nu(U_i)} \\ \vdots \end{pmatrix},$$

$l_\nu(U_i)$ being the length of the ν-th orbit of U_i on X.

Example: For the graphs on $v = 4$ vertices we obtain

$$B(S_4) \cdot \begin{pmatrix} (y_0+y_1)^6 \\ (y_0+y_1)^2(y_0^2+y_1^2)^2 \\ (y_0+y_1)^2(y_0^2+y_1^2)^2 \\ (y_0^3+y_1^3)^2 \\ (y_0+y_1)^2(y_0^4+y_1^4) \\ (y_0^4+y_1^4)(y_0^2+y_1^2) \\ (y_0^2+y_1^2)^3 \\ (y_0^3+y_1^3)^2 \\ (y_0^4+y_1^4)(y_0^2+y_1^2) \\ y_0^6+y_1^6 \\ y_0^6+y_1^6 \end{pmatrix} = \begin{pmatrix} 0 \\ y_0^4y_1^2+y_0^2y_1^4 \\ y_0^3y_1^3 \\ 0 \\ y_0^5y_1+y_0y_1^5 \\ 0 \\ 0 \\ 2y_0^3y_1^3 \\ y_0^4y_1^2+y_0^2y_1^4 \\ 0 \\ y_0^6+y_1^6 \end{pmatrix}.$$

Recall what we have now, for graphs on 4 vertices:

- Their total number is 11.
- The generating function by number of edges is

$$1 + y + 2y^2 + 3y^2 + 2y^4 + y^5 + y^6.$$

- The numbers of such graphs by stabilizer class, and by stabilizer class and number of edges were just given.

Summary of this section: The Lemma of Cauchy-Frobenius and its refinements (the Lemma of Burnside, and the weighted forms of both)

- allow to enumerate the total number of equivalence classes as well as
- the numbers of equivalence classes (= finite structures) that have additional properties like a prescribed weight or a prescribed symmetry group (= stabilizer class)!

4 Construction of finite structures

Again we assume that our favourite finite structure is an unlabeled structure, for example, the unlabeled graphs with v vertices, and that it is already defined as an orbit $\omega \in G\backslash\backslash Y^X$, say.

We should like now to put our hands on these unlabeled structures, say by getting them drawn on a piece of paper, after having them generated by the computer.

There are sometimes too many such labeled graphs, of course, and so we better cut our problem into smaller pieces as follows.

4.1 Reduce complexity by restricting attention to subset

In the graph case, as in very many others, we can restrict attention to the unlabeled structures which have a given parameter value, for example a prescribed number v of vertices *and* a prescribed number e of edges. Such a labeled graph is a

$$\gamma \in \{0,1\}^{\binom{v}{2}}$$

where

$$|\gamma^{-1}(1)| = e.$$

We should like to apply the Basic Lemma, and so we note the following facts:

1. $S_{\binom{v}{2}}$ is transitive on the set of graphs with v vertices and e edges.
2. The stabilizer of γ is

$$S_{\gamma^{-1}(0)} \oplus S_{\gamma^{-1}(1)} = S_{(\binom{v}{2}-e,e)} \simeq S_{\binom{v}{2}-e} \oplus S_e.$$

3. Hence, the action of S_v on this set is the same as on the set of left cosets

$$S_{\binom{v}{2}}/S_{(\binom{v}{2}-e,e)}.$$

4. Therefore the set of orbits of S_v on this set of graphs is bijective to the set of *double cosets*

$$S_v \backslash S_{\binom{v}{2}} / S_{(\binom{v}{2}-e,e)}.$$

5. Thus it suffices to evaluate a transversal of this set of double cosets and to retranslate its elements into graphs. The result will be the complete set of unlabeled graphs with v vertices and e edges!

This motivates the following quite approach to a construction of finite structures:

A general strategy for the evaluation of a transversal of $U \backslash\backslash X$:

- Restrict attention to a subset of $U \backslash\backslash X$, say, to the orbits of given weight.
- Choose a (usually bigger) group G which acts *transitively* on this set of orbits.
- Pick an element x of this subset of orbits, evaluate its stabilizer G_x.
- Evaluate a transversal of the set of double cosets $U \backslash G / G_x$ and use the inverse of the bijection

$$U \backslash\backslash G(x) \longrightarrow U \backslash G / G_x$$

in order to get a complete system of unlabeled structures on X via retranslation.

Example, the evaluation of the 22 isomers of dioxine. They can be obtained from any transversal of

$$V_4 \backslash\backslash_{(4,4)} \{H, Cl\}^8,$$

which is bijective to the set of double cosets

$$V_4 \backslash S_8 / S_4 \oplus S_4.$$

Representatives can be evaluated using the following *subgroup folder* in S_8 :

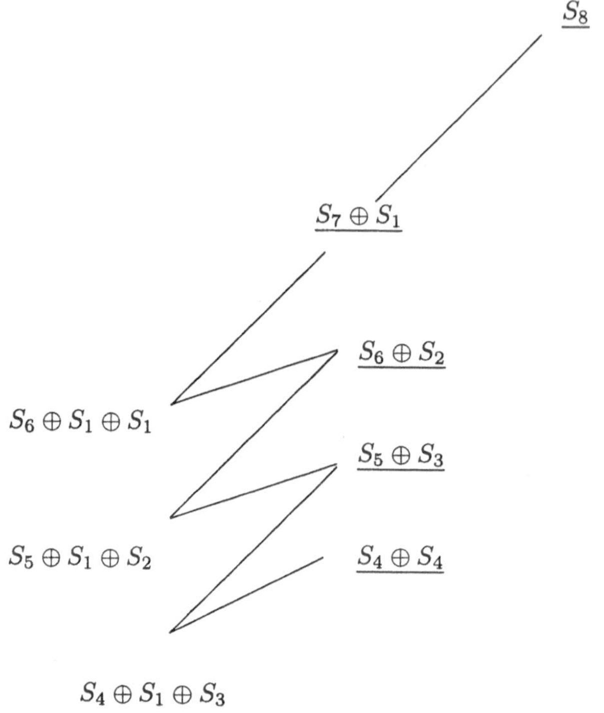

This means that we start from the set of left cosets S_8/S_8 (which is, of course, the trivial one element set $\{S_8\}$) and evaluate a transversal of the set of double cosets $V_4\backslash S_8/S_8 = \{S_8\}$, and so we can take any element of S_8 as a representative of this single class).

In the second step we pass to the subgroup $S_7 \oplus S_1$, i.e. to the set of left cosets $S_8/S_7 \oplus S_1$, in order to evaluate a transversal of the orbits of the Kleinian 4-group V_4 on this set, or, in other words, a transversal of the set of double cosets $V_4\backslash S_8/S_7 \oplus S_1$.

Steps downwards (from smaller to bigger sets!) can be managed using the method that we are going to describe next:

Assume *epimorphic* actions ${}_GX$ und ${}_GY$, i. e. a surjective $\theta\colon X \to Y$ with

$$\theta(gx) = g\theta(x),$$

for each $x \in X$, $g \in G$. Moreover, we assume a transversal T_G of $G\backslash\backslash Y$ at hand. Then the following is true:

- Each orbit $\omega \in G\backslash\backslash X$ is represented in exactly one of the inverse images $\theta^{-1}(y)$ of the $y \in T_G$,

$$\forall\, \omega \in G\backslash\backslash X\ \exists_1\ y \in T_G\colon\ \omega \cap \theta^{-1}(y) \neq \emptyset.$$

- The action of G on the inverse image of y can be replaced by the action of the stabilizer of y,

$$G\backslash\backslash\theta^{-1}(y) = G_y\backslash\backslash\theta^{-1}(y).$$

Hence we can obtain a transversal $\mathcal{T}_G$ of $G\backslash\backslash X$ by simply taking the (disjoint) union of the transversals of these inverse images:

$$\mathcal{T}_G := \bigcup_{y\in T_G} T(y), \text{ where } T(y) \in \mathcal{T}(G_y\backslash\backslash\theta^{-1}(y)).$$

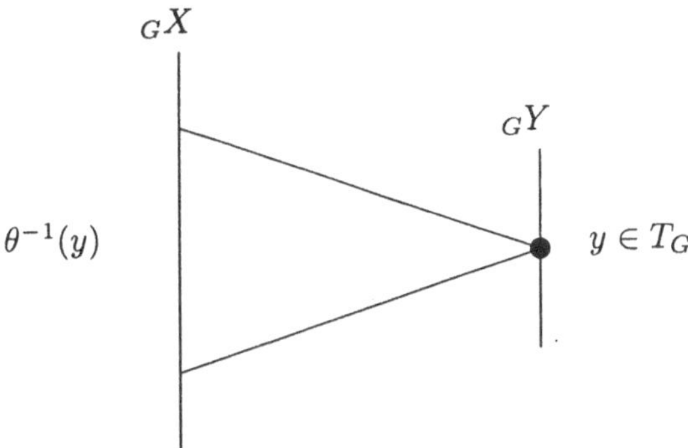

We use this principle (and double cosets) for the construction of

- permutational isomers of molecules,
- unlabeled graphs by number of vertices and edges,
- isometry classes of linear codes,
- designs with prescribed stabilizer.

This method of going from a smaller set to a bigger one is called the **Homomorphism Principle.** It can be used in many cases.

Another way of reducing the complexity is by restricting attention to the set of structures with prescribed symmetry group:

4.2 Reducing complexity by prescribing an automorphism group

This method was successfully used for a considerable push of the constructive theory of t-designs (see [1],[2]), and so we are going to describe it by this particular example.

A $t-(v,k,\lambda)$-*design* is a pair $(V,\mathcal{B})$, where V is a set of vertices, $\mathcal{B}$ a set of blocks, such that

$$|V| = v, \ \mathcal{B} \subseteq \binom{V}{k},$$

and

$$\forall\, T \in \binom{V}{t}\colon \ |\{B \in \mathcal{B} \mid T \subseteq B\}| = \lambda.$$

In order to describe the designs on V we consider the matrix

$$M_{t,k}^{v} = (m_{TB}^{V}),\ T \in \binom{V}{t}, B \in \binom{V}{k},$$

where

$$m_{TB}^{V} := \begin{cases} 1 & \text{if } T \subseteq B \\ 0 & \text{otherwise.} \end{cases}$$

A $t-(v,k,\lambda)$-design $(V,\mathcal{B})$ is a selection of columns (=blocks) or, equivalently, a 0-1-vector x which solves the system of linear equations

$$M_{t,k}^{v} \cdot x = \begin{pmatrix} \lambda \\ \vdots \\ \lambda \end{pmatrix}$$

$M_{t,k}^{v}$ is $\binom{v}{t} \times \binom{v}{k}$-matrix. People tried for many years to find 7-designs. Suspicion: there should be 7-designs with $v = 33$ and $k = 8$. The matrix has about $6 \cdot 10^{13}$ entries, λ is unknown!

The trick: *prescribe a subgroup* G of the symmetric group S_V on the set of vertices *to be contained in the automorphism group* of the designs $D = (V,\mathcal{B})$ in question: $G \leq \mathrm{Aut}(D)$. (This means that the elements of G permute the blocks of the designs!) $M_{t,k}^{v}$ then can be replaced by

$$M_{t,k}^{G} := (m_{T,K}^{G}), T \in \mathcal{T}(G\backslash\!\backslash \binom{V}{t}), K \in \mathcal{T}(G\backslash\!\backslash \binom{V}{k}),$$

where

$$m_{T,K}^{G} := |\{K' \in G(K) \mid T \subseteq K'\}|.$$

We note that the labels T of the rows and the labels k of the columns are representatives of orbits of G on the set $\binom{V}{t}$ of t-subsets and on the set $\binom{V}{k}$ of k-subsets, respectively, and therefore, by the Basic Lemma, these T and K can be obtained from transversals of the following sets of double cosets:

$$G\backslash S_V/S_{v-t} \oplus S_t,\ G\backslash S_V/S_{v-k} \oplus S_k,$$

respectively. Hence *the subgroup folder method (also called* ladder game*) applies again!*

Theorem of Kramer and Mesner *The $t-(v,k,\lambda)$-designs $(V,\mathcal{B})$ with $G \leq Aut(V,\mathcal{B})$ come from the set of all the 0-1-solutions x of*

$$M_{t,k}^{G} \cdot x = \begin{pmatrix} \lambda \\ \vdots \\ \lambda \end{pmatrix}.$$

Proof:

A 0-1-solution of the system means just picking a subset of the set of orbits of G on $\binom{V}{k}$, i. e. to pick orbits of blocks.

□

Example $t := \lambda := 1$, $v := 4$, $k := 2$, $V := \{0,1,2,3\}$,

$$G := \langle (0123), (13) \rangle,$$

$V := \{0,1,2,3\}$,

$$\binom{V}{2} = \{\{0,1\},\{0,2\},\{0,3\},\{1,2\},\{1,3\},\{2,3\}\},$$

$$\binom{V}{t} = \{\{0\},\{1\},\{2\},\{3\}\},$$

This yields the matrix

$$M^v_{t,k} = M^4_{1,2} = \begin{pmatrix} 1\,1\,1\,0\,0\,0 \\ 1\,0\,0\,1\,1\,0 \\ 0\,1\,0\,1\,0\,1 \\ 0\,0\,1\,0\,1\,1 \end{pmatrix}.$$

Clearly

$$G\backslash\!\backslash\binom{V}{k} = \{\{\{0,1\},\{0,3\},\{1,2\},\{2,3\}\},\{\{0,2\},\{1,3\}\}\}, G\backslash\!\backslash\binom{V}{t} = \{\{0,1,2,3\}\}.$$

These orbits parametrize the rows and columns of the incidence matrix, which, according to its definition, has the following entries:

$$M^G_{t,k} = (2\,1)\,.$$

This is a much smaller matrix, its number of entries is by the *factor* 12 smaller than that of the original matrix $M^4_{1,2}$. Moreover, we see immediately that there is obviously exactly one solution of the system of equations, namely

$$x = \begin{pmatrix} 0 \\ 1 \end{pmatrix}.$$

The historically first 7-design was a $7-(33,8,10)$-design with prescribed group G of automorphisms, generated by

$$\begin{aligned} \alpha = {} & (1\,2\,4\,8\,16)(3\,6\,12\,24\,17)(5\,10\,20\,9\,18)(7\,14\,28\,25\,19) \\ & (11\,22\,13\,26\,21)(15\,30\,29\,27\,23)(31)(32)(33) \\ \beta = {} & (1\,18\,30)(2\,21\,12)(3\,10\,28)(4\,31\,32)(5\,24\,14)(6\,7\,17)(8\,25\,27) \\ & (9\,19\,20)(11\,15\,13)(16\,23\,29)(22\,33\,26). \end{aligned}$$

The Kramer–Mesner matrix, which was already known to Magliveras and Leavitt, and which was recalculated by A. Betten, using double coset methods, is

```
2222222222222000000000000000000000000000000000000000000000000000000000000000000000000000000000000
2101110000000211112111111111110000000000000000000000000000000000000000000000000000000000000000000
1110000000100010020000000000010001111121121111111000000000000000000000000000000000000000000000000
0011000000000002100100000000000000100011000000010001213111111111000000000000000000000000000000000
0001200000000000000000100000100111000000000100100100110001000011211111111000000000000000000000000
0000110000000001000000010002000001001000001100111000000101001000000110021111110000000000000000000
0000013000110100100000101001001000100020001000100101010010000000100010001001000100000000000000000
0000004200000000020000200220000000002000000200000000000002000002000000000000020002000000000000000
0000000220000000000200000200000200000200002000000002000000000002000020000220200000000000000000000
0001000121000000110010000001010000000001000100010100100001000000010011100000101001011100000000000
0000002001200010011011010000110000000110100011001000001000000000010000000001000010100110000000000
0000000200020200000000220002000020000000220000000200000200000000000000000000020000000002000000000
0000000001012000000000010100002200000000000000000111000000110010110000100100100010200101000000000
0200000000000221002000010100000111100000000001000000000000001000110010100000000100001011100000000
0010100000000100100120010000000100010000020000000001001011011000100000000000000101000000001021100
0000110011000010100000102010000000000000000101000000101000000010100100000000001000000001021100111
0011020000000010010000000010010000000100010000000000001010000010000010001100001000101021200000100
0000000220000000000024000200000000000000000000000000000000000000020022000000000220200000000000200
0000010010011000001011001001100111000010000000010000010100010000000001000000010110001000000000100
0200000200000200000200000200020002000000000202000000000000002000000000000000200020000000000000020
0001000000101001100000000000000100001100010001000010000100000100001100000001000000000002001003101
0010100000011000102001001000000000100011000010110010010011000000000000200000100000000000000100000
0200000000000000100000010000000000010220100000011000000000010010010011000000010001101002100001000
0011000000000000010000100000001010001101000000010001000100000001010000000201000000101000000010100
0000000200000000000020020000000000000000200000000200000000020000000002002022000000000000002000200
0010000000000000000110000101000000000000100000010010100101000020010000001101110011000000000111000
0000000000100000010000001101000000000000100010200000000200110000200000010010000011100010011000100
0000000000000000000001001010001011000010000001000010101010110000000100010100020100110000000000003
0000000010000000100000000000010100011010001001000011100111001000100000001001000110001000000000011
0000000005000000000000000000000000050000000000000500050000000000000000000000000000000000050000001
0000000000000000000000000000000000000000000050000000000500000000050000000050005000000000000050001
0000000000000000000000000000000000000000000000000500000000000000500000000000000000000500500000001
```

Compared with the size $\binom{33}{7} \cdot \binom{33}{8}$ of the matrix $M^{33}_{7,8}$ we have obtained a reduction of the number of entries by the *factor* $2.3 \cdot 10^{10}$, and this is what brings this problem into the reach of computers! The first 0-1-solutions of the system of linear equations with this matrix where found for $\lambda = 10$ and $\lambda = 16$:

```
0011100010100100110001101000001001000000001011011000101111000000010010010100001101110110010101110
1100011101011011001110010111101101111111101001001110100001111111011011010111100100010011010100011
```

They were obtained using the implementation of an improved version (by A. Wassermann) of the LLL-algorithm for the evaluation of short vectors.

The interested reader can obtain the computer algebra package **DISCRETA** where these methods are implemented, and which is devoted to the constructive theory of discrete structures and its applications, from

http://www.mathe2.uni-bayreuth.de/ discreta/

5 Generating orbit representatives

Sometimes we cannot or we do not want to generate of to store the complete catalog of unlabeled structures that we examine, but we nevertheles do want toThey must be distributed over the orbits uniformly at random!

5.1 The Dixon/Wilf Algorithm *If ${}_GX$ denotes a finite action, then we can generate orbit representatives uniformly at random in the following way:*

- *Choose a conjugacy class C of G with probability*

$$p(C) := \frac{|C||X_g|}{|G||G \backslash\backslash X|}, \text{ where } g \in C.$$

- *Pick any $g \in C$ and generate a fixed point x of g, uniformly at random.*

Then

$$p(x \in \omega) = 1/|G \backslash\backslash X|,$$

i.e. x is uniformly distributed over the orbits of G.

□

Corollary 4. *We generate $f \in Y^X$, distributed over the G–classes uniformly at random:*

- *Choose a conjugacy class C of G with the probability*

$$p(C) := \frac{|C||Y|^{c(\bar{g}')}}{\sum_g |Y|^{c(\bar{g})}}, g' \in C.$$

- *Pick any $g \in C$, evaluate its cycle decomposition and construct an $f \in Y^X$ that takes values $y \in Y$ on these cycles which are distributed uniformly at random over Y.*

This can be used, for example, in order to test hypotheses!

Example Generate *unlabeled* graphs on 4 vertices uniformly at random!

1. The conjugacy classes are parametrized by the proper partitions $(4), (3,1), (2^2), (2,1^2), (1^4)$ of 4.
2. Their orders are 6,8,3,6 and 1.
3. The numbers of cyclic factors of the permutations induced on $\binom{4}{2}$ are 2,2,4,4,6, so the numbers of fixed points on $2^{\binom{4}{2}}$ amount to 4,4,16,16,64.
4. This yields for the probabilities $p(C)$ the values

$$3/33, 4/33, 6/33, 12/33, 8/33.$$

5. These numbers may be multiplied by the common denominator 33 in order to get the natural numbers 3,4,6,12,8, which we accumulate obtaining

$$3, 7, 13, 25, 33.$$

6. A generator that yields natural numbers between 1 and 33 uniformly at random is now used in order to choose C. Assume,say, that it generates 12,

then the conjugacy class C_2 is picked, an element of which is the permutation (01)(23). It induces on the set of pairs of vertices

$$\{a = \{0,1\}, b = \{0,2\}, c = \{0,3\},$$

$$d = \{1,2\}, e = \{1,3\}, f = \{2,3\}\}$$

the permutation

$$(a)(be)(cd)(f).$$

A random generator of zeros and ones is now used in order to associate values 0 or 1 with the cyclic factors of $(a)(be)(cd)(f)$. If it generates the sequence 1,0,0,1,say, we obtain the labeled graph that has edges joining the elements of the pairs a and f,

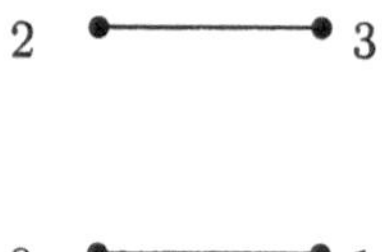

Its orbit can be identified (by just erasing the labels of the vertices) with the unlabeled graph

This is the unlabeled graph that we generated! The Dixon/Wilf Algorithm allows in general to provide large sets of examples without prejudice! Therefore it can be used for *pattern recognition in the search for hypotheses on unlabeled structures.*

An application we did was to test graph invariants, for example generalized matrix functions of the adjacency matrix, in order to show that these functions do not separate the unlabeled graphs in general.

References

1. A. Betten, A. Kerber, A. Kohnert, R. Laue, A. Wassermann: Es gibt 7-Designs mit kleinen Parametern! *Bayreuther Math. Schr.* **49**, p. 213, 1995.
2. Anton Betten, Adalbert Kerber, Reinhard Laue, and Alfred Wassermann: Simple 8-Designs with Small Parameters. *Designs, Codes, Cryptography,* 15:5–27, 1998.
3. A. Betten, H. Fripertinger, A. Kerber, A. Wassermann, K.-H. Zimmermann: Codierungstheorie, Konstruktion und Anwendung linearer Codesl *Springer-Verlag 1999.*
4. H. Fripertinger: Enumeration of isometry classes of linear (n, k)-codes over $GF(q)$ in SYMMETRICA. *Bayreuther Mathematische Schriften,* **49**, 215-223, 1995.

5. H. FRIPERTINGER, A. KERBER: Isometry Classes of Indecomposable Linear Codes *Proc. of AAECC 11, Springer LN in Computer Science* **948**, 194-204, 1995.
6. TH. GRÜNER: Strategien zur Konstruktion diskreter Strukturen und ihre Anwendung auf molekulare Graphen. *MATCH* **39**, 39-126, 1999.
7. A. KERBER: *Algebraic Combinatorics Via Finite Group Actions.* BI-Wissenschaftsverlag, Mannheim, Wien, Zürich, 1991.
8. A. KERBER: *Applied Finite Group Actions.* Springer-Verlag 1999.
9. A. KERBER, R. LAUE; Group actions, double cosets, and homomorphisms: unifying concepts for the constructive theory of discrete structures. *Acta Applicandae Mathematica,* **52**, No. 1-3, 115-126 (1998).
10. A. KERBER, R. LAUE, TH. WIELAND: Discrete Mathematics for Combinatorial Chemistry. *DIMACS Series in Discrete Mathematics and Theoretical Computer Science,* **51**, *225-234, 2000.*
11. R. LAUE: Construction of combinatorial objects – a tutorial. *Bayreuther Mathematische Schriften,* **43**, pp. 53–96, 1993.

Heat Invariant E_2 for Nonminimal Operator on Manifolds with Torsion

Vladimir V. Kornyak

Laboratory of Computing Techniques and Automation
Joint Institute for Nuclear Research
141980 Dubna, Russia

Abstract. Computer algebra methods are applied to the investigation of spectral asymptotics of elliptic differential operators on curved manifolds with torsion and in the presence of a gauge field. In this paper we present complete expressions for the second coefficient (E_2) in the heat kernel expansion for *nonminimal* operators on manifolds with nonzero torsion. The expressions were computed for the general case of manifolds of arbitrary dimension n and also for the most important for E_2 case $n = 2$. The calculations have been carried out on PC with the help of a program written in C.

1 Introduction

Determination of the internal structure of an object via the spectra of different radiations and waves around the object is one of the archetypal problems in physics. More restricted mathematical version of this problem may be formulated as follows. A manifold (bundle) equipped with such structures as metric, curvature, torsion, gauge fields etc. and an elliptic (pseudo)differential operator acting on this manifold are given. What information about the manifold can one obtain by studying the spectral properties of the operator? M. Kac phrases the problem in the evocative title of his paper "Can one hear the shape of a drum?" [1]. The answer to this radical question is negative. In 1964, J. Milnor [2] found a pair of isospectral (with respect to the Laplace operator) but non-isometric tori in dimension sixteen, and recently, in 1999, D. Schüth [3] constructed continuous isospectral families of metrics on the product of spheres $S^4 \times S^3 \times S^3$.[1] Nevertheless, many global geometric invariants, such as dimension, volume, and total scalar curvature, are known to be spectrally determined. Moreover, various manifolds such as round spheres of dimension less than or equal to six and 2-dimensional flat tori are uniquely determined by the spectra of the Laplacian acting on them.

One of the most constructive approaches to study the spectral properties of operators on manifolds is investigation of the *heat kernel expansion*. This

[1] After the Milnor's result many examples of *multiply connected* isospectral manifolds have been constructed, but the Schüth's construction is the first example of closed *simply connected* isospectral but non-isometric Riemannian manifolds.

approach can be described briefly as follows. Starting with an elliptic operator A of the order $2r$, acting on a bundle whose base is a compact closed n-dimensional manifold M, and introducing an additional "time" variable t one can construct the *heat* operator $A - \frac{\partial}{\partial t}$. Then one can compute the short-time asymptotic expansion of the diagonal elements of the kernel of this heat operator:

$$\langle x|e^{-tA}|x\rangle \sim \sum_{m\geq 0} \mathrm{E_m(x|A)t^{\frac{m-n}{2r}}}, \qquad \mathrm{t} \to +0\ . \tag{1}$$

The coefficients $E_m(x|A)$ in this expansion are spectral invariants of the operator A, and encode information about the asymptotic properties of the spectrum. These coefficients are called the *heat invariants* or *heat kernel coefficients.* They are also widely known under the names Hadamard coefficients [4][2], Hadamard-Minakshisundaram-DeWitt-Seeley, DeWitt-Seeley-Gilkey (HMDS or HAMIDEW, DWSG) coefficients, according to papers of these authors [5–8]. The heat invariants E_m are of fundamental importance in quantum field theory, quantum gravity, spectral geometry and topology of manifolds. Many quantities of interest (such as the effective action, Green function, anomalies in quantum field theory [6–10], the indices of elliptic operators and the invariants of manifolds in spectral geometry [11, 5, 1, 12]) are expressed in terms of the heat invariants.

Most papers devoted to computation of the heat invariants deal with so-called *minimal* operators whose leading term is a power of the Laplacian and symbol is a scalar (w.r.t space-time indices). A typical example of such an operator is

$$A = -\Box + X.$$

Here $\Box = g^{\mu\nu}D_\mu D_\nu$, D_μ is a covariant derivative including generally different connections (affine and spinor connections, gauge fields), X is a matrix in internal space, i.e., an operator acting in sections of the bundle. For minimal operators there are rather efficient methods for computing the heat invariants.

In this paper we consider *nonminimal* operators of the form

$$A^{\mu\nu} = -g^{\mu\nu}\Box + aD^\mu D^\nu + X^{\mu\nu}, \tag{2}$$

where $X^{\mu\nu}$ is a tensor field (bundle indices are assumed implicitly), a is a scalar parameter which should satisfy the condition $a < 1$ for positive definiteness and, hence, for the ellipticity of operator (2). In recent years special cases of operator (2) have been encountered by physicists studying the quantization of gauge and gravitational fields in arbitrary gauges [10, 13]. For example, the quantization of Yang-Mills field in an arbitrary covariant background gauge leads to the operator

$$A^{ab}_{\mu\nu} = -\delta_{\mu\nu}\Box^{ab} - \left(\frac{1}{\alpha} - 1\right) D^{ac}_\mu D^{cb}_\nu - 2f^{acb}G^c_{\mu\nu},$$

where D_μ is a covariant derivative containing the external field potential A_μ, $G_{\mu\nu}$ is a corresponding field strength, f^{abc} are the structure constants of a corresponding Lie algebra and α is a scalar (gauge) parameter. Another example:

[2] It was Hadamard who introduced these coefficients for scalar operator A already in 1923 and established their essential properties.

the quantization of electro-magnetic field in an external gravitational field leads to the operator [14, 15]

$$A_{\mu\nu} = -g_{\mu\nu}\Box - \left(\frac{1}{\alpha} - 1\right) D_\mu D_\nu + R_{\mu\nu}$$

(for an analogous operator in quantum gravity see [14]).

The torsion is defined as antisymmetric part of affine (or linear) connection

$$T^\lambda{}_{\mu\nu} = \Gamma^\lambda{}_{\nu\mu} - \Gamma^\lambda{}_{\mu\nu},$$

where $\Gamma^\lambda{}_{\mu\nu}$ are connection coefficients. The Einstein's General Relativity (see e.g. [16]) is based on a special connection called *Levi-Civita* connection, i.e., *symmetric* and compatible with metric affine connection. This connection can be expressed completely in terms of metric and is torsionless. The General Relativity well describes the interaction of the matter with the gravity as far as macroscopic bulk matter is considered. However, on the microscopic level, where the elementary particles possess such quantum property as spin, it seems necessary to take into account the influence of spin on the geometry of space-time. To describe the interaction of spinning particles with the gravitation, a gravitation theory should include the non-vanishing torsion. In 1922 Elie Cartan first pointed out [17, 18] that there is no *a priori* reason to assume an affine connection to be symmetric in the context of General Relativity. He proposed also a theory of gravitation with torsion which development is known now as the Einstein-Cartan theory. The torsion arises naturally in the different (based on Poincaré and affine groups) gauge theories of gravity developed in the recent years (see Refs. [19, 20]). Moreover, all kinds of modern superstring theories [21] (for the recent review see e. g. [22]), which allow to deduce the properties of space-time, also predict, along with the metric, the existence of torsion.

In [23] we computed the heat invariants for operator (2) up to E_4 but for manifolds without torsion. In this paper we consider more general (and computationally much more difficult) case of manifolds with torsion.

2 Algorithm and Implementation

The algorithm we use was developed by V. Gusynin [24, 25]. This algorithm is based on the covariant generalization of the *pseudodifferential calculus* given by Widom [26]. The main advantage of this algorithm is its universality. It can be applied to the wide class of pseudodifferential operators, in particular, to the nonminimal and higher-order operators intractable by other methods such as the *DeWitt ansatz* [6] for heat kernel matrix elements.

The algorithm has the following main features. For a positive elliptic operator A the spectrum of which lies inside a contour C, the heat operator $\exp(-tA)$ can be expressed in terms of the resolvent $(A-\lambda)^{-1}$ via the formula

$$e^{-tA} = \int_C \frac{id\lambda}{2\pi} e^{-t\lambda}(A-\lambda)^{-1}. \qquad (3)$$

The pseudodifferential calculus method uses the following representation for the matrix elements of the resolvent

$$G(x,x',\lambda) \equiv \langle x|\frac{1}{A-\lambda}|x'\rangle = \int \frac{d^n k}{(2\pi)^n \sqrt{g(x')}} e^{il(x,x',k)} \sigma(x,x',k;\lambda), \qquad (4)$$

where $\sigma(x,x',k;\lambda)$ is an amplitude, $l(x,x',k)$ is a (real) phase function which is a biscalar with respect to general coordinate transformations, k is a wave vector.

The resolvent satisfies the equation $(A-\lambda)G = 1$ which leads to the equation for the amplitude:

$$(A(x, D_\mu + iD_\mu l) - \lambda)\sigma(x,x',k;\lambda) = I(x,x'), \qquad (5)$$

where $I(x,x')$ is a transport function having both bundle and Lorentz indices.

In the pseudodifferential calculus, it is assumed that in the flat space the phase function has the form $l = (x-x')_\mu k^\mu$. The covariant analogue of the linearity of the function l is based on the requirement that all higher-order symmetrized covariant derivatives of $l(x,x',k)$ vanish at the points $x = x'$, i. e., satisfy the infinite set of relations [26]:

$$[\{D_{\mu_1} \dots D_{\mu_m}\} l] = 0, \qquad m > 1, \qquad (6)$$

where $\{\dots\}$ means symmetrizing in all indices, and $[\dots]$ means transition to coincidence limit ($x = x'$). In an analogous way, the covariant transport function should satisfy the relations:

$$[\{D_{\mu_1} \dots D_{\mu_m}\} I] = 0, \quad m \geq 1. \qquad (7)$$

Equations (6) and (7) together with the "initial conditions" $[l] = 0, [D_\mu l] = k_\mu$ and $[I] = \mathbb{1}$ (unit operator) allow one to compute the coincidence limits for nonsymmetrized covariant derivatives $[D_{\mu_1} \dots D_{\mu_m} l]$ and $[D_{\mu_1} \dots D_{\mu_m} I]$. These nonsymmetrized derivatives are obtained directly from (6) and (7) by reducing all terms to a unified index ordering with the help of the Ricci identity. The resulting expressions are universal polynomials in the torsion $T^\lambda{}_{\mu\nu}$, curvature tensor $R^\lambda{}_{\mu\nu\eta}$, gauge curvature $W_{\mu\nu}$ and their covariant derivatives. In fact, once computed and stored the coincidence limits $[D_{\mu_1} \dots D_{\mu_m} l]$ and $[D_{\mu_1} \dots D_{\mu_m} I]$ can be used in many calculations for different operators A. The functions $l(x,x',k)$ and $I(x,x')$, introduced with the help of formulas (6) and (7),[3] play an important role in the covariant pseudodifferential calculus called also *intrinsic symbolic calculus* [26]. In fact, just these universal functions manifest the geometric properties of a base manifold and a bundle.

Expanding the amplitude σ in degrees of homogeneity of k:

$$\sigma = \sum_{m=1}^{\infty} \sigma_m(x,x',k;\lambda),$$

[3] The existence of these functions has been proved in [26].

we obtain the recursion equations for σ_m from equation (5). For example, for operator (2) these recursion expressions take the form

$$
\begin{aligned}
&A^{\mu\lambda}\sigma_{0\lambda\nu} = I^{\mu}_{\nu}, \\
&A^{\mu\lambda}\sigma_{1\lambda\nu} + i\left[-g^{\mu\lambda}(\Box l + 2D^{\eta}lD_{\eta}) + a(D^{\mu}D^{\lambda}l + D^{\mu}lD^{\lambda} + D^{\lambda}lD^{\mu})\right] \\
&\quad\times \sigma_{0\lambda\nu} = 0, \\
&\vdots \\
&A^{\mu\lambda}\sigma_{m\lambda\nu} + i\left[-g^{\mu\lambda}(\Box l + 2D^{\eta}lD_{\eta}) + a(D^{\mu}D^{\lambda}l + D^{\mu}lD^{\lambda} + D^{\lambda}lD^{\mu})\right] \\
&\quad\times \sigma_{(m-1)\lambda\nu} + (-g^{\mu\lambda}\Box + aD^{\mu}D^{\lambda} + X^{\mu\lambda})\sigma_{(m-2)\lambda\nu} = 0, \quad m \geq 2,
\end{aligned}
$$

where the matrix

$$A^{\mu\nu} = g^{\mu\nu}(D^{\eta}lD_{\eta}l - \lambda) - aD^{\mu}lD^{\nu}l$$

is the principal symbol for operator (2). Solving the recursion equations we obtain σ_m. The heat invariants are expressed in terms of integrals of the coincidence limits $[\sigma_m]$:

$$E_m(x|A) = \int \frac{d^n k}{(2\pi)^n\sqrt{g}} \int_C \frac{i d\lambda}{2\pi} e^{-\lambda}[\sigma_m](x,k,\lambda) \equiv J([\sigma_m]). \tag{8}$$

The integrals in (8) can be expressed in terms of gamma and Gauss hypergeometric functions for a wide class of operators A. The typical integral of terms of the coincidence limit $[\sigma_m]$ takes the form

$$
\begin{aligned}
&J\Big(\frac{k^{2p}k_{\mu_1}\ldots k_{\mu_{2s}}}{(k^{2r}-\lambda)^l[(1-a)k^{2r}-\lambda]^m}\Big) = \\
&g_{\{\mu_1\ldots\mu_{2s}\}}\frac{\Gamma((p+s+n/2)/r)}{(4\pi)^{n/2}2^s r\Gamma(n/2+s)\Gamma(l+m)}F(m,(p+s+n/2)/r;l+m;a),
\end{aligned}
$$

where $g_{\{\mu_1\ldots\mu_{2s}\}}$ is a symmetrized sum of products of metric tensors. Using the fact that m and l are integer numbers, one can express the hypergeometric function in (9) in terms of elementary functions with the help of the Gauss relation

$$a(1-z)F(a+1,b;c;z) = (c-a)F(a-1,b;c;z) + (2a-c-az+bz)F(a,b;c;z), \tag{9}$$

and using then the formula [27]

$$
\begin{aligned}
F(1,b;m;z) = (m-1)!\frac{(-z)^{1-m}}{(1-b)_{m-1}}\Big[(1-z)^{m-b-1} - \sum_{k=0}^{m-2}\frac{(b-m+1)_k}{k!}z^k\Big], \\
m = 1,2,\ldots, \quad m-b \neq 1,2,\ldots,
\end{aligned}
$$

where $(a)_k = a(a+1)\ldots(a+k-1)$ is the Pochhammer symbol (shifted factorial). During simplification of tensor expressions we use various symmetry properties of the tensors $R^{\lambda}{}_{\eta\mu\nu}, T^{\lambda}{}_{\mu\nu}, W_{\mu\nu}$, and also the Ricci identity

$$[D_\mu, D_\nu]\varphi^{\eta_1\ldots\eta_l}_{\lambda_1\ldots\lambda_k} = \sum_{i=1}^{l} R^{\eta_\iota}{}_{\alpha\mu\nu}\varphi^{\eta_1\ldots\eta_{\iota-1}\alpha\eta_{\iota+1}\ldots\eta_l}_{\lambda_1\ldots\lambda_k}$$

$$-\sum_{i=1}^{k} R^{\alpha}{}_{\lambda_\iota\mu\nu}\varphi^{\eta_1\ldots\eta_l}_{\lambda_1\ldots\lambda_{\iota-1}\alpha\lambda_{\iota+1}\ldots\lambda_k} + T^{\alpha}{}_{\mu\nu}D_\alpha\varphi^{\eta_1\ldots\eta_l}_{\lambda_1\ldots\lambda_k} + W_{\mu\nu}\varphi^{\eta_1\ldots\eta_l}_{\lambda_1\ldots\lambda_k},$$

the Bianchi identities for both affine and gauge curvatures

$$D_\alpha R^{\beta}{}_{\gamma\delta\epsilon} + D_\delta R^{\beta}{}_{\gamma\epsilon\alpha} + D_\epsilon R^{\beta}{}_{\gamma\alpha\delta}$$
$$+T^{\lambda}{}_{\alpha\delta}R^{\beta}{}_{\gamma\epsilon\lambda} + T^{\lambda}{}_{\delta\epsilon}R^{\beta}{}_{\gamma\alpha\lambda} + T^{\lambda}{}_{\epsilon\alpha}R^{\beta}{}_{\gamma\delta\lambda} = 0,$$
$$D_\alpha W_{\beta\gamma} + D_\beta W_{\gamma\alpha} + D_\gamma W_{\alpha\beta}$$
$$+W_{\alpha\lambda}T^{\lambda}{}_{\beta\gamma} + W_{\beta\lambda}T^{\lambda}{}_{\gamma\alpha} + W_{\gamma\lambda}T^{\lambda}{}_{\alpha\beta} = 0,$$

and the cyclic identity

$$R^{\alpha}_{\beta\gamma\delta} + R^{\alpha}_{\gamma\delta\beta} + R^{\alpha}_{\delta\beta\gamma} + D_\beta T^{\alpha}{}_{\gamma\delta} + D_\gamma T^{\alpha}{}_{\delta\beta} + D_\delta T^{\alpha}{}_{\beta\gamma}$$
$$+T^{\alpha}{}_{\beta\lambda}T^{\lambda}{}_{\gamma\delta} + T^{\alpha}{}_{\gamma\lambda}T^{\lambda}{}_{\delta\beta} + T^{\alpha}{}_{\delta\lambda}T^{\lambda}{}_{\beta\gamma} = 0.$$

The above algorithm has been implemented in the C language. The C code of total length about 11000 lines contains about 250 functions for different manipulations with tensors and scalars. These functions are gathered into two programs DWSGCOEF and COLIM.

The COLIM program computes coincidence limits of the $l(x, x'; k)$ and $I(x, x')$ functions and writes them to the disk. Once computed and stored[4] the coincidence limits, being universal functions, can be used in many calculations for different operators A.

The DWSGCOEF program computes E_m coefficients by the following steps:

1. *Reading input information (operator, order m, etc.)*
2. *Computing a set of asymptotic operators for constructing recursion equations.*
3. *Computing σ_m with the help of the recursion equations.*
4. *Taking the coincidence limit $[\sigma_m]$.*
5. *Integrating $[\sigma_m]$ to obtain the coefficient E_m.*
6. *Substituting tensor expressions for $[D_{\mu_1}\ldots D_{\mu_k}l]$ and $[D_{\mu_1}\ldots D_{\mu_k}I]$ into E_m.*
7. *Reducing hypergeometric to elementary functions in the scalar coefficients (C_i in the formulas of Section 3) included in the heat invariants in the case of nonminimal or higher-order operator. Eliminating possible linear dependencies among these scalar coefficients[5] to make the resulting formulas as compact as possible.*

[4] For the operators of different tensor ranks $A, A^{\mu\nu},\ldots$ the coincidence limits for the functions $I, I^{\mu\nu},\ldots$ should be computed separately.

[5] Usually there are many dependencies among the scalar coefficients which are not seen in terms of hypergeometric functions, i.e., quite different hypergeometric expressions may be reduced sometimes to the same elementary function.

8. *Output E_m (and its Lorentz trace in the nonminimal case).*

To cut down the swelling of the intermediate expressions, we use *term-by-term* strategy, i.e., the most cumbersome Steps 4-6 are applied consecutively to single terms of σ_m generated during the execution of Step 3.

3 Heat Invariant E_2

We present here the full expression for the coefficient E_2 and also its trace with respect to Lorentz indices for nonminimal operator (2) on a curved manifold with the torsion and gauge field manifested itself in the gauge curvature $W_{\mu\nu}$. We consider the case of arbitrary dimension n and also the most important[6] for E_2 case $n = 2$. In the below formulas the indices α, β, γ and μ, ν are dummy and free, correspondingly. We use the following definition for the torsion trace: $T_\mu = T^\alpha{}_{\alpha\mu}$.

3.1 Full Expression

$$
\begin{aligned}
E_2 = (4\pi)^{-\frac{n}{2}} \Big\{ & -C_1 X^{\mu\nu} - C_2 \left(X^{\nu\mu} + g^{\mu\nu} X_\alpha{}^\alpha\right) + C_3 \left(W^{\mu\nu} + \frac{8}{3} D_\alpha T^{\alpha\mu\nu} \right. \\
& \left. + \frac{19}{6} T_\alpha T^{\alpha\mu\nu} \right) + C_4 R^{\mu\nu} - C_5 R^{\nu\mu} + C_6 \left(D_\alpha T^{\mu\alpha\nu} + D_\alpha T^{\nu\alpha\mu}\right) \\
& + C_7 T_{\alpha\beta}{}^\mu T^{\alpha\beta\nu} + C_8 T_{\alpha\beta}{}^\mu T^{\beta\alpha\nu} + C_9 \left(T_{\alpha\beta}{}^\mu T^{\nu\alpha\beta} + T_{\alpha\beta}{}^\nu T^{\mu\alpha\beta}\right) \\
& - C_{10} T^\mu{}_{\alpha\beta} T^{\nu\alpha\beta} + C_{11} D^\mu T^\nu - C_{12} D^\nu T^\mu + C_{13} T_\alpha \left(T^{\mu\alpha\nu} + T^{\nu\alpha\mu}\right) \\
& - C_{14} T^\mu T^\nu + C_{15} g^{\mu\nu} \left(R + D_\alpha T^\alpha\right) + C_{16} g^{\mu\nu} T_\alpha T^\alpha \\
& + \left(C_{16} - C_{15}\right) g^{\mu\nu} T_{\alpha\beta\gamma} T^{\beta\alpha\gamma} - C_{17} g^{\mu\nu} T_{\alpha\beta\gamma} T^{\alpha\beta\gamma} \Big\} .
\end{aligned}
$$

Coefficients C_i in arbitrary dimension n:

$$
\begin{aligned}
C_1 &= \frac{1}{a(n-2)n(n+2)} \left\{ (1-a)^{-\frac{n}{2}} \left(-3an - 6a + 4n + 4\right) + an^3 - 2an^2 \right. \\
&\quad \left. -3an + 6a - 4n - 4 \right\} , \\
C_2 &= \frac{1}{a(n-2)n(n+2)} \left\{ (1-a)^{-\frac{n}{2}} (an + 2a - 4) + an - 2a + 4 \right\} , \\
C_3 &= \frac{1}{a(n-2)n} \left\{ (1-a)^{1-\frac{n}{2}} (an - 8) + 3an - 8a + 8 \right\} , \\
C_4 &= \frac{1}{6a(n-2)n(n+2)} \left\{ (1-a)^{-\frac{n}{2}} \left(17a^2n^2 + 34a^2n - 17an^2 - 168an \right. \right. \\
&\quad \left. \left. -268a + 140n + 256\right) - 53an^2 + 40an + 268a - 140n - 256 \right\} ,
\end{aligned}
$$

[6] The *Atiyah–Singer index* of an elliptic operator on a manifold of dimension n can be expressed in terms of an integral of E_n over the manifold.

$$C_5 = \frac{1}{6a(n-2)n(n+2)}\left\{(1-a)^{-\frac{n}{2}}\left(15a^2n^2+30a^2n-15an^2-152an\right.\right.$$
$$\left.\left.-244a+116n+256\right)-43an^2+24an+244a-116n-256\right\},$$

$$C_6 = \frac{1}{6a^2(n-2)n(n+2)}\left\{(1-a)^{-\frac{n}{2}}\left(a^3n^2+2a^3n-a^2n^2-20a^2n-36a^2\right.\right.$$
$$\left.\left.+12an+96a-48\right)+a^2n^2-16a^2n+36a^2+12an-96a+48\right\},$$

$$C_7 = \frac{1}{6a^2(n-2)n(n+2)(n+4)}\left\{(1-a)^{-\frac{n}{2}}\left(-a^3n^3-6a^3n^2+a^2n^3-\right.\right.$$
$$8a^3n+30a^2n^2+152a^2n-24an^2+192a^2-240an-576a+144n$$
$$\left.+288\right)-7a^2n^3+18a^2n^2+64a^2n-48an^2-192a^2+96an+576a$$
$$\left.-144n-288\right\},$$

$$C_8 = \frac{1}{6a^2(n-2)n(n+2)}\left\{(1-a)^{1-\frac{n}{2}}\left(a^2n^2+2a^2n-24an-48a+144\right)\right.$$
$$\left.-7a^2n^2+34a^2n-48a^2-48an+192a-144\right\},$$

$$C_9 = \frac{1}{a^2(n-2)n(n+2)(n+4)}\left\{(1-a)^{-\frac{n}{2}}\left(a^2n^2+6a^2n+8a^2-12an\right.\right.$$
$$\left.\left.-48a+48\right)-a^2n^2+6a^2n-8a^2-12an+48a-48\right\},$$

$$C_{10} = \frac{1}{12a^2(n-2)n(n+2)(n+4)}\left\{(1-a)^{-\frac{n}{2}}\left(-a^3n^3-6a^3n^2+a^2n^3-\right.\right.$$
$$\left.8a^3n+30a^2n^2+152a^2n+192a^2-192an-768a+576\right)-a^2n^3$$
$$\left.-6a^2n^2+88a^2n-192a^2-96an+768a-576\right\},$$

$$C_{11} = \frac{1}{3a^2(n-2)n(n+2)}\left\{(1-a)^{1-\frac{n}{2}}\left(-9a^2n^2-18a^2n+76an+152a\right.\right.$$
$$\left.\left.-24\right)+2\left(-13a^2n^2+6a^2n+76a^2-32an-88a+12\right)\right\},$$

$$C_{12} = \frac{1}{3a^2(n-2)n(n+2)}\left\{2(1-a)^{1-\frac{n}{2}}\left(-5a^2n^2-10a^2n+38an+76a\right.\right.$$
$$\left.\left.+12\right)-31a^2n^2+26a^2n+152a^2-88an-128a-24\right\},$$

$$C_{13} = \frac{1}{6a(n-2)n(n+2)(n+4)}\left\{(1-a)^{-\frac{n}{2}}\left(a^3n^3+6a^3n^2-a^2n^3+8a^3n\right.\right.$$
$$\left.-18a^2n^2-80a^2n+12an^2-96a^2+72an+96a-48n+96\right)+a^2n^3$$
$$\left.-18a^2n^2+8a^2n+12an^2+96a^2-120an-96a+48n-96\right\},$$

$$C_{14} = \frac{1}{2a^2(n-2)n(n+2)}\left\{(1-a)^{1-\frac{n}{2}}\left(-a^2n^2-2a^2n+8an+16a-16\right)\right.$$
$$\left.-a^2n^2-2a^2n+16a^2-32a+16\right\},$$

$$C_{15} = \frac{1}{6a(n-2)n(n+2)}\left\{(1-a)^{-\frac{n}{2}}\left(-a^2n^2-2a^2n+an^2+8an+12a-24\right)\right.$$
$$\left.+an^3-an^2-12a+24\right\},$$

$$C_{16} = \frac{1}{12a^2(n-2)n(n+2)(n+4)}\left\{(1-a)^{-\frac{n}{2}}\left(-a^3n^3-6a^3n^2+a^2n^3\right.\right.$$

$$-8a^3n + 6a^2n^2 + 8a^2n + 48an + 192a - 288\big) + a^2n^4 + 3a^2n^3$$
$$+2a^2n^2 - 48a^2n + 96an - 192a + 288\big\},$$
$$C_{17} = \frac{1}{24a^2(n-2)n(n+2)(n+4)} \big\{(1-a)^{-\frac{n}{2}} \big(-a^3n^3 - 6a^3n^2 + a^2n^3$$
$$-8a^3n + 30a^2n^2 + 152a^2n + 192a^2 - 192an - 768a + 576\big) + a^2n^4$$
$$+3a^2n^3 - 10a^2n^2 + 72a^2n - 192a^2 - 96an + 768a - 576\big\}.$$

Coefficients C_i in the dimension $n = 2$:

$$C_1 = -\frac{3\ln(1-a)}{4a} - \frac{2-a}{8(1-a)},$$
$$C_2 = \frac{\ln(1-a)}{a} + \frac{2-a}{2(1-a)},$$
$$C_3 = -\frac{(a-4)\ln(1-a)}{2a} + 2,$$
$$C_4 = \frac{(17a-67)\ln(1-a)}{12a} + \frac{137a-134}{24(1-a)},$$
$$C_5 = \frac{(15a-61)\ln(1-a)}{12a} + \frac{119a-122}{24(1-a)},$$
$$C_6 = \frac{(a^2-9a+6)\ln(1-a)}{12a^2} + \frac{(3a-2)(a-2)}{8a(1-a)},$$
$$C_7 = -\frac{(a^2-12a+12)\ln(1-a)}{12a^2} - \frac{2a^2-9a+6}{6a(1-a)},$$
$$C_8 = -\frac{(a^2-12a+18)\ln(1-a)}{12a^2} + \frac{a-6}{4a},$$
$$C_9 = \frac{(a-2)\ln(1-a)}{4a^2} - \frac{a^2-12a+12}{24a(1-a)},$$
$$C_{10} = -\frac{(a^2-12a+12)\ln(1-a)}{24a^2} - \frac{2a^2-9a+6}{12a(1-a)},$$
$$C_{11} = \frac{(9a^2-38a+3)\ln(1-a)}{6a^2} - \frac{73a-6}{12a},$$
$$C_{12} = \frac{(10a^2-38a-3)\ln(1-a)}{6a^2} - \frac{79a+6}{12a},$$
$$C_{13} = \frac{(a-6)\ln(1-a)}{12a} + \frac{2a-3}{6(1-a)},$$
$$C_{14} = \frac{(a^2-4a+2)\ln(1-a)}{4a^2} - \frac{3a-2}{4a},$$
$$C_{15} = -\frac{(a-3)\ln(1-a)}{12a} - \frac{7a-10}{24(1-a)},$$
$$C_{16} = -\frac{(a^2-6)\ln(1-a)}{24a^2} - \frac{3a^2+a-6}{24a(1-a)},$$

$$C_{17} = -\frac{(a^2 - 12a + 12)\ln(1-a)}{48a^2} - \frac{3a^2 - 10a + 6}{24a(1-a)}.$$

3.2 Lorentzian Trace

$$\mathrm{tr_L} E_2 = (4\pi)^{-\frac{n}{2}} \{ -C_1 X_\alpha{}^\alpha + C_2 R - C_3 T_{\alpha\beta\gamma} T^{\alpha\beta\gamma} + C_4 T_{\alpha\beta\gamma} T^{\beta\alpha\gamma} \\ + C_5 D_\alpha T^\alpha + (C_4 + C_5) T_\alpha T^\alpha \}$$

C_i for the trace in arbitrary dimension n:

$$C_1 = \frac{(1-a)^{-\frac{n}{2}} + n - 1}{n}, \quad C_2 = \frac{(1-a)^{-\frac{n}{2}}(-an + n + 6) + n^2 - n - 6}{6n},$$

$$C_3 = \frac{1}{24an(n+2)} \{ -(1-a)^{-\frac{n}{2}} (a^2n^2 + 2a^2n - an^2 - 26an - 48a + 96) \\ + an^3 + an^2 + 22an - 48a + 96 \},$$

$$C_4 = \frac{1}{12an(n+2)} \{ (1-a)^{-\frac{n}{2}} (a^2n^2 + 2a^2n - an^2 - 14an - 24a + 48) \\ - an^3 - an^2 - 10an + 24a - 48 \},$$

$$C_5 = \frac{1}{6a(n-2)n} \{ -(1-a)^{-\frac{n}{2}} (a^2n^2 + 4a^2n - an^2 - 10an - 36a + 48) \\ + an^3 - 3an^2 + 14an - 36a + 48 \}.$$

C_i for the trace in the dimension $n = 2$:

$$C_1 = \frac{2-a}{2(1-a)}, \quad C_2 = \frac{2+a}{6(1-a)}, \quad C_3 = \frac{1}{12}, \quad C_4 = -\frac{1}{6},$$

$$C_5 = -\frac{(a-4)\ln(1-a)}{2a} - \frac{11a - 14}{6(1-a)}.$$

4 Conclusion

The program computes E_2 with torsion for operator (2) rather easily (about 10 sec on a Pentium-75 PC). Unfortunately, computational complexity of the problem under consideration is very high. For example, the timings for torsionless computations of E_2 and E_4 for the same operator (2) are < 1 sec and 4 h 5 min, correspondingly. It is clear that the inclusion of the torsion increases the computational efforts considerably and the computation of E_4 with torsion may take too much time. Another problem is the volume of the resulting expressions. There are two ways to handle this problem. First of all, some work is needed for developing of algorithms for further reduction of large tensor expressions. However, due to the natural complexity of the heat invariants, one can not hope to make the higher-order invariants tractable by hand. Thus, the methods for automatic usage of these invariants should be elaborated.

Acknowledgements

I would like to thank V. Gusynin for initiating this work and helpful communications. This work was supported in part by INTAS project No. 96-0842 and RFBR project No. 98-01-00101.

References

1. Kac, M.: Can one hear the shape of a drum? *Amer. Math. Monthly* **73**, No. 4, Part II (1966) 1–23.
2. Milnor, J.: Eigenvalues of the Laplace operator on certain manifolds. *Proc. Nat. Acad. Sci. U.S.A.*, **51** (1964) 542.
3. Schüth, D.: Continuous families of isospectral metrics on simply connected manifolds. *Annals of Mathematics* **149** (1999) 287–308.
4. Hadamard, J.: *Lectures on Cauchy's Problem in Linear Partial Differential Equations,* Yale University Press, New Haven, 1923.
5. Minakshisundaram, S., Pleijel, A.: Some properties of the eigenvalues of the Laplace operator on Riemannian manifolds. *Can. J. Math.* **1** (1949) 242–256.
6. DeWitt, B.: *Dynamical Theory of Groups and Fields,* Gordon and Breach, New York, 1965.
7. Seeley, R.T.: Complex powers of an elliptic operator. *Singular Integrals (Proc. Symp. Pure Math.,Providence), Amer. Math. Soc.* **10** (1967) 288–307.
8. Gilkey, P.B.: The spectral geometry of a Riemannian manifold. *J. Diff. Geom.* **10** (1975) 601–618.
9. Birrell, N.D., Davies, P.C.W.: *Quantum Fields in Curved Space,* Cambridge University Press, Cambridge, 1982.
10. Barvinsky, A.O., Vilkovisky, G.A.: The generalized Schwinger–DeWitt technique in gauge theories and quantum gravity. *Phys. Repts.* **119** (1985) 1–74.
11. Atiyah, M.A., Bott, R., Patodi, V.K.: On the Heat Equation and the Index Theorem. *Invent. Math.* **19** (1973) 279–330.
12. McKean, M.P., Singer, I.M.: Curvature and eigenvalues of the Laplacian *J. Diff. Geom.* **1** (1967) 43–69.
13. Barvinsky, A.O., Vilkovisky, G.A.: Beyond the Schwinger–DeWitt technique: Converting loops into trees and in-in currents. *Nucl. Phys. B.* **282** (1987) 163–188.
14. Barth, N.H., Christensen, S.M.: Quantizing fourth order gravity theories: The functional integral. *Phys. Rev. D.* **28** (1983) 1876–1893.
15. Cho, H.T., Kantowski, R.: Gauge independent conformal anomaly for gravitons. *Phys. Rev. D.* **52** (1995) 4600–4608.
16. Misner, C.W., Thorne, K.S., and Wheeler, J.A.: *Gravitation,* W.H. Freeman and Company, 1973.
17. Cartan, E.: *Sur une généralisation de la notion de courbure de Riemann et les espaces à torsion.* Comptes Rendus Acad. Sci. **174** (1922) 593; English translation by G.D. Kerlick in *Cosmology and Gravitation: Spin, Torsion Rotation and Supergravity,* Eds.: P.G. Bergman and V. De Sabbata, Plenum Press, New York, 1980.
18. Cartan, E.: *Sur les variétés à connection affine et la théory de la relativisteé généralisée I, II,* Ann. Ec. Norm. Sup., **40**, 325 (1923), **41**, 1 (1924), **42**, 17 (1925); English Translation by A. Magnon, A Ashtekar and A. Trautmann, *On Manifolds with Affine Connection and the Theory of General Relativity*, Bibliopolis, Naples, 1985.

19. Hehl, F.W., von der Heyde, P., Kerlick, G.D., Nester, J.M.: *General Relativity with Spin and Torsion: Foundations and Prospects,* Rev. Mod. Phys. **48** (1976) 393.
20. Hehl, F.W.: *Four Lectures on Poincaré Gauge Field Theory,* in Proceedings of the 6th Course of the School of Cosmology and Gravitation on *Spin, Torsion, Rotation and Supergravity,* P. Bergman, V. de Sabbata (Eds.), Plenum, New York, 1980.
21. Green, M., Schwarz, J., Witten, E.: *Superstring Theory,* Vols. I and II, Cambridge University Press, 1987.
22. Kiritsis, E.: *Introduction to Superstring Theory,* hep-th/9709062.
23. Gusynin, V.P., Kornyak, V.V.: DeWitt-Seeley-Gilkey Coefficients for Nonminimal Differential Operators in Curved Space, *Fundamental and Applied Mathematics (Fundamental'naya i prikladnaya matematika)* **5** (1999) 649-674 (in Russian); Complete Computation of DeWitt-Seeley-Gilkey Coefficient E_4 for Nonminimal Operator on Curved Manifolds, E-print math.SC/9909145.
24. Gusynin, V.P.: New algorithm for computing the coefficients in the heat kernel expansion, *Phys. Lett.* **B225** (1989) 233–239.
25. Gusynin, V.P.: Seeley-Gilkey coefficients for the fourth-order operators on a Riemannian manifold, *Nucl. Phys.* **B333** (1990) 296–316.
26. Widom, H.: A complete symbolic calculus for pseudodifferential operators, *Bull. Sci. Math.* **104** (1980) 19–63.
27. Prudnikov, A.P., Brychkov, Yu.A., and Marichev, O.I.: *Integrals and Series.*, Vol. III, Nauka, Moscow, 1986.

Computer Algebra for Automated Performance Modeling of Fortran Programs

Hermann Mierendorff and Helmut Schwamborn

Institute for Algorithms and Scientific Computing (SCAI)
GMD - German National Research Center for Information Technology
Schloss Birlinghoven, D-53754 Sankt Augustin, Germany

Abstract. Time complexity of sequential programs or segments of parallel programs can be estimated using the dynamic frequency of statements and their execution time. The execution time of single statements can be estimated by counting basic operations. The present paper deals with a method of determining the global dynamic frequency of statements by transient analysis of a Markov model. For defining the model automatically and solving the related equations, we use AUGUR which is a research tool for performance modeling of Fortran programs. While static program analysis and model generation are implemented by routine compile techniques, AUGUR uses MAPLE for evaluating expressions, defining transition matrices, and transient analysis. Routines of LAPACK are considered to demonstrate the achievable results.

Key Words: time complexity of programs, performance analysis, automated model generation, parametrized models, computer algebra.

1 Introduction

There is a variety of tools for estimating the performance of computer systems. For a rich collection of those tools, we refer to [5]. Most of these tools offer a possibility for describing hardware and software models under consideration by the use of a specific language. Once the user has described the model it can be transformed into an internal form for evaluation by a simulator. In contrast to these systems, a research tool called AUGUR (AUtomatic model Generation with User Response) is under development which assists the user in generating performance models of existing Fortran programs. AUGUR subdivides a given program source automatically into large sequential blocks. The runtime of these blocks is estimated by a time-formula. If required, e.g. for parallel programs, a model of the whole program is generated for evaluation by a simulator. For a detailed description, we refer to [8].

For estimating the *dynamic frequency* of statements of a sequential program segment (i.e. number of executions of a statement during a single run), we follow a method which has been mentioned first in [9]. We assume that a sequential block of statements has exactly one entry in the beginning and one exit at the end. Furthermore, we exclude non-terminating loops, jumps of a statement to itself, and jumps from the last statement into the considered code. This can easily be satisfied by appropriate program transformations.

Let $A = (a_{ik})_{i,k=1,\ldots,n}$ be the matrix of transition probabilities for a block of n statements, i.e. $a_{i,k}$ is the probability that statement i is executed next if k is the current statement. For a run of the considered program, this probability is defined as the ratio of the number of executions of statement i immediately after statement k over the number of executions of statement k. The vector x of dynamic frequencies of statements satisfies the equation (see [9, 8]):

$$(I - A)x = \begin{pmatrix} 1 \\ 0 \\ \vdots \\ 0 \end{pmatrix} \tag{1}$$

where I denotes the unit matrix. Because of our assumptions on the considered programs, there is always a unique solution. The execution time of statements is estimated by accumulating runtime of its basic operations. At present, the user of AUGUR can specify classes of arbitrary arithmetic, Boolean, and relational operations of Fortran for basic operations. Let the $n \times m$ matrix $B = (b_{kl})$ contain operation counters, i.e. b_{kl} is the number of operations of type l in statement number k, and $t = (t_l)_{l=1,\ldots,m}$ the vector of timing values of the m basic operations which has to be estimated by the user, then

$$x^T Bt \tag{2}$$

is the time-formula of the considered block of statements. This complexity function can contain size parameters which come via A from the program code and timing parameters which come from t. Those parts of a program which cannot be treated in this way (mainly the parallel program structure) are translated into a model which can be evaluated by simulation.

There is a relation to performance estimation in the optimizing run of parallelizing compilers where also models of parallel programs are developed and evaluated automatically (see [4]). In contrast to those methods, we use simulation of complex models and we allow user interaction for definition of unknown control variables during model generation. Here we mean variables which determine the control flow of the program. Such a method has already been described in [3]. Furthermore, we try to generate highly accurate parametrized models. The existence of parameters in these models requires the use of computer algebra for defining the matrix A as well as for solving the equation (1). Because the number of user interactions is a crucial point in particular with large programs, we try to define control variables automatically wherever possible.

While the principles of our method and numerical examples are already known (see [8]), the present paper deals with engineering problems of implementing this method using computer algebra which have not been reported before.

An overview of the components of AUGUR and their collaboration is given in section 2. Section 3 deals with the definition of matrix A and solution of

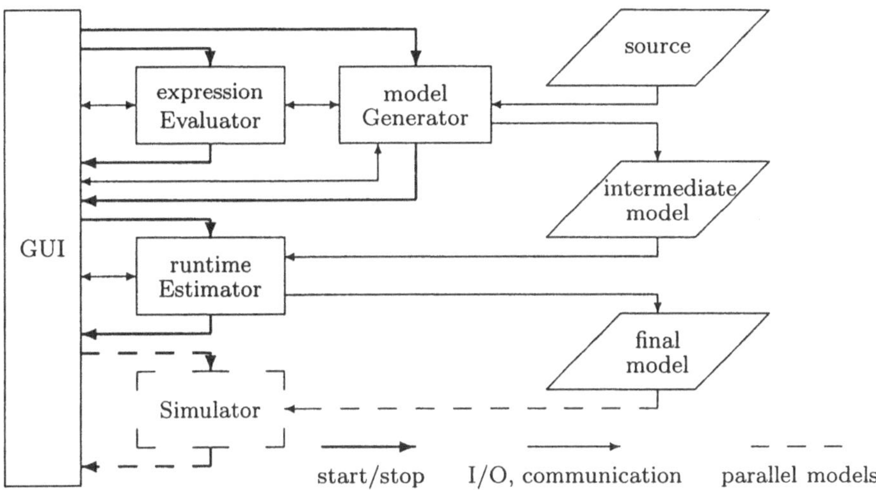

Fig. 1. Main components of AUGUR and their collaboration.

the related linear equations. The challenge is to find transformations of the original problems for making higher level MAPLE constructs applicable. The applicability to Fortran programs is demonstrated in section 4 with a BLAS routine of LAPACK (see [2]) for example. Finally, we summarize the results and announce future work in the concluding remarks.

2 Components of AUGUR and Their Collaboration

AUGUR is a tool which assists the user during generation and evaluation of performance models for FORTRAN77 programs. Fig. 1 describes the main components of AUGUR and their collaboration. All components run under the control of a Graphical User Interface (GUI). At the beginning of an analysis the *Generator* reads the program source, defines the sequential code segments which have to be replaced by a time-formula, and generates the intermediate model. The Generator tries to evaluate all required control variables by static program analysis. If algebraic transformation is required, the Generator collaborates with the *Evaluator*. This is a MAPLE program which runs in parallel to the Generator and provides facilities for evaluation of Boolean expression. If the value of an `IF` expression is always true or always false for the considered run of a program, early evaluation of these constructs can help to suppress analysis of the `ELSE` or the `THEN` clause and this might avoid unnecessary user interactions.

The estimation of time-formulas for sequential blocks is postponed to the run of the *Estimator*. This is also a MAPLE program which produces the final performance model. Evaluation of several functions during the final definition of equation (1) and their solution using the MAPLE routine `linsolv` is the task of the Estimator. Computer algebra is required because many calculations have to be done with multivariate expressions. 2 to 6 size parameters for the application and 1 to 20 timing parameters for the basic operations are absolutely normal cases. Up to 1000 statements per block, the solution of the related linear equations is no problem with standard MAPLE routines. The definition of the equations, however, requires strong improvements of some MAPLE functions.

If the final model contains more than a single time-formula only (e.g. for parallel programs), the model can be evaluated by simulation. At present, AUGUR is connected to Modline/Arch which is a commercial tool of Simulog for simulation of arbitrary parallel systems.

3 Definition of Transition Probabilities

The current section deals with estimating the transition probabilities of statements in order to define the equation (1).

3.1 Simple Cases

Non-branching Statements. The static and the dynamic order is identical for non-branching statements which is the majority of statements (after some beautifying transformations, the LAPACK routine DSYMM had 106 non-branching statements and 96 others). This is valid for assignment statements, IO statements, statements which are not executable, etc. We estimate $a_{k+1,k} = 1$ and $a_{jk} = 0$ for $j \neq k+1$, if statement k is of this type.

GOTO. A `GOTO` from statement k to j leads to $a_{jk} = 1$ and $a_{ik} = 0$ for $i \neq j$.

Subroutine or Function Call. Nested calls of procedures are analyzed in a hierarchical way. We recall that the expression $x^T Bt$ has to be evaluated (see (2)), where Bt represents a vector containing the complexity of single statements for elements. AUGUR interrupts the evaluation of a code segment when a procedure call is detected. Then it evaluates the procedure first and inserts the resulting complexity expression in the already existing part of Bt. Because the result might depend on current values of parameters, it cannot be reused. The call itself is treated like non-branching statements.

RETURN and STOP. `RETURN` statements are treated like a jump to the end of the subroutine. `STOP` statements in the main program can be handled like a jump to the end of the program. In general, `STOP` statements are considered model statements and the program is intersected at this place.

3.2 Handling of `IF-THEN-ELSE` Statements

The transition probability for continuing after an `IF` statement is a crucial point. If this problem is left to the user, the number of user interactions can be unacceptably high with large programs. Therefore, we try to evaluate an `IF` expression automatically whenever possible. Let us briefly discuss the three main classes of `IF` statements which cover all cases occurring in application programs.

Mode Definition. An IF statement which selects the mode the program is currently running in contains constants and parameters and can be evaluated by static analysis. This might require an advanced *def-use* analysis of variables and the use of computer algebra including assumption handling for parameters.

Typical for mode definition by an IF statement is that for all executions during the current run always the same is happening, either the THEN or the ELSE block is entered and the other block represents currently *bypassed* code. Here we mean code with execution probability 0. Bypassed code is not necessarily *dead code* because it could be executed if another run and different parameters are considered. AUGUR tries to detect bypassed code automatically. In this way, detailed analysis of critical statements belonging to bypassed code can be avoided.

We consider a code segment of the LAPACK/BLAS routine DSYMM for example. Checking of mode parameters for this routine (these are the variables SIDE, UPLO, M, N, LDA, LDB) is done in the following way:

```
      IF( LSAME( SIDE, 'L' ) )THEN
         NROWA = M
      ELSE
         NROWA = N
      END IF
      UPPER = LSAME( UPLO, 'U' )
      INFO = 0
      IF(      ( .NOT.LSAME( SIDE, 'L' ) ).AND.
     $         ( .NOT.LSAME( SIDE, 'R' ) )       )THEN
         INFO = 1
      ELSE IF( ( .NOT.UPPER                ).AND.
     $         ( .NOT.LSAME( UPLO, 'L' ) )       )THEN
         INFO = 2
      ELSE IF( M  .LT.0               )THEN
         INFO = 3
      ELSE IF( N  .LT.0               )THEN
         INFO = 4
      ELSE IF( LDA.LT.MAX( 1, NROWA ) )THEN
         INFO = 7
          :
      END IF
      IF( INFO.NE.0 )THEN
         CALL XERBLA( 'DSYMM ', INFO )
         RETURN
      END IF
```

We consider a simple case assuming SIDE='L' and UPLO='U' which leads to LSAME = true in all essential cases where LSAME is a logical function checking case insensitively for equality of its arguments. Using the assumption handling capability of MAPLE and providing appropriate assumptions on the indeterminates M,N, the above code segment can be evaluated automatically. In particular the last IF has to be evaluated already during the Generator run. The last IF decides whether the error routine (XERBLA) of LAPACK is called or the main part of DSYMM is entered. This decision is possible only if all preceding IFs have already been evaluated before. This is an example where the symbolic computation facility has also to be available during the model generation.

There is a *def-use* analysis for control variables in AUGUR which guarantees reliability of values obtained by constant propagation. Statements are scanned in their static order for this purpose. In each place where a control variable is used, it has to be clarified if the last definition is still reliable. The *def-use* analysis has to detect that definitions using reliable arguments are also reliable when located within code which is executed with probability 1. Furthermore, the *def-use* analysis has to ignore definitions in bypassed code. Code is in particular bypassed code if it is located in a subroutine behind a RETURN which is executed with probability 1. This is important for analyzing the function calls LSAME(UPPER,'U') and LSAME(SIDE,'L') in the above code segment.

The evaluation of Boolean expressions can be optimized using

$$\texttt{FALSE AND } \textit{any boolean} = \texttt{FALSE}, \quad \texttt{TRUE OR } \textit{any boolean} = \texttt{TRUE}$$

While evaluating the AND expressions of the above example, AUGUR analyzes the relational expressions from left to right. Evaluation of the first relational expression will already do in both cases.

At the end of analyzing the above piece of code, the Generator knows that INFO = 0 for a correct call of DSYMM, i.e. the parameter control part is executed automatically and the evaluation of the main part can be started.

Dynamic Selection of the Following Code Segment. An IF statement decides dynamically which one of the alternative continuations should be selected. If the decision depends on results which are available only by running the program, the required value has to be estimated by external definition e.g. by user interaction. The user can find a numerical estimate for the probability that a specific IF expression is true by evaluating a run of a properly instrumented version of the code. As an alternative a user of AUGUR can define this probability by a parameter which had to be declared in the beginning. This parameter will appear in the final time-formula and can be studied afterwards.

If the decision depends on values of loop variables, however, an automatic estimation is possible as pointed out in [7] or later in the present paper.

Iterations. An IF statement is used for termination control of an iterative method. There is no way of estimating the probability automatically but it could be handled as an undetermined parameter. If definition is required, user interaction is necessary. Only few cases of this kind can be expected in a program.

3.3 Handling of DO-Loops

Single Loops. Let us consider a single run of a simple DO loop as an example:

```
DO i = b,e,d
    loop body
ENDDO
```

Assuming the loop is entered, the number of executions of the loop body is

$$w := \max\left(1, \left\lfloor \frac{e-b+d}{d} \right\rfloor\right)$$

The probability of leaving the loop at the end of the loop body is $1/w$. The probability of entering the loop is

$$\min\left(1, \max\left(0, \left\lfloor \frac{e-b+d}{d} \right\rfloor\right)\right)$$

which is either 0 or 1. Therefore, the frequency of executing the loop body is

$$\max\left(1, \left\lfloor \frac{e-b+d}{d} \right\rfloor\right) \min\left(1, \max\left(0, \left\lfloor \frac{e-b+d}{d} \right\rfloor\right)\right) \tag{3}$$

This expression can be simplified considerably using the following assumptions: b, e, and d are integers, $e \geq b$, $d > 0$, and $\frac{e-b}{d}$ is integer. Then we obtain

$$\frac{e-b+d}{d}$$

instead of (3) as usually expected. If expressions like (3) remain in the form of (1), its evaluation takes much more effort and the final resulting expression will not be satisfactory. Therefore, we have to pay attention to the functions used in (3). It is not worthwhile to discuss `min` or `max` because their implementation is straight forward. The `Floor` function we need, however, has to evaluate much harder problems than MAPLE's `floor` is able to handle.

Evaluation of the `floor` Function. Let us consider `Floor(expr)` where the expression is a multivariate polynomial $P_{r,s}(n_1, \ldots, n_m)$ of degree r having s terms of degree r depending on m variables n_i, $i = 1, \ldots, m$. For formal reasons we also use $P_{r,0}$ in which case we mean $P_{r-1,s}$ with some s. For n_i we consider the indeterminates of the numerators of the expanded form of `expr` which do not occur in the denominators. The coefficients of that polynomial can still contain further indeterminates. We assume that appropriate assumptions have been provided on the size of all indeterminates and on the properties of divisibility of the n_i. In terms of MAPLE we allow `AndProp`'s like `RealRange` and `LinearProp(`α`,integer,`β`)` as well as `property/objects` like `((`$n_i-\beta$`)/`α`,integer)`, i.e. assumptions which mean $n_i \equiv \beta \pmod{\alpha}$.

Because it is not clear which of several `property/objects` or `LinearProps` for a single variable should be applied, we try to find a linear property which represents all of them. We consider `LinearProp(`α_i`,integer,`β_i`)` for this purpose. If all $\alpha_i^{(k)}$ are numeric, β_i is solution of $\beta_i \equiv \beta_i^{(k)} \bmod \alpha_i^{(k)}$ for $k = 1, \ldots, l_i$. If not

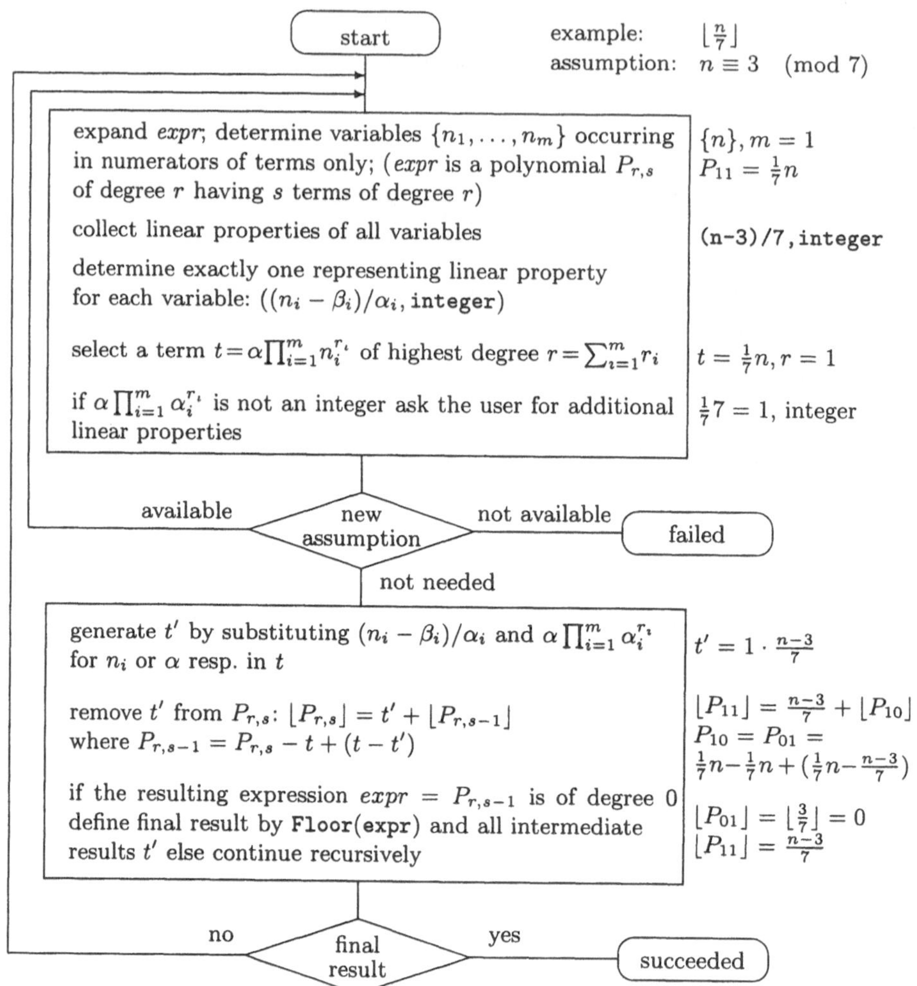

start
example: $\lfloor \frac{n}{7} \rfloor$
assumption: $n \equiv 3 \pmod 7$
expand *expr*; determine variables $\{n_1, \ldots, n_m\}$ occurring in numerators of terms only; (*expr* is a polynomial $P_{r,s}$ of degree r having s terms of degree r)
$\{n\}, m = 1$
$P_{11} = \frac{1}{7}n$
collect linear properties of all variables
(n-3)/7,integer
determine exactly one representing linear property for each variable: $((n_i - \beta_i)/\alpha_i,$ integer)
select a term $t = \alpha \prod_{i=1}^{m} n_i^{r_i}$ of highest degree $r = \sum_{i=1}^{m} r_i$
$t = \frac{1}{7}n, r = 1$
if $\alpha \prod_{i=1}^{m} \alpha_i^{r_i}$ is not an integer ask the user for additional linear properties
$\frac{1}{7}7 = 1$, integer
available
new assumption
not available
failed
not needed
generate t' by substituting $(n_i - \beta_i)/\alpha_i$ and $\alpha \prod_{i=1}^{m} \alpha_i^{r_i}$ for n_i or α resp. in t
$t' = 1 \cdot \frac{n-3}{7}$
remove t' from $P_{r,s}$: $\lfloor P_{r,s} \rfloor = t' + \lfloor P_{r,s-1} \rfloor$ where $P_{r,s-1} = P_{r,s} - t + (t - t')$
$\lfloor P_{11} \rfloor = \frac{n-3}{7} + \lfloor P_{10} \rfloor$
$P_{10} = P_{01} =$
$\frac{1}{7}n - \frac{1}{7}n + (\frac{1}{7}n - \frac{n-3}{7})$
if the resulting expression *expr* $= P_{r,s-1}$ is of degree 0 define final result by Floor(expr) and all intermediate results t' else continue recursively
$\lfloor P_{01} \rfloor = \lfloor \frac{3}{7} \rfloor = 0$
$\lfloor P_{11} \rfloor = \frac{n-3}{7}$
no
final result
yes
succeeded

Fig. 2. Evaluating `Floor(expr)` where `expr` is a multivariate polynomial.

all $\alpha_i^{(k)}$ are numeric, we use $\alpha_i = lcm\{\alpha_i^{(k)} : k = 1, \ldots, l_i\}$. This works if $l_i = 1$ or if all known linear properties of n_i, say `LinearProp(`$\alpha_i^{(k)}$`, integer, `$\beta_i^{(k)}$`)` for $k = 1, \ldots, l_i$, have a remainder $\beta_i^{(k)} = 0$. In the latter case, we use $\beta_i = 0$. At present no automatic solution can be presented for other cases. Therefore, the user has to spend some attention when defining the assumptions.

For evaluation of MAPLE's `floor` function, the following facts are used:

$$\lfloor xy \rfloor = xy \quad \text{if} \quad x, y \in \mathbb{Z} \tag{4}$$

$$\lfloor x + y \rfloor = x + \lfloor y \rfloor \quad \text{if} \quad x \in \mathbb{Z} \tag{5}$$

Our improved version `Floor` makes use of (4) and (5) very intensively and we use in addition

$$\lfloor x_1 x_2 \cdots x_k \rfloor = x_1 x_2 \cdots x_k \text{ if } x_1 x_2 \cdots x_k \in \mathbb{Z} \tag{6}$$

$$\lfloor x \rfloor = k \text{ if } k \le x < k+1 \text{ for } k = -2, -1, 0, 1, 2 \tag{7}$$

(6) might be known by a particular assumption. (7) seems to be very artificial. But after splitting off the numeric part using (5) and reducing the degree of indeterminates in the numerators of all terms by the algorithm of Fig. 2, the remainder is mostly a small number according to appropriate assumptions which can, therefore, be tackled in this way. The algorithm of Fig. 2 is able to evaluate $\lfloor P_{r,s}(n_1, \ldots, n_m) \rfloor$ if the user provides for appropriate assumptions.

In the present paper, all examples need only evaluation of `Floor(expr)` where the `expr` is a polynomial of the indeterminates (parameters). Rational expressions occur for example for sequential code segments of parallel programs. If a grid of size $m \times n$ in a grid oriented application is partitioned and mapped onto a $r \times s$ process system, then m/r and n/s might appear with loop parameters.

3.4 Evaluation of Nested DO-Loops

Handling of DO loops has already been discussed in [4, 7]. The probability of leaving a loop at the end of the loop body is $\frac{1}{n}$ if n is the number of executions of the loop body. If nested loops are independent of each other, then the same estimation can be used for each loop. If loop parameters of the inner loop depend on loop variables of the outer loops as in

$$\begin{array}{l} \texttt{DO } i_1 = b_1, e_1, d_1 \\ \quad \texttt{DO } i_2 = b_2(i_1), e_2(i_1), d_2(i_1) \\ \quad\quad \vdots \end{array}$$

the probability of leaving the j-th loop at the end of their loop body is

$$\frac{\sum\limits_{i_1} \sum\limits_{i_2} \cdots \sum\limits_{i_{j-1}} 1}{\sum\limits_{i_1} \sum\limits_{i_2} \cdots \sum\limits_{i_{j-1}} \sum\limits_{i_j} 1} \tag{8}$$

where

$$\begin{aligned} i_k &= b_k(i_1, \ldots, i_{k-1}) + l_k \cdot d_k(i_1, \ldots, i_{k-1}) \\ i_k &\le e_k(i_1, \ldots, i_{k-1}) \\ l_k &= 0, 1, 2, \ldots \end{aligned}$$

for $k = 0, 1, \ldots, j$.

Automated evaluation of (8) was a tantalizing engineering problem which has in principle been solved for all 28 types of nested loops found in BLAS routines of LAPACK. Here we distinguish cases of depending nested loops where

- the depth of nesting is 2 or 3,
- the outer loop index appears positive or negative in the inner loop,
- functions like *min* or *max* are used with loop parameters,
- different IF statements depending on loop variables occur.

In a simple case like

```
DO i = 1,n,1
   DO k = 1,i,1
      ⋮
```

we could simply apply the MAPLE function

```
sum(sum(1,k=1..i),i=1..n)
```

which delivers $\frac{1}{2}n^2 + \frac{1}{2}n$ after simplification for the number of executing the inner loop body. This approach works well only if the increment is 1 and if each value for the loop variable of the outer loop leads to a non-empty inner loop.

In a more general case like

```
DO i1 = b1, e1, d1
   IF (f1(i1)) THEN
      DO i2 = b2(i1), e2(i1), d2(i1)
         IF (f2(i1, i2)) THEN
            ⋮
```

however, the above approach fails. Before MAPLE's sum can be applied, we have to transform the code properly. Here we mean transformations which guarantee globally identical number of executions of the innermost loop (see [7]).

Transforming Increments. Using the substitution $i_k = b_k + d_k i'_k$ the k-th loop of the above loop nest becomes

$$\texttt{DO}\ i'_k = 0, Floor((e_k - b_k)/d_k), 1$$

The above expression for i_k has to be substituted in all loop parameters and IF expressions for each level $k' > k$. From now on, we always assume loop increments of 1.

Removing IF Statements. An IF statement can be removed from a loop nest without any effect to the number of executing the innermost loop body if the relational elements of the IF expression can be integrated in the corresponding loop parameters.

We are able to solve this problem for IF expressions which use operations like AND, OR, or NOT to combine simple relational expressions. If there is a loop variable i within a relational expression $f(i,x) \prec_1 0$ then there should be an equivalent relation $i \prec_2 g(x)$ where $\prec_1, \prec_2 \in \{<, \leq, >, \geq\}$. This is the case if f is linear in i but also in some other cases.

The first step is building the disjunctive normal form (DNF) of the IF expression using the logic package of MAPLE. It is only important that the terms are disjoint. In this case, we can split the loop into a collection of loops where each loop handles one of the terms and summarize the results afterwards. The algorithm could be optimized by simplifying the DNF using for example Quine-MacClusky algorithm but this is only a matter of saving time and will not be discussed here. Within a term we replace factors beginning with NOT by the opposite relation without NOT, i.e. NOT$(i \prec g(x))$ is replaced by $i \succ g(x)$ where

opposite symbols	represented pairs of operations			
$\prec$	$<$	$\leq$	$>$	$\geq$
$\succ$	$\geq$	$>$	$\leq$	$<$

Because of formal problems in handling expressions like `x=y` and `x<>y` with MAPLE, we have to replace these expressions by `'x<=y AND x>=y'` or `'x<y OR x>y'` resp. All resulting relations in level k are combined with the loop parameters of level $k' < k$ if k is the highest level for which the loop variable $i_{k'}$ occurs in the IF expression of the k-th level. Depending on the current relational expression of level k we replace in level k'

relation	replace	substitute
$i_{k'} < g_k(x)$	$e_{k'}$	$\min(e_{k'}, b_{k'} + \lceil g_k(x) - b_{k'} \rceil - 1)$
$i_{k'} \leq g_k(x)$	$e_{k'}$	$\min(e_{k'}, b_{k'} + \lfloor g_k(x) - b_{k'} \rfloor)$
$i_{k'} > g_k(x)$	$b_{k'}$	$\max(b_{k'}, b_{k'} + \lfloor g_k(x) - b_{k'} \rfloor + 1)$
$i_{k'} \geq g_k(x)$	$b_{k'}$	$\max(b_{k'}, b_{k'} + \lceil g_k(x) - b_{k'} \rceil)$

We do not need to discuss *ceiling* function separately because of $\lceil x \rceil = -\lfloor -x \rfloor$.

Handling of Case Functions within the Loop Parameters. We call a function $f(x_1, \ldots, x_m)$ a *case function* if there is a well defined sequence $r_1, \ldots, r_m$ of disjoint and exhaustive Boolean expressions (i.e. in each case exactly one of them is true) and $f(x_1, \ldots, x_m) = x_k$ iff $r_k = true$. At present we implemented in AUGUR the functions $\texttt{max}(x_1, x_2)$ and $\texttt{min}(x_1, x_2)$ using $r_1 := x_1 \geq x_2$, $r_2 := x_1 < x_2$ or $r_1 := x_1 \leq x_2$, $r_2 := x_1 > x_2$ respectively and restricted the analysis to cases with linear dependency of r_j on loop variables.

If a case function f appears at level k, e.g.

```
DO i_k' = b_k', e_k'
  :
    DO i_k = b_k(f(x_1,...,x_m)), e_k(f(x_1,...,x_m))
      :
```

and the related Boolean expressions $r_1, \ldots, r_m$ depend on $i_{k'}$ but not on any $i_{k''}$ where $k'' > k'$ then we consider the loop nests

```
DO i_k' = b_k', e_k'
  :
  IF (r_j = true) THEN
    DO i_k = b_k(x_j), e_k(x_j)
      :
```

for $j = 1, \ldots, m$ separately and summarize the results at the end. This kind of loop nest, however, has already been treated within the latter subsection.

Avoiding Empty Loops. After having removed all `IF` statements and case functions which depend on loop variables, there is a last problem with nested loops. In the case

```
⋮
DO i_{k'} = b_{k'}, e_{k'}
⋮
    DO i_k = b_k(i_{k'}), e_k(i_{k'})
    ⋮
```

$b_k(i_{k'}) > e_k(i_{k'})$ might happen for certain values of $i_{k'}$. In order to satisfy all requirements for using the `sum` routine of MAPLE successfully, we have to transform the above loop nest into

```
⋮
DO i_{k'} = b_{k'}, e_{k'}
⋮
IF (b_k(i_{k'}) ≤ e_k(i_{k'})) THEN
   DO i_k = b_k(i_{k'}), e_k(i_{k'})
   ⋮
```

Again we can proceed as before in order to complete the required transform.

Real Examples of Nested `DO`-Loops. We consider some loop nests as they can be found in BLAS routines of LAPACK (see Table 1). These results have been generated automatically where the user had to provide for appropriate assumptions only. Using the expression p_l in the last column of this table, we can define $a_{i+1,i} = \min(1, p_l)$ if i is the line number of the last statement of the inner loop. For the probability of entering the inner loop, we have to use $\min(1, 1/p_l)$. By the way, $1/p_l$ is the average number of executing the loop body for a single execution of the inner loop. In particular the last example of the table shows that a time-formula cannot easily be found by hand even for simple examples.

The assumptions have to be found by the user. They represent in general all useful cases of the application. Sometimes, it might be necessary to split the set of all cases into subsets which are described by appropriate collections of assumptions and which have to be analyzed separately (see `DGBMV` in Table 1).

In particular the assumptions for the latter example can be found by repeated call of the tool. If not all required assumptions are provided, a few expressions like $\min(1 + k_u, m - k_l)$ for example remain unevaluated. In this example, the user has to decide which value the min-function should take and after that he can provide the additional assumption, e.g. $1 + k_u \leq m - k_l$.

4 Analysis of LAPACK Routine `DSYMM` as an Example

In principal AUGUR is able to analyze all BLAS routines of LAPACK. We refer to [2] for more details of LAPACK. We use `DSYMM` for example. The original code

LAPACK routine	code	assumptions	probability p_l of leaving the inner loop at the end

DSYMM	`DO 170, J=1,N`
	`DO 140, K=1,J-1`
	⋮
		n,integer
		$n > 1$
			$\frac{2}{n-1}$

DSBMV	`DO 60, J=1,N`
	`DO 50, I=MAX(1,J-K),J-1`
	⋮
		n,integer
		k,integer
		$n > 1,\ k \geq 1$
		$1 + k < n$
			$-2\frac{n}{k(-2n+1+k)}$

DTPSV	`DO 80, J=1,N`
	`DO 70, K=KK,KK+J-2`
	⋮
	`70 CONTINUE`
	`KK=KK+J`
	⋮
		n,integer
		k_k,integer
		$n > 1$
		$k_k \geq 0$
			$\frac{2}{n-1}$

DGBMV	`DO 100, J=1,N`
	`DO 90, I=MAX(1,J-KU),`
	`* MIN(M,J+KL)`
	⋮
		n,integer, m,integer
		k_u,integer, k_l,integer
		$n > 2,\ k_u > 1,\ k_l > 1$
		$1 + k_u \leq m - k_l$
		$1 + k_u \leq n$
		$n \lesseqgtr m-k_l,\ n \lesseqgtr m+k_u$
			1) in the case of $m-k_l \leq n \leq m+k_u$
			2) in the case of $n < m - k_l$
			3) in the case of $n > m + k_u$

1) $\frac{-2n}{k_u-n+k_l-m+k_u^2+k_l^2-2mk_l+m^2-2k_un+n^2-2mn}$

2) $\frac{-2n}{-2nk_l-2n-2k_un+k_u+k_u^2}$ 3) $\frac{-2n}{k_l-2m-2mk_u-2mk_l+k_l^2}$

Table 1. Results of automated analysis for nested loops in BLAS routines of LAPACK

has been used as is. We only followed the recommendation of LAPACK to adapt the `STOP` statement of the error routine `XERBLA` to the user specific purposes. We removed the `STOP` in order to get a single time-formula for the complete routine.

The basic functionality of `DSYMM` is a certain matrix product. The original code consists of 147 executable statements and is rather complex because of several modes the routine can be used for. The kernel of `DSYMM` shows nested loops where the loop parameter of the inner loop depend on the outer loop. We refer to the original LAPACK library for the complete code but we list all values for mode variables which have been used and we show some parts of the code for illustration.

We consider the following call

```
CALL DSYMM(SIDE,UPLO,M,N,ALPHA,A,M,B,M,BETA,C,M)
```

With mode parameters `SIDE='L'`, `UPLO='U'`, `ALPHA`$\neq$`0`, and `BETA`$\neq$`0` the routine calculates the matrix product

```
C := ALPHA*A*B + BETA*C
```

where `A` is an upper triangular $m \times m$ matrix and `B` and `C` are $m \times n$ matrices.

The performance estimation is generated automatically by AUGUR after declaration of two parameters m, n for the size of the matrices and after definition of simple assumptions on these parameters like $m, n > 2$. In our case, 29 IF expressions were to be handled because they represent the evaluation of modes the routine is running in. All these expressions have been evaluated or bypassed automatically. Such a good result can not always be achieved. Mostly a few user interactions for defining IF expressions have to be accepted if other modes are selected or another way of encoding parameters is used.

At present, AUGUR supports counting of arbitrary classes of basic operations in which way the user is able to define his own performance metric. The user can select these operations and group subsets of them. Each subset of basic operations is accompanied by a timing parameter. For the following example, we assume that arithmetic integer operations (including index operations) take T1, arithmetic floating point operations take T2, and logical operations or relational operations of any type take T3 time units.

In this case, the kernel part of the routine is

```
      DO 70, J = 1, N
         DO 60, I = 1, M
            TEMP1 = ALPHA*B( I, J )
            TEMP2 = ZERO
            DO 50, K = 1, I - 1
               C( K, J ) = C( K, J ) + TEMP1     *A( K, I )
               TEMP2     = TEMP2     + B( K, J )*A( K, I )
   50       CONTINUE
            IF( BETA.EQ.ZERO )THEN
               C( I, J ) = TEMP1*A( I, I ) + ALPHA*TEMP2
            ELSE
               C( I, J ) = BETA *C( I, J ) +
     $                     TEMP1*A( I, I ) + ALPHA*TEMP2
            END IF
   60    CONTINUE
   70 CONTINUE
```

We obtain automatically by AUGUR the time-formula

```
n*m*T1+(4*n*m+2*n*m**2)*T2+(17+n*m)*T3
```

for the complete routine. Counting arithmetic operations of the above piece of code, the reader could find the T1 and T2 part of the time-formula. The remaining part would need consideration of the whole routine. As usual with members of a very general library, many other sets of mode parameters could be considered. If the parameters are changed, a slightly different result has to be expected due to the executed part of the routine.

5 Concluding Remarks

Runtime of sequential Fortran programs is predicted by a product of single statement analysis and an estimation of the frequency of statements. The method has been implemented in AUGUR which is a research tool for automatic generation of performance models for Fortran programs. The frequency vector is defined as the solution of a system of linear equations. AUGUR uses routine compile technology and the computer algebra tool MAPLE to define all required matrices and to solve the related equations.

The evaluation with MAPLE is concentrated on functions like `min`, `max`, `floor`, and `sum`. Because parametrized evaluation is required, the user has to provide for an appropriate collection of assumptions which simplify the resulting time-formulas. For this purpose the functions `min`, `max`, `floor` have been improved and the function `sum` has been generalized considerably. In addition to the kernel of MAPLE, the linear algebra and the logic package are used.

The considered method is able to assist the user in estimating the performance of a given code segment. However, it cannot replace external interactions in a situation where the control flow depends on the history, i.e. on numerical results having been obtained at the corresponding point in the original program.

The current version of AUGUR is still a prototype. Mainly the robustness has to be improved and the subset of Fortran which is accepted has to be extended. Future work will also include extensions of the tool set for automatic generation of time formulas for simple parallel programs.

References

1. A. O. Allen, *Introduction to Computer Performance Analysis with Mathematica*, Harcourt Brace & Company, Boston, 1994.
2. E. Anderson, Z. Bai, C. Bischof, J. Demmel, J. Dongarra, J. Du Croz, A. Greenbaum, S. Hammarling, A. McKenney, S. Ostroushov, and D. Sorensen, *Lapack User's Guide*, SIAM, Philadelphia, PA, 1992.
3. J. Cohen, *Computer-Assisted Microanalysis of Programs*, Comm. of the ACM, Vol. 25, 1982, pp. 724-733.
4. T. Fahringer, *Automatic Performance Prediction of Parallel Programs*, Kluwer Academic Publishers, Norwell, Massachusetts, 1996.
5. B. R. Haverkort and I. G. Niemegeers, *Performability modelling tools and techniques*, Performance Evaluation, Vol. 25, 1996, pp. 17-40.
6. M. Hofri, *Probabilistic Analysis of Algorithms*, Springer Verlag, New York, 1987.
7. H. Mierendorff and H. Schwamborn, Definition of control variables for automatic performance modeling, in: *Proceedings of the 1997 Conference on Advances in Parallel and Distributed Computing*, IEEE Comp. Soc. Press, Los Alamitos, CA, 1997, pp. 42-49.
8. H. Mierendorff and H. Schwamborn, *Automatic Model Generation for Performance Estimation of Parallel Programs*, Parallel Computing, vol. 25, 1999, pp. 667-680.
9. C. V. Ramamoorthy, *Discrete Markov Analysis of Computer Programs*, Proceedings of the 20th National Conference, ACM, August 1965, pp. 386-392.

A Parallel Symbolic-Numerical Approach to Algebraic Curve Plotting*

Christian Mittermaier, Wolfgang Schreiner, and Franz Winkler

Research Institute for Symbolic Computation (RISC-Linz)
Johannes Kepler University, Linz, Austria
FirstName.LastName@risc.uni-linz.ac.at
http://www.risc.uni-linz.ac.at

Abstract. We describe a parallel hybrid symbolic-numerical solution to the problem of reliably plotting a plane algebraic curve. The original sequential program is implemented in the software library CASA on the basis of the computer algebra system Maple. The new parallel version is based on Distributed Maple, a distributed programming extension written in Java. We describe the mathematical foundations of the algorithm, give sequential algorithmic improvements and discuss our parallelization approach.

1 Introduction

All modern computer algebra systems provide functions for plotting and visualizing the real affine part of the graph of curves which are given in implicit representation. These algorithms work well for curves defined by bivariate polynomials of reasonably low degree, however, as soon as the defining polynomial becomes somewhat complicated or the curve has singular points, these methods usually fail and return a picture from which one can only vaguely guess the shape of the curve.

Consider, for example, the *tacnode* curve, defined by the polynomial $f(x,y) := 2x^4 - 3x^2y + y^4 - 2y^3 + y^2$, a curve of degree four. In Figure 1 we see the graph of this curve produced by two different plotting algorithms. The picture on the left hand side was produced by Maple's `implicitplot` algorithm, which, as almost all plotting algorithms, uses numerical methods such as Newton's method in order to generate the visualization of the curve. The picture on the right hand side was produced by CASA's `pacPlot` algorithm, which is a hybrid symbolic numerical algorithm, combining both, the exactness of symbolic methods and the speed and efficiency of numerical approximation.

The difference in the quality of the output of the two algorithms is quite obvious but can be explained easily. Purely numerical methods use linearization in order to get an approximation. For this, the Jacobian of the defining polynomial has to be computed; but at those points on the curve, where two or

* Supported by projects P11160-TEC (HySaX) and SFB F013/F1304 of the Austrian Fonds zur Förderung der wissenschaftlichen Forschung.

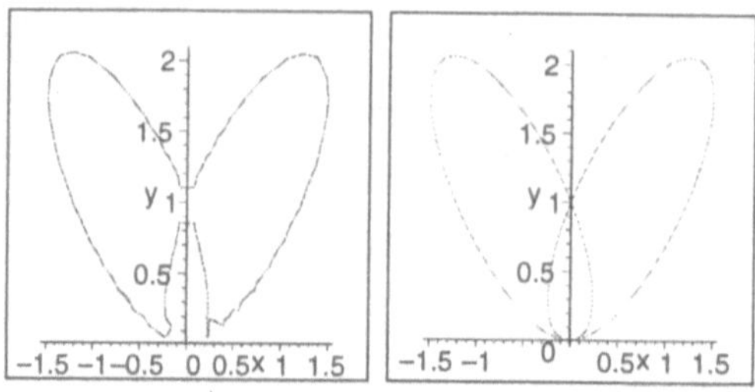

Fig. 1. Comparing Maple's `implicitplot` and CASA's `pacPlot`

more branches intersect each other, the Jacobian becomes singular as the partial derivatives with respect to both variables vanish. Increasing the granularity in the numerical plotting algorithm makes the picture smoother, but nevertheless the singular points are still missing.

A "good" plotting algorithm should fulfill the following informal requirements:

- **Draw topologically correct plots.** The course of the branches of the curve through singular points should be represented correctly, singular points must not be missed, in particular even isolated singularities, originating from two complex branches intersecting in one real point, should be represented by exactly one point.
- **Use an efficient method for rendering the curve.** Computing time should be as low as possible without sacrificing correctness. Increasing efficiency and creating smoother pictures is not acceptable if it means a compromise on topological correctness.

In the following sections we will analyze CASA's `pacPlot` algorithm which fulfills the requirements stated above and we show how computing time can be cut down by finding ways of employing parallel computation. Also the sequential version of `pacPlot` can be improved a lot by avoiding unnecessary computations in algebraic extension fields.

The program system CASA is based on Maple and contains a collection of algorithms for designing, analyzing and rendering algebraic curves and surfaces. CASA has been developed by a research group under the direction of the third author at RISC-Linz [MW96].

In a companion paper [SMW00] we describe the `pacPlot` algorithm from the point of view of parallelization while in this work we concentrate on the mathematical and algorithmic details.

The structure of this paper is as follows: In Section 2 we describe the mathematical and algorithmic ideas of the `pacPlot` algorithm and in Section 3 we list the most time consuming parts of the program. Section 4 deals with sequential algorithmic improvements of the original algorithm, Section 5 describes our parallelization approach, Section 6 presents experimental results which demonstrate the efficiency of the algorithm and the performance of the underlying distributed system. Finally, in Section 7 we summarize our results.

2 The pacPlot Algorithm

The `pacPlot` algorithm (plane algebraic curve plot) uses a hybrid symbolic numerical approach in order to plot plane algebraic curves defined by bivariate polynomials over $\mathbb{Q}$. It was developed in 1995 by Quoc-Nam Tran [Tra96].

The hybrid approach combines the exactness of symbolic methods with the speed and efficiency of numerical approximations. `pacPlot` uses a symbolic preprocessing step where the points determining the topological structure of the curve are computed. For tracing the branches between these points Nam's Extended Newton Method [Tra94] is used. The general structure of the algorithm looks as follows:

1. Determine all critical points of the curve.
2. Find out starting points for visualization.
3. Trace the simple branches of the curve by using numerical approximation.

Definition 1. *Let C be an affine plane algebraic curve defined by the polynomial $f \in \mathbb{Q}[x, y]$. $(\xi, \eta) \in \overline{\mathbb{Q}}^2$ is a* critical point *of C iff $f(\xi, \eta) = \frac{\partial f}{\partial x}(\xi, \eta) = 0$. ($\overline{\mathbb{Q}}$ is the algebraic closure or $\mathbb{Q}$, i.e. the field of algebraic numbers.)*

In order to clarify how critical points determine the topological structure of the curve, we present the following picture (Figure 2) of a curve defined by the polynomial $f(x, y) := 8y^2 - 8x^2 - 8y^3 - 8x^3 + 2y^4 + x^5$.

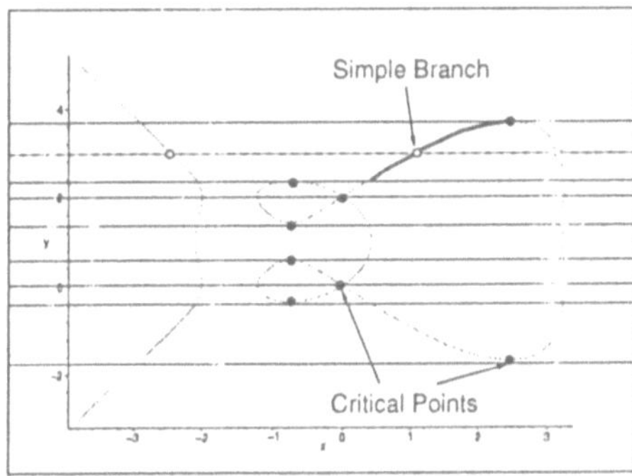

Fig. 2. Critical Points and the Topological Structure of an Algebraic Curve

The critical points are marked by black dots (we do not distinguish between singular points and extremal points). The curve is partitioned into horizontal stripes, each of them bounded by two critical points. Within such a stripe, the curve has no self intersecting branches and the Jacobian is regular in every point. The picture of the curve is generated by tracing all simple branches for each stripe. For every stripe, starting points for the visualization of all simple branches have to be computed. This is done by intersecting the curve in the middle of every stripe with a horizontal line. The intersection points form the set of starting points for the rendering process. In Figure 2, the starting points are marked by grey dots. Next, all simple branches are traced in both directions using the Extended Newton method starting at the initial points of visualization.

Before we describe critical point computation in detail, we mention two important observations:

1. We are interested in the real solutions of the system $f(x,y) = \frac{\partial f}{\partial x}(x,y) = 0$ only, as we want to plot the real part of the algebraic curve.
2. We do not need the exact solutions of the system. It is sufficient to have isolating intervals for the x and y values of reasonably small length as the resolution of the output device is limited to pixel level.
   By this approach, introduction of and computation with real algebraic numbers and their numerical evaluation is avoided.

For computing critical points we first transform the system $f = \frac{\partial f}{\partial x} = 0$ into a set of triangular systems of the form $p(y) = q(x,y) = 0$ ($p(y)$ irreducible over $\mathbb{Q}$) such that the solutions of all triangular systems are exactly the solutions of the original system. Isolating intervals for the real zeros of a triangular system are obtained by the following steps:

1. Find isolating intervals $Y_1, \ldots, Y_n \subset \mathbb{Q}$ for the real roots $\eta_1, \ldots, \eta_n$ of $p(y)$.
2. From p_i and q_i compute a polynomial $r(x)$ among whose real roots $\alpha_1, \ldots, \alpha_m$ we can also find for each real root η_j ($j \in \{1, \ldots, n\}$ of p_i all α_k's such that (α_k, η_j) is a zero of $p_i = q_i = 0$.
3. Find isolating intervals $X_1, \ldots, X_m \subset \mathbb{Q}$ for all real roots $\alpha_1, \ldots, \alpha_m$ of $r(x)$.
4. Check which combinations $X_k \times Y_j$, ($k \in \{1, \ldots, m\}$, $j \in \{1, \ldots, n\}$) contain a common zero of $p{=}q{=}0$.

Theorem 1 guarantees that we can always construct a polynomial $r(x)$ with the desired property.

Theorem 1. *Let $\mathbb{K}$ be an algebraically closed field and let $f(x,y) = \sum_{i=0}^{m} f_i(x)y^i$, $g(x,y) = \sum_{i=0}^{n} g_i(x)y^i$ be elements of $\mathbb{K}[x,y]$ of positive degrees m and n in y, respectively. Let $r(x) := \mathrm{res}_y(f,g)$. If $(\xi,\eta) \in \mathbb{K}^2$ is a common zero of f and g, then $r(\xi) = 0$. Conversely, if $r(\xi) = 0$, then one of the following holds:*

1. $f_m(\xi) = g_n(\xi) = 0$,
2. $\exists \eta \in \mathbb{K}: \ f(\xi,\eta) = g(\xi,\eta) = 0$.

Steps 2 and 3 are repeated until all intervals Y_j, $j \in \{1, \ldots, n\}$ are processed. In order to check which combinations $X_k \times Y_j$ contain a common zero of the system $p = q = 0$ we proceed as follows:

1. Factor p over $\mathbb{Q}$ and apply the following steps to each factor. W.l.o.g. we assume that p is the only irreducible factor.
2. Compute the greatest square free divisor $q^*(x,\eta)$ of $q(x,\eta)$ over $\mathbb{Q}(\eta)[x]$, where η is an arbitrary real root of $p(y)$.
$$q^*(x,\eta) = q(x,\eta)/\gcd(q(x,\eta), \frac{\partial}{\partial x}q(x,\eta)).$$
3. Take an interval Y_j, $j \in \{1,\dots,n\}$ and consider p as the minimal polynomial for the real root η_j represented by Y_j.
4. Use a real root isolation algorithm for polynomials with real algebraic number coefficients to determine the appropriate intervals X_k . Such an algorithm requires as input the real algebraic number η_j represented by the isolating interval Y_j, the minimal polynomial $p(y)$, the squarefree polynomial $q(x,y) \in (\mathbb{Q}[y]_{/\langle p_i \rangle})[x]$ representing $q(x,\eta_j)$, for which we want to isolate the real roots and a list of candidates $X_1,\dots,X_m$ for the isolating intervals (isolating intervals for the real roots of the resultant $r(x)$).

The algorithm used in step 4 computes the sign of the real algebraic numbers $q(\xi_1,\eta)$ and $q(\xi_2,\eta)$, where ξ_1 and ξ_2 are the end points of an interval X_k. If the signs are different or at least one of of the signs is zero then we have found an isolating interval. For determining the signs, Algorithm 1 is used. For details we refer to [Loo83].

Algorithm 1 Computing the sign of a real algebraic number β

Input: $p(y)$, minimal polynomial for the real algebraic number α,
$I =]\eta_1,\eta_2[$, isolating interval for α,
$q(y) \in \mathbb{Q}[y]_{/\langle p \rangle}$ representing the real algebraic number $\beta \in \mathbb{Q}(\alpha)$.

Output: s, the sign of β.

Compute q^* and b such that $(1/b)q^* = q$, $q^* \in \mathbb{Z}[y]$ and $b \in \mathbb{Q} - \{0\}$. $I^*(=]\eta_1^*,\eta_2^*[) := I$.
Set $\overline{q}$ to the greatest squarefree divisor of q^*.
loop
  Set n to the number of roots of $\overline{q}$ in I^*.
  if $n = 0$ **then**
    $s := \operatorname{sgn}(b) \cdot \operatorname{sgn}(q^*(\eta_2^*))$.
    return(s).
  end if
  $w := (\eta_1^* + \eta_2^*)/2$.
  if $p(\eta_1^*)p(w) < 0$ **then**
    $\eta_2^* := w$.
  else
    $\eta_1^* := w$.
  end if
end loop

Critical point computation for a curve C defined by the bivariate polynomial $f(x,y)$ is done by the following steps:

1. $f_x := \frac{\partial f}{\partial x}$; $L := 0$
2. From f and f_x compute a set T of triangular systems $(p(y), q(x,y))$ such that the union of the solutions of all triangular systems is the solution set of $f = f_x = 0$. This can be done using Kalkbrener's `bsolve` algorithm [Kal90].
3. For each triangular system $(p(y), q(x,y)) \in T$ do the following:
   (a) Factor p over $\mathbb{Q}$ giving a set $P := \{p_j | \prod_j p_j = p\}$.
   (b) For each non-constant factor $p_j \in P$ do the following operations:
      - If p_j is linear, then compute the exact solution η of $p_j = 0$. Substitute η for y in q and compute isolating intervals $]\xi_{1i}, \xi_{2i}[$ for the real roots of $q(x, \eta)$ whose length is smaller then the resolution of the output device. $L := L \cup \{]\xi_{1i}, \xi_{2i}[\times [\eta, \eta]\}$.
      - If $\deg(p_j) > 1$ then
         i. Let $I_1 := \{]\eta_{1i}, \eta_{2i}[\mid \eta_{1i}, \eta 2i \in \mathbb{Q}\}$ be a set of isolating intervals for the real roots of p_j of the required length.
         ii. Let α be a root of p_j and $q' := q(x, \alpha)$. Compute q'' as the quotient $q' / \gcd(q', \frac{\partial}{\partial x} q')$ over $\mathbb{Q}(\alpha)$. $q^* := q''(x, y)$.
         iii. $r(x) := \mathrm{res}_y(pj(y), q^*(x, y))$.
         iv. Let $I_2 := \{]\xi_{1i}, \xi_{2i}[\mid \xi_{1i}, \xi_{2i} \in \mathbb{Q}\}$ be a set of isolating intervals for the real roots of r whose length is smaller than the resolution of the output device.
         v. For every interval $]\eta_{1i}, \eta_{2i}[\in I_1$ consider p_j as minimal polynomial for the real root η represented by the isolating interval and the algebraically square free polynomial $q^*(x, y) \in (\mathbb{Q}[y]_{/\langle p_j \rangle})[x]$ as representation of $q_i(x, \eta)$. Use a real root isolation algorithm for polynomials with real algebraic number coefficients in order to check which intervals from I_2 represent a real algebraic number ξ such that $p_j(\eta) = q(\xi, \eta) = 0$. (Such an algorithm requires q to be square free for every root of p). As a result we get a list of Cartesian product of intervals of the form $L' := \{X_i \times Y_j \mid X_i \in I_2,\ Y_j \in I_1\}$.
         vi. $L := L \cup L'$.

3 Profiling `pacPlot`

We have analyzed the most time consuming parts of `pacPlot` by doing sample executions for randomly generated high degree curves. The sample runs were executed on a PIII@450MHz Linux machine.

Critical point computation is dominated by the gcd computation in the algebraic extension field of $\mathbb{Q}$ used for computing the greatest squarefree divisor. Resultant computation and real root isolation for polynomials with real algebraic number coefficients also require a large part of the overall computing time. Ordinary real root isolation is time consuming in only some of the cases, but still it is worth to put some effort in developing a parallel algorithm for this sub-problem.

	Example 1 deg = 14	Example 2 deg = 13	Example 3 deg = 18	Example 4 deg = 21
real root	137	187	7	171
resultant	542	455	82	4175
algebraic squarefree	[1]	7213	1204	[1]
real root over $\mathbb{Q}(\alpha)$	6525	1807	252	20060
$\sum$ critical points	>7213	9666	1546	>24484
$\sum$ `pacPlot`	>7267	9689	1552	>24518

Table 1. Profiling the `pacPlot` algorithm (all timings in seconds)

The overall impression obtained from this execution profile perfectly meets ones expectations, namely that the symbolic part of `pacPlot` is the toughest and most time consuming one.

4 Sequential Algorithmic Improvements

Also the sequential version of `pacPlot` can be improved a lot by avoiding unnecessary computations in algebraic extension fields. This section is based on the first author's diploma thesis [Mit00].

4.1 Replacing gcd-Computation over $\mathbb{Q}(\alpha)$

Critical point computation is mainly dominated by a gcd computation in an algebraic extension field of $\mathbb{Q}$. However, this gcd operation does not provide any information in the critical point computing step because it only ensures that the polynomial $q \in (\mathbb{Q}[y]_{/\langle p \rangle})[x]$, which is passed to the real root isolation algorithm for polynomials with algebraic number coefficients as a parameter is square free.

Algebraic real root isolation is used for checking which combinations of isolating intervals contain a critical point. By making this check more explicit instead of hiding it behind a real root isolation algorithm, we get completely rid of this time consuming operation and additionally generate lots of independent tasks for which employing parallel computation becomes easy. Algorithm 2 describes the idea in pseudo code.

For this approach, it is no longer necessary to compute the squarefree divisor of the polynomial $q(x, \alpha)$ for every real root α of $p(y)$ in advance. Hence, the gcd computation over $\mathbb{Q}(\alpha)$ becomes obsolete. However, the algorithm for computing the sign of a real algebraic number requires a squarefree polynomial representation of this number. By first substituting the left and right endpoint of the interval for the x values, we reduce the gcd computation over $\mathbb{Q}(\alpha)$ to a gcd computation over $\mathbb{Q}$ which is much more faster. From a mathematical point of view, the original approach is more elegant as we have to do several gcd computations here. But when efficiency is investigated carefully, our approach is superior to the former.

[1] Maple fails with an `object too large` error in this position but computation runs through when the gcd computation is omitted.

Algorithm 2 A more explicit and efficient algorithm for computing isolating rectangles for the real solutions of the triangular system $p(y) = q(x, y) = 0$.

Input: $p(y) \in \mathbb{Q}[y]$, irreducible over $\mathbb{Q}$,
$q(x, y) \in \mathbb{Q}[x, y]$, a bivariate polynomial,
$\{Y_1, \ldots, Y_m\}$, a set of isolating intervals for the real roots of p,
$\{X_1, \ldots, X_n\}$, list of isolating intervals for the real roots of $r(x) = \mathrm{res}_y(p, q)$ of the form $]\xi_{1i}, \xi_{2i}[$.
Output: L, set of isolating rectangles for the real solutions of $p = q = 0$.

$L := \emptyset$.
for i from 1 to n **do**
  $q_l(y) := q(\xi_{1i}, y) / \gcd(q(\xi_{1i}, y), \frac{\partial}{\partial y} q(\xi_{1i}, y))$.
  $q_r(y) := q(\xi_{2i}, y) / \gcd(q(\xi_{2i}, y), \frac{\partial}{\partial y} q(\xi_{2i}, y))$.
  {q_l, q_r are the greatest squarefree divisors of $q(\xi_{1i}, y)$ and $q(\xi_{2i}, y)$). }
  $q_l^* := q_l \bmod p$.
  $q_r^* := q_r \bmod p$.
  for j from 1 to m **do**
    $l := \mathrm{sgn}(q_l^*, p, Y_j)$.
    $r := \mathrm{sgn}(q_r^*, p, Y_j)$.
    if $l = 0 \vee r = 0 \vee l \cdot r < 0$ **then**
      $L := L \cup \{X_i \times Y_j\}$.
    end if
  end for
end for
return(L).

4.2 Speeding Up the Sign Computations

In the algorithm for computing the sign of a real algebraic number Moebius-transformations are used in order to transform the real roots of $\overline{q}$ in the interval I^* to the interval $]0, \infty[$ in order to apply Descartes' rule of signs. These transformations involve the left and right end-point η_1^* and η_2^*, which are huge rational numbers because all intervals obtained from real root isolation in the critical point computation algorithm are refined to at least the resolution of the output device. The huge rational numbers make the Moebius-transformations time consuming and furthermore, for some isolating intervals refinement is superfluous as they do not occur in any isolating rectangle. In order to make sign computation more efficient and simultaneously avoiding extra work, we pursue the following strategy, which has successfully been implemented in our version of the `pacPlot` algorithm:

1. Do not refine the isolating intervals in the real root isolation process.
2. Apply sign computation to the unrefined intervals only, in order to check which intervals actually form isolating rectangles.
3. Having determined the isolating rectangles, do refinement for all occurring intervals.

4.3 Execution Time Comparisons

Comparing execution times of the improved sequential algorithm (Table 2) with execution times of the original method (Table 1) is not possible in the level of the individual sub-algorithms, however, comparing the total computing time for Step 1, we see that our improved version is significantly better. Moreover, if we replace the gcd computation over the algebraic extension field of $\mathbb{Q}$ by our equivalent but less complex approach, the big examples (Example 1 and 4) executable in the sense that Maple does not fail with an `'object too large'` error.

	Example 1 deg = 14	Example 2 deg = 13	Example 3 deg = 18	Example 4 deg = 21
real root	83	97	3	64
resultant	386	150	55	3121
checking	6180	166	86	8279
refinement	127	52	8	49
$\sum$ critical points	6870	470	155	11748

Table 2. Execution times of the improved sequential `pacPlot` algorithm

5 The Parallel Plotting Algorithm

Parallelization of the plotting algorithm is based on the sequentially improved version. Parallelism is applied on several levels:

1. *Parallel Resultant Computation* Our parallelization approach applies a modular method to compute resultants in multiple homomorphic images $\mathbb{Z}_{p_i}[x, y]$, where the p_i are prime numbers. We get a divide and conquer structure where both, the divide phase (the modular resultant computations) and the conquer phase (the application of the Chinese Remainder Theorem) can be parallelized obviously. For details, see [HL94, Sch99].
2. *Parallel Real Root Isolation* We use Uspensky's method [CL83], a divide and conquer search algorithm, for computing isolating intervals. A naive parallelization of this algorithm typically yields poor speedups due to the narrowness of the highly unbalanced search trees associated with the isolation process [CJK90].
   Our approach is to broaden and flatten the search trees up to a certain extent such that in every step each processor can participate in the computation of isolating intervals. Our algorithm is based on speculative parallelism yielding better results in many of the cases. For details, see [Mit00].
3. *Parallel Solution Test and Interval Refinement* The tests which intervals $X \times Y$ do indeed contain a solution of the given triangular system can be performed in parallel in a straight-forward fashion. Likewise we can refine all isolating intervals in parallel to the desired accuracy.

Algorithm 3 (Parallel Critical Points) A parallel algorithm for critical point computation

---

Input: $f(x,y) \in \mathbb{Q}[x,y]$, irreducible
Output: L, list of critical points of f with respect to y-direction.

$f_x := \frac{\partial f}{\partial x}(x,y)$; $L := \emptyset$.
Apply `bsolve` to get a set T of triangular systems from f and f_x.
for all $(p,q) \in T$ **do**
  $F := \{p_i \mid \text{the } p_i \text{ are the irreducible factors of } p \text{ over } \mathbb{Q}\}$
  for all $p \in F$ **do**
    $Y := \{]\eta_{i1}, \eta_{i2}[\}$ set of isolating intervals for the real roots of p obtained by the **parallel real root isolation** algorithm.
    $r := \mathrm{res}_y(p,q)$ resultant of p and q computed by the **parallel modular approach.**
    $X := \{]\xi_{i1}, \xi_{i2}[\}$ set of isolating intervals for the real roots of r computed by the **parallel real root isolation** algorithm.
    do in parallel
      For each $]\eta_1, \eta_2[\in Y$ and each $]\xi_1, \xi_2[\in X$ compute the sign of $q(\xi_1, \alpha)$ and $q(\xi_2, \alpha)$, where α is the real root of p represented by the isolating interval $]\eta_1, \eta_2[$. Add $]\xi_1, \xi_2[\, \times\,]\eta_1, \eta_2[$ to L if the signs are different or at least one sign is 0.
    end do
  end for
end for
do in parallel
  Refine each interval occurring in L to the desired granularity.
end do

The various ideas developed above are incorporated in the parallel version of the `pacPlot` algorithm, the `dpacPlot` algorithm (distributed `pacPlot`). Parallelization affects the critical point computation step of `pacPlot`, only. Obviously, the various subalgorithms simply have to be replaced by their parallel counterpart. In Algorithm 3 (Parallel Critical Points) we give a high level description of how critical points are computed in `dpacPlot`.

We used the *Distributed Maple* environment [Sch98], a system for writing parallel programs on the basis of Maple, for implementing our algorithms. It allows to create tasks and execute them on Maple kernels running on various machines in a network.

Each node of a *Distributed Maple* session comprises two components, a Java based scheduler coordinating the interaction between nodes and the Maple interface, which is a link between kernel and scheduler. Both components use pipes to exchange messages (which may employ any Maple objects). The parallel programming paradigm is essentially based on functional principles which is sufficient for many kinds of algorithms in computer algebra.

The basic operations representing the core programming interface consist of methods for executing a Maple command on every machine connected to the session, creating a task for evaluating an arbitrary Maple expression (tasks are represented by task handles) and waiting for the result of a task represented by its handle.

6 Experimental Results

We have systematically benchmarked the `dpacPlot` algorithm implemented in *Distributed Maple* with four randomly generated algebraic curves for which the sequentially improved program takes 6870s, 470s, 155s and 11748s respectively. These times refer to a PIII@450MHz PC. The parallel algorithm has been executed in three system environments consisting of 24 processors each:

- A cluster which comprises 4 Silicon Graphics Octanes (2 R10000@250Mhz each) and 16 Linux PCs (various Pentium processors) linked by a 100 Mbit switched (point-to-point) Ethernet;
- a Silicon Graphics Origin multiprocessor (64 R12000@300 Mhz, 24 processors used);
- a mixed configuration consisting of our 4 dual-processor Octanes and 16 processors of the Origin multiprocessor interconnected via a high-bandwidth ATM line.

The raw total computational performance of cluster, Origin, and mixed configuration is 18.3, 17.1, and 18.7, respectively; these numbers (measured by a representative Maple benchmark) denote the sum of the performances of all processors relative to a PIII@450MHz processor. In the cluster and in the mixed configuration, the initial (frontend) Maple kernel was executed on an Octane.

The top diagram in Figure 3 generated from a Distributed Maple profile illustrates the execution in the cluster configuration with 16 processors listed on the vertical axis (8 Octane processors above 8 Linux PCs) and each line denotes a task executed on a particular machine; we see the real root isolation phase followed by the phases for resultant computation, the second real root isolation, the solution tests and the solution refinements. This visualization is generated from a run with Example 2 and illustrates the dynamic behavior, i.e. the number of tasks generated in each step of the algorithm. The length of these intervals does not reflect the execution time of these parts.

The table in Figure 3 lists the execution times measured in each system environment for each input with varying numbers of processors. The subsequent row of diagrams visualizes the absolute speedups $\frac{T_s}{T_n}$ (where T_s denotes the sequential execution time and T_n denotes the parallel execution time with n processors), the second row visualizes these speedups multiplied with $\frac{n}{\sum_{i=1}^{n} p_i}$ (where p_i denotes the relative performance of processor i), the third row visualizes the scaled efficiency, i.e., $\frac{T_s}{T_n \sum_{i=1}^{n} p_i}$, which compares the speedup we actually got to an upper bound of the speedup we could have got. The markers $+$, $\times$, and $\Box$ denote execution on cluster, Origin, and in the mixed configuration, respectively.

Analyzing the experimental data gives some interesting results. Most obviously, the speedups for larger examples is better than with smaller ones; for instance, in Example 1 the Cluster/mixed configuration gives an absolute speedup

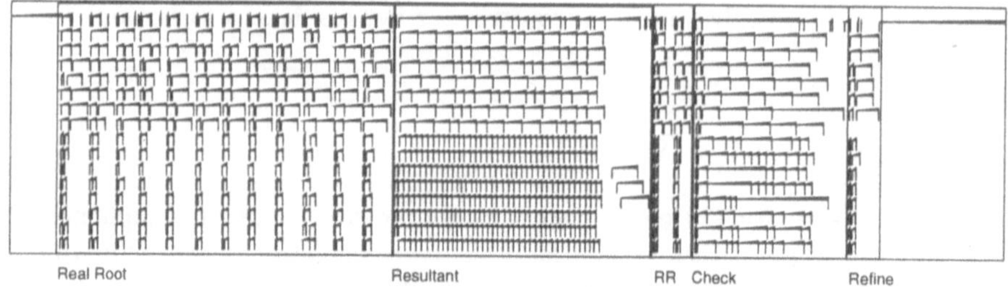

Execution Times (s)							
Example	Environment	1	2	4	8	16	24
1 (6870s)	Cluster	14992	8035	2732	1186	552	488
	Origin	7290	4217	1789	872	446	513
	Mixed	14992	8035	2732	1368	597	519
2 (470s)	Cluster	810	648	328	173	95	108
	Origin	667	541	297	166	112	116
	Mixed	810	648	328	210	116	127
3 (155s)	Cluster	267	191	112	67	46	45
	Origin	196	147	90	63	54	54
	Mixed	267	191	112	74	58	56
4 (11748s)	Cluster	25178	15559	6820	3562	1915	1563
	Origin	13397	8223	4009	2281	1726	1420
	Mixed	25178	15559	6820	4042	2004	1599

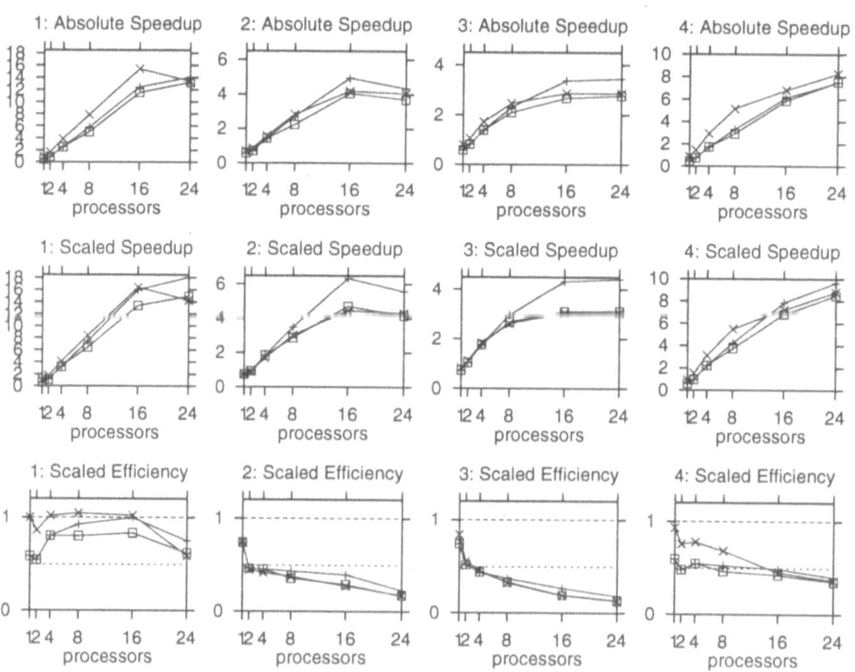

Fig. 3. Experimental Results

of 16 but only a speedup of 5 for Example 2. The Origin operates in Example 1 with scaled efficiencies close to 1 and gives in Example 4 (which has very large intermediate data) due to its high-bandwidth interconnection fabric for smaller processor numbers significantly better results than the other environments. Especially with 16 to 24 processors, however, in all examples the scaled speedups/efficiencies of the cluster compete with (are equal or higher than) those of the Origin. Moreover, the cheap Linux PCs in the cluster give overall better performance than the much more expensive Silicon Graphics machines.

When we consider the execution times of the parallel subalgorithms (not listed due to lack of space) individually, we realize that the speedups are partially much higher than the speedup of the overall algorithm. In Example 2 with 24 processors, the parallelization of resultant computation gives absolute speedups of 10.2 (cluster), 10.2 (Origin), and 7.9 (mixed). The parallelization of the checking phase gives absolute speedups of 12.3, 14.1, and 16.1 of the respective configurations. Although both phases together account for almost 80% of the total work, the less efficient parallelization of the remaining (much shorter) phases limits the overall speedup.

7 Conclusions

We have described an algorithm for reliably plotting plane algebraic curves which is based on both, symbolic and numerical methods. Analyzing the algorithm carefully, we have found sequential algorithmic improvements significantly reducing computing time.

Our parallelization of the symbolic part of the `pacPlot` algorithm demonstrates that also in the area of symbolic computation significant absolute speedups can be achieved. This was only possible after careful analysis and redesign of the original sequential algorithms. Also we see that computer networks can give speedups that are comparable to those on a massively parallel multiprocessor. Subtle algorithmic differences between the parallel and sequential version of the program give super-linear speedups in certain situations.

References

[CJK90] G. E. Collins, J. R. Johnson, and W. Küchlin. Parallel Real Root Isolation Using the Coefficient Sign Variation Method. In R. E. Zippel, editor, *Computer Algebra and Parallelism*, LNCS, pages 71–87. Springer Verlag, (1990).

[CL83] G. E. Collins and R. Loos. Real Zeros of Polynomials. In B. Buchberger, G. E. Collins, and R. Loos, editors, *Computer Algebra, Symbolic and Algebraic Computation*, pages 83–94. Springer Verlag Wien New York, 2nd ed. edition, (1983).

[HL94] H. Hong and H. W. Loidl. Parallel Computation of Modular Multivariate Polynomial Resultants on a Shared Memory Machine. In B. Buchberger and J. Volkert, editors, *Parallel Processing: CONPAR94-VAPP VI – Third*

Joint International Converence on Vector and Parallel Processing, number 854 in LNCS, pages 325–336. Springer, Berlin, (1994).

[Kal90] M. Kalkbrener. Solving Systems of Bivariate Algebraic Equations by Using Primitive Polynomial Remainder Sequences. Technical report, Research Institute for Symbolic Computation (RISC), 1990.

[Loo83] R. Loos. Computing in Algebraic Extensions. In B. Buchberger, G. E. Collins, and R. Loos, editors, *Computer Algebra, Symbolic and Algebraic Computation*, pages 173–187. Springer Verlag Wien New York, 2nd ed. edition, (1983).

[Mit00] Christian Mittermaier. Parallel Algorithms in Constructive Algebraic Geometry. Master's thesis, Johannes Kepler University, Linz, Austria, 2000.

[MW96] M. Mňuk and F. Winkler. CASA - A System for Computer Aided Constructive Algebraic Geometry. In *DISCO'96 — International Symposium on the Design and Implementation of Symbolic Computation Systems*, volume 1128 of *LNCS*, pages 297–307, Karsruhe, Germany, 1996. Springer, Berlin.

[Sch98] W. Schreiner. Distributed Maple — User and Reference Manual. Technical Report 98-05, Research Institute for Symbolic Computation (RISC-Linz), Johannes Kepler University, Linz, Austria, May 1998.

[Sch99] W. Schreiner. Developing a Distributed System for Algebraic Geometry. In B.H.V. Topping, editor, *EURO-CM-PAR'99 Third Euro-conference on Parallel and Distributed Computing for Computational Mechanics*, pages 137–146, Weimar, Germany, March 20-25, (1999). Civil-Comp Press, Edinburgh.

[SMW00] W. Schreiner, C. Mittermaier, and F. Winkler. On Solving a Problem in Algebraic Geometry by Cluster Computing. Submitted for publication, 2000.

[Tra94] Q.-N. Tran. Extended Newton's Method for Finding the Roots of an Arbitrary System of Equations and its Applications. Technical Report 94-49, Research Institute for Symbolic Computation (RISC), Johannes Kepler University Linz, Austria, 1994.

[Tra96] Q.-N. Tran. *A Hybrid Symbolic-Numerical Approach in Computer Aided Geometric Design (CAGD) and Visualization.* PhD thesis, RISC-Linz, Johannes Kepler University Linz, Austria, (1996).

Parametric Analysis for a Nonlinear System

M.A.Novickov

Institute of Systems Dynamics and Control Theory, Siberian Branch of the Russian Academy of Sciences, 134 Lermontov st., Irkutsk, 664033, Russia, E-mail: irteg@icc.ru

Abstract. The paper suggests an investigation of one system of differential equations with parameter carried out on the basis of the author's theorem on signdefiniteness of nonuniform structures [1, 2].

Consider a nonuniform structure of $(n+1)$ variables

$$V(x) = V_{2m}(x_1, \ldots, x_n) + V_*(x_1, \ldots, x_{n+1}), \tag{1}$$

where $n, m \geq 1$ are integer; $x \in R^{n+1}$; $V_{2m}(x_1, \ldots, x_n)$ is a positive uniform structure with respect to its variables; $V_*(x)$ is a polynomial of degree higher than $2m$ of $(n+1)$ variables. Perform the parametric substitution of variables [3, 4]:

$$x_i = \sum_{j=L}^{\infty} a_{ij} t^j, \quad (i = 1, \ldots, n), \quad a_{ij} \in R, \quad L > M, \quad x_{n+1} = t^M \tag{2}$$

and substitute these expressions into (1). As a result, we have

$$V(x(t)) = A_Q(a_{ij}, L, M)\, t^Q + \ldots , \tag{3}$$

where the coefficient A_Q affecting t^Q is dependent on both the coefficients of form (1) and parametrization (2); $Q \geq 2mL$; the dots ... denote the terms of higher orders with respect to t . Basing on the structure of the expression A_Q in (3), it is possible to prove the following

Theorem 1. *In the case when*

1° : *a)* $Q = 2\gamma + 1$ *(*γ *is integer) , or*

b) $Q = 2\gamma$ *and* A_Q *assumes negative values for some real* a_{ij}, *then* $V(x)$ *is signvariable;*

2° : $Q = 2\gamma$ *and* $A_Q > 0$ *for all* $a_{ij} \in R$, *then* $V(x)$ *is positive definite;*

3° : $Q = 2\gamma$ *and* $A_Q \geq 0$ *for all* $a_{ij} \in R$,

then $V(x)$ *can be either signdefinite or signvariable. This is determined with the aid of the terms of orders higher than* L *in expansion (2), and, respectively, terms of orders higher than* Q *in (3).*

Let us employ this theorem in the analysis of the following problem. Let disturbed motions of controlled system be described by the differential equations:

$$\begin{cases} \dot{x}_1 = -x_1^3 - 2\,x_1x_2^2 - 2.5\,x_1^2x_2x_3^2 + x_3^6\,, \\ \dot{x}_2 = -x_1^2x_2 - 2\,x_2^3 + x_1^3x_3^2 + 1/2\,x_3^6\,, \\ \dot{x}_3 = 2\,x_1^3x_2x_3 + (2\,x_1 + x_2)\,x_3^5 - s\,x_3^7\,, \end{cases} \tag{4}$$

where s is a real control parameter.

Consider the problem of finding the largest set of values of s, for which the trivial solution of (4) is asymptotically stable.
One can encounter such systems in the process of modeling (under definite reservations (conditions)) some problems of mechanics, e.g., rolling of a 3-wheel vehicle, which has a controlled (drived) forward wheel, along some absolutely rough, horizontal surface [5].
In accordance with Lyapunov's theorem of asymptotic stability [6] let us choose the negative definite quadratic form

$$V(x) = -\frac{1}{2}\,(x_1^2 + ax_2^2 + bx_3^2) \tag{5}$$

in the capacity of the Lyapunov function. The indeterminate weighting coefficients a and b will be found from the condition of the "best" choice of the function (5) corresponding to asymptotic stability of the trivial solution of the system (4). The derivative of $V(x)$ due to differential equations (4) has the form:

$$\dot{V}(x) = (x_1^2 + 2x_2^2)\,(x_1^2 + ax_2^2) - [(1 + 2b)\,x_1 + (a + 2b)\,\frac{x_2}{2}]\,x_3^6 \\ + (2.5 - a - 2b)\,x_1^3x_2x_3^2 + s\,b\,x_3^8\,. \tag{6}$$

In accordance with [6], in order to obtain asymptotic stability of the trivial solution of (1) we must impose the requirement $\dot{V}(x) \gg 0$. The goal of the present investigation implies the finding of the positive parameters a and b such that the domain S of values of the parameter s be the largest.
In the expression (6) the form of lower 4th order for $a > 0$ is sign-constant. For the purpose of investigation of positive definiteness of the nonuniform form $\dot{V}(x)$ let us apply the sign-definiteness criterion [1, 2] based on the parametric representation of the variables [3, 4]. From the very outset, the parametrization in our problem may be chosen in the form:

$$x_1 = a_{12}t^2 + a_{13}t^3 + \ldots,\; x_2 = a_{22}t^2 + a_{23}t^3 + \ldots,\; x_3 = t\,. \tag{7}$$

After the substitution of (7) into (6) we have

$$\dot{V}(x(t), a, b, s) = A_8(a, b, s, a_{12}, a_{22})\,t^8 + \ldots\,,$$

where

$$A_8(a, b, s, a_{12}, a_{22}) = (a_{12}^2 + 2\,a_{22}^2)\,(a_{12}^2 + a\,a_{22}^2) - [(1 + 2b)\,a_{12} \\ + (a + 2b)\,a_{22}/2] + sb\,,$$

and the dots (...) denote the terms of orders higher than 8 with respect to t. For the purpose of analysis of the sign of $A_8(a, b, s, a_{12}, a_{22})$ let us first introduce the denotations:

$$c = \frac{a_{12}}{a_{22}}, \ \alpha = \frac{1}{a_{22}}, \ r_1 = (1 + 2b)\, c + a/2 + b,$$
$$r_2 = (c^2 + 2)\,(c^2 + a)\ ,$$

hence we obtain

$$A_8 = a_{22}^4\, sb \left[\alpha^4 - \frac{r_1\, \alpha^3}{sb} + \frac{r_2}{sb}\right] .$$

For positive A_8 and for all real α the polynomial

$$\varphi = \alpha^4 - \frac{r_1\, \alpha^3}{sb} + \frac{r_2}{sb}$$

will assume only positive values (i.e. the polynomial assumes the value of the same sign as that of the coefficient of the highest 4th order). Hence, in case of positive values φ and the parameter b naturally follows the requirement $s > 0$. Therefore, the condition $A_8 > 0$ for $a, b \in R^+$; $c, \alpha, s \in R$ is reduced to the following two inequalities

$$s > 0\ , \ f(\alpha) = sb\, \alpha^4 - r_1\, \alpha^3 + r_2 > 0\ . \tag{8}$$

The last inequality in (8) expresses the absence of real roots for the equation $f(\alpha) = 0$. For this result, in accordance with Theorem 2 [7] on sign-definiteness of uniform fourth-order forms of two variables $sb\, u^4 - r_1\, u^3 v + r_2\, v^4$, it is necessary and sufficient that

$$\frac{r_2}{sb} > \frac{3^3\, r_1^4}{4^4\, s^4\, b^4} > 0\ .$$

By resolving the latter inequality with respect to the parameter s we obtain

$$s^3 > \frac{3^3\, r_1^4}{4^4\, r_2\, b^3} = P(a, b, c)\ . \tag{9}$$

Hence the initial problem is reduced to finding the smallest value of s_0 for $a \in R^+$, $b \in R^+$, $c \in R$.
Since inequality (9) must hold for all real values of c, it is natural that

$$s^3 > P_*(a, b) = \max_{c \in R} P(a, b, c)\ .$$

The variable parameters a and b will be chosen to satisfy the condition of reaching the minimum for the function $P_*(a, b)$. So the estimation of the set S leads to the extremum problem

$$P_0 = \min_{a \in R^+} \min_{b \in R^+} \max_{c \in R} P(a, b, c)\ . \tag{10}$$

When there are several stationary points (a,b,c) for the problem (10), the best estimate s_0 is obviously provided by the values which are the smallest for the function $P(a,b,c)$.
Stationary points of problem (10) are realized at the points (a,b,c), which are among the solutions:

$$\frac{\partial P}{\partial a} = 0, \quad \frac{\partial P}{\partial b} = 0, \quad \frac{\partial P}{\partial c} = 0\,, \quad a > 0\,,\ b > 0\,, \tag{11}$$

$$\frac{\partial P}{\partial b} = 0, \quad \frac{\partial P}{\partial c} = 0\,, \quad a = 0\,,\ b > 0\,, \tag{12}$$

where

$$\begin{aligned}
\frac{\partial P}{\partial a} &= \frac{r_1^3}{b^3\, r_2^2}\,[2\,c^2 - (1+2b)\,c + \frac{3}{2}\,a - b]\,,\\
\frac{\partial P}{\partial b} &= \frac{r_1^3}{b^4\, r_2}\,[2(3 - 2b) + 3a - 2b]\,,\\
\frac{\partial P}{\partial c} &= -\frac{r_1^3}{b^3\, r_2^2}\,[2(a+2b)\,c^3 - 2(1+2b)\,(a+2)\,c^2\\
&\quad + (a+2)\,(a+2b)\,c - 8a(1+2b)]\,.
\end{aligned}$$

Other possible variants of the constrained extremum

$$1)\quad \frac{\partial P}{\partial a} = 0,\ b = 0,\ \frac{\partial P}{\partial c} = 0\,,\ a > 0\,;$$

$$2)\quad a = 0,\ b = 0,\ \frac{\partial P}{\partial c} = 0$$

for $b \to 0,\ r_1 \neq 0$ provide that $P \to +\infty$ and, hence, do not provide any solutions of (10).
The solution $r_1 = 0$ for $a \in R^+,\ b \in R^+$ is one of possible solutions for system (11). Analysis of expressions of first derivatives of $P(a,b,c)$ shows that in case of variations of the parameter a only there can be the values both greater and smaller than $P = 0$ in the neighbourhood of the set $r_1 = 0$. So, the set $r_1 = 0$ cannot contain solutions of (10).
For the system (11) we have obtained two possible solution sets:

$$1)\ a_1 = 1\,,\ b_1 = 1.5\,,\ c_1 = 2.951021337$$

(here c_1 has been found approximately with precision of 10^{-11} from the equation $2\,c^3 - 6\,c^2 + 3\,c - 8 = 0$),

$$2)\ a_2 = -2\,,\ b_2 = 9/8\,,\ c_2 = 11/2.$$

The latter set of solutions is exempt from analysis in virtue of the necessary requirement $a > 0$. For the first solution our computations have given

$P(a_1, b_1, c_1) = 10.9143403327$, $s_1 = (P(a_1, b_1, c_1))^{1/3} = 2.21819215$. Analysis of the quadratic form $P(a, b, c)$ at point (a_1, b_1, c_1) allows to define the "saddle point" with respect to the variables a, b.

The following solutions are possible for the system (12):

1. $a_3 = 0,\ b_3 = (11 - \sqrt{41})/4,\ c_3 = (13 - \sqrt{41} - \sqrt{274 - 26\sqrt{41}})/8,$
2. $a_4 = 0,\ b_4 = (11 - \sqrt{41})/4,\ c_4 = (13 - \sqrt{41} + \sqrt{274 - 26\sqrt{41}})/8,$
3. $a_5 = 0,\ b_5 = (11 + \sqrt{41})/4,\ c_5 = (13 + \sqrt{41} - \sqrt{274 + 26\sqrt{41}})/8,$
4. $a_6 = 0,\ b_6 = (11 + \sqrt{41})/4,\ c_6 = (13 + \sqrt{41} + \sqrt{274 + 26\sqrt{41}})/8,$
5. $a_7 = 0,\ b_7 = 1.220042044132934,\ (4\, b_7^3 - 10\, b_7^2 + 21\, b_7 - 18 = 0),$
   $c_7 = 2.178973696879423.$

At the first point (a_3, b_3, c_3) we have:

$$P(a_3, b_3, c_3) = 0.00382477070995,\ s_3 = 0.1563874152069742.$$

The second derivatives $\frac{\partial^2 P}{\partial b^2}$, $\frac{\partial^2 P}{\partial c^2}$ help to define the minimum condition at the point

$$\min_{b \in R^+} \min_{c \in R} P(0, b, c)\ .$$

Since the maximum condition with respect to the variable c is not satisfied here, the point (a_3, b_3, c_3) is not the solution of (10).
At the second point we have:

$$P(a_4, b_4, c_4) = 10.462059750521,\ s_4 = 2.187118925763814,$$

and the stationary point corresponds to

$$\min_{b \in R^+} \min_{c \in R} P(0, b, c)\ .$$

For the next point we have:

$$P(a_5, b_5, c_5) = 0.557471072718,\ s_5 = 0.823014421944817,$$

and here

$$\max_{b \in R^+} \min_{c \in R} P(0, b, c)\ .$$

For the point (a_6, b_6, c_6) the following values have been obtained:

$$P(a_6, b_6, c_6) = 14.78484186349014, \quad s_6 = 2.454363561376983.$$

In this case, the form of the extremum point writes:

$$\min_{b \subset R^+} \max_{c \in R} P(0, b, c).$$

For the last solution we have:

$$P(a_7, b_7, c_7) = 10.4609954377932, \; s_7 = 2.187044757531\,,$$

and so the point (a_7, b_7, c_7) does not satisfy the problem (7) because of

$$\min_{b\in R^+} \min_{c\in R} P(0, b, c)\,.$$

Finally, for (10) we have only one solution

$$P(a_6, b_6, c_6) = 14.78484186349014,$$

and the desired value of $s_0 = 2.454363561376983$.

In the critical case of three zero roots of the characteristic equation of the system (4), it is impossible to obtain necessary stability conditions in the first approximation. So, let us compare the obtained sufficient conditions of asymptotic stability with the conditions of instability. To this end let us apply Khamenkov's theorem [8]. In accordance with this theorem, to provide instability of the trivial solution for the system $\dot{x}_j = X_j(x)$ it is sufficient that on any real solution

$$F_{1j}(x) = 0, \;\ldots,\; F_{(k-1)j}(x) = 0\,, F_{(k+1)j}(x)\,,\;\ldots,\; F_{nj}(x) = 0$$
$$\text{for } j, k = 1, \ldots, n; \; F_{ij}(x) = X_i\, x_j - X_j\, x_i$$

the expression $L(x) = X_1\, x_1 + \ldots + X_n\, x_n$ would assume positive sign. For the system (4) the Khamenkov functions assume the form:

$$\begin{aligned}
F_{12}(x) &= (x_2 - x_1/2)\, x_3^6 - 2.5\, x_1^2 x_2^2 x_3^2 - x_1^4 x_3^2\,;\\
F_{13}(x) &= (-x_1^3 - 2x_1 x_2^2 + x_3^6)\, x_3 - (2\, x_1 + x_2)\, x_1 x_3^5 + s\, x_1 x_3^7\\
&\quad - 2.5\, x_1^2 x_2 x_3^3 - 2\, x_1^4 x_2 x_3\,;\\
F_{23}(x) &= (-x_1^2 x_2 - 2x_2^3 + x_3^6/2)\, x_3 - (2\, x_1 + x_2)\, x_2 x_3^5\\
&\quad + s\, x_2 x_3^7 + x_1^3 x_3^3 - 2\, x_1^4 x_2^2 x_3\,;\\
L(x) &= -(x_1^2 + x_2^2)\,(x_1^2 + 2x_2^2) + 3\,(2\, x_1 + x_2)\, x_3^6/2 - s\, x_3^8\,.
\end{aligned}$$

For finding real solutions construct the systems:

$$\begin{aligned}
F_{12}(x) = 0\,, \quad F_{13}(x) &= 0\,,\\
F_{12}(x) = 0\,, \quad F_{23}(x) &= 0\,,\\
F_{13}(x) = 0\,, \quad F_{23}(x) &= 0\,.
\end{aligned} \tag{13}$$

Here we also applied the technique of parametric representation of the curves [3, 4]: $x_1 = \alpha_2\, t^2 + \alpha_3\, t^3 + \ldots$, $x_2 = \beta_2\, t^2 + \beta_3\, t^3 + \ldots$, $x_3 = t$. The initial terms of the expansions are

$$\begin{aligned}
G_{12} &= (\beta_2 - \alpha_2/2)\, t^8 + \ldots\,,\\
G_{13} &= (\alpha_2^3 - 2\alpha_2\, \beta_2^2 + 1)\, t^7 + \ldots\,,\\
G_{23} &= (-\alpha_2^2\, \beta_2 - 2\beta_2^3 + 1/2)\, t^7 + \ldots\,,\\
L &= [-(\alpha_2^2 + \beta_2^2)\,(\alpha_2^2 + 2\beta_2^2) + 3/2\,(2\,\alpha_2 + \beta_2) - s]\, t^8 + \ldots\,.
\end{aligned}$$

Each of the systems in (10) has a unique solution $\alpha_2 = \beta_2/2$, $12\,\beta_2^3 = 1$. Having substituted $\beta_2 = (1/12)^{1/3}$ into the expression L , we finally obtain $L = 5\,\beta_2 - s$. For providing instability it is necessary to impose the requirement $L > 0$, and hence, $s < 5\,\beta_2 = 2.1839116185$.

Consequently, for $s < s_* = 2.1839116185$ the trivial solution of the system (4) is unstable, and for $s > s_0 = 2.4543635613775$ it is asymptotically stable.

There is no need to conduct an analysis of the boundaries $s = s_*$ and $s = s_0$ since the coefficients for the 8th order of t in $\dot{V}(x)$ and $L(x)$ vanish, and the higher-order terms with respect to t in this problem can take on the values with any signs.

The solution of the extremum problem is reached at the boundary $a = 0$.

As obvious from the sequence of computational operations, in the process of solving the considered control problem there appeared many operations related to different special computations. A personal computer (PC) was used for (i) obtaining expressions of the first- and second-order derivatives of the function $P(a,b,c)$; (ii) analytical transformations in the process of solving algebraic equations and systems of equations (11), (12). For such a type of problems it is possible to apply the well-known systems "Maple" or "Mathematica". In our case we employed the system "Mathematica".

References

1. Novickov, M.A.: On sign-definiteness of analytical functions. In: *The Method of Lyapunov Functions in the Analysis of Systems Dynamics*, Nauka, Novosibirsk, 1987, pp. 256–261 (in Russian).
2. Novickov, M.A.: An Investigation into Stability of Conservative Mechanical Systems Using Analytic Calculations. In: *Computer Algebra in Scientific Computing / CASC'99*, V.G. Ganzha, E.W. Mayr and E.V. Vorozhtsov (Eds.), Springer-Verlag, Berlin, 1999, pp. 317–322.
3. Walker, R.: *Algebraic Curves*, Inostr. Lit. Publ., Moscow, 1952 (in Russian).
4. Bruno, A.D.: *A Local Method of Nonlinear Analysis of Differential Equations*, Nauka, Moscow, 1979 (in Russian).
5. Letov, A.M.: *Stability of Nonlinear Controlled Systems*, FIZMATGIZ, Moscow, 1962 (in Russian).
6. Lyapunov, V.M.: *A General Problem of Stability of Motion*, Gostekhizdat, Moscow and Leningrad, 1950 (in Russian).
7. Irtegov, V.D., Novickov, M.A.: Sign-definiteness of 4th order forms of two variables. In: *The Method of Lyapunov Functions and Its Applications*, Nauka, Novosibirsk, 1984, pp. 87–93 (in Russian).
8. Khamenkov, G.V.: *Selected Works. Stability of Motion. Oscillations. Aerodynamics*, Nauka, Moscow, 1971 (in Russian).

Extended Splicing System and Semigroup of Dominoes

Pethuru Raj, Naohiro Ishii

Department of Intelligence and Computer Science
Nagoya Institute of Technology
Nagoya, 466 8555, Japan
{peter,ishii}@ics.nitech.ac.jp

Abstract. Tom Head[5] introduced the novel idea of splicing system as a generative device for representing the string restructuring that takes place during the interactions of linear biopolymers in the presence of precisely specified enzymatic activities and thus established a new relationship between formal language theory and the study of informational macromolecules. In this paper, we discuss the relationship existing between the extended splicing system and the semigroup of dominoes, a special algebraic structure acting on linked strings.

1 Introduction

One of the most recent suggestions in developing new types of computers consists of considering computers based on molecular interactions, which under some circumstances, might be an alternative to the classical Turing/Von Neumann notion of computing. The concept of splicing system serves as one of the proposals for achieving universal programmable molecular computer[7].

DNA(deoxyribonucleic acid) is found in all living organisms as the storage medium for genetic information. It consists of polymer chains, customarily referred to as DNA strands. A chain is composed of nucleotides, also referred to as bases. The four DNA nucleotides or bases are denoted by A(adenine), C(cytosine), G(guanine), and T(thymine) and hence the DNA alphabet is $\Sigma_{DNA} = \{A, C, G, T\}$.

Thus, DNA strands may be viewed as words over the DNA alphabet. According to a chemical convention, each strand has a $5'$ end and a $3'$end, for instance, $5'$ $ATTAGCAT$ $3'$ or $3'$ $TAATCGTA$ $5'$ making the words oriented. Bonding happens by the pairwise attraction of bases: The base A bonds with T, and C bonds with G. Thus the two strands mentioned above will form the double strand:

$$\begin{array}{c} 5'\ ATTAGCAT\ 3' \\ 3'\ TAATCGTA\ 5' \end{array}$$

The effect of a restriction enzyme is to cut the DNA molecule into two pieces at a specific pattern. For example, as quoted in [9], the enzyme EcoRI represented by the pair of strings(also called its recognition sequence)

GAATTC
CTTAAG

acts on a hypothetical DNA molecule

...GCTACTAGAATTCGCGCTA...
...CGATGATCTTAAGCGCGAT...

The enzyme works on the molecule by finding a substring pair in the DNA molecule identical to its corresponding recognition sequence and cuts the molecule into two staggered pieces

...GCTACTAG   AATTCGCGCTA...
...CGATGATCTTAA   GCGCGAT...

The staggered ends of a DNA molecule recombine with others if they match. For example, if there exists the following piece

AATTACATT...
..........TGTAA...

cut from some other molecule by the same enzyme or different enzyme, a new hybrid DNA molecule is generated by combining this piece with the first half of the fragments cut by EcoRI. That is,

...GCTACTAGAATTACATT...
...CGATGATCTTAATGTAA...

Thus, a combination of enzymes and ligases with the DNA sequences in a test tube can create new hybrid DNA molecules under certain situations.

To abstract this chemical process and bring out a formal model mathematically, each pair is viewed as a single symbol in an alphabet Σ. An enzyme is represented by a triple (a, x, b) for $a, x, b \in \Sigma^*$. The cut and recombination takes place at x. In the above example, $a = G/C$,and x is

AATT
TTAA

and $b = C/G$. Thus, if (c, x, d) represents another enzyme and $uaxbv$ and $wcxdz$ are the original DNA molecules, then $uaxdz$ and $wcxbv$ are said to have been formed by splicing at x. Here is one another biological complexity. In the above example, the DNA was cut as below.

...G   AATTC...
...CTTAA   G...

But by a different enzyme, the same DNA could be cut as follows:

...GAATT   C...
...C   TTAAG...

Because of the orientation of the molecules, cuts of the former type cannot recombine with cuts of the latter type. Consequently, two sets of patterns are introduced, called left and right-handed.

2 Splicing System

This section formally defines splicing system, which was first introduced in [5].

2.1 Definition

A splicing system is a quadruple $S = (\Sigma, I, B, C)$, where
Σ : a finite alphabet,
I : a set of initial strings in Σ^*, and
B, C : finite sets of triples (a, x, b), for $a, x, b \in \Sigma^*$.

The size of I can be infinite. The sets B and C are called left-hand patterns and right-hand patterns respectively. String x is called the crossing of the site axb. (In a biological interpretation, one may take I as the initial set of DNA molecule sequences, B and C as the sets of splicing rules specified by restriction enzymes and a ligase).

2.2 Definition

For a splicing system $S = (\Sigma, I, B, C)$, $L(S)$ is the set of strings generated by S which is formally defined as follows:

1. $I \subseteq L(S)$.
2. If $uaxbv$ and $pcxdq$ are in $L(S)$, and axb and cxd are sites of the same hand, then $uaxdq$ and $pcxbv$ are also in $L(S)$.

In other words, $L(S)$ is the minimal subset of Σ^* which contains I and is closed under the operation of splicing.

3 Semigroup of Dominoes

3.1 Definition

Let S be a semigroup and $\Delta \subseteq S \times S$, a semigroup of the direct product, and let S' be the extension of S to a monoid with the identity element denoted by 1. A Δ-domino α is a triple $(l(\alpha), m(\alpha), r(\alpha))$, usually written as $\alpha = l(\alpha).m(\alpha).r(\alpha)$ where $l(\alpha), r(\alpha) \in (dom\Delta \times 1) \cup (1 \times ran\Delta) \cup (1 \times 1), m(\alpha) \in \Delta$, and the products are in $S' \times S'$. The components of α are said to be its left, middle and right parts.

If $\alpha = m(\alpha)$, then α is called blunt domino. Letting $\tau_i : S' \times S' \to S'$ be the projection for $i = 1, 2$, the pair $(\tau_1(\alpha), \tau_2(\alpha))$ is the underlying pair of the Δ-domino α. Blunt domino α is called an equal domino if its underlying pair is (x, x) for some $x \in S'$, that is, α is an equal domino whenever $l(\alpha) = (1, 1) = r(\alpha)$ and $\tau_1(\alpha) = \tau_2(\alpha)$. The graphical representation of Δ-domino α with $l(\alpha) \in S' \times 1, r(\alpha) \in 1 \times S'$ is depicted in fig 1.

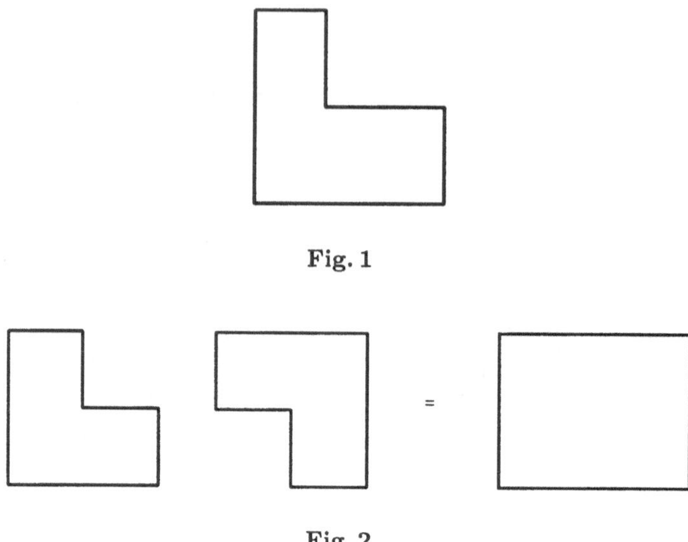

Fig. 1

Fig. 2

Two Δ-dominoes α and β over S can be matched together in this order to form a larger Δ-domino $\alpha \otimes \beta$, as illustrated in Fig 2, if $r(\alpha).l(\beta) \in \Delta$, that is, if the right part of α can be joined with the left part of β.

In this case, $l(\alpha\otimes\beta) = l(\alpha), m(\alpha\otimes\beta) = m(\alpha)r(\alpha)l(\beta)m(\beta), r(\alpha\otimes\beta) = r(\beta)$. Note that above $m(\alpha\otimes\beta) \in \Delta$, since Δ is a subsemigroup of the direct product. The product $\alpha\otimes\beta$ is undefined if $r(\alpha).l(\beta) \notin \Delta$. In this case, we write $\alpha\otimes\beta = 0$, where 0 is a special element for this purpose. If we agree that 0 is a Δ- domino, then the Δ- dominoes form a semigroup with a zero element.Let this semigroup be denoted by D_Δ. Also, $D_\Delta(I)$ denotes the subsemigroup of D_Δ generated by the subset I.

The set $BD_\Delta(I) = \{\alpha \in D_\Delta(I) : \alpha = m(\alpha)\}$, is the blunt language generated by the subset $I \subseteq D_\Delta$, and $ED_\Delta(I) = \{\alpha \in BD_\Delta(I) : \tau_1(\alpha) = \tau_2(\alpha)\}$, is the equality language generated by I.

3.2 Example

Let $\Sigma = \{A, T, C, G\}$ and Δ be the semigroup of $\Sigma^* \times \Sigma^*$ generated by the relation $\{(A,T), (T,A), (C,G), (G,C)\}$. The double string

$$\alpha = (AG, 1)(GTC, CAG)(T, 1),$$

is a Δ-domino.

The Δ-domino α can be matched with the Δ - domino $\beta = (1, A)(A, T)$ and the result is $\alpha\otimes\beta = (AG, 1)(GTCTA, CAGAT)$, where $r(\alpha\otimes\beta) = r(\beta) = (1, 1)$. The product $\beta \otimes \alpha$ is undefined: $\beta \otimes \alpha = 0$. Also the product $\beta \otimes \gamma$ is undefined when $\gamma = (TC, AG)(1, A)$ since $r(\beta)l(\gamma) = (1, 1) \notin \Delta$.

3.3 Definition

Let P be a set of Δ-dominoes over a semigroup S such that each $\sigma \in P$ is oriented: $l(\sigma) \in (S-1) \times 1$ implies $r(\sigma) \in 1 \times S'$ and $l(\sigma) \in 1 \times (S-1)$ implies $r(\sigma) \in S' \times 1$.

The splicing product $\gamma = \alpha|_p\beta$ of two Δ-dominoes α, β is $\alpha \otimes \beta$ if $\alpha \otimes \beta \neq 0$ and there exists $\sigma \in P$ such that
i. $m(\sigma) = r(\alpha)l(\beta)$,
ii. $l(\sigma) = (1, x)$ for some suffix x of $\tau_2(\alpha)$,
iii. $r(\sigma) = (y, 1)$ for some prefix y of $\tau_1(\beta)$,
or
i. $m(\sigma) = r(\alpha)l(\beta)$,
ii. $l(\sigma) = (x, 1)$ for some suffix x of $\tau_1(\alpha)$,
iii. $r(\sigma) = (1, y)$ for some prefix y of $\tau_2(\beta)$.

Also, $\alpha|_p\beta = 0$ if $\alpha|_p\beta$ is undefined. A subsemigroup T of D_Δ is called a splicing semigroup with respect to P if $\alpha|_p\beta \in T - \{0\}$ implies $\alpha, \beta \in T$.

For example, if $S = (\Sigma, I, B, C)$ be a splicing system with $\Sigma = \{c, x\}, I = \{cxcxc\}, B = \{(c, x, c)\}$ and $C = \emptyset$, then the splicing language $L(S) = \{c(xc)^n xc : n \geq 0\}$ [5]. Using the domino terminology, we can get $\alpha = (cxcxc, cxcxc) = (c, c)(1, x)|_P(x, 1)(cxc, cxc) = \alpha_1|_p\alpha_2$ and $\alpha = (cxc, cxc)(1, x)|_p(x, 1)(c, c) = \alpha_3|_p\alpha_4$, where $\alpha_1 = (c, c)(1, x), \alpha_2 = (x, 1)(cxc, cxc), \alpha_3 = (cxc, cxc)(1, x)$, and $\alpha_4 = (x, 1)(c, c)$.

Also, $\alpha_1|_p\alpha_2 = \alpha_3|_p\alpha_4$.

Now $\alpha_1 \otimes \alpha_4 = (c, c)(1, x) \otimes (x, 1)(c, c) = (cxc, cxc) = \alpha_5$ and $\alpha_3 \otimes \alpha_2 = (cxc, cxc)(1, x) \otimes (1, x)(cxc, cxc) = (cxcxcxc, cxcxcxc) = \alpha_6$.

On continuing this process on α_6 and matching the flexible pieces, we get $\alpha_7 = (c(xc)^3xc, c(xc)^3xc)$. As the dominoes $\alpha, \alpha_5, \alpha_6, \alpha_7, \cdots$, are blunt, being generated by $\alpha \in I$ with respect to P, and no other elements can be generated from I , $BD^P_\Delta(I) = \{(c(xc)^n xc, c(xc)^n xc) : n \geq 0\}$.

Thus, $L(S) = BD^P_\Delta(I)$.

3.4 Theorem

Let $S = (\Sigma, I, B, C)$ be a splicing system.There is a splicing semigroup of dominoes $BD^P_\Delta(I)$ such that $L(S) = BD^P_\Delta(I)$.

Proof: Let $S = (\Sigma, I, B, C)$ be a splicing system, where Σ is the alphabet, I is the set of initial strings, B is the set of left patterns, C is the set of right patterns.

The existence of the splicing semigroup of dominoes $BD^P_\Delta(I)$ is true by taking Δ as the semigroup generated by the pairs (a, a) for all $a \in \Sigma$, the alphabet of the splicing system,and the set P of oriented dominoes is chosen as mentioned in the beginning.

Now to prove $L(S) = BD^P_\Delta(I)$, that is ,to prove $L(S) \subseteq BD^P_\Delta$.

Let $\alpha \in L(S)$, then α is in I_0 or $\notin I_0$.

Case 1: If $\alpha \in I_0$, then it is in $BD^P_\Delta(I)$ as $I = I_0$ and I is in $BD^P_\Delta(I)$.
Case 2: If $\alpha \notin I_0$, then α should be in some I_i for $i > 0$.

Then there exists two strings $uaxbv$ and $pcxdq$ in I_{i-1} such that on splicing and recombination of these two strings, we get $\alpha = uaxdq$ or $pcxbv$.

Now to prove α in $BD^P_\Delta(I)$, the oriented dominoes corresponding to the patterns (c, x, d) and (e, x, f) in B are $(c, 1)(x, x)(1, d)$ and $(e, 1)(x, x)(1, f)$. The strings $ucxdv$ and $pexfq$ in I_{i-1} are written as $(ucxdv, ucxdv)$ and $(pexfq, pexfq)$. Now the strings in I_{i-1} can be spliced with respect to P as below:

$(ucxdv, ucxdv) = (uc, uc)(1, x)|_p(x, 1)(dv, dv) = \alpha_1|_p\alpha_2$.
$(pexfq, pexfq) = (pe, pe)(1, x)|_p(x, 1)(fq, fq) = \alpha_3|_p\alpha_4$.

Thus, α should be $\alpha_1 \otimes \alpha_4$ or $\alpha_3 \otimes \alpha_2$ as $(x, x) \in \Delta$.
Thus $\alpha \in BD^P_\Delta(I)$ and hence $L(S) \subseteq BD^P_\Delta(I)$.

To prove the other way, let $\alpha \in BD^P_\Delta(I)$. Suppose $\alpha \in I$, then $\alpha \in L(S)$ since $I = I_0$ and $I \subseteq L(S)$.
Suppose $\alpha \notin I$.

We prove this by induction on the number of times $|_P$,the splicing operation, is performed.

Suppose, α is in $BD^P_\Delta(I)$ as a result of one splicing,then there exists $\alpha_1, \alpha_2 \in I_0 = I$ such that $\alpha_1 = \alpha_{11}|_p\alpha_{12}, \alpha_2 = \alpha_{21}|_p\alpha_{22}$ and of course $\alpha = \alpha_{11} \otimes \alpha_{22}$ or $\alpha_{21} \otimes \alpha_{12}$. Then, $\alpha \in I_1$ and hence in $L(S)$ as $L(S) = I_0 \cup I_1 \cup \cdots$.

Suppose $\alpha \in BD^P_\Delta(I)$ after $\leq k$ splicing operations on the strings in I, then $\alpha \in L(S)$.

If α is obtained after $k+1$ operations, there exists $\alpha_1, \alpha_2 \in BD^P_\Delta(I)$ which are obtained after $\leq k$ splicing operations, then $\alpha_1 = \alpha_{11}|_P\alpha_{12}, \alpha_2 = \alpha_{21}|_P\alpha_{22}$ and of course $\alpha = \alpha_{11} \otimes \alpha_{22}$ or $\alpha_{21} \otimes \alpha_{12}$. Therefore, $\alpha \in L(S)$ and $BD^P_\Delta(I) \subseteq L(S)$ and hence the result. Now we generalize the definition of splicing system.

4 Extended Splicing System

4.1 Definition

Let $S = (\Sigma, I, B, C)$ be an extended splicing system, then, $L(S)$ consists of the set of initial strings I and strings of the form

$$\begin{matrix} u_1c_1x_1f_1q_1 & p_1e_1y_1d_1v_1 \\ u_2c_2y_2f_2q_2 & p_2e_2x_2d_2v_2 \end{matrix}$$

where

$$\begin{pmatrix} c_1 & x_1 & d_1 \\ c_2 & , x_2 & , d_2 \end{pmatrix} \begin{pmatrix} e_1 & y_1 & f_1 \\ e_2 & , y_2 & , f_2 \end{pmatrix}$$

are patterns of the same hand and

$$\begin{pmatrix} x_1 \\ x_2 \end{pmatrix}, \begin{pmatrix} y_1 \\ y_2 \end{pmatrix}, \begin{pmatrix} x_1 \\ y_2 \end{pmatrix}, \begin{pmatrix} y_1 \\ x_2 \end{pmatrix} \in \Delta$$

and Δ is generated by $E \subseteq \Sigma^* \times \Sigma^*$.

The following lemma follows from this definition.

4.2 Lemma

Let $S = (\Sigma, I, B, C, E)$ be an extended splicing system and $L(S)$ be its language, then, $L(S)$ can be partitioned into an infinite pair-wise disjoint family of finite subsets $I_0, I_1, I_2 \cdots$. That is, $L(S) = \cup I_j : j \geq 0$ and $I_i \cap I_j = \emptyset$ when $i \neq j$.

4.3 Theorem

Given $S = (\Sigma, I, B, C, E)$ is an extended splicing system, then there is a splicing semigroup $BD_\Delta^P(I)$ such that $L(S) = BD_\Delta^P(I)$.

Proof: Let the semigroup $\Delta \subseteq \Sigma^* \times \Sigma^*$ be generated by E. The existence of $BD_\Delta^P(I)$ is as found in the last theorem. To show that $BD_\Delta^P(I) = L(S)$.

Let $\alpha \in L(S)$.
Case 1: If $\alpha \in I_0$, then $\alpha \in BD_\Delta^P(I)$ as $I = I_0$ and $I \subseteq BD_\Delta^P(I)$.

Case 2: $\alpha \notin I_0$, If $\alpha \in I_1$, then there exists two strings

$$\frac{u_1c_1x_1d_1v_1}{u_2c_2x_2d_2v_2}, \frac{p_1e_1y_1f_1q_1}{p_2e_2y_2f_2q_2} \in I_0$$

with the patterns

$$\begin{pmatrix} c_1 \; x_1 \; d_1 \\ c_2 \; x_2 \; f_2 \end{pmatrix}, \begin{pmatrix} e_1 \; y_1 \; f_1 \\ e_2 \; y_2 \; f_2 \end{pmatrix}$$

of the same hand such that

$$\alpha = \frac{u_1c_1x_1f_1q_1}{u_2c_2y_2f_2q_2}, \frac{p_1e_1y_1d_1v_1}{p_2e_2x_2d_2v_2}$$

where

$$\begin{pmatrix} x_1 \\ x_2 \end{pmatrix}, \begin{pmatrix} x_1 \\ y_2 \end{pmatrix}, \begin{pmatrix} y_1 \\ x_2 \end{pmatrix}, \begin{pmatrix} y_1 \\ y_2 \end{pmatrix} \in \Delta$$

Now to prove $\alpha \in BD_\Delta^P(I)$.
Now

$$\frac{u_1c_1x_1d_1v_1}{u_2c_2x_2d_2v_2} \in I$$

can be written as $(u_1c_1x_1d_1v_1,\ u_2c_2x_2d_2v_2)$ etc..
Thus, $(u_1c_1x_1d_1v_1, u_2c_2x_2d_2v_2) = (u_1c_1, u_2c_2)(x_1, 1)|_p(1, x_2)(d_1v_1, d_2v_2) = \alpha_1|_p\alpha_2$
Similarly, $(p_1e_1y_1f_1q_1, p_2e_2y_2f_2q_2) = (p_1e_1, p_2e_2)(y_1, 1)|_p(1, y_2)(f_1q_1, f_2q_2) = \alpha_3|_p\alpha_4$.

Then, α must be $\alpha_1 \otimes \alpha_4$ or $\alpha_3 \otimes \alpha_2$ as

$$\begin{pmatrix} x_1 \\ y_2 \end{pmatrix}, \begin{pmatrix} y_1 \\ x_2 \end{pmatrix} \in \Delta$$

Thus, $\alpha \in BD_{\Delta}^{P}(I)$.

Assume that if $\alpha \in I_0 \cup I_1 \cup \cdots \cup I_{n-1}$, then $\alpha \in BD_{\Delta}^{P}(I)$.

Suppose $\alpha \in I_n$, then there exists strings α_1, α_2 in $I_0 \cup ... \cup I_{n-1}$ and by assumption, $\alpha_1, \alpha_2 \in BD_{\Delta}^{P}(I)$.

On splicing $\alpha_1 = \alpha_{11} \otimes \alpha_{12}$ and $\alpha_2 = \alpha_{21} \otimes \alpha_{22}$, we get $\alpha = \alpha_{11} \otimes \alpha_{22}$ or $\alpha_{21} \otimes \alpha_{12}$. Thus, $\alpha \in BD_{\Delta}^{P}(I)$. That is, $L(S) \subseteq BD_{\Delta}^{P}(I)$.

The converse part is as similar as the proof of theorem 3.4.

4.4 Example

Let $\Sigma = \{a, b\}$ and $\Delta \subseteq \Sigma^* \times \Sigma^*$ be generated by the finite set $\{(a,a), (b,b), (a,a^2), (b^2,b), (b,a)\}$ and let $P = \{(ab^2,ab^2)(b,1), (a^2b,a^2b)(b,1), (ab^2,a^2)(b,1), (a^2b,a^3)(b,1)\}$ and $I = \{\alpha\}$, where $\alpha = (a^3b^3, a^3b^3)$.

Now α can be spliced only in the following two ways:

$\alpha = (a^2,b^2)(ab^2,1)|_p(1,ab^2)(b,b) = \alpha_{11}|_p\alpha_{12}$ by the presence of the oriented domino $\sigma = (ab^2,ab^2)(b,1)$.

$\alpha = (a,a)(a^2b,1)|_p(1,a^2b)(b^2,b^2)$ by the oriented domino $\sigma = (a^2b,a^2b)(b,1)$.

Here, of course, $\alpha_{11} \otimes \alpha_{12} = \alpha = \alpha_{13} \otimes \alpha_{14}$. Also, $\alpha_{13} \otimes \alpha_{12} = 0$, since $(a^2b, ab^2) \notin \Delta$ and $\alpha_4 = \alpha_{11} \otimes \alpha_{14} = (a^3b^4, a^4b^3) \in BD_{\Delta}^{P}(I)$ as $(ab^2, a^2b) \in \Delta$.

Also, α_4 can be further spliced as follows:

$\alpha_4 = (a^2,b^2)(ab^2,1)|_p(1,a^2b)(b^2,b^2) = \alpha_{21}|_p\alpha_{22}$ by the presence of the oriented domino $\sigma = (ab^2,ab^2)(b,1)$.

Now $\alpha_5 = \alpha_{21} \otimes \alpha_{24} = (a^3b^5, a^5b^3) \in BD_{\Delta}^{P}(I)$ as $(ab^2, a^3) \in \Delta$.

Further, $\alpha_{23} \otimes \alpha_{22} = (a^3b^3, a^3b^3) \in BD_{\Delta}^{P}(I)$.

Thus, $BD_{\Delta}^{P}(I) \supseteq \{\alpha, \alpha_4, \alpha_5\} = \{(a^3b^3, a^3b^3), (a^3b^4, a^4b^3), (a^3b^5, a^5b^3)\}$.

Suppose α_n is obtained from α_{n-1} by splicing and recombination,

that is, $\alpha_n = (a^3b^n, a^nb^3) \in BD_{\Delta}^{P}(I)$.

Now to prove α_{n+1} is obtained from α_n,

$\alpha_n = (a^3b^n, a^nb^3) = (a^2,a^2)(ab^2,1)|_p(1,a^3)(b^{n-2}a^{n-2}, a^{n-5}b^3) = \alpha_{n1}|_p\alpha_{n2}$.

Also, $\alpha_n = (a,a)(a^2b,1)|_p(1,a^3)(b^{n-1}, a^{n-4}b^3) = \alpha_{n3}|_p\alpha_{n4}$.

Now, $\alpha_{n1} \otimes \alpha_{n4} = (a^3b^{n+1}, a^{n+1}b^3) = \alpha_{n+1}$, and $\in BD_{\Delta}^{P}(I)$.

Also, $\alpha_{n3} \otimes \alpha_{n2} = (a^3b^{n-1}, a^{n-1}b^3) = \alpha_{n-1}$.

Thus, $BD_{\Delta}^{P}(I) = \{(a^3b^n, a^nb^3) : n \geq 0\}$. Take this as (1).

Now to construct the corresponding extended splicing system $S = (\Sigma, I, B, C, E)$ from the splicing semigroup of dominoes. We have $\Sigma = \{a, b\}$,

$$I = \begin{pmatrix} a^3b^3 \\ a^3b^3 \end{pmatrix} \text{ and } E = \begin{pmatrix} a \\ a \end{pmatrix}, \begin{pmatrix} b \\ b \end{pmatrix}, \begin{pmatrix} a \\ a^2 \end{pmatrix}, \begin{pmatrix} b^2 \\ b \end{pmatrix}$$

From P, the set of oriented dominoes, for the oriented domino $\sigma = (ab^2, ab^3)(b, 1)$, we construct the corresponding pattern

$$\begin{pmatrix} 1 & ab^2 & b \\ 1, & ab^2, & 1 \end{pmatrix} \in C$$

If any oriented domino is of the form $(ab^2, ab^2)(1, b)$, we would have constructed

$$\begin{pmatrix} 1 & ab^2 & 1 \\ 1, & ab^2, & 1 \end{pmatrix} \in B$$

and no chance for any right pattern arises. Thus

$$C = \begin{pmatrix} 1 & ab^2 & b \\ 1, & ab^2, & 1 \end{pmatrix}, \begin{pmatrix} 1 & a^2b & b \\ 1, & a^2b, & 1 \end{pmatrix}, \begin{pmatrix} 1 & ab^2 & b \\ 1, & a^2b, & 1 \end{pmatrix}, \begin{pmatrix} 1 & a^2b & b \\ 1, & a^3, & 1 \end{pmatrix}, \begin{pmatrix} 1 & ab^2 & b \\ 1, & a^3, & 1 \end{pmatrix}$$

$B = \emptyset$

Now the initial string α can be written as

$$\alpha = \begin{matrix} u_1c_1x_1d_1v_1 & = aa\ 1\ abb\ b\ 1 \\ u_2c_2x_2d_2v_2 & = aa\ 1\ abb\ 1\ b \end{matrix}$$

$$\alpha = \begin{matrix} p_1e_1y_1f_1v_1 & = a\ 1\ aab\ b\ b \\ p_2e_2y_2f_2v_2 & = a\ 1\ aab\ 1\ bb \end{matrix}$$

On splicing and recombination of these two strings according to the definition of extended splicing system, we get,

$$\begin{matrix} u_1c_1x_1f_1q_1 & = & aa\ 1\ abb\ b\ 1 & = & a^3b^4 \\ u_2c_2y_2f_2q_2 & = & aa\ 1\ abb\ 1\ b & = & a^4b^3 \end{matrix}$$

and

$$\begin{matrix} p_1c_1y_1d_1v_1 & = & a\ 1\ aab\ b\ b & = & a^3b^2 \\ p_2c_2x_2d_2v_2 & = & a\ 1\ abb\ 1\ bb & = & a^2b^4 \end{matrix}$$

As

$$\begin{pmatrix} a^2b \\ ab^2 \end{pmatrix} \notin \Delta, \begin{pmatrix} a^3b^3 \\ a^2b^4 \end{pmatrix} \notin L(S)$$

But,

$$\alpha_4 = \begin{pmatrix} a^3b^4 \\ a^4b^3 \end{pmatrix} \in I_1 \text{ since } \begin{pmatrix} ab^2 \\ a^2b \end{pmatrix} \in \Delta$$

On repeating the same procedure on the string α_4, we get

$$\alpha_5 = \begin{pmatrix} a^3b^5 \\ a^5b^3 \end{pmatrix}$$

Suppose, $I_0, I_1, I_2, \cdots I_n$ have been generated. As $I_{n+1} = \{s \in \sigma^* \backslash (I_0 \cup I_1 \cup, \cdots I_n)$: there exists two strings

$$\frac{u_1c_1x_1d_1v_1}{u_2c_2x_2d_2v_2}, \frac{p_1e_1y_1f_1q_1}{p_2e_2y_2f_2q_2} \in I_0 \cup I_1 \cup \cdots I_n$$

and patterns for which

$$s = \frac{u_1c_1x_1f_1q_1}{u_2c_2y_2f_2q_2}, \; and \frac{p_1e_1y_1d_1v_1}{p_2e_2x_2d_2v_2}$$

As $L(S) = \cup I_j : j \geq 0$

$$= \left\{ \frac{a^3b^3}{a^3b^3}, \frac{a^3b^4}{a^4b^3}, \frac{a^3b^5}{a^5b^3}, \cdots \right\}$$

$$= \left\{ \frac{a^3b^n}{a^nb^3} : n \geq 3 \right\}$$

Take this as (2).
Hence $L(S) = BD^P_\Delta(I)$ from 1 and 2.
We generalize this phenomenon in the next theorem.

4.5 Theorem

Let $\Delta \subseteq \Sigma^* \times \Sigma^*$ be generated by a relation E. Let $BD^P_\Delta(I)$ be the smallest semigroup with respect to P, generated by $I \subseteq D_\Delta$, the semigroup of Δ-dominoes. Then there exists an extended splicing system $S = (\Sigma, I, B, C, E)$ such that $L(S) = BD^P_\Delta(I)$.

Proof: Let Σ be the alphabet, I, the set of initial Δ-blunt dominoes. Consider $\sigma \in P$ with

$$l(\sigma) = \begin{pmatrix} c_1 \\ 1 \end{pmatrix}, m(\sigma) = \begin{pmatrix} x_1 \\ x_2 \end{pmatrix}, r(\sigma) = \begin{pmatrix} 1 \\ d_2 \end{pmatrix}$$

then introduce $(c_1, x_1, 1)$ in B or with $l(\sigma) = (1)$, $m(\sigma) = (y_1)$, $r(\sigma) = (f_1)$ then

$$\begin{pmatrix} 1 & y_1 & f_1 \\ e_2, & y_2, & 1 \end{pmatrix} \in C$$

Thus, the extended splicing system $S = (\Sigma, I, B, C, E)$ is specified. The proof for $L(S) \subseteq BD^P_\Delta(I)$ is the same as theorem 3.4 and for $L(S) \supseteq BD^P_\Delta(I)$ is the same as theorem 4.3. Hence the proof.

4.6 Persistent Splicing Languages

In this section, we discuss about the persistent splicing languages[8] and its relationships with other formal languages.

1. Consider a language L consisting of all finite-length strings over $\Sigma = \{a, b\}$ that begin with a, end with a and contain no occurrence of a as a strictly interior subword, that is, L is the language denoted by the regular expression $a + ab^*a$.
   Here, $A = B = \{a\}$, and $C = \{b\}$, then, it is easily seen that for all $w \in \Sigma^*$ with length of $w \geq 1, w \in L$ if and only if $L_1(w) \in A, R_1(w) \in B$, and $I_1(w) \subseteq C$. Hence, L is strictly 1-testable.
   We can form the corresponding persistent splicing system $S = (\Sigma, I, B, C)$, which can generate this language. Let the alphabet set $\Sigma = \{a, b\}$, and the set of initial strings $I = \{a, aa, abba\}$. The set of patterns $B = \{(1, b, 1)\}$ and $C = \emptyset$.
2. Consider a language L containing of all finite-length strings over $\Sigma = \{a, b\}$ that begin with aa, end with aa and contain no occurrence of words of length two other than aa as a strictly interior subword, that is, L is the language denoted by the regular expression $aa + aa(b^+a)^*a$. Here $A = B = C = \{aa\}$. Hence, L is strictly 2-testable.
   The corresponding persistent splicing system $S = (\Sigma, I, B, C)$, where $I = \{aa, aaa, aabbaa, aababaa\}$ and $B = \{(1, b, 1), (1, ba, 1)\}$.
3. Let $S = (\Sigma, I, B, C)$ where $\Sigma = \{a, b, c, x\}, I = \{axbxc\}$, $B = \{(a, x, b), (b, x, c), (b, x, b), (a, x, c)\}$ and $C = \emptyset$ be a persistent splicing system. The language generated by this system is $L(S) = \{a(xb)^n xc : n \geq 0\}$ and is persistent. This language is 2-strictly locally testable with $S_2 = \emptyset, L_2 = \{ax\}, R_2 = \{xc\}, M_2 = \{xb, bx\}$.
4. Let $S = (\Sigma, I, B, C)$ where $\Sigma = \{a, b, c\}, I = \{baa, bb, aaac, cc, ac\}$, $B = \{(b, a, \wedge), (ba, a, \wedge), (\wedge, b, \wedge), (\wedge, aa, c), (\wedge, aa, ac)\}$, $C = \emptyset$, is a non-persistent splicing system. The language generated by this $L(S) = \{bb^*a^*\} \cup \{a^*c^*c\}$ is not strictly locally testable.

4.7 Results

A reduced automaton $M = (Q, \Sigma, \delta, q_0, q_f)$ recognizing a persistent splicing language $L(S)$, $S = (\Sigma, I, B, C)$ with only one crossing 'x' satisfies the following:

1. $\delta(q_0, a) = q_{1x}$, $\delta(q_{1x}, x) = q_2x$, $\delta(q_{2x}, c) = q_f$ for all $a \in A_x$, $c \in C_x$, q_{1x}, q_{2x} are the states forming the vertices of a loop with an edge 'x'.
2. $\delta(q_{1x}, xb) = \delta(q_{1x}, xbxb)$ for all $b \in B_x$.

If a reduced automaton M is the union of automaton with input sets $\Sigma_x \cup \Sigma_y \cup \cdots$, where $\Sigma_x = A_x \cup B_x \cup C_x \cup \{x\}$, $\Sigma_y = A_y \cup B_y \cup C_y \cup \{y\}$ etc., and the transition function satisfying the properties of the previous one, then the language recognized is 2−strictly locally testable and vice versa.

4.8 Notes

Let Σ_1 and Σ_2 be two disjoint subsets of the alphabet Σ, and let $L_1(S) \subset \Sigma_1^*$ and $L_2(S) \subset \Sigma_2^*$ be two persistent splicing languages. Then the languages $L_1 \cup L_2$ and $L_1 L_2$ are also persistent splicing languages.

Let L be a persistent splicing language, then the language L^* is also a persistent splicing language. Let Σ_1 and Σ_2 be two disjoint subsets of the alphabet Σ, and let $L_1(S) \subset \Sigma_1^*$ and $L_2(S) \subset \Sigma_2^*$ be two persistent splicing languages. Then the languages $L_1 \cup L_2$ and $L_1 L_2$ are also persistent splicing languages.

5 Conclusion

In this paper, we have brought out the similarities that exist between the two mathematical models available to study the recombinant behavior of DNA molecules. Having extended the splicing system to accommodate the splicing and recombination for the case of unequal crossing too, we have established that every splicing semigroup of blunt dominoes gives rise to a splicing language and vice versa.

Acknowledgments

We would like to thank Japan Society for Promotion of Science (JSPS) for their continued financial support.

References

1. Culik II,K., and Harju,T., Dominoes over a free monoid, Vol. 18, Theoretical Computer Science,(1998) 279 - 300
2. Culik II,K., and Harju,T., Dominoes and the regularity of DNA splicing languages, Vol. 182. LNCS Springer-Verlag,(1998) 222 - 233
3. Culik II,K., and Harju,T., Splicing semigroups of dominoes and DNA, Vol. 31. Discrete Applied Mathematics,(1991) 261-277
4. Denninghoff,K.L., and Gatterdam,R.W., On the undecidability of splicing systems, Vol. 27. International Journ. of Computer Mathematics, (1989) 133-145
5. Head,T., Formal language theory and DNA:An analysis of the generative capacity of specific recombinant behaviors, Vol. 49. Bulletin of Mathematical Biology, (1987) 737 - 759.
6. Lewin,B., Genes IV, (1987) Wiley, New York.
7. Paun,G.,and Salomaa,A., DNA computing based on the Splicing operation, Math. Japonica, (1996) 607 - 632
8. Pethuru Raj,C., Contributions to the studies on Persistent Splicing Languages, (1997) Ph.D thesis, Anna University, India.
9. Watson,J.D., Tooze,J.,and Kurtz,D.T., Recombinant DNA: A Short Course, Freeman,(1983) New York.

Newton Polyhedra and the Reversible System of Ordinary Differential Equations*

A. Soleev and I. Yarmukhamedov

Faculty of Mathematics, Samarkand State University
15, University bld.,Samarkand, Uzbekistan
soleev@samuni.silk.org

Abstract. We investigate a reversible system of fourth-order ODEs by using the method of Newton polyhedron. We discuss the program for IBM PC, which computes the Newton polyhedron of the system under study and all of its corresponding objects. The results of the program runs are presented as a table of correspondences.

1 Introduction

An apparatus was developed in [1] for local analysis of a system of ordinary differential equations

$$dX/dt \stackrel{\text{def}}{=} \dot{X} = \Phi(X), \quad X \in \mathbb{R}^n \quad \text{or} \quad \mathbb{C}^n.$$

It is based on two methods:

1. The method of normal forms. It can be used for a system with nonzero linear part.
2. The method of truncated systems, which employs the Newton polygon. It can be used for systems with zero linear part.

It was shown that by combining the two methods it is possible to find both the asymtotics of solutions of the system as $t \to \infty$ and $t \to 0$, and its periodic and quasi-periodic solutions, in arbitrary complicated cases. The application of these methods to the specific multidimensional problems, however, was hindered by the absence of an efficient algorithm for computing the Newton polyhedron. Such an algorithm has been worked out in [2] and realized as a computer program [3, 4].

In present work, the computer program given in [4] is generalized and applied for the local investigation of solutions of reversible fourth-order system

* The work is supported by the State Committee of Science and Technology of the Republic of Uzbekistan.

depending on two small parameters $X'' = (x_5, x_6)$:

$$\dot{X}' = \Phi(X', X''), \quad X' \in \mathbb{R}^4,$$
$$\dot{X}'' = 0, \quad X'' \in \mathbb{R}^2. \tag{1}$$

Here $X = (X', X'')$, $X' = (x_1, x_2, x_3, x_4)$, and the number of coordinates of the problem is $n = 4 + 2 = 6$. Namely, we consider the system (1), where $\Phi(0, X'') = 0$ and for $X = 0$

$$\frac{\partial \Phi}{\partial X'} = \begin{pmatrix} 0 & 1 & 0 & 0 \\ 0 & 0 & 1 & 0 \\ 0 & 0 & 0 & 1 \\ 0 & 0 & 0 & 0 \end{pmatrix} \stackrel{\text{def}}{=} L, \tag{2}$$

that is, the point $X' = 0$ is a fixed point, and the matrix L of the unperturbed linear part is a Jordan block with zero eigenvalue of multiplicity four. Moreover, the system (1) passes to itself under the substitution $X', X'', t \to SX', X'', -t,$, where

$$S = \begin{pmatrix} 1 & 0 & 0 & 0 \\ 0 & -1 & 0 & 0 \\ 0 & 0 & 1 & 0 \\ 0 & 0 & 0 & -1 \end{pmatrix}, \tag{3}$$

that is, the reversibility property of (1) means that

$$S\Phi(X', X'') = -\Phi(SX', X''). \tag{4}$$

See [5] and [6] for surveys of the general theory of reversible systems.

Such a system arises in hydrodynamics. For example, the problem of surface waves on water for Bond number b close to 1/3 and Froude number $\tilde{\lambda}$ close to 1 passes to such a system (1) as a result of center-manifold reduction (see [7], [8]–[13])

2 Computation of Newton Polyhedron

According to [1], we can write the system (1) in the form

$$d(\ln X')/dt = F(X) \stackrel{\text{def}}{=} \sum F_Q X^Q, \quad \dot{X}'' = 0, \tag{5}$$

where $X^Q = x_1^{q_1} x_2^{q_2} x_3^{q_3} x_4^{q_4} \mu^{q_5} \mu_2^{q_6}$, that is $Q = (q_1, q_2, q_3, q_4, q_5, q_6) \in \mathbb{R}^6$, $\ln X' = (\ln x_1, \ldots, \ln x_4)$ and $F_Q = (f_{1Q}, \ldots, f_{4Q})$. In R^6 we consider the set $\mathbf{D} \stackrel{\text{def}}{=} \mathbf{D}(F) = \{Q : F_Q \neq 0\}$, called the *support* of the system (5). The closure $\Gamma = \Gamma(F)$ of its convex hull is called the *Newton polyhedron* of this system (see [14, 4]).

The boundary $\partial\Gamma$ of Γ consists of the *faces* $\Gamma_j^{(d)}$ of various dimensions d. The *boundary subset* $\mathbf{D}_j^{(d)} = \mathbf{\Gamma}_j^{(d)} \cap \mathbf{D}$ corresponds to each face $\Gamma_j^{(d)}$ of the polyhedron Γ, and the *truncated system*

$$d(\ln X)/dt = \widehat{F}_j^{(d)}(X) \stackrel{\text{def}}{=} \sum F_Q X^Q \quad \text{over} \quad Q \in \mathbf{D}_j^{(d)} \tag{6}$$

corresponds to the above subset. In the space R_*^6 dual to R^6 each face $\mathbf{\Gamma}_j^{(d)}$ has its corresponding *normal cone* $\mathbf{U}_j^{(d)}$, that is, the set of vectors $P \in R_*^6$ for which the supporting hyperplane $\langle P, Q\rangle = \text{const}$ of $\mathbf{\Gamma}$ normal to P intersects $\mathbf{\Gamma}$ in the face $\mathbf{\Gamma}_j^{(d)}$. If $X \to 0$ along curves of the form $x_i = b_i \tau^{p_i}(1 + o(1)), \quad i = 1, \ldots, 6$, where the $b_i \in \mathbb{C} \setminus \{0\}$ are constants, $\tau \to \infty$, and $P = (p_1, \ldots, p_6) \in \mathbf{U}_j^{(d)}$, then the truncated system (6) is the first approximation of system (5). The brackets $\langle ., .\rangle$ denote the inner product.

Suppose that in R_*^6 dual to R^6 we are given the polyhedral convex *cone* $\mathbf{K}$ *of the problem*, that is, we are interested only in the faces $\mathbf{\Gamma}_j^{(d)}$ (as well as the truncated systems (6)) whose normal cones $\mathbf{U}_j^{(d)}$ intersect the cone $\mathbf{K}$ of the problem. The intersection $\mathbf{U}_j^{(d)} \bigcap \mathbf{K}$ of two convex cones is also a polyhedral convex cone and can be given as the convex conic hull of a minimal finite number of vectors $V_1, \ldots, V_s$ forming the *frame* of this cone. We let N_j stand for the normals to the hyperfaces $\mathbf{\Gamma}_j^{(5)}$. The frame of the cone $\mathbf{U}_j^{(5)}$ obviously consists of a single vector N_j. On the other hand, the boundary $\partial\mathbf{K}$ of the cone $\mathbf{K}$ of the problem consists of finitely many cones $\mathbf{K}_j^{(d)}$ of smaller dimension; the intersection

$$K_k^{(d)} \cap \mathbf{U}_j^{(d_1)}$$

of each of them with a normal cone is a polyhedral cone and can be given by some finite frame $V_1^*, \ldots, V_t^*$. It has been proved that the frame of the intersection $\mathbf{U}_j^{(d)} \bigcap \mathbf{K}$ consists of the normal vector $N_j \in \mathbf{K}$ to the hyperface $\mathbf{\Gamma}_j^{(5)}$ and vectors D_j lying on the boundary $\partial\mathbf{K}$ of the $\mathbf{K}$ ([5], [8]).

We assume that the linear part of the system (5) has normal form, and that the nonlinear part is not normalized. Then (5) has the form

$$\begin{aligned}
d(\ln x_1)/dt &= x_1^{-1}x_2 && + \textstyle\sum f_{1Q}X^Q,\\
d(\ln x_2)/dt &= x_2^{-1}x_3 + \mu x_2^{-1}x_1 && + \textstyle\sum f_{2Q}X^Q,\\
d(\ln x_3)/dt &= x_3^{-1}x_4 + \mu x_3^{-1}x_2 && + \textstyle\sum f_{3Q}X^Q,\\
d(\ln x_4)/dt &= \mu_2 x_4^{-1}x_1 + \mu x_4^{-1}x_3 && + \textstyle\sum f_{4Q}X^Q,
\end{aligned} \tag{7}$$

where the integers $q_1, \ldots, q_4$ are ≥ -1, $q_1 + \ldots + q_4 \geq 1$, and $q_2 + q_4$ is odd by the property (4). For this system we single out its support $\mathbf{D}$ in R^6 and compute all the accompanying objects up to the determination of the set of the truncations.

3 About a Program for IBM PC

The algorithm of determination of all elements $\mathbf{\Gamma}_j^{(d)}$ of the minimal convex hull of the set of the points $\mathbf{D}$ described above was realized as a program for IBM PC.

This program reads the file with input data of the problem. The file contains the dimension of space, the number of points in the investigated set $\mathbf{D}$, and the number of the inequalities determining the cone of the problem. The file must contain the coordinates of the points Q_i of the set $\mathbf{D}$ and coefficients of the inequalities of the problem cone.

The results of running the programm are (see [4]):

1) Dimension of the subspace in which an investigated set of the points lies;

2) The set of the vectors of normals to the hyperfaces (the coordinates of the vectors have the greater common divisor 1) marked by "N";

3) Vectors of intersection of the normal cones with cone of the problem marked by "D";

4) The table of correspondences;

5) The columns of the table of the problem remained the same. Thus, to analyze the system (7) with the help of truncations of the power series expansions of the right-hand sides of the equations it suffices to keep the terms through third order. The result is given as the Table of Correspondences.

The computational results showed that for the Newton polyhedron corresponding to the system (7) the cone $\{P \leq 0\}$ of the problem is intersected by the normal cones of 12 hyperfaces, 54 faces of dimension 4, 114 faces of dimension 3, 128 faces of the dimension 2, 74 edges and 16 vertices.

It is interesting to consider only those truncations that correspond to hyper faces (in the Table of Correspondences 1 these rows are labelled by "N" in the first column). There are 12 such truncations. However, for 7 of the hyperfaces the normal vectors have some zero components, that is, they lie on the boundary of the cone of the problem. From the truncations corresponding to these hyperfaces it is possible to analyze the behavior of the solutions of the system (7) in a neighborhood of zero (see [1, 2]).

For the remaining five hyperfaces the normal vectors N_j do not contain zero components and lie within the cone of the problem. The truncated system corresponding to the hyperface $\mathbf{\Gamma}_1^{(5)}$ with normal vector $N_1 = (-4, -5, -6, -7, -2, -4)$ (the 12th row of the Table of Correspondences has the form

$$(\ln \dot{x}_1) = x_1^{-1} x_2, \; (\ln \dot{x}_2) = x_2^{-1} x_3 + \mu x_2^{-1} x_1, \; (\ln \dot{x}_3) = x_3^{-1} x_4 + \mu x_3^{-1} x_2, \\ (\ln \dot{x}_4) = \mu_2 x_4^{-1} x_1 + \mu x_4^{-1} x_3 + a x_4^{-1} x_1^2. \tag{8}$$

After power transformation:

$$x_1 = y_1\mu^2,\ x_2 = y_2|\mu|^{\frac{5}{2}},\ x_3 = y_3|\mu|^3\ x_4 = y_4|\mu|^{\frac{7}{2}}\ \mu = \mu,\ \mu_2 = \nu\mu^2$$

and the time change $t_1 = |\mu|^{\frac{1}{2}}$ the system (8) turns to the system

$$\frac{dy_1}{dt_1} = y_2,\ \frac{dy_2}{dt_1} = \sigma y_1 + y_3,\ \frac{dy_3}{dt_1} = \sigma y_2 + y_4,\ \frac{dy_4}{dt_1} = \nu y_1 + \sigma y_3 + a y_1^2, \qquad (9)$$

where $\sigma = \operatorname{sgn}\mu,\ \nu = \mu_2/\mu^2$. The differential equation corresponding to the last system has the first integral of the form:

$$J_1 \stackrel{\text{def}}{=} \nu y_1^2 + 2\sigma y_1 y_3 + y_3^2 - \sigma y_2^2 - 2y_2y_4 + \frac{a}{3}y_1^3 = C.$$

Theorem 1. *The system (9) is a Hamiltonian system with Hamiltonian function* $H = (1/2)J_1$.

(For detailed proof see [15]).

In [15] the system (9) was completely investigated by finding the locating of the eigenvalues of the linear part of system (9) with respect to different values of parameters $\sigma = \pm 1$ and $\nu \in \mathbb{R}^1$.

In the truncated systems corresponding to the remaining four hyperfaces one of the equations of the truncated system has a right-hand side which identically equals zero. We write these truncated systems and find their solutions. The truncated system corresponding to the hyperface $\mathbf{\Gamma}_2^{(5)}$ with normal vector $N_2 = (-1,-1,-2,-2,-1,-1)$ (the 9th row of the Table of Correspondences 1) has the form

$$(\ln x_1)^{\cdot} = x_1^{-1}x_2, \quad (\ln x_2)^{\cdot} = 0, \quad (\ln x_3)^{\cdot} = x_3^{-1}x_4 + \mu x_3^{-1}x_2 + a_3x_3^{-1}x_1x_2,$$
$$(\ln x_4)^{\cdot} = \mu_2 x_4^{-1}x_1 + a x_4^{-1}x_1^2 + b x_4^{-1}x_2^2,$$

or, in usual notation

$$\dot{x}_1 = x_2,\ \dot{x}_2 = 0,\ \dot{x}_3 = \mu x_2 + x_4 + a_3x_1x_2, \quad \dot{x}_4 = \mu_2 x_1 + a x_1^2 + b x_2^2. \qquad (10)$$

Integrating the system (10), we obtain its explicit solutions

$$x_1 = \sigma_2 t + c_1, \quad x_2 = c_2, \quad x_3 = \frac{1}{12}ac_2t^4 + \frac{1}{6}(\mu_2c_2 + 2ac_1c_2)t^3$$
$$+ \frac{1}{2}(ac_1^2 + bc_2 + \mu_2c_1 + a_3c_3^2)t^2 + (c_1c_2 + \mu c_2 + c_4)t + c_3,$$
$$x_4 = \frac{1}{3}ac_2^2t^3 + \frac{1}{2}(\mu^2c_2 + 2ac_1c_2)t^2 + (\mu_2c_1 + ac_1^2 + bc_2)t + c_4,$$

where c_1, c_2, c_3, c_4 are arbitrary constants. This notation is also used below.

We now consider the truncated system corresponding to the hyperface $\mathbf{\Gamma}_3^{(5)}$ with normal vector $N_3 = (-1,-1,-1,-2,-1,-1)$ (the 8th row of Table of Correspondences 1), and we write it at once in the usual form:

$$\dot{x}_1 = x_2, \quad \dot{x}_2 = x_3, \quad \dot{x}_3 = 0, \quad \dot{x}_4 = \mu_2 x_1 + \mu x_3 + a x_1^2 + b x_2^2 + a_{13} x_1 x_3 + d x_3^2.$$

The solutions are

$$x_1 = \frac{1}{2} c_3 t^2 + \sigma_2 t + c_1, \quad x_2 = c_3 t + c_2, \quad x_3 = c_3,$$

$$x_4 = \frac{1}{20} a c_3^3 t^5 + \frac{1}{4} a c_2 c_3 t^4 + \frac{1}{6}(\mu_2 c_3 + 2 a c_2^2 + 2 a c_1 c_3 + a_{13} c_3^2 + b c_3^2) t^3 + \frac{1}{2}(\mu_2 c_2 + 2 a c_1 c_2 + a_{13} c_2 c_3 + 2 b c_3) t^2 + (\mu_2 c_1 + a c_1^2 + a_{13} c_1 c_3 + b c_2^2 + d c_3^2) t + c_4.$$

The truncated system

$$\dot{x}_1 = x_2, \quad \dot{x}_2 = 0, \quad \dot{x}_3 = \mu x_2, \quad \dot{x}_4 = \mu_2 x_1 + \mu x_3 + a x_1^2 + b x_2^2.$$

corresponds to the hyperface $\mathbf{\Gamma}_4^{(5)}$ with normal vector $N_4 = (-2,-2,-3,-4,-1,-2)$ (the 10th row of the Table of Correspondences 1). Its solutions are

$$x_1 = \sigma_2 t + c_1, \quad x_2 = \sigma_2, \quad x_3 = \mu c_2 t + c_3,$$

$$x_4 = \frac{1}{3} a c_2^2 t^3 + \frac{1}{2}(\mu^2 c_2 + \mu_2 c_2 + 2 a c_1 c_2) t^2 + (\mu_2 c_1 + \mu c_3 + a c_1^2 + b c_2^2) t + c_4.$$

Finally, the truncated system

$$\dot{x}_1 = 0, \quad \dot{x}_2 = \mu x_1 + x_3 + a_2 x_1^2, \quad \dot{x}_3 = x_4, \quad \dot{x}_4 = \mu_2 x_1 + a x_1^2.$$

corresponds to the hyperface $\mathbf{\Gamma}_5^{(5)}$ with normal vector $U_5 = (-1,-2,-2,-2,-1,-1)$ (11th row of the Table of Correspondences 1). Its solutions are

$$x_1 = c_1, \quad x_2 = \frac{1}{6}(a c_1^2 + \mu_2 c_1) t^3 + \frac{1}{2} c_4 t^2 + (c_3 + \mu c_1 + a_2 c_1^2) t + c_2,$$

$$x_3 = \frac{1}{2}(a c_1^2 + \mu_2 c_1) t^2 + c_4 t + c_3, \quad x_4 = (a c_1^2 + \mu_2 c_1) t + c_4.$$

The program described in Section 2 was also applied by authors to investigate the following symmetric-reversible system of ODEs:

$$\begin{cases} \dot{x}_1 = x_3 + \Theta_1(X) \\ \dot{x}_2 = x_4 + \Theta_2(X) \\ \dot{x}_3 = -x_1 + \Theta_3(X) \\ \dot{x}_4 = -x_2 + \Theta_4(X), \end{cases} \tag{11}$$

where Θ_i $(\overline{i,4})$ are the analytic functions having two automorphisms:

the reversibility: $x_1, x_2, x_3, x_4, t \to x_1, x_2, -x_3, -x_4, -t$;

the symmetry: $x_1, x_2, x_3, x_4, t \to -x_1, x_2, -x_3, -x_4, t$.

It is worth to be noted that by combining the method described in Section 2 with the method of Belitski's normal form [16] one can consider the complicated cases of system (1). The special interest is represented by conditional-periodic and homoclinic solutions of systems (1) and (11), which did not manage to be found out by other methods until now.

Table of Correspondences

		j	1	2	3	4	5	6	7	8	9	10	11	12	13	14	15	16	17	18	19	20	21
		T_j	V	V	V	V	V	V	V	V	d	V	V	d	d	d	d	V	d	d	V	V	V
		Q_j	-1	0	0	1	0	0	1	2	2	0	1	0	1	1	-1	0	0	0	1	-1	0
			1	-1	0	-1	1	0	0	-1	0	1	1	2	-1	0	1	-1	0	0	0	0	-1
			0	1	-1	0	-1	1	0	0	0	0	-1	0	1	1	1	2	2	0	-1	1	0
			0	0	1	0	0	-1	-1	0	-1	0	0	-1	0	-1	0	0	-1	1	1	1	2
						1	1	1	0														
i	t_i	B_i, N_i, D_i				0	0	0	1														
1	N	0 0 0 -1 0 0	-	-	-	-	-	+	+	-	+	-	-	+	-	+	-	-	+	-	-	-	-
2	N	0 -1 0 -1 0 0	-	+	-	+	-	+	+	+	+	-	-	-	+	+	-	+	+	-	-	-	-
3	N	0 0 -1 -1 0 0	-	-	-	-	+	-	+	-	+	-	+	+	-	-	-	-	-	-	-	-	-
4	N	0 -1 -2 -1 0 0	-	-	+	+	+	-	+	+	+	-	+	-	-	-	-	-	-	-	+	-	-
5	N	0 0 0 -1 0 0	+	+	+	-	-	-	+	+	+	+	+	+	+	+	+	+	+	+	+	+	+
6	N	-2 -1 -4 -3 -2 0	+	-	+	-	+	-	+	-	-	-	+	+	-	-	-	-	-	-	-	-	-
7	N	0 0 0 0 0 -1	+	+	+	+	+	+	-	+	+	+	+	+	+	+	+	+	+	+	+	+	+
8	N	-1 -1 -1 -2 -1 -1	+	+	-	-	-	+	+	-	+	-	-	+	-	+	-	-	+	-	-	-	-
9	N	-1 -1 -2 -2 -1 -1	+	-	+	-	+	-	+	-	+	-	+	+	-	-	-	-	-	-	-	-	-
10	N	-2 -2 -3 -4 -1 -2	+	-	-	-	+	+	+	-	+	-	-	+	-	-	-	-	-	-	-	-	-
11	N	-1 -2 -2 -2 -1 -1	-	+	+	+	-	-	+	+	+	-	-	-	-	-	-	-	-	-	-	-	-
12	N	-4 -5 -6 -7 -2 -4	+	+	+	+	+	+	+	-	+	-	-	-	-	-	-	-	-	-	-	-	-
13	D	-2 -1 -2 -3 0 0	+	-	-	-	+	+	+	-	-	-	-	+	-	-	-	-	-	-	-	-	-
14	D	-1 -1 -1 -1 0 0	+	+	+	+	+	+	+	-	-	-	-	-	-	-	-	-	-	-	-	-	-
15	D	0 -1 0 0 0 0	-	+	-	+	-	-	-	+	-	-	-	-	+	-	-	+	-	-	-	-	+
16	D	0 0 -1 0 0 0	-	-	+	-	+	-	-	-	-	-	+	-	-	-	-	-	-	-	+	-	-
17	D	-1 0 -1 0 0 0	+	-	+	-	+	-	-	-	-	-	-	-	-	-	-	-	-	-	-	-	-
18	D	0 -1 -1 0 0 0	-	-	+	+	-	-	-	+	-	-	-	-	-	-	-	-	-	-	+	-	+
19	D	-1 0 0 -1 0 0	+	-	-	-	-	+	-	-	-	-	-	+	-	-	+	-	+	-	-	-	-
20	D	-2 -1 0 -1 0 0	+	+	-	-	-	+	-	-	-	-	-	-	-	-	+	+	+	-	-	+	-
21	D	-1 0 -1 -1 0 0	+	-	-	-	-	+	-	-	-	-	-	-	+	-	-	-	-	-	-	-	-
22	D	-1 0 0 0 0 0	-	-	-	-	-	-	-	-	-	-	-	-	-	-	+	-	-	-	-	+	-
23	D	-2 -1 -1 -3 -1 0	+	-	-	-	-	+	+	-	-	-	-	+	-	-	-	-	+	-	-	-	-
24	D	-4 -3 -2 -5 -2 0	+	+	-	-	-	+	+	-	-	-	-	-	-	-	-	-	+	-	-	-	-
25	D	-2 -5 -4 -3 -2 0	-	+	+	+	-	-	+	+	-	-	-	-	-	-	-	-	-	-	-	-	-
26	D	-1 -1 -2 -3 0 -1	-	-	-	-	+	+	+	-	+	-	-	+	-	-	-	-	-	-	-	-	-
27	D	-2 -3 -4 -5 0 -2	-	-	-	+	+	+	+	-	+	-	-	-	-	-	-	-	-	-	-	-	-
28	D	-1 -1 0 0 0 0	-	+	-	-	-	-	-	-	-	-	-	-	-	-	-	+	-	-	-	+	+
29	D	-1 -1 -1 0 0 0	-	-	+	-	-	-	-	-	-	-	-	-	-	-	-	-	-	-	-	-	+
30	D	-4 -3 -2 -1 0 0	+	+	+	-	-	-	-	-	-	-	-	-	-	-	-	-	-	-	-	+	+
31	D	-2 -3 -2 -1 0 0	-	+	+	+	-	-	-	-	-	-	-	-	-	-	-	-	-	-	-	-	+
32	D	-1 -3 -2 -1 -1 0	-	+	+	+	-	-	-	+	-	-	-	-	-	-	-	-	-	-	-	-	+

Table of Correspondences – *continued*

i	t_i \ j	22	23	24	25	26	27	28	29	30	31	32	33	34	35	36	37	38	39	40	41	42	43	44	45	46	47	48	49
	T_j	V	V	d	V	V	V	V	d	d	d	V	d	d	V	V	V	d	V	d	d	d	V	V	d	d	V	V	V
		3	3	1	2	1	-1	0	2	2	0	0	1	1	-1	0	0	1	2	-1	0	0	-1	1	-1	0	0	-1	0
		-1	0	1	1	2	3	3	-1	0	1	2	-1	0	1	-1	0	0	0	2	2	0	0	-1	1	1	-1	0	0
	Q_j	0	0	0	-1	0	0	-1	1	1	1	1	2	2	2	3	3	0	-1	0	-1	1	2	0	0	-1	1	0	-1
		0	-1	0	0	-1	0	0	0	-1	0	-1	0	-1	0	0	-1	1	1	1	1	1	1	2	2	2	2	3	3
1	N	-	-	-	-	-	-	+	-	+	-	+	-	-	+	-	-	-	-	-	-	-	-	-	-	-	-	-	-
2	N	+	+	-	-	-	-	-	+	+	-	-	+	+	-	+	+	-	-	-	-	-	-	-	-	-	-	-	-
3	N	-	+	-	+	+	-	+	-	-	-	-	-	-	-	-	-	-	-	-	-	-	-	-	-	-	-	-	-
4	N	+	+	-	+	-	-	-	-	-	-	-	-	-	+	-	-	-	+	-	-	-	-	-	-	-	-	-	-
5	N	+	+	+	+	+	+	+	+	+	+	+	+	+	+	+	+	+	+	+	+	+	+	+	+	+	+	+	+
6	N	-	-	-	-	-	-	+	-	-	-	-	-	-	-	-	-	-	-	-	-	-	-	-	-	-	-	-	-
7	N	+	+	+	+	+	+	+	+	+	+	+	+	+	+	+	+	+	+	+	+	+	+	+	+	+	+	+	+
8	N	-	-	-	-	-	-	-	-	-	-	-	-	-	-	-	-	-	-	-	-	-	-	-	-	-	-	-	-
9	N	-	-	-	-	-	-	-	-	-	-	-	-	-	-	-	-	-	-	-	-	-	-	-	-	-	-	-	-
10	N	-	-	-	-	-	-	-	-	-	-	-	-	-	-	-	-	-	-	-	-	-	-	-	-	-	-	-	-
11	N	-	-	-	-	-	-	-	-	-	-	-	-	-	-	-	-	-	-	-	-	-	-	-	-	-	-	-	-
12	N	-	-	-	-	-	-	-	-	-	-	-	-	-	-	-	-	-	-	-	-	-	-	-	-	-	-	-	-
13	D	-	-	-	-	-	-	-	-	-	-	-	-	-	-	-	-	-	-	-	-	-	-	-	-	-	-	-	-
14	D	-	-	-	-	-	-	-	-	-	-	-	-	-	-	-	-	-	-	-	-	-	-	-	-	-	-	-	-
15	D	+	-	-	-	-	-	-	+	-	-	-	+	-	-	+	-	-	-	-	-	-	-	+	-	-	+	-	-
16	D	-	-	-	+	-	-	+	-	-	-	-	-	-	-	-	-	-	+	-	+	-	-	-	-	+	-	-	+
17	D	-	-	-	-	-	+	+	-	-	-	-	-	-	-	-	-	-	-	+	+	-	-	-	+	+	-	+	+
18	D	+	-	-	-	-	-	-	-	-	-	-	-	-	-	-	-	-	-	+	-	-	-	-	+	-	-	-	+
19	D	-	-	-	-	-	+	-	-	-	-	+	-	-	+	-	+	-	-	-	-	-	-	-	-	-	-	-	-
20	D	-	-	-	-	-	-	-	-	-	-	-	-	-	+	+	+	-	-	-	-	-	+	-	-	-	-	-	-
21	D	-	-	-	-	-	+	+	-	-	-	-	-	-	-	-	-	-	-	-	-	-	-	-	-	-	-	-	-
22	D	-	-	-	-	-	+	-	-	-	-	-	-	-	+	-	-	-	-	+	-	-	+	-	+	-	-	+	-
23	D	-	-	-	-	-	-	-	-	-	-	-	-	-	-	-	-	-	-	-	-	-	-	-	-	-	-	-	-
24	D	-	-	-	-	-	-	-	-	-	-	-	-	-	-	-	-	-	-	-	-	-	-	-	-	-	-	-	-
25	D	-	-	-	-	-	-	-	-	-	-	-	-	-	-	-	-	-	-	-	-	-	-	-	-	-	-	-	-
26	D	-	-	-	-	-	-	-	-	-	-	-	-	-	-	-	-	-	-	-	-	-	-	-	-	-	-	-	-
27	D	-	-	-	-	-	-	-	-	-	-	-	-	-	-	-	-	-	-	-	-	-	-	-	-	-	-	-	-
28	D	-	-	-	-	-	-	-	-	-	-	-	-	-	-	+	-	-	-	-	-	-	+	-	-	-	+	+	-
29	D	-	-	-	-	-	-	-	-	-	-	-	-	-	-	-	-	-	-	-	-	-	-	-	-	-	-	+	+
30	D	-	-	-	-	-	-	-	-	-	-	-	-	-	-	-	-	-	-	-	-	-	-	-	-	-	-	+	-
31	D	-	-	-	-	-	-	-	-	-	-	-	-	-	-	-	-	-	-	-	-	-	-	-	-	-	-	-	-
32	D	-	-	-	-	-	-	-	-	-	-	-	-	-	-	-	-	-	-	-	-	-	-	-	-	-	-	-	-

References

1. Bruno, A.D.: *The Local Method of Nonlinear Analysis of Differential Equations*, Nauka, Moscow, 1979.
2. Bruno, A.D., Soleev A.: Local uniformization of branches of a space curve and Newton polyhedra, *Algebra i Anal.* **3**, No. 1 (1991) 67-102.
3. Soleev, A.: An algorithm for computing Newton polyhedra. *DAN UzSSR*, No. 5 (1982) 14–16.
4. Soleev, A., Aranson, A.: *Computation of a Polyhedron and Normal Cones of its Faces*. Preprint No. 36, Inst. Prikl. Mat. Ross. Akad. Nauk, 1994, pp. 1–25.
5. Roberts, J. A. G., Quispel, G. R. W.: Chaos and time-reversal symmetry. Order and chaos in reversible dynamical systems. *Phys. Rep.* **216** (1992) 63–177.
6. Sevryuk, M. B.: Reversible systems. In: *Lecture Notes in Math.* **1211**, 1986.
7. Soleev, A., Aranson, A.: *First Approximations of a Certain Revesible System of ODEs*. Preprint No 28, Inst. Prikl. Mat. Ross. Akad. Nauk, 1995, pp. 1–20.
8. Amick, C., McLeod, J.: A singular perturbation problem in water waves, *Stab. and Appl. Anal. of Cont. Media* **1** (1991) 127–148.
9. Iooss, G.: A codimension 2 bifurcation for reversible vector fields. *Fields Institute Communications* **4** (1995) 201–217.
10. Iooss, G.: Capillary-gravity water-waves problem as a dynamical system. In: *Structure and Dynamics of Nonlinear Waves in Fluids*, Mielke A., Kirchgässner K. (Eds.), Advanced series in nonlinear dynamics. V. 7, World Scientific, Singapore, 1995, pp. 42–57.
11. Iooss, G., Kirchgässner, K.: Bifurcation d'ondes solitaires en présence d'une faible tension superficielle. *C.R. Acad. Sci. Paris* **311**, Ser. I (1990) 265–268.
12. Iooss, G., Kirchgässner, K. Water waves for small surface tension: an approach via normal form. *Proceedings of the Royal Society of Edinburgh* **122A** (1992) 267–299.
13. Iooss, G., Pérouème, M.–C.: Perturbed homoclinic solutions in reversible 1:1 resonance vector fields. *J. Diff. Equations* **102** (1993) 62–88.
14. Bruno, A. D., Soleev A.: *Local Analysis of Singularities of Reversible System of ODEs. Complicated Cases*. Preprint No. 47, Inst. Prikl. Mat. Ross. Akad. Nauk, 1995, pp. 1–30.
15. Bruno, A.D.: *The Restricted Three-Body Problem: Plane Periodic Orbits*, Nauka, Moscow, 1990.
16. Belitski, G. R.: *Normal Forms, Invariants and Local Maps*, Naukova Dumka, Kiev, 1979.

Condition Analysis of Overdetermined Algebraic Problems

Hans J. Stetter*

Tech.Univ. Vienna, A-1040 Vienna, Austria
stetter@aurora.anum.tuwien.ac.at

Abstract. In analogy to numerical linear algebra, the evolving *numerical polynomial algebra* studies the modifications of classical polynomial algebra necessary to accomodate inaccurate data and inexact computation. A standard part of this endeavor is a *condition analysis* of the problems to be solved, i.e. an assessment of the *sensitivity* of their results with respect to *variations in their data.* In this paper, we extend this condition analysis to *overdetermined problems,* like greatest common divisors, multivariate factorization etc. Here we must consider the fact that results exist only for data from restricted low-dimensional domains. The discontinuous dependence of the results on the data, however, is "smoothed" by their limited accuracy so that a condition analysis becomes meaningful. As usual, the condition numbers indicate the accuracy with which results may be specified.

1 Introduction

Many mathematical problems in Scientific Computing are distinguished by the fact that some of their data are only known to a *limited accuracy.* This may reflect the limited accuracy of measured or observed data or express the simplifications of the underlying model; in any case, it has far-reaching consequences for the mathematical and computational treatment of such problems. In *linear algebra,* it has led to the enormous growth of *numerical linear algebra* into one of the supporting pillars of Scientific Computing. In an analogous fashion, the presently evolving *numerical polynomial algebra* studies the modifications and extensions of classical polynomial algebra necessary to accomodate the presence of inaccurate data and inexact computation.

Except in some very special situations, data to which "point values" cannot reasonably be assigned must be assumed to vary in $\mathbb{R}$ or $\mathbb{C}$. This makes it possible to consider the *analytic aspects* of classical algebraic problems which opens the way for new approaches in their computational solution. The formal introduction of *tolerances* for algebraic objects with values in $\mathbb{C}^n$ meets with no great difficulties; cf., e.g., [1], [18]: A tolerance defines a *family of neighborhoods* for the specified value of the object. E.g., a polynomial with a tolerance (or an "empiric polynomial") has (a few) coefficients whose values are only known

* partly joint work with Huang Yuzhen, Wu Wenda, Zhi Lihong

to lie in some small interval. Correspondingly, the concept of a zero of such a polynomial – or, more generally, of a result of an empiric algebraic problem, – extends into that of a *pseudozero* (*pseudoresult*): z is a pseudoresult if it is the *exact result* of a problem within the tolerance neighborhood of the specified problem.

For the computational solution of such problems, this poses two fundamental tasks: Given an approximate solution $\tilde{z}$, from whatever computational source,

1) check whether $\tilde{z}$ is a pseudoresult;

2) if yes, assess how many digits of $\tilde{z}$ are meaningful.

In Numerical Analysis, where problems are generally assumed to have data of limited accuracy, these two tasks are associated with "Backward Error Analysis" and "Condition Analysis" resp.; cf. any textbook of Numerical Analysis and [1], [4].

Task 1 requires that we consider the position of the tolerance neighborhoods relative to the manifold of all data which lead to $\tilde{z}$ as an *exact* result; this can be done and effectively implemented for a wide variety of algebraic problems; cf. [18]. Task 2 requires that we analyze the effect of the data variations in the tolerance neighborhoods on the exact results of the problem. While this task has received a good deal of attention in numerical linear algebra, it has only been considered for some standard problems in polynomial algebra so far.

In this paper, we systematically address this task of finding quantitative expressions for the "*condition*" of polynomial algebraic problems, i.e. for the *sensitivity* of their results with respect to *variations in their data*; we will outline an approach which generalizes easily to non-standard situations. Moreover, we will extend this condition analysis to *overdetermined problems*, like greatest common divisors, multivariate factorization, and the like. In the classical context, this appears impossible since the results depend discontinuously on their data and exist only for specially restricted data. With the pseudoresult concept, however, these discontinuities are "smoothed" so that a condition analysis becomes meaningful.

In section 2 of this paper, we summarize the formalization of empiric algebraic problems which we use. In section 3, we consider the analytic basis of a quantitative condition analysis and apply it to clusters of polynomial zeros. The main part of the paper is devoted to overdetermined algebraic problems: In section 5, we introduce our approach to a condition analysis for such problems. Then we consider three particular problem classes of this kind: Multiple zeros of polynomials, greatest common divisors of univariate polynomials, and the factorization of a multivariate polynomial. To our knowledge, no sensitivity analysis for the last two of these has so far been proposed. These sections are supplemented by numerical examples. Some conclusions complete the paper.

2 Preparations

When we wish to assess the sensitivity of a result of an algebraic problem with respect to variations in the data of the problem, we must consider the mapping

F which assigns a particular result value z to a particular data quantity a from a suitably restricted domain A.

Definition 1: Consider an algebraic problem $\mathcal{P}$, with data in a domain A of the data space $\mathcal{A} = \mathbb{C}^n$ and results in the result space $\mathcal{Z} = \mathbb{C}^m$. The mapping $F : \mathcal{A} \to \mathcal{Z}$ which assigns to each data quantity $a \in A$ the exact result $z \in \mathcal{Z}$ of $\mathcal{P}$ will be called *data→result mapping* of $\mathcal{P}$. □

The data a of a polynomial $p \in \mathbb{C}[x_1, \ldots, x_s]$, with support $J \subset \mathbb{N}_0^s$, consist of its coefficients α_j, $j \in J$. However, there may be two kinds of coefficients: Some may be *intrinsic* in the given context, i.e. they have a well-known *fixed* value (e.g. 1) in the model under consideration; such coefficients may be considered as part of the problem formulation (of the data→result mapping F) rather than as data in the strict sense. All non-intrinsic data will be considered as *empiric*, i.e. of limited accuracy.

Definition 2: For $\mathcal{A} = \mathbb{C}^n$, the specification of an *empiric data quantity* a consists of a *specified value* $\bar{a} = (\bar{\alpha}_1, \ldots, \bar{\alpha}_n) \in \mathbb{C}^n$ and of a *tolerance* $e = (\varepsilon_1, \ldots, \varepsilon_n) \in \mathbb{R}_+^n$ which define the *tolerance neighborhood*

$$N(\bar{a}, e) \; := \; \{\tilde{a} \in \mathcal{A} \, : \, \| \; \ldots, \frac{|\tilde{\alpha}_\nu - \bar{\alpha}_\nu|}{\varepsilon_\nu}, \ldots \; \|^* \leq 1\} \; \subset \; \mathcal{A} . \tag{1}$$

Each value $\tilde{a} \in N(\bar{a}, e)$ is an *equally valid* value for the data quantity. A problem with at least one empiric data quantity is called an *empiric problem.* □

The norm $\|..\|^*$ in (1) is a vector norm in $(\mathbb{R}^n)^*$; for $\|b^T\|^* = \max_\nu |\beta_\nu|$, e.g., (1) requires

$$|\tilde{\alpha}_\nu - \bar{\alpha}_\nu| \; \leq \; \varepsilon_\nu \, , \quad \nu = 1(1)n \, . \tag{2}$$

More generally, we consider an *absolute norm* on $\mathbb{R}^n$, i.e. a norm for which $\||u|\| = \|u\|$. Then $\|..\|^*$ denotes the associated *dual norm* (operator norm), i.e.

$$\|v^T\|^* \; := \; \sup_{u \neq 0} \frac{|v^T u|}{\|u\|} \, . \tag{3}$$

The following norms are most common:

$$\begin{aligned} \|u\|_1 &:= \textstyle\sum_j |u_j| \, , & \|v^T\|_1^* &:= \max_j |v_j| \, ; \\ \|u\|_\infty &:= \max_j |u_j| \, , & \|v^T\|_\infty^* &:= \textstyle\sum_j |v_j| \, ; \\ \|u\|_2 &:= (\textstyle\sum_j |u_j|^2)^{\frac{1}{2}} \, , & \|v^T\|_2^* &:= (\textstyle\sum_j |v_j|^2)^{\frac{1}{2}} \, . \end{aligned}$$

With (1), the $\|..\|_1^*$ norm appears most appropriate as shown by (2); it is the only norm where the requirements on the individual components of the empiric data quantity a remain *separated.*

It is natural that the "result" of an empiric problem is not a specific point $z \in \mathcal{Z}$ but rather *some* point in the set of all results for data in $N(\bar{a}, e)$.

Definition 3: A quantity $\tilde{z} \in \mathbb{C}^m$ is a *pseudoresult* of the problem $\mathcal{P}$ with empiric data $(\bar{a}, e)$ if it is the exact result $F(\tilde{a})$ of $\mathcal{P}$ for some $\tilde{a} \in N(\bar{a}, e)$. The set of all pseudoresults

$$Z(\bar{a}, e) \; := \; \{\tilde{z} \in \mathbb{C}^m \; : \; \tilde{z} = F(\tilde{a}) \;\; \text{for some} \;\; \tilde{a} \in N(\bar{a}, e)\} \, , \tag{4}$$

where F is the data→result mapping of $\mathcal{P}$, is the *pseudoresult set* of $\mathcal{P}$ for data $(\bar{a}, e)$. Each $\tilde{z} \in Z(\bar{a}, e)$ is an *equally valid* approximate solution of $\mathcal{P}$ for $(\bar{a}, e)$. □

An important tool in dealing with empiric problems is the set of all data such that a given $\tilde{z}$ is the exact result of $\mathcal{P}$:

Definition 4: For a problem $\mathcal{P}$ with data→result mapping $F : A \subset \mathcal{A} \to \mathcal{Z}$ and some $\tilde{z} \in \mathcal{Z}$, the manifold

$$\mathcal{M}(\tilde{z}) := \{\tilde{a} \in A : F(\tilde{a}) = \tilde{z}\} \subset \mathcal{A} \tag{5}$$

is called the *equivalent-data-manifold* for $\tilde{z}$. □

The following criterion is an immediate consequence of this definition:

Theorem 1:

$$\tilde{z} \in Z(\bar{a}, e) \iff \mathcal{M}(\tilde{z}) \cap N(\bar{a}, e) \neq \emptyset . \tag{6}$$

This criterion can be checked algorithmically for many situations; cf. e.g. [18] for further details. This solves task 1 described in section 1.

For many purposes, it would appear sufficient to have an approximate result $\tilde{z}$ which has been verified to be a pseudoresult. But even in the specification of such a $\tilde{z}$, there arises a question: *How many digits* of $\tilde{z}$ should be stated? If both 1.23 and 1.24 are pseudoresults of a problem for the same data $(\bar{a}, e)$, it is certainly not meaningful to write $\tilde{z} = 1.24786$; in fact, this could falsely indicate that these digits are well-defined in the given context. One should rather specify $\tilde{z}$ as either one of 1.23, 1.24, 1.25 or as $1.24 \pm .01$ or the like. Obviously, this question is closely related to task 2 of section 1.

While it might be desirable to know details about the size and shape of a pseudoresult set $Z(\bar{a}, e)$, the associated computational effort is far too high (except for demonstration purposes); cf., e.g., [19], [20]. Also, how should one display a nontrivial domain in $\mathbb{C}^s$ for $s > 1$? To test the situation by checking a few neighboring result values may work in one real dimension; but it will be unreliable to worthless in several complex dimensions.

On the other hand, all that is generally needed in a real-life problem is an *order-of-magnitude* information about the largest extension of $Z(\bar{a}, e)$, and – for $m > 1$ – perhaps some directional information. For sufficiently small tolerances, this information can be generated with a moderate computational effort, as we will explain in the remainder of this paper.

3 The Condition of an Algebraic Problem

Formally, an analysis of the effect of data variations on the result of an algebraic problem $\mathcal{P}$ is straightforward: If the associated data→result mapping F is continuously diferentiable in the neighborhood $N(\bar{a}, e)$ of the specified data, we can use the Frechet derivative $F' : A \to \mathbb{C}^{m \times n}$ to express the effect Δz of a data perturbation Δa. The following result from multidimensional analysis (and functional analysis) is well-known (cf., e.g., [10]):

Proposition 2: Let $F : \mathbb{C}^n \to C^m$ have a continuous Frechet derivative in some convex domain $A \subset \mathbb{C}^n$; then, for $a, a + \Delta a \in A$,

$$F(a + \Delta a) = F(a) + \int_0^1 F'(a + \lambda\,\Delta a)\,d\lambda \cdot \Delta a\,, \tag{7}$$

which implies

$$\|F(a + \Delta a) - F(a)\| \le \sup_{\lambda \in (0,1)} \|F'(a + \lambda\,\Delta a)\| \cdot \|\Delta a\|^* \tag{8}$$

and

$$\|F(a + \Delta a) - F(a) - F'(a)\,\Delta a\| \le \sup_{\lambda \in (0,1)} \|F'(a + \lambda\,\Delta a) - F'(a)\| \cdot \|\Delta a\|^*\,, \tag{9}$$

where the norm on derivatives is the operator norm for linear maps from $\mathcal{A}$ with norm $\|..\|^*$ to $\mathcal{Z}$ with norm $\|..\|$.

Corollary 3: In the situation of Proposition 2, if F' is Lipschitz continuous with Lipschitz constant L', then

$$F(a+\Delta a) - F(a) = F'(a)\,\Delta a + r(a, \Delta a) \quad \text{with } \|r(a, \Delta a)\| \le \frac{L'}{2}(\|\Delta a\|^*)^2\,. \tag{10}$$

Thus, in a reasonably regular situation with a Lipschitz continuous Frechet derivative and a moderate L', the linear mapping $F'(a)$ describes the effect of *small perturbations* of the data well. By (10), a perturbation Δa may be considered as small if $L'\|\Delta a\|^*/2 \ll \|F'(a)\|$. In the following, we will always assume this condition to hold so that we can neglect quadratic terms in $\|\Delta a\|^*$. If necessary, we will use $\doteq$ and $\dot{\le}$ for relations in which a quadratic term has been omitted. As a consequence we have

Proposition 4: In the situation just described, the pseudoresult set $Z(\bar{a}, e)$ satisfies, for sufficiently small tolerances e,

$$Z(\bar{a}, e) \doteq \{\tilde{z} \in \mathcal{Z} : \tilde{z} = F(a) + F'(a)\,(\tilde{a} - \bar{a})\,,\ \tilde{a} \in N(\bar{a}, e)\}\,. \tag{11}$$

In the normed finite-dimensional spaces $\mathcal{A}$ and $\mathcal{Z}$, the *diameter* of a closed and bounded set M is defined by

$$\text{diam } M := \max_{m_1, m_2 \in M} \|m_1 - m_2\|\,.$$

For the max-norm in the data space $\mathcal{A}$, equ. (2) implies immediately

$$\text{diam } N(\bar{a}, e) = 2 \max_{\nu} \varepsilon_\nu = 2\,\|e\|^*\,; \tag{12}$$

for other norms, an equally simple expression arises only if the ε_ν are all equal.

Corollary 5: In the situation of Proposition 4,

$$\text{diam } Z(\bar{a}, e) \;\dot{\le}\; 2\,\|F'(\bar{a})\|\,\|e\|^*\,, \tag{13}$$

with the operator norm for F' from $\mathcal{Z}$ to $\mathcal{A}$ with the max-norm.

A large diameter of $Z(\bar{a}, e)$ may be due to only one stretched direction in the result space $\mathcal{Z}$. Therefore, it may be interesting also to look at the *singular values* of $F'(\bar{a})$ which represent the half axes of the ellipsoidal image of a Euclidean ball in $\mathcal{A}$ under the linear map $F'(\bar{a})$. A small singular value indicates that the sensitivity of the result z is low for variations of the data in a subspace defined by the associated singular vector so that an approximate solution is more reliable in the corresponding direction in the result space. This effect may become dramatic for near-singular algebraic problems; cf. e.g. [20], [1], [17].

So far, we have tacitly assumed that the Jacobian F' of the data→result mapping F is explicitly available. In almost all cases, F is only implicitly defined by the formulation of the algebraic problem $\mathcal{P}$; but, generally, that does not prevent the (approximate) evaluation of $F'(\bar{a})$ after an approximate solution for that data has been obtained. Assume that $\mathcal{P}$ may be formulated in terms of M *equations*:

$$G(x;\, a) \;=\; 0\,, \tag{14}$$

with a mapping $G:\; \mathbb{C}^m \times \mathbb{C}^n \to \mathbb{C}^M$. Assume that (14) defines a data→result mapping F in a suitable domain A such that

$$G(F(a);\, a) \;=\; 0 \tag{15}$$

holds for all a in a sufficiently large neighborhood of the specified data $\bar{a}$. Assume that G is differentiable w.r.t. to both arguments in that neighborhood; then differentiation of (15) w.r.t. a yields, with $\bar{z} = F(\bar{a})$,

$$\frac{\partial G}{\partial x}(\bar{z}, \bar{a}) \cdot F'(\bar{a}) + \frac{\partial G}{\partial a}(\bar{z}, \bar{a}) \;=\; 0\,, \tag{16}$$

or

$$F'(\bar{a}) \;=\; -\left[\frac{\partial G}{\partial x}(\bar{z}, \bar{a})\right]^{-1} \frac{\partial G}{\partial a}(\bar{z}, \bar{a})\,. \tag{17}$$

An obvious prerequisite for that last transformation is $M = m$ or "as many equations as unknowns"; this excludes consistent overdetermined systems.

But consistent overdetermined systems abound in polynomial algebra. Note that most Groebner bases for a total degree order consist of more polynomials than there are variables. Many other interesting problems like greatest common divisors or multivariate factorization may also be written in the form (14), with $M > m$. It is true that such problems possess solutions only if their data lie on a specific manifold $\mathcal{S}$ in the data space; but it can often be assumed that this is the case or that – in empiric problems – the tolerance neighborhood $N(\bar{a}, e)$ of the specified data has a nonempty intersection with $\mathcal{S}$ so that the pseudoresult set of the given problem is not empty; cf. Definition 3. Thus, the analysis of the perturbation sensitivity of the (pseudo)results of such (near-)consistent overdetermined problems appears particularly important.

4 Condition of Zero Clusters

Expressions for the condition of an isolated zero of a univariate polynomial $p(x,a) = \sum_{\nu=0}^{n} \alpha_\nu x^\nu \in \mathbb{C}[x]$ have been known for a long time; cf. textbooks of Numerical Analysis and [1], [4]. From (17), we obtain for a particular zero $\bar{z} = z(\bar{a})$ of $p(x,\bar{a})$

$$\frac{\partial}{\partial a_\nu} z(\bar{a}) = -\frac{1}{p'(\bar{z})}\,\bar{z}^\nu ; \tag{18}$$

for the corresponding pseudozero set of the empiric polynomial $p(x,(\bar{a},e))$; this implies (cf. (11) and (13))

$$\operatorname{diam} Z(\bar{a},e) \mathrel{\dot{\leq}} \frac{2}{|p'(\bar{z})|}\sum_{\nu=0}^{n} \varepsilon_\nu |\bar{z}|^\nu \leq \frac{2}{|p'(\bar{z})|}\sum_{\nu=0}^{n} |\bar{z}|^\nu \max_\nu \varepsilon_\nu , \tag{19}$$

for sufficiently small tolerances ε_ν. If we interpret "sufficiently small" as indicated in section 3, we have $|p'(z)| \gg 0$ in $Z(\bar{a},e)$ so that the evaluation of (19) at a computed pseudozero $\tilde{z}$ in place of the exact zero $\bar{z}$ leaves the estimate valid.

The actual size of the maximal tolerance level for which (19) is valid depends on the size of $p'(\bar{z})$ relative of $\|\bar{a}\|^*$ and thus, indirectly, on the closeness of further zeros of p. This is made more obvious by the following result:

Theorem 6 (Mosier [12]): If a pseudozero set $Z(\bar{a},e)$ for an empiric polynomial p contains more than one zero of $p(x,\bar{a})$, it contains at least one zero of $p'(x,\bar{a})$.

Thus, it is a necessary condition for p' to be bounded away from 0 in the pseudozero sets of the individual (isolated simple) zeros of p that these pseudozero sets *remain isolated* at the specified tolerance level e.

Due to the continuous dependence of polynomial zeros on the coefficients, if $p(x,\bar{a})$ has exactly m zeros (counting multiplicities) in some $Z(\bar{a},e)$ then *each* polynomial $p(x,\tilde{a})$, $\tilde{a} \in N(\bar{a},e)$, has exactly m zeros in $Z(\bar{a},e)$.

Definition 5: If some polynomial $p(x,\tilde{a})$, $\tilde{a} \in N(\bar{a},e)$, has m zeros in $Z(\bar{a},e)$, the associated empiric polynomial is said to have an *m-cluster* of zeros in $Z(\bar{a},e)$. □

For an empiric polynomial, the zeros in an m-cluster have *no individuality* in the following sense: If we mark a certain simple zero of $p(x,\bar{a})$ in $Z(\bar{a},e)$ and then vary the coefficients of $p(x,\tilde{a})$ continuously over $N(\bar{a},e)$, we cannot follow that zero in a unique fashion since there occurs at least one confluence of zeros in this process. Generally (though not necessarily), it will even happen that there exists an $\tilde{a} \in N(\bar{a},e)$ so that $p(x,\tilde{a})$ has an m-fold zero in $Z(\bar{a},e)$. Thus for a polynomial with coefficients of limited accuracy, a zero cluster must be considered as *one mathematical quantity* (cf. also [7]) !

The appropriate tool for the description of an m-cluster is the monic polynomial $s(x,c)$ of degree m which vanishes at the zeros of the cluster; cf. [16]. The *coefficients of s* depend *Lipschitz continuously* and smoothly on the coefficients of p.

Theorem 7: Let

$$p(x,a) \;=\; q(x,b)\cdot s(x,c)\,, \tag{20}$$

with monic p, q, s of degrees $n, n-m, m$ resp. Assume that s and q are coprime. Then there exists an $n \times m$-matrix C such that $p(x, \bar{a} + \Delta a) = q(x, \bar{b} + \Delta b) \cdot s(x, \bar{c} + \Delta c)$ implies

$$(\ldots \Delta c^T \ldots) = (\ldots \Delta a^T \ldots)\cdot C + \; \mathrm{O}(\|\Delta a\|^2) \quad \text{or} \quad C = \left(\frac{dc}{da}\right)_{a=\bar{a}} . \tag{21}$$

Proof: Let the $n \times m$-matrix X represent the normal forms of the monomials x^ν mod $\langle s(x,\bar{c})\rangle$:

$$\begin{pmatrix} 1 \\ x \\ \vdots \\ \vdots \\ x^{n-1} \end{pmatrix} \equiv \begin{pmatrix} 1 & & \\ & \ddots & \\ & & 1 \\ \cdots & & \cdots \\ \cdots & & \cdots \end{pmatrix} \begin{pmatrix} 1 \\ x \\ \vdots \\ x^{m-1} \end{pmatrix} \quad \text{mod } \langle s\rangle\,.$$

Let the $m \times m$-matrix A_q represent multiplication by $q(x,\bar{b})$ in the quotient ring $\mathbb{C}[x]/\langle s\rangle$:

$$q(x,\bar{b}) \begin{pmatrix} 1 \\ x \\ \vdots \\ x^{m-1} \end{pmatrix} \equiv \begin{pmatrix} \; A_q \; \end{pmatrix} \begin{pmatrix} 1 \\ x \\ \vdots \\ x^{m-1} \end{pmatrix} \quad \text{mod } \langle s\rangle\,.$$

Since the eigenvalues of A_q are the values of q at the zeros of s, the coprimality of s and q implies the regularity of A_q. The normal form mod $\langle s\rangle$ of

$$p(x,\bar{a}+\Delta a)-p(x,a) = p(x,\Delta a) = q(x,\bar{b})\, s(x,\Delta c)+q(x,\Delta b)\, s(x,\bar{c})+\mathrm{O}(\|\Delta b\|\,\|\Delta c\|)$$

yields

$$(\ \ldots \Delta a^T \ldots\ )\, X \begin{pmatrix} 1 \\ x \\ \vdots \\ x^{m-1} \end{pmatrix} = (\ \ldots \Delta c^T \ldots\ )\, A_q \begin{pmatrix} 1 \\ x \\ \vdots \\ x^{m-1} \end{pmatrix} + \ \mathrm{O}(\|\Delta b\|\,\|\Delta c\|)\,.$$

From the above, we have $\|\Delta c\| = \mathrm{O}(\|\Delta a\|) + \mathrm{O}(\|\Delta b\|\,\|\Delta c\|)$ and $\|\Delta b\| = \mathrm{O}(\|\Delta a\|) + \mathrm{O}(\|\Delta b\|\,\|\Delta c\|)$ which implies $\|\Delta b\|\,\|\Delta c\| = \mathrm{O}(\|\Delta a\|^2)$. Thus we have (21) with

$$C \;=\; X\, A_q^{-1}\,. \qquad \square \tag{22}$$

Remark: The coprimality assumption about q, s requires that s must contain the zeros of the cluster with their full multiplicity. A_q^{-1} will have a moderate norm if the cluster represented by s is well separated from the remaining zeros of p. □

Corollary 8: The *moments* of the zeros in a cluster depend Lipschitz continuously and smoothly on the coefficients of p.

Proof: By Vieta's Theorem, the coefficients of s are the fundamental symmetric functions of the cluster zeros, and the moments $Z_\lambda = \frac{1}{m}\sum_{\mu=1}^{m} z_\mu^\lambda$ of a set of m point $z_\mu \in \mathbb{C}$ may be expressed as polynomials in their fundamental symmetric functions. □

For the *arithmetic mean* Z_1 of the zeros in a cluster, the above result has been known for a long time; cf., e.g., the seminal paper [7] by W.Kahan. Here,

$$\left(\frac{dZ_1}{da^T}\right)_{a=\bar{a}} = -\frac{1}{m}\left(\frac{dc_{m-1}}{da^T}\right)_{a=\bar{a}} = -\frac{1}{m}\cdot(\text{ last column of } C). \tag{23}$$

Our result shows that *all* moments of the cluster zeros (up to Z_m) vary only linearly with a small change of the coefficients of p. E.g., $Z_2 = \frac{1}{m}\sum z_\mu^2 = \frac{1}{m}(c_{m-1}^2 - 2\,c_{m-2})$ implies

$$\Delta Z_2 \doteq \frac{2}{m}(c_{m-1}\,\Delta c_{m-1} - \Delta c_{m-2}). \tag{24}$$

This suggests that – for a polynomial with coefficients of limited accuracy – the *moments* of the zeros of a cluster should be computed and specified rather than the locations of the individual zeros. These moments lie in pseudoresult sets whose approximate diameters can be estimated from (21) and (22).

Example 1 (cf. also [18], section 7.3): Assume the polynomial

$$p(x,\bar{a}) := x^4 - 2.83088\,x^3 + .00347\,x^2 + 5.66176\,x - 4.00694 \tag{25}$$

to be empiric with a tolerance of 10^{-5} for each coefficient. There is a 3-cluster of zeros near $\sqrt{2}$ which cannot be separated at this tolerance level; actually, for polynomials within the tolerance neighborhood, there may be either 3 real or 1 real and 2 conjugate complex zeros, and the real zeros can vary in the interval $[1.385826, 1.444818]$. Thus, the pseudozero set of the cluster has a diameter of $\approx .06$.

On the other hand, the cluster polynomial (cf. (20))

$$s(x,\bar{c}) = x^3 - 4.24509\,x^2 + 6.00694\,x - 2.83333 \qquad \text{(to 5 digits)}$$

is *insensitive* to the variations in p. All elements of the matrix C in (21) are below 1 in modulus; from the columns of C (cf. (22)), one finds that the pseudoresult sets for the coefficients c_0, c_1, c_2 of s have diameters of $\approx (2.3, 3.6, 2.1) \times 10^{-5}$. Thus, it is meaningful so specify the cluster polynomial s to 5 digits for the above tolerance level of p.

Moreover, the moments of the cluster zeros are, for $\bar{a}$,

$$Z_1 \approx 1.41503, \quad Z_2 \approx 2.00231, \quad Z_3 \approx 2.83334.$$

From (24) and the analogous relations for other moments, one obtains that the pseudoresult sets for these moments have approximate diameters $(0.7, 5.4, 26.7) \times 10^{-5}$ so that it is meaningful to specify Z_1 and Z_2 to 5 and Z_3 to 4 digits. Note that the *arithmetic mean* Z_1 of the three cluster zeros varies only by $\approx 7 \times 10^{-6}$ within the tolerance neighborhood while the zeros vary by up to 6×10^{-2} ! □

For isolated zeros of *systems P of multivariate equations*, the univariate results (18), (19) generalize immediately, with the inverse of the Jacobian of P at $\bar{z}$ in place of $1/p'(\bar{z})$; cf. (14) - (17). As we have remarked there, this requires the regularity of P; (near-) consistent overdetermined systems will be considered in the next section. *Zero clusters* of multivariate systems are likewise defined by Definition 6; as in one dimension, the individual zeros in a cluster have no true identity. However, it is now not so clear how the cluster should be described in a well-conditioned fashion.

In [16], it has been shown that a multivariate m-cluster can be represented by a system S of polynomials which is a small perturbation of a system with exactly one m-fold zero at the center of the cluster. In analogy to (20), we must have $P \in \langle S \rangle$; [16] deals with the construction of this "cluster system" S. A condition analysis for S can, in principle, be conducted as for the cluster polynomial s in the univariate case; but the technical details are much more involved and will be treated in a separate paper. If no singular situation is present (cf. [17]), the coefficients of S depend smoothly on the coefficients of P, and one can define generalized moments of the zeros which also depend on P in a well-conditioned manner.

5 Condition of Pseudoresults of Overdetermined Algebraic Problems

Pseudoresults for systems of $n > s$ linear equations in s unknowns have been introduced by the "surveyor" C.F. Gauss nearly 200 years ago. Gauss's least squares approach has remained dominant in "smoothing" because of its structural and computational simplicity. In the full-rank case, it defines a unique data→result mapping so that the data sensitivity of the least-squares solution is well-defined. This has been analyzed in much detail in numerical linear algebra; cf. e.g. [3].

In classical polynomial algebra, with its paradigms of exact data and exact computation, overdetermined problems either have a solution or not; the chief interest is in relevant criteria. Since solutions exist only for data from some lower-dimensional manifold $\mathcal{S}$ in the data space $\mathcal{A}$, generic data perturbations lead to non-existence and a sensitivity analysis appears out of place.

With the consideration of data of limited accuracy, the dichotomy between existence and non-existence of solutions of overdetermined problems is *smoothed*: When the specified data $\bar{a}$ of such a problem move away from the manifold $\mathcal{S}$, the tolerance neighborhoods $N(\bar{a}, e)$ continue to intersect $\mathcal{S}$ so that pseudoresults continue to exist. The associated pseudoresult sets $Z(\bar{a}, e)$ of (4) decreases in size with an increasing distance of $\bar{a}$ from $\mathcal{S}$ and tends to $\emptyset$; but in a set topology, the transition of a set to $\emptyset$ is a *continuous* process. Thus, for pseudoresults, the data sensitivity problem returns.

In order to take the right approach, we must remember our purpose: We want to assess the *meaningful accuracy* of pseudoresults of such problems. In section 3, we had argued that a valid *indicator* of this accuracy level was the potential change in the *exact* solution upon a change of the data within the tolerance level. We had chosen to evaluate the relevant derivatives at the specified data $\bar{a}$: Under our smoothness assumptions and with a sufficiently small tolerance e, this cannot seriously affect the sensitivity analysis.

For overdetermined problems, we wish to follow the same route, with the necessary modifications: Since exact solutions exist only for data on $\mathcal{S}$, we analyze the result sensitivity with respect to data variations in $\mathcal{S}$. This is justified because we are only interested in the sensitivity analysis *when pseudoresults exist*, i.e. when the manifold $\mathcal{S}$ passes through the tolerance neighborhood $N(\bar{a}, e)$ of the specified data $\bar{a}$.

For a formal description, we assume again (cf. (14)) that the algebraic problem may be formultated as

$$G(x, a) = 0, \qquad \text{with } G : \mathbb{C}^m \times \mathbb{C}^n \rightarrow \mathbb{C}^M . \tag{26}$$

But now we admit $M > m$ so that generally no solution exists for a generic data set a. We assume that a must lie on an algebraic manifold $\mathcal{S}$ in the data space $\mathcal{A}$ for (26) to be *consistent*, i.e. a must satisfy a system of k polynomial equations $S(a) = 0$; this covers the cases in which we are interested. Naturally, there may be domain restrictions and other technical details which must be observed. By Definition 3, z^* is a pseudoresult of the *empiric* overdetermined problem $G(x, (\bar{a}, e)) = 0$ if there exist data $a^* \in N(\bar{a}, e)$ with $S(a^*) = 0$ and $G(z^*, a^*) = 0$.

Now we note that, in a neighborhood of a^* *on the consistency manifold* $\mathcal{S}$, there must exist a data$\rightarrow$result mapping F_0 which maps data $a^* + \Delta a \in \mathcal{S}$ into the respective results $z^* + \Delta z$ of the *consistent* system $G(x, a^* + \Delta a) = 0$. As in the regular case (cf. section 3), F_0 expresses the result sensitivity of small data variations in $\mathcal{S}$ near z^* through its Jacobian $F_0'(a^*)$; the differentiability assumption is generally no restriction in our algebraic context. Accordingly, we assume differentiability of G as in the regular case. If we disregard quadratic terms in the increments, we obtain, from $G(z^*, a^*) = G(z^* + \Delta z, a^* + \Delta a) = 0$,

$$\frac{\partial G}{\partial x}(z^*, a^*)\, \Delta z + \frac{\partial G}{\partial a}(z^*, a^*)\, \Delta a = 0 . \tag{27}$$

This is now an *overdetermined linear system* for Δz; but, as explained above, we assume that Δa is such that (27) is *consistent*. Due to the preceeding linearization, this now requires that

$$\sum_{\nu=1}^{n} \frac{\partial S}{\partial \alpha_\nu}(a^*)\, \Delta \alpha_\nu =: \mathbf{S} \cdot \Delta a = 0 , \tag{28}$$

where $S(a) = 0$ is the description of the consistency manifold $\mathcal{S}$. More intuitively, we replace $\mathcal{S}$ by its tangential hyperplane $d\mathcal{S}$ at a^*. With this assumption about Δa, we may use the Moore-Penrose pseudo-inverse to solve (27): With $G_x := \frac{\partial G}{\partial x}(z^*, a^*)$, $G_a := \frac{\partial G}{\partial a}(z^*, a^*)$, we obtain

$$\Delta z \;=\; -\,(G_x^T \cdot G_x)^{-1}\, G_x^T\, G_a\, \Delta a \;=\; F_0'(a^*) \cdot \Delta a\,. \tag{29}$$

This relation takes the place of (17) in the regular case. If G_x is rank-deficient, (27) contains some duplicated information; in such cases, one must reduce (27) accordingly.

An alternate and often simpler route to an expression for $F_0'(a^*)$ comes from the observation that – on $\mathcal{S}$ – it must suffice to solve a *regular consistent subsystem* $G_0(x, a) = 0$ of (26), with $M_0 = m$ equations, to obtain the solution $z = F_0(a)$ of (26). Furthermore, with variations Δa of the data which keep $a^* + \Delta a$ in $\mathcal{S}$, the corresponding result variations $\Delta z^* = F_0(a^* + \Delta a) - z^*$ must also follow from $G_0(z^* + \Delta z, a^* + \Delta a) = 0$. Since G_0 is regular, we can proceed as in section 3 to determine derivatives of F_0 at a^* with respect to the data components α_ν, for $a \in \mathcal{S}$.

However we have obtained $F_0'(a^*)$, to find the maximal variation of Δz for a variation of Δa within the tolerance neighborhood, we cannot simply take the appropriate operator norm of $F_0'(a^*)$ but we must respect the restriction of the variation of the Δa to the linearized consistency manifold $d\mathcal{S}$ represented by (28). Therefore, we must solve the *restricted linear optimization problem*

$$\|\Delta z_{\max}\| \;=\; \max_{\mathbf{S}\Delta a = 0,\; a^* + \Delta a \in N(\bar a, e)} \|\, F_0'(a^*) \cdot \Delta a \,\|\,, \tag{30}$$

and use the size of the maximal variation $\Delta z_{\max}$ as an assessment of the accuracy level in our pseudoresult like in (13). For problems of type (30), there are standard algorithmic procedures for any combinations of norms; in this paper, we have always used the max-norm in $\mathcal{A}$ (cf. (2)).

Naturally, if we omit the restriction $\mathbf{S}\Delta a = 0$ in (30) and use the respective operator norm of $F_0'(a^*)$ only, we obtain an *upper bound* for $\|\Delta z_{\max}\|$. This may be good enough for order-of-magnitude results, particularly in well-conditioned situations.

We claim that our approach covers the worst case for *arbitrary* small perturbations of *near-consistent* overdetermined problems. Consider, at first, a specified problem with data $a^* \in \mathcal{S}$ under a generic data perturbation Δa with $a^* + \Delta a \in N(a^*, e)$. The pseudoresult $z^* + \Delta z$ for $a^* + \Delta a$ must refer to data $a^* + \Delta a^* \in \mathcal{S} \cap N(a^*, e)$; these are precisely the data which we have considered above.

If we have specified data $\bar a$ *near* $\mathcal{S}$, let $\bar a^*$ be the normwise closest data point on $\mathcal{S}$. The intersection of $N(\bar a, e)$ with $\mathcal{S}$ is contained in the intersection of $N(\bar a^*, e)$ with $\mathcal{S}$; hence the variation of consistent data inside $N(\bar a, e)$ is at most as large as for $\bar a^*$, which implies the same for the variation of the associated pseudoresults. Therefore, (30) is a valid and meaningful condition estimate for the pseudoresults in every case.

There remains one question: How do we obtain $\bar{a}^* \in \mathcal{S}$ for the specified empiric data $(\bar{a}, e)$ and a computed pseudoresult z^*? For this, we refer to the checking procedure (task 1 in section 1) which we must perform for a pseudoresult in any case: As explained in [18], a suitable $\bar{a}^*$ may be obtained as a part of the checking procedure at virtually no extra cost.

5.1 Condition of a Multiple Zero of a Univariate Polynomial

Consider a univariate empiric polynomial $p(x, (\bar{a}, e))$ of degree n with an m-cluster of zeros so that the associated pseudozero set Z has a diameter of $O(\|e\|^{\frac{1}{m}})$. Let

$$Z^*(\bar{a}, e) := \{\tilde{z}^* \in Z : \tilde{z}^* \text{ is an } m\text{-fold zero of } p(x, \tilde{a}^*), \text{ with } \tilde{a}^* \in N(\bar{a}, e)\} \subset \mathbb{C} \tag{31}$$

be the set of all *m-fold pseudozeros* of p in Z and assume that $Z^* \neq \emptyset$; this fact can be established computationally as shown in [18] (cf. also [13]). We want to assess diam Z^* or the potential variation of an m-fold pseudozero of p, at tolerance level e. Note that each $\tilde{z}^* \in Z^*$ satisfies

$$p(x, a) = p'(x, a) = \ldots = p^{(m-1)}(x, a) = 0 \tag{32}$$

for some $a = \tilde{a}^* \in N(\bar{a}, e)$ so that Z^* is the pseudoresult set of the *overdetermined system* (32) of m univariate polynomials.

The algebraic conditions $S(a) = 0$ which the $\tilde{a}^*$ have to satisfy may be obtained from the determinants of the $m-1$ pairwise Sylvester matrices of $p^{(\mu-1)}, p^{(\mu)}$, $\mu = 1(1)m-1$. But according to (28), we are only interested in the Jacobian of S; therefore we may proceed more directly: For a^*, $a^* + \Delta a \in \mathcal{S}$ and z^* the associated zero in Z^*, we have

$$p(x, a^*) = q(x)\,(x - z^*)^m \quad \text{and} \quad p(x, a^* + \Delta a) = (q(x) + \Delta q(x))\,(x - z^* - \Delta z^*)^m,$$

which implies

$$p(x, \Delta a) = m\,q(x)\,(x - z^*)^{m-1} Deltaz^* + (x - z^*)^m \Delta q + \text{ quadratic terms in } \Delta z^*, \Delta q. \tag{33}$$

Taking remainders modulo $\langle (x - z^*)^m \rangle$, we obtain a polynomial relation of degree $m-1$ in x, linear and homogeneous in the $\Delta \alpha_\nu$ and Δz^*:

$$\sum_{\mu=0}^{m-1} \rho_{0\mu}(\Delta a)\, x^\mu = \left(\sum_{\mu=0}^{m-1} \rho_{1\mu}\, x^\mu \right) \cdot \Delta z^*, \tag{34}$$

or

$$\rho_{00}(\Delta a) : \rho_{01}(\Delta a) : \ldots : \rho_{0,m-1}(\Delta a) = \rho_{10} : \rho_{11} : \ldots : \rho_{1,m-1}, \tag{35}$$

which is equivalent to a system of $m-1$ linear homogeneous equations

$$\mathbf{S} \cdot \begin{pmatrix} \Delta\alpha_0 \\ \vdots \\ \Delta\alpha_{m-1} \end{pmatrix} = 0; \tag{36}$$

the $(m-1) \times n$ matrix $\mathbf{S}$ is the Jacobian at a^* of S; cf. (28).

For $a^* \in \mathcal{S}$, the *consistent subsystem* $G_0(x, a^*) = 0$ of (32) which yields the solution $z^* = F_0(a^*) = F(a^*)$ of (32), is simply

$$p^{(m-1)}(x, a^*) = 0, \tag{37}$$

which has a unique zero z^* in Z. From (17), we have

$$p^{(m)}(z^*, a^*)\, F_0'(a^*) + \frac{\partial}{\partial a} p^{(m-1)}(z^*, a^*) = 0,$$

$$\begin{aligned} F_0'(a^*) &= \left(\ldots \tfrac{\partial z^*}{\partial \alpha_\nu}(a^*) \ldots\right) \\ &= \tfrac{1}{p^{(m)}(z^*, a^*)}\, (0, \ldots, 0, (m-1)!, m!\, z^*, \ldots, (n-1)\ldots(n-m+1)(z^*)^{n-m}). \end{aligned} \tag{38}$$

The zeros in (38) reflect the fact that $\alpha_0, \ldots, \alpha_{m-2}$ no longer appear in $p^{(m-1)}$. But $\Delta\alpha_0, \ldots, \Delta\alpha_{m-2}$ are uniquely determined by the $\Delta\alpha_{m-1}, \ldots, \Delta\alpha_{n-1}$ through (36) since one can show that the first $m-1$ columns of $\mathbf{S}$ are linearly independent.

Therefore, the linear optimization problem (30) takes the form

Find the maximum of $|\sum_{\nu=m-1}^{n-1} \frac{\partial z^*}{\partial \alpha_\nu}(a^*)\Delta\alpha_\nu|$ for $|\Delta\alpha_\nu| \le \varepsilon_\nu$, $\nu = m-1(1)n-1$, $|\Delta\alpha_\mu(\Delta\alpha_{m-1} \ldots \Delta\alpha_{n-1})| \le \varepsilon_\mu$, $\mu = 0(1)m-2$,
which is a standard linear program.

Example 2: We use the empiric polynomial (25) from Example 1 for which we expect 3-fold pseudozeros near $\sqrt{2}$. The respective zero $z^* \approx 1.415031$ of $p''(x, \tilde{a})$ (cf. (37)) satisfies not only the requirements for a 3-fold pseudozero (cf. [18], section 5.1) but there are $\tilde{a}^*$ in (31) which differ from $\tilde{a}$ by less than 10^{-6} so that we may use $a^* = \tilde{a}$.

Eq. (34) takes the form (to 5 digits)

$$\begin{aligned} &(\Delta\alpha_0 + 2.83333\,\Delta\alpha_3) + (\Delta\alpha_1 - 6.00694\,\Delta\alpha_3)\,x + (\Delta\alpha_2 + 4.24509\,\Delta\alpha_3)\,x^2 \\ &\quad = (16.99509 - 24.02081\,x + 8.48773\,x^2)\,\Delta z \end{aligned}$$

which, via (35), leads to the representation

$$\begin{pmatrix} 24.02081 & 16.99509 & 0 & -34.02949 \\ 0 & 8.48773 & 24.02081 & 50.98528 \end{pmatrix} \begin{pmatrix} \Delta\alpha_0 \\ \Delta\alpha_1 \\ \Delta\alpha_2 \\ \Delta\alpha_3 \end{pmatrix} = 0,$$

of the linear manifold dS of (28).

With (38) and $p'''(z^*,\bar{a}) \approx 16.975$, we must maximize $|\Delta z_{\max}| \approx |.11782\,\Delta\alpha_2 + .50014\,\Delta\alpha_3|$ for $|\Delta\alpha_2|, |\Delta\alpha_3| \leq 10^{-5}$ and $|\Delta\alpha_0|, |\Delta\alpha_1| \leq 10^{-5}$ in terms of $\Delta\alpha_2$, $\Delta\alpha_3$. $|\Delta z_{\max}|$ is found as $< .15 \cdot 10^{-5}$ for $\Delta\alpha_0 = -\Delta\alpha_2 = \pm 10^{-5}$, $\Delta\alpha_1 \approx .35\,\Delta\alpha_2$, $\Delta\alpha_3 \approx -.53\,\Delta\alpha_2$. Thus the pseudoresult set Z^* for the 3-fold pseudozeros of p has a diameter of $< .3 \cdot 10^{-5}$(!) and it is meaningful to specify z^* to 6 digits as it has been done above. Experimentally, the extension of Z^* along the real axis was found as [1.4150300, 1.4150320] which matches well with the result of our linearized analysis. The omission of the restriction in (30) leads to $|\Delta z_{\max}| \approx .62 \cdot 10^{-5}$ which still displays the very good condition. □

The fact that the consideration of m-fold pseudozeros in place of clusters of m zeros improves the condition a great deal has been stressed by W.Kahan in his paper [7].

5.2 Condition of a GCD of Univariate Polynomials

Consider two empiric polynomials $p(x, (\bar{a}_i, e_i))$, monic, of degree $n_i, i = 1, 2$, with a pseudo-g.c.d. of degree $m \geq 1$. This means that the set $Z^*((\bar{a}_1, \bar{a}_2), (e_1, e_2)) \subset \mathbb{C}^m$ containing the coefficients c of monic polynomials $s(x, c)$ of degree m which satisfy

$$s(x, \tilde{c}^*) \;=\; \text{(exact) g.c.d. of } p(x, \tilde{a}_1^*),\, p(x, \tilde{a}_2^*) \;\text{ with } \tilde{a}_i^* \in N(\bar{a}_i, e_i),\, i = 1, 2\,, \tag{39}$$

is not empty; this fact can be established computationally, cf. [6]. Again, we want to assess diam Z^* as a measure for the potential variation of the pseudo-g.c.d.'s within the data tolerance level. Note that each $\tilde{c}^* \in Z^*$ satisfies, for the appropriate $a_i \in N(\bar{a}_i, e_i)$,

$$p(x, a_1) \;=\; q(x, b_1)\, s(x, c)\,, \quad p(x, a_2) \;=\; q(x, b_2)\, s(x, c)\,; \tag{40}$$

this is an *overdetermined* bilinear system of $n_1 + n_2$ equations for the $n_1 + n_2 - m$ coefficients b_1, b_2, c. Efforts to determine pseudoresults of this system have been reported by various authors; cf., e.g., [2], [8], [11].

In the following, we will – as in section 4 – denote coefficient vectors by *row* vectors (with a superscript ^{T}). By (21), there exist $n_i \times m$-matrices C_i such that small changes in the coefficients of p_1, p_2 and s in (40), locally at $\tilde{a}_1^*, \tilde{a}_2^*$, are related by

$$(\ldots \Delta c^T \ldots) \doteq (\;\ldots \Delta a_i^T \ldots\;)\, C_i\,, \quad i = 1, 2\,. \tag{41}$$

This assumes coprimality of s and q_i in (40); but this must be satisfied for at least one of the q_i (say q_1) because otherwise there would exist a g.c.d. of higher degree.

According to (21) and (41), variations of the p_i which preserve the existence of the g.c.d of degree m must satisfy (in linearized form)

$$(\ldots \Delta a_1^T \ldots, \ldots \Delta a_2^T \ldots) \begin{pmatrix} C_1 \\ -C_2 \end{pmatrix} \;=\; 0\,; \tag{42}$$

this system of m homogeneous linear equations represents the linear manifold dS in the coefficient space $\mathcal{A}$.

Naturally, either part ($i = 1$ or 2) of (40) constitutes a consistent subsystem, with (41) as the associated linearized data→result mapping F_0' at a coefficient set $\tilde{a}_i^*$ from (39); if s and q_2 have a common factor, one must use $i = 1$. This permits us to evaluate the maximal potential variation $\|\Delta c_{\max}\|$ of the coefficients of s for (say) $\tilde{a}_1^* + \Delta a_1 \in N(\tilde{a}_1^*, e_1)$; this maximum is reached at some Δa_1^* on the *boundary* of $N(\tilde{a}_1^*, e_1)$. If it is possible to choose Δa_2^* with $\tilde{a}_1^* + \Delta a_2^* \in N(\tilde{a}_2^*, e_2)$ such that (42) is satisfied for the Δa_i^*, then $\|\Delta c_{\max}\|$ is the overall optimum; otherwise we have to restrict the optimization suitably (cf. (30)).

Example 3: We form p_1, p_2 such that they have a 2nd degree common divisor:

$$\begin{aligned} p_1(x) &:= x^5 - 7.258\,x^4 + 8.910416\,x^3 + 4.457116\,x^2 - 15.603204\,x + 5.596536 \\ &= (x^2 - 2.364\,x + 1.892)\,(x^3 - 4.894\,x^2 - 4.551\,x + 2.958)\,; \end{aligned}$$

$$\begin{aligned} p_2(x) &:= x^4 - 5.482\,x^3 + 13.064952\,x^2 - 14.887184\,x + 7.193384 \\ &= (x^2 - 2.364\,x + 1.892)\,(x^2 - 3.118\,x + 3.802)\,. \end{aligned}$$

The matrices C_1, C_2 of (41) which relate local changes of the p_i and s are found as (rounded to 5 digits)

$$C_1 = \begin{pmatrix} -.17362 & .09975 \\ -.18872 & .06218 \\ -.11764 & -.04173 \\ .07895 & -.21629 \\ .40922 & -.43236 \end{pmatrix} \; ; \quad C_2 = \begin{pmatrix} .09668 & .57154 \\ -1.08136 & 1.44781 \\ -2.73925 & 2.34126 \\ -4.42966 & 2.79549\,. \end{pmatrix}$$

From (42) and these C_i, we obtain the representation of dS by two linear homogeneous equations in the Δa_i. Under this relation between Δa_1 and Δa_2, we obtain the linearized data→result mapping F_0' at a_1^{*T}, a_2^{*T} also from (41), with either $i = 1$ or 2. Without solving the optimization problem (30) explicitly, we may deduce from the elements of the C_i that the condition of the g.c.d. in this particular case is below 10. The restricted maximum is approx. 2, which is confirmed by systematic tests.

Thus, if we had been confronted with the task of computing a pseudo-g.c.d. of two empiric polynomials $p_i(x, \bar{a}_1) \approx p_i(x)$ above, with a tolerance of (say) 10^{-4} in their coefficients, it would have been meaningful to specify the pseudo-g.c.d. to 3-4 digits. □

5.3 Condition of a Multivariate Factorization

In contrast to a univariate polynomial, a multivariate polynomial $p \in \mathbb{C}[x_1, \ldots, x_s]$, $s > 1$, generally does not factor over $\mathbb{C}$. For a factorizable p, we have

$$p(x) \;=\; u(x) \cdot v(x) \quad \text{or} \quad \sum_{j \in J_p} \alpha_j\, x^j \;=\; \sum_{j \in J_u} \beta_j\, x^j \cdot \sum_{j \in J_v} \gamma_j\, x^j\,, \qquad (43)$$

with multi-indices $j := (j_1, \ldots, j_s)$ and multi-exponents $x^j := x_1^{j_1} \ldots x_s^{j_s}$, and with supports $J_p, J_u, J_v \subset \mathbb{N}_0^s$ which may be sparse. (Except in simple situations, it is difficult to decide what the sparsity of p implies for the supports of u and v; cf. [9].) For the sake of simplicity, we assume total degrees $n, m, n-m$ for p, u, v, without a particular sparsity structure. To keep the symmetry in the variables, we assume $p(0) \neq 0$ and normalize by $\alpha_0 = \beta_0 = \gamma_0 = 1$. Then, p, u, v have $\binom{n+s}{s} - 1, \binom{m+s}{s} - 1, \binom{n-m+s}{s} - 1$ coefficients and (43) represents an *overdetermined* bilinear system for the coefficient vectors b^T, c^T of u, v. An effort to determine pseudoresults of this system directly has been reported by [8]; but the iterative approach of [6] appears to be more flexible and efficient.

According to our approach, we assume to have a factorizable polynomial p^* (in the tolerance neighborhood of some empiric polynomial $p(\bar{a}, e)$), with coefficients a^{*T}, and factors u^*, v^*, with coefficients b^{*T}, c^{*T}. We want to assess the sensitivity of the coefficients of u^*, v^* with respect to small perturbations Δa of the coefficients of p^* which *leave* $p^* + \Delta p$ *factorizable.* Upon subtraction of (43) from $p(x, a^* + \Delta a) = (u^*(x) + \Delta u(x))(v^*(x) + \Delta v(x))$, we have (with linearization)

$$p(x, \Delta a) \doteq v^*(x)\, \Delta u(x) + u^*(x)\, \Delta v(x)\,; \tag{44}$$

we may write this linear relation between the coefficient vectors $\Delta a^T, \Delta b^T, \Delta c^T$ of dimensions N_a, N_b, N_c resp. as

$$(\ \ldots \Delta a^T \ldots\ ) = (\ldots \Delta b^T \ldots)\, C^* + (\ldots \Delta c^T \ldots)\, B^*\,, \tag{45}$$

where the $N_c \times N_a$-matrix B^* contains the coefficients of u^*, in a well-known arrangement, and the $N_b \times N_a$-matrix C^* those of v^*. E.g., for $s = 2, n = 2, m = n - m = 1$, (45) takes the form

$$(\Delta\alpha_{10}, \Delta\alpha_{01}, \Delta\alpha_{20}, \Delta\alpha_{11}, \Delta\alpha_{02}) =$$

$$(\Delta\beta_{10}, \Delta\beta_{01}) \begin{pmatrix} 1 & 0 & \gamma_{10}^* & \gamma_{01}^* & 0 \\ 0 & 1 & 0 & \gamma_{10}^* & \gamma_{01}^* \end{pmatrix} + (\Delta\gamma_{10}, \Delta\gamma_{01}) \begin{pmatrix} 1 & 0 & \beta_{10}^* & \beta_{01}^* & 0 \\ 0 & 1 & 0 & \beta_{10}^* & \beta_{01}^* \end{pmatrix}.$$

The relation (45) can only be satisfied if the row vector Δa^T lies in the row space of the combined rows of B^* and C^*, i.e. if the matrix

$$BC(\Delta a) := \begin{pmatrix} C^* \\ B^* \\ \Delta a^T \end{pmatrix}$$

has a rank deficiency 1. The rows of the numeric matrices B^* and C^* have full rank due to our normalization assumption; thus, upon Gaussian elimination of $BC(\Delta a)$, with partial pivoting and column exchanges where necessary, there remain $N_a - (N_b + N_c)$ non-zero elements in the last row of BC which are homogeneous linear functions in the $\Delta\alpha_j$ and represent the components of the linearized restrictions (28).

In the situation of (43), the selection of such a subsystem is not so obvious; but the bilinearity of (43) makes the approach via (27) straightforward: With the joint coefficient vector $(\dots b^T \dots\dots c^T \dots)$ of u, v taking the place of the result z, (45), written as

$$(\dots \Delta a^T \dots) = (\dots \Delta b^T \dots\dots \Delta c^T \dots) \begin{pmatrix} C^* \\ B^* \end{pmatrix} , \tag{46}$$

is immediately identified with (27). With the assumption of a consistent Δa, we obtain the Moore-Penrose pseudosolution of (45) (observing the change from column to row vectors) as

$$(\dots \Delta b^T \dots \Delta c^T \dots) = (\dots \Delta a^T \dots) \left(C^{*T} \; B^{*T} \right) \cdot \left(\begin{pmatrix} C^* \\ B^* \end{pmatrix} \left(C^{*T} \; B^{*T} \right) \right)^{-1} ; \tag{47}$$

the right-hand matrix is the representation of $F_0'(a^*)$. If we are satisfied with an upper bound, we can deduce our condition estimate directly from this matrix; otherwise, we have to go through the restricted optimization procedure of (30).

Example 4, cf. [6]: The empiric bivariate polynomial $p(x, (\bar{a}, e))$ specified by

$$\bar{p} = 1 - .82\,x + .91\,y + .15\,x^2 + 11.22\,xy - 8.71\,y^2 + 4.69\,x^3 - .65\,x^2y - 12.08\,xy^2 + 7.14\,y^3 ,$$

with tolerances .005 for all coefficients (except the normalizing constant term 1), has pseudofactors (to 3 digits)

$$\begin{aligned} u^*(x,y) &= 1 + 1.415\,x - 1.733\,y , \\ v^*(x,y) &= 1 - 2.235\,x + 2.644\,y + 3.314\,x^2 + 3.601\,xy - 4.123\,y^2 . \end{aligned}$$

The exactly factorizable polynomial $p(x, a^*) \in N(\bar{p}, e)$ with these factors is (to 4 digits)

$$\begin{aligned} &1 - .8200x + .9112y + .1501x^2 + 11.2172xy - 8.7052y^2 \\ &+4.6899x^3 - .6460x^2y - 12.0757xy^2 + 7.1448y^3 . \end{aligned}$$

The relation (46) becomes

$$(..\Delta a^T..) = \\ (\Delta b^T \Delta c^T) \begin{pmatrix} 1 & 0 & -2.2353 & 2.6442 & 0 & 3.3137 & 3.6011 & -4.1228 & 0 \\ 0 & 1 & 0 & -2.2353 & 2.6442 & 0 & 3.3137 & 3.6011 & -4.1228 \\ 1 & 0 & 1.4153 & -1.7330 & 0 & 0 & 0 & 0 & 0 \\ 0 & 1 & 0 & 1.4153 & -1.7330 & 0 & 0 & 0 & 0 \\ 0 & 0 & 1 & 0 & 0 & 1.4153 & -1.7330 & 0 & 0 \\ 0 & 0 & 0 & 1 & 0 & 0 & 1.4153 & -1.7330 & 0 \\ 0 & 0 & 0 & 0 & 1 & 0 & 0 & 1.4153 & -1.7330 \end{pmatrix} .$$

This gives us, from (47), the Moore-Penrose representation of $F_0'(a^*)$ (to 4 digits) as

$$\begin{pmatrix} .2123 & -.3034 & .6437 & .1265 & -.4801 & -.4216 & .8224 \\ -.0504 & .2007 & -.1265 & .2252 & .2161 & .0644 & -.4550 \\ -.1621 & .0757 & .1185 & .0742 & .3407 & .5678 & -.0920 \\ -.0099 & -.1132 & -.1088 & .1336 & .0012 & .2204 & .3995 \\ -.0372 & .0233 & -.1618 & -.3380 & .1257 & .2172 & .0637 \\ .1283 & .1405 & -.1044 & -.0018 & .3796 & -.2349 & -.3200 \\ .0113 & .1585 & -.0169 & .0414 & -.0705 & .1358 & -.3144 \\ .0035 & .0641 & -.0765 & .1108 & -.0569 & -.3389 & -.0262 \\ -.0186 & .0658 & -.1559 & -.1045 & .0261 & -.1515 & -.5617 \end{pmatrix} .$$

By Gauss elimination in the matrix $BC(\Delta a)$, we obtain two linear homogeneous equations as representation of the linear manifold dS (to 4 digits):

$$(\Delta\alpha_{10}, \Delta\alpha_{01}, \Delta\alpha_{20}, \Delta\alpha_{11}, \Delta\alpha_{02}, \Delta\alpha_{30}, \Delta\alpha_{21}, \Delta\alpha_{12}, \Delta\alpha_{03}) \begin{pmatrix} -.7583 & 1.1427 \\ -2.7984 & 2.1854 \\ .8346 & -.0997 \\ .2443 & .5779 \\ -1.4153 & 1.7330 \\ .6981 & -.4295 \\ 1.0519 & -.4083 \\ 1. & 0 \\ 0 & 1. \end{pmatrix}$$

$$= 0 .$$

Without a restricted optimization, the first two columns of F_0' show that the maximal variation of the coefficients of u^* upon a variation of the coefficients of $\bar{p}$ must remain well below $\max|\Delta\alpha_j|$; the sensitivity of the coefficients of v^* may be slightly larger but O(1). These bounds are further decreased when we consider the restriction of the $\Delta\alpha_j$ to dS. Thus, it is meaningful in this case to specify the pseudofactors to 3 digits, as we have done above.

6 Conclusions

The admission of data with limited accuracy and of pseudoresults for algebraic problems has widened the scope and the applicability of computer algbraic procedures for real-life problems. In this context, it is natural to ask for the *meaningful accuracy* of computed (pseudo) results, under the specified tolerance of the data. An answer to this question requires an analysis of the sensitivity of the results of algebraic problems to variations in their data, or of the *condition* of the results.

While this a standard exercise in numerical analysis for regular problems with a well-behaved data→result mapping, it is a less transparent task for problems which have solutions in the classical sense only for data which *satisfy certain algebraic restrictions* because they are *overdetermined.* Such problems include standard algebraic problems like multiple zeros of polynomials, polynomial greatest common divisors, and the factorization of multivariate polynomials. In Scientific Computing, the deviation from consistency may only be due to a perturbation of the data and pseudoresults may be of great interest because they correspond to the actual modelled situation.

In this paper, we have suggested an approach to the assessment of the sensitivity of pseudoresults of such problems with respect to variations in their data. We have explained an algorithmic implementation of this approach in connection with the above-mentioned special situations and also with numerical examples. These algorithms for the condition analysis of pseudoresults of overdetermined algebraic problems should become standard tools of computational computer algebra. They would help to raise the usefulness and reliability of computer algebra software in Scientific Computation.

The application of our approach to problems with systems of multivariate polynomial equations presents some technical difficulties; these will be addressed in a forthcoming separate paper.

References

1. F. Chaitin-Chatelin, V. Frayssé: Lectures on Finite Precision Computations, Software - Environments - Tools, SIAM, Philadelphia, 1996
2. R.M. Corless, P.M. Gianni, B.M. Trager, S.M. Watt: The singular value decomposition for polynomial systems, in: Proceed. ISSAC'95 (Ed. A.H.M.Levelt) (1995) 195-207
3. G.H. Golub, C.F. Van Loan: Matrix Computations, John Hopkins Univ. Press, Baltimore, 1996 (3rd.ed.)
4. N.J. Higham: Accuracy and Stability of Numerical Algorithms, SIAM, Philadelphia, 1996
5. V. Hribernig, H.J. Stetter: Detection and Validation of Clusters of Polynomial Zeros, J.Symb.Comp. **24** (1997) 667-681
6. Y.Huang, H.J. Stetter, W.Wu, L.Zhi: Pseudofactors of Multivariate Polynomials, submitted to ISSAC 2000
7. W. Kahan: Conserving Confluence Curbs Ill-condition; Comp.Science UC Berkeley Tech.Rep. 6, 1972
8. M.A. Hitz, E. Kaltofen, Y.N. Lakshman: Efficient Algorithms for Computing the Nearest Polynomial with a Real Root and Related Problems, in: Proceed. ISSAC'99 (Ed. S.Dooley) (1999) 205-212
9. E. Kaltofen: Polynomial Factorization 1987-1991, in: Proceed. LATIN'92 (Ed. I.Simon), Lect.Notes Comp.Sci. **583** (1992) 294-313
10. L.V. Kantorovich, G.P. Akhilov: Functional Analysis in Normed Spaces, Pergamon Press
11. N.K. Karmarkar, Y.N. Lakshman: On Approximate GCDs of Univariate Polynomials, J. Symb.Comp. **26** (1986) 653-666
12. R.G. Mosier: Root neighbourhoods of a polynomial, Math.Comput. **47** (1986) 265–273
13. A. Neumaier: An Existence Test for Root Clusters and Multiple Roots, ZAMM **68** (1988) 256–257

14. T. Sasaki, M. Suzuki, M. Kolar, M. Sasaki: Approximate Factorization of Multivariate Polynomials and Absolute Irreducibility Testing, Japan J.Industr.Appl. Math. **8** (1991), 357-375
15. T. Sasaki, T. Saito, T. Hilano: Analysis of Approximate Factorization Algorithm I, Japan J.Industr.Appl.Math. **9** (1992), 351-368
16. H.J. Stetter: Analysis of Zero Clusters in Multivariate Polynomial Systems, in: Proceed. ISSAC 96 (Ed. Y.N. Lakshman), 127 - 136
17. H.J. Stetter, G.H. Thallinger: Singular Polynomial Systems, in: Proceed. ISSAC 98 (Ed. O. Gloor), 9 - 16
18. H.J. Stetter: Polynomials with Coefficients of Limited Accuracy, in: Computer Algebra in Scientific Computing (Eds. V.G.Ganzha, E.W.Mayr, E.V.Vorozhtsov), Springer, 1999, 409-430
19. K.-C. Toh, L.N. Trefethen: Pseudozeros of Polynomials and Pseudospectra of Companion Matrices, Numer.Math. **68** (1994) 403 - 425
20. L.N. Trefethen: Pseudospectra of Matrices, in: Numerical Analysis (Eds: D.F. Griffiths, G.A. Watson), Longman, 1991

An Algebraic Approach to Offsetting and Blending of Solids

Thomas Sturm

Department of Mathematics and Computer Science, University of Passau, Germany
sturm@uni-passau.de, http://www.fmi.uni-passau.de/~sturm

Abstract. We propose to broaden the framework of CSG to a representation of solids as boolean combinations of polynomial equations and inequalities describing regular closed semialgebraic sets of points in 3-space. As intermediate results of our operations we admit arbitrary semialgebraic sets. This allows to overcome well-known problems with the computation of blendings via offsets. The operations commonly encountered in solid modelers plus offsetting and constant radius blending can be reduced to quantifier elimination problems, which can be solved by exact symbolic methods. We discuss the general properties of such offsets and blendings for arbitrary regular closed semialgebraic sets in real n-space. Our computational examples demonstrate the capabilities of the REDLOG package for the discussed operations on solids within our framework.

1 Introduction

Solid modeling knows two major representation schemata for solid objects: *boundary representation* (B-rep) and *constructive solid geometry* (CSG). B-rep is very close to the image rendered on the screen. CSG, in contrast, resembles the inherent structure of the described solid by combining primitive solids via regularized boolean operations. Today's modelers are usually dual representation modelers, which combine both schemata to make use of the particular advantages of each of them [8].

We propose a generalization of CSG representing solids as certain boolean combinations of polynomial equations and inequalities, where the solids are derived as the corresponding *semialgebraic sets* of points in 3-space satisfying these *formulas*. Among all semialgebraic sets, we define regular closed sets to be valid solids. It is not hard to see that all operations commonly encountered in today's solid modelers do not lead outside the realm of semialgebraic sets. On the other hand, our notion of a solid obviously extends that of CSG. In particular, it includes common types of offsetted and blended solids.

The treatment of various problems from computational geometry within our framework has already been discussed elsewhere [13]. This includes parallel and central projections, corresponding reconstruction problems, offsets of algebraic surfaces, Voronoi diagrams, and collision problems. This article focuses on offsets and blendings of solids. Among the numerous variants of blending [7, 15] we restrict ourselves here to circular constant radius blending that can be visualized as an envelope of a ball rolling along parts of the surfaces or solids. We discuss the general properties of such blendings reducing them to an iterated formation of certain offsets, where our approach differs from the usual concatenations of shrinking and expanding [11, 12]. Though restricting the notion of a solid to regular closed sets, our approach allows to naturally handle arbitrary semialgebraic sets as *intermediate* results of our operations. It shall turn out that this freedom, together with a new concept of defining blendings in terms of offsets, allows us to understand and to overcome the well-known problems with the definition of blendings via shrinking and expanding mentioned above.

All necessary operations can be straightforwardly performed if we allow the resulting formulas to involve *quantifiers* "$\exists x$" and "$\forall x$" as they are used, e.g., in definitions of elementary calculus. Maybe a little surprisingly, it is possible to map each formula φ containing such quantifiers to an equivalent formula φ' not containing any quantifier. Here, *equivalent* means that both φ and φ' describe the same semialgebraic set. The step from φ to φ' is known as *quantifier elimination.* There are various software packages performing quantifier elimination available [5].

In contrast to purely algebraic approaches discussed elsewhere [8, 9] quantifier elimination can handle polynomial inequalities both in the input and in the output. Besides this, it is another crucial feature of our approach that it can solve problems parametrically. We can, e.g., compute the blend of a solid with respect to a parametric radius r. The resulting description will contain case distinctions that make it correct for any substitution of a positive real number for r.

It is a well-known experience in computational real algebra that real elimination methods are very limited in practice. We mainly use a *test point method* originally due to Weispfenning [10, 16], which has stretched these limits considerably under the restriction that the polynomials to be handled are of low degree. This method is well-suited for parallelization [2]. The present limitations of our approach are due to the degree restrictions and the possible size of the output description. The test point method is included in the the package REDLOG by the author together with A. Dolzmann [4, 3], which is part the computer algebra system REDUCE.

REDLOG has been successfully applied to a large variety of areas in science and engineering; see, e.g., [5] for a survey. The procedures are currently by no means tuned to the particular type of input discussed here. All our example computations have been performed on a Sun Ultra 10 Sparc Workstation using 32 MB of memory.

2 Basic Definitions and the Tarski Principle

A *semialgebraic set* is a boolean combination of sets of the form $\{\, \mathrm{a} \in \mathbb{R}^n \mid f(\mathrm{a}) \geqslant 0 \,\}$ for $f \in \mathbb{R}[\mathrm{x}]$. For our practical computations we will restrict our attention to semialgebraic sets in $\mathbb{R}^n$ with defining polynomials $f \in \mathbb{Q}[\mathbf{x}]$ over the rational numbers. The following fact, due to A. Tarski [14], is fundamental.

Theorem 1 (Tarski). *Let $\pi : \mathbb{R}^n \to \mathbb{R}^{n-1}$ denote the projection along the last coordinate. Whenever $A \subseteq \mathbb{R}^n$ is semialgebraic, then the image $\pi(A)$ is semialgebraic in $\mathbb{R}^{n-1}$.*

From the viewpoint of logic, this fact can be rephrased as quantifier elimination for the elementary theory of the reals: *Atomic formulas* are polynomial equations $f = 0$, negated polynomial equations $f \neq 0$, and polynomial ordering inequalities $f \geqslant 0$, $f > 0$ where $f \in \mathbb{Q}[\mathbf{x}]$. *Quantifier-free formulas* are obtained from atomic formulas by the logical operators "$\wedge$" (and), "$\vee$" (or), and "$\neg$" (not). Arbitrary *formulas* are obtained by these operations together with existential quantification "$\exists x$" (there exists an x) and universal quantification "$\forall x$" (for all x) over real variables. Any universal quantification $\forall x \varphi$ can be rewritten as an existential quantification $\neg\exists x \neg\varphi$. Existential quantification, in turn, corresponds geometrically to a projection. Consequently, Tarski's principle can be rephrased as quantifier elimination, which is effective over the reals:

Theorem 2 (Tarski). *Every first-order formula φ in the theory of the reals is equivalent to a quantifier-free formula φ', and there is an algorithm computing φ' from φ.*

So for any formula $\varphi(\mathbf{x})$, including quantified formulas, the set $S(\varphi) = \{ \mathbf{a} \in \mathbb{R}^n \mid \mathbb{R} \models \varphi(\mathbf{a}) \}$ is a semialgebraic set. We call φ a *description* of $S(\varphi)$. Accordingly, a quantifier-free formula φ' is called a *quantifier-free description* of $S(\varphi')$.

Although our formulas formally range over the real numbers, we allow ourselves to briefly write down constraints involving vectors. Quantifications such as "$\exists \mathbf{x}$" have to be read as "$\exists x_1 \ldots \exists x_n$."

For formally writing down formulas involving applications of a metric, we have to ensure that the graph of this metric is a semialgebraic set. Luckily, all metrics commonly encountered have this property. Throughout this paper we exclusively use the Euclidean metric denoted by d. For points $\mathbf{a} \in \mathbb{R}^n$ and sets $B \subseteq \mathbb{R}^n$, we define as usual the *distance* of $\mathbf{a}$ to B as $d(\mathbf{a}, B) = \inf_{\mathbf{b} \in B} d(\mathbf{a}, \mathbf{b})$. We denote by $\mathrm{B}(\mathbf{a}, r)$ and $\mathrm{B}[\mathbf{a}, r]$ the *open* and *closed ball*, respectively, with center $\mathbf{a} \in \mathbb{R}^n$ and radius $0 \leqslant r \in \mathbb{R}$. Balls are semialgebraic, and so are all objects commonly considered in solid modeling, in particular cylinders, cuboids, and polyhedrons.

3 Solids as Regular Closed Semialgebraic Sets

Our intuition of a solid in $\mathbb{R}^3$ is that it is a closed semialgebraic set that is "everywhere 3-dimensional." It is thus reasonable and compatible with the more specialized definitions commonly used [9] to define a solid as a non-empty *regular closed* semialgebraic subset of $\mathbb{R}^3$.

For $A \subseteq \mathbb{R}^n$ the *regular closure* A^* is defined as the closure of the interior $\overline{A^\circ}$ of A, and A is called regular closed if $A^* = A$. The dual concept of regular open sets has been studied intensively in the theory of boolean algebras, where it is the crucial tool for the completion of boolean algebras [6]. The results on regular open sets can easily be transferred to regular closed sets.

For open sets A we have $A^* \supseteq A$, and for closed sets A we have $A^* \subseteq A$. If A is neither open nor closed, then there need not hold any inclusion. Since both interior and closure are monotone, so is the regular closure, i.e., $X \subseteq A$ implies $X^* \subseteq A^*$. It follows that if $X \subseteq A$ is already regular closed, then $X \subseteq A^*$. Finally, it is not hard to see that the regular closure is idempotent, i.e. $A^{**} = A^*$.

For closed sets A the formation of A^* will—roughly speaking—kill lower dimensional parts of A. This is exactly what is intended in the well-known formation of the normalized boolean operations such as the normalized intersection

$$A \cap^* B = (A \cap B)^*.$$

In the literature, this regularization is usually carried out with all boolean operations. From a computational point of view, it might be noteworthy that this is a certain overkill. For regular closed input solids A and B we have:

$$(A \cup B)^* = A \cup B, \quad (\sim A)^* = \overline{\sim A}, \quad (A \setminus B)^* = \overline{A \setminus B},$$

i.e., the union requires no regularization at all, and for regularizing complement and difference one need not take the interior. Note that, at least in our framework, it cannot be efficiently recognized whether a set is already open. So this external knowledge will save superfluous computations.

4 Boolean and Topological Operations on Solids

Taking unions, intersections, complements, interiors, and closures leads us from semialgebraic sets to semialgebraic sets: $S(\varphi) \cup S(\psi) = S(\varphi \vee \psi)$, $S(\varphi) \cap S(\psi) = S(\varphi \wedge \psi)$, $\sim S(\varphi) = S(\neg\varphi)$, and

$$\begin{aligned} S(\varphi(\mathbf{x}))^\circ &= S(\exists r(r > 0 \wedge \forall \mathbf{u}(d(\mathbf{x}, \mathbf{u}) < r \longrightarrow \varphi(\mathbf{u})))), \\ \overline{S(\varphi(\mathbf{x}))} &= S(\forall r(r > 0 \longrightarrow \exists \mathbf{u}(d(\mathbf{x}, \mathbf{u}) < r \wedge \varphi(\mathbf{u})))). \end{aligned}$$

So does any composition of these, such as the regular closure and regularized boolean operations or the boundary $\partial S(\varphi) = \overline{S(\varphi)} \cap \overline{\sim S(\varphi)}$.

5 Offsets

The central geometrical construction for our purposes is the formation of *offsets*. Given *arbitrary* $A \subseteq \mathbb{R}^n$ and $0 < r \in \mathbb{R}$ we define:

$$\text{off}_{\leqslant r}(A) = \{\, \mathbf{x} \in \mathbb{R}^n \mid \exists(\mathbf{a} \in A) : d(\mathbf{x}, \mathbf{a}) \leqslant r \,\}.$$

The offset of a semialgebraic set is again semialgebraic:

$$\text{off}_{\leqslant r}(S(\varphi(\mathbf{x}))) = S(\exists \mathbf{u}(\varphi(\mathbf{u}) \wedge d(\mathbf{x}, \mathbf{u}) \leqslant r)),$$

and for regular closed A, we have that $\mathrm{off}_{\leqslant \mathrm{r}}(A)$ is again regular closed.

This notion of an offset, also called *expanding*, and its application to sets that are not necessarily regular closed is quite standard [11, 12]. In the corresponding literature, there is also an "opposite" operation of *shrinking* a set. The result of this is not necessarily regular closed even when applied to regular closed input sets. Thus shrinking has usually been used in combination with regularization similar to the regularized boolean operations discussed in Section 3. Compositions of expanding and shrinking have been used to perform blending operations on solids [12], which can lead to very unintuitive results. We will analyze these problems in Section 7, and exhibit that they are actually caused by the regularization of shrinkings.

To avoid the problems, we choose an alternate approach to defining blendings in terms of offsets. Therefore, we generalize our notion of an offset by defining also the following offsets, which just like $\mathrm{off}_{\leqslant \mathrm{r}}$ lead from semialgebraic sets to semialgebraic sets:

$$\begin{aligned}
\mathrm{off}_{\geqslant \mathrm{r}}(A) &= \{\, \mathbf{x} \in \mathbb{R}^n \mid \forall(\mathbf{a} \in A) : d(\mathbf{x}, \mathbf{a}) \geqslant r \,\} \\
\mathrm{off}_{< \mathrm{r}}(A) &= \{\, \mathbf{x} \in \mathbb{R}^n \mid \exists(\mathbf{a} \in A) : d(\mathbf{x}, \mathbf{a}) < r \,\} \\
\mathrm{off}_{> \mathrm{r}}(A) &= \{\, \mathbf{x} \in \mathbb{R}^n \mid \forall(\mathbf{a} \in A) : d(\mathbf{x}, \mathbf{a}) > r \,\}.
\end{aligned}$$

Note that these definitions are not necessarily conform with the notion of *distance* between points and sets. For closed $A \subseteq \mathbb{R}^n$, we can, however, assert that

$$\mathrm{off}_{\leqslant \mathrm{r}}(A) = \{\, \mathbf{x} \in \mathbb{R}^n \mid d(\mathbf{x}, A) \leqslant r \,\} \quad \text{and} \quad \mathrm{off}_{\geqslant \mathrm{r}}(A) = \{\, \mathbf{x} \in \mathbb{R}^n \mid d(\mathbf{x}, A) \geqslant r \,\}.$$

The remainder of this section is devoted to a careful study of the properties of our various offsets and their interaction with boolean and topological operations. It follows immediately from the definitions that $\mathrm{off}_{\leqslant \mathrm{r}}(A) = {\sim}\mathrm{off}_{> \mathrm{r}}(A)$ and $\mathrm{off}_{< \mathrm{r}}(A) = {\sim}\mathrm{off}_{\geqslant \mathrm{r}}(A)$. The following lemma is concerned with offsets of unions.

Lemma 3. *Let* $A \subseteq \mathbb{R}^n$ *and* $0 < r \in \mathbb{R}$. *Then*

1. $\mathrm{off}_{\leqslant \mathrm{r}}(A \cup B) = \mathrm{off}_{\leqslant \mathrm{r}}(A) \cup \mathrm{off}_{\leqslant \mathrm{r}}(B)$
2. $\mathrm{off}_{\geqslant \mathrm{r}}(A \cup B) = \mathrm{off}_{\geqslant \mathrm{r}}(A) \cap \mathrm{off}_{\geqslant \mathrm{r}}(B)$.

Analogous results hold when replacing "$\leqslant$" by "$<$" and "$\geqslant$" by "$>$," respectively.

Proof. 1. $\mathbf{x} \in \mathrm{off}_{\leqslant \mathrm{r}}(A \cup B)$ iff there exists $\mathbf{y} \in A \cup B$ such that $d(\mathbf{x}, \mathbf{y}) \leqslant r$ iff there exists $\mathbf{y} \in A$ such that $d(\mathbf{x}, \mathbf{y}) \leqslant r$ or there exists $\mathbf{y} \in B$ such that $d(\mathbf{x}, \mathbf{y}) \leqslant r$ iff $\mathbf{x} \in \mathrm{off}_{\leqslant \mathrm{r}}(A)$ or $\mathbf{x} \in \mathrm{off}_{\leqslant \mathrm{r}}(B)$ iff $\mathbf{x} \in \mathrm{off}_{\leqslant \mathrm{r}}(A) \cup \mathrm{off}_{\leqslant \mathrm{r}}(B)$. The same argument holds for "$<$" instead of "$\leqslant$."

2. $\mathbf{x} \in \mathrm{off}_{\geqslant \mathrm{r}}(A \cup B)$ iff for all $\mathbf{c} \in A \cup B$ we have $d(\mathbf{x}, \mathbf{c}) \geqslant r$ iff for all $\mathbf{a} \in A$ we have $d(\mathbf{x}, \mathbf{a}) \geqslant r$ and for all $\mathbf{b} \in B$ we have $d(\mathbf{x}, \mathbf{b}) \geqslant r$ iff $\mathbf{x} \in \mathrm{off}_{\geqslant \mathrm{r}}(A)$ and $\mathbf{x} \in \mathrm{off}_{\geqslant \mathrm{r}}(B)$ iff $\mathbf{x} \in \mathrm{off}_{\geqslant \mathrm{r}}(A) \cap \mathrm{off}_{\geqslant \mathrm{r}}(B)$. The same argument holds for "$>$" instead of "$\geqslant$." □

Next, the closure properties of our offsets are quite surprising. For some types of offsets they depend on the set considered, while for other offsets they do not.

Lemma 4. *Let $A \subseteq \mathbb{R}^n$ and $0 < r \in \mathbb{R}$.*

1. *$\mathrm{off}_{<\mathrm{r}}(A)$ is open and $\mathrm{off}_{\geqslant\mathrm{r}}(A)$ is closed.*
2. *If A is open, then $\mathrm{off}_{\leqslant\mathrm{r}}(A)$ is open and $\mathrm{off}_{>\mathrm{r}}(A)$ is closed.*
3. *If A is closed, then $\mathrm{off}_{\leqslant\mathrm{r}}(A)$ is closed and $\mathrm{off}_{>\mathrm{r}}(A)$ is open.*

Proof. 1. Let $\mathbf{x} \in \mathrm{off}_{<\mathrm{r}}(A)$, pick $\mathbf{a} \in A$, $0 < \varepsilon \in \mathbb{R}$ such that $d(\mathbf{x}, \mathbf{a}) = r - \varepsilon < r$. For $\mathbf{y} \in \mathrm{B}(\mathbf{x}, \varepsilon)$ we have $d(\mathbf{y}, \mathbf{a}) \leqslant d(\mathbf{y}, \mathbf{x}) + d(\mathbf{x}, \mathbf{a}) < \varepsilon + r - \varepsilon = r$, i.e., $\mathrm{B}(\mathbf{x}, \varepsilon) \subseteq \mathrm{off}_{<\mathrm{r}}(A)$.

2. For open A let $\mathbf{x} \in \mathrm{off}_{\leqslant\mathrm{r}}(A)$, say $\mathbf{a} \in A$ such that $d(\mathbf{x}, \mathbf{a}) \leqslant r$. Choose $0 < \varepsilon \in \mathbb{R}$ such that $\mathrm{B}(\mathbf{a}, \varepsilon) \subseteq A$. For $\mathbf{y} \in \mathrm{B}(\mathbf{x}, \frac{\varepsilon}{2})$ pick $\mathbf{z}$ on the line segment $\overline{\mathbf{ax}}$ with $d(\mathbf{a}, \mathbf{z}) = \frac{\varepsilon}{2}$. Then $d(\mathbf{y}, \mathbf{z}) \leqslant d(\mathbf{y}, \mathbf{x}) + d(\mathbf{x}, \mathbf{z}) < \frac{\varepsilon}{2} + r - \frac{\varepsilon}{2} = r$, i.e., $\mathrm{B}(\mathbf{x}, \frac{\varepsilon}{2}) \subseteq \mathrm{off}_{\leqslant\mathrm{r}}(A)$.

3. For closed A, we show that $\overline{\mathrm{off}_{\leqslant\mathrm{r}}(A)} \subseteq \mathrm{off}_{\leqslant\mathrm{r}}(A)$: Let $\mathbf{x} \in \overline{\mathrm{off}_{\leqslant\mathrm{r}}(A)}$, and assume for a contradiction that $\mathbf{x} \in \mathrm{off}_{>\mathrm{r}}(A) = {\sim}\mathrm{off}_{\leqslant\mathrm{r}}(A)$. Since A is closed, there is $\mathbf{a} \in A$ and $0 < \varepsilon \in \mathbb{R}$ such that $d(\mathbf{x}, \mathbf{a}) = d(\mathbf{x}, A) = r + \varepsilon > r$. Pick $\mathbf{y} \in \mathrm{B}(\mathbf{x}, \varepsilon) \cap \mathrm{off}_{\leqslant\mathrm{r}}(A) \neq \emptyset$, then $d(\mathbf{x}, \mathbf{a}) \leqslant d(\mathbf{x}, \mathbf{y}) + d(\mathbf{y}, \mathbf{a}) < \varepsilon + r$, a contradiction. □

Finally, we turn to the interplay between offsets and topological operations and regularizations. The proof is again not too hard, but long and tedious:

Lemma 5. *1. Let $A \subseteq \mathbb{R}^n$, and let $0 < r \in \mathbb{R}$. Then*

$$\begin{aligned}\mathrm{off}_{>\mathrm{r}}(A)^\circ &= \overset{\circ}{\overline{\mathrm{off}_{>\mathrm{r}}(A)}} = \mathrm{off}_{\geqslant\mathrm{r}}(A)^\circ = \overset{\circ}{\overline{\mathrm{off}_{\geqslant\mathrm{r}}(A)}} = \mathrm{off}_{>\mathrm{r}}(\overline{A})\\ &\subseteq \mathrm{off}_{>\mathrm{r}}(A) \subseteq \overline{\mathrm{off}_{>\mathrm{r}}(A)} \subseteq \mathrm{off}_{\geqslant\mathrm{r}}(\overline{A}) = \mathrm{off}_{\geqslant\mathrm{r}}(A) = \overline{\mathrm{off}_{\geqslant\mathrm{r}}(A)}\end{aligned}$$

and $\mathrm{off}_{>\mathrm{r}}(A)^* = \mathrm{off}_{\geqslant\mathrm{r}}(A)^*$.

2. If A is open, then

$$\begin{aligned}\mathrm{off}_{>\mathrm{r}}(A)^\circ &= \overset{\circ}{\overline{\mathrm{off}_{>\mathrm{r}}(A)}} = \mathrm{off}_{\geqslant\mathrm{r}}(A)^\circ = \overset{\circ}{\overline{\mathrm{off}_{\geqslant\mathrm{r}}(A)}} = \mathrm{off}_{>\mathrm{r}}(\overline{A}) \subseteq \mathrm{off}_{>\mathrm{r}}(A)^*\\ &= \mathrm{off}_{\geqslant\mathrm{r}}(A)^* \subseteq \mathrm{off}_{>\mathrm{r}}(A) = \overline{\mathrm{off}_{>\mathrm{r}}(A)} = \mathrm{off}_{\geqslant\mathrm{r}}(\overline{A}) = \mathrm{off}_{\geqslant\mathrm{r}}(A) = \overline{\mathrm{off}_{\geqslant\mathrm{r}}(A)}.\end{aligned}$$

3. If A is closed, then

$$\begin{aligned}\mathrm{off}_{>\mathrm{r}}(A)^\circ &= \overset{\circ}{\overline{\mathrm{off}_{>\mathrm{r}}(A)}} = \mathrm{off}_{\geqslant\mathrm{r}}(A)^\circ = \overset{\circ}{\overline{\mathrm{off}_{\geqslant\mathrm{r}}(A)}} = \mathrm{off}_{>\mathrm{r}}(\overline{A}) = \mathrm{off}_{>\mathrm{r}}(A)\\ &\subseteq \mathrm{off}_{>\mathrm{r}}(A)^* = \overline{\mathrm{off}_{>\mathrm{r}}(A)} = \mathrm{off}_{\geqslant\mathrm{r}}(A)^* \subseteq \mathrm{off}_{\geqslant\mathrm{r}}(\overline{A}) = \mathrm{off}_{\geqslant\mathrm{r}}(A) = \overline{\mathrm{off}_{\geqslant\mathrm{r}}(A)}.\end{aligned}$$

A similar result for $<$-offsets and $\leqslant$-offsets is derived in the following way: We take the complement of each offset in Lemma 5, which inverts all inclusion relations. Then the complements are encoded into offsets using rules as discussed above together with the topological rules ${\sim}\overline{X} = ({\sim}X)^\circ$ and ${\sim}(X^\circ) = \overline{{\sim}X}$:

Lemma 6. *1. Let $A \subseteq \mathbb{R}^n$, and let $0 < r \in \mathbb{R}$. Then*

$$\begin{aligned}\mathrm{off}_{<\mathrm{r}}(A)^\circ &= \mathrm{off}_{<\mathrm{r}}(A) = \mathrm{off}_{<\mathrm{r}}(\overline{A}) \subseteq \mathrm{off}_{\leqslant\mathrm{r}}(A)^\circ \subseteq \mathrm{off}_{\leqslant\mathrm{r}}(A) \subseteq \mathrm{off}_{\leqslant\mathrm{r}}(\overline{A})\\ &= \mathrm{off}_{<\mathrm{r}}(A)^* = \overline{\mathrm{off}_{<\mathrm{r}}(A)} = \mathrm{off}_{\leqslant\mathrm{r}}(A)^* = \overline{\mathrm{off}_{\leqslant\mathrm{r}}(A)},\end{aligned}$$

and $\overset{\circ}{\overline{\mathrm{off}_{<\mathrm{r}}(A)}} = \overset{\circ}{\overline{\mathrm{off}_{\leqslant\mathrm{r}}(A)}}$.

2. If A is open, then

$$\begin{aligned}\mathrm{off}_{<\mathrm{r}}(A)^\circ &= \mathrm{off}_{<\mathrm{r}}(A) = \mathrm{off}_{<\mathrm{r}}(\overline{A}) = \mathrm{off}_{\leqslant\mathrm{r}}(A)^\circ = \mathrm{off}_{\leqslant\mathrm{r}}(A) \subseteq \overline{\overset{\circ}{\mathrm{off}_{<\mathrm{r}}(A)}} \\ &= \overline{\overset{\circ}{\mathrm{off}_{\leqslant\mathrm{r}}(A)}} \subseteq \mathrm{off}_{\leqslant\mathrm{r}}(\overline{A}) = \mathrm{off}_{<\mathrm{r}}(A)^* = \overline{\mathrm{off}_{<\mathrm{r}}(A)} = \mathrm{off}_{\leqslant\mathrm{r}}(A)^* = \overline{\mathrm{off}_{\leqslant\mathrm{r}}(A)}.\end{aligned}$$

3. If A is closed, then

$$\begin{aligned}\mathrm{off}_{<\mathrm{r}}(A)^\circ &= \mathrm{off}_{<\mathrm{r}}(A) = \mathrm{off}_{<\mathrm{r}}(\overline{A}) \subseteq \overline{\overset{\circ}{\mathrm{off}_{<\mathrm{r}}(A)}} = \mathrm{off}_{\leqslant\mathrm{r}}(A)^\circ = \overline{\overset{\circ}{\mathrm{off}_{\leqslant\mathrm{r}}(A)}} \\ &\subseteq \mathrm{off}_{\leqslant\mathrm{r}}(A) = \mathrm{off}_{\leqslant\mathrm{r}}(\overline{A}) = \mathrm{off}_{<\mathrm{r}}(A)^* = \overline{\mathrm{off}_{<\mathrm{r}}(A)} = \mathrm{off}_{\leqslant\mathrm{r}}(A)^* = \overline{\mathrm{off}_{\leqslant\mathrm{r}}(A)}.\end{aligned}$$

Table 1 illustrates by example that already in $\mathbb{R}$ all inclusions in Lemma 5 are proper in general. Replacing the intervals by corresponding balls, they can be lifted to any dimension. Since all the inclusions are proper, it follows that $\mathrm{off}_{>\mathrm{r}}(A)^*$ cannot be positioned into the given chain of inclusions in Lemma 5 (1). These results can certainly be transferred to $</\leqslant$-offsets.

Table 1. All inclusions in Lemma 5 are proper in general.

A	$\mathrm{off}_{>1}(\overline{A})$	$\mathrm{off}_{>1}(A)^*$	$\mathrm{off}_{>1}(A)$	$\overline{\mathrm{off}_{>1}(A)}$	$\mathrm{off}_{\geqslant 1}(\overline{A})$
$]-\infty,0[$	$]1,\infty[$	$[1,\infty[$			
$\sim[-1,1]$		$\emptyset$	$\{0\}$		
$]-\infty,0]$			$]1,\infty[$	$[1,\infty[$	
$\sim]-1,1[$				$\emptyset$	$\{0\}$

6 Blendings

Among the numerous types of blendings of solids in the literature [7, 15] we consider global fixed-radius rolling ball roundings in the classification of Vida et al. [15].

Our blending operations are by definition *global* on the involved objects. In practice, however, one often wishes to formate *local* blendings, i.e., blendings restricted to certain edges or corners. This can be achieved by equivalently modifying the boolean structure of the quantifier-free representation of a given solid in such a way that the target parts are isolated. Then the operations are applied only to the subformulas describing these parts. Afterwards, the boolean structure can once more be rearranged to a more canonical form.

6.1 Roundings from Inside

Let A be a closed set in $\mathbb{R}^n$, and let $0 \leqslant r \in \mathbb{R}$. Then the *r-rounding from inside* of A is the union of all closed balls completely contained in A:

$$\mathrm{round}_r(A) = \bigcup_{\mathrm{B}[\mathbf{a},r]\subseteq A} \mathrm{B}[\mathbf{a},r] = \{\, \mathbf{x} \in \mathbb{R}^n \mid \exists(\mathbf{a} \in A) : \mathbf{x} \in \mathrm{B}[\mathbf{a},r] \subseteq A \,\}.$$

Accordingly, A is called *r-rounded from inside* if $A = \mathrm{round}_r(A)$.

From the definition it follows that $\mathrm{round}_r(A) \subseteq A$, and for $A \subseteq B$ we may conclude that $\mathrm{round}_r(A) \subseteq \mathrm{round}_r(B)$.

The following theorem characterizes roundings from inside in terms of offsets. From this characterization, it follows that r-roundings from inside are, though *infinite* unions of closed balls, again closed. In fact, they are even regular closed.

Theorem 7. *Let $A \subseteq \mathbb{R}^n$ be closed, and let $0 \leqslant r \in \mathbb{R}$. Then*

$$\mathrm{round}_r(A) = \mathrm{off}_{\leqslant \mathrm{r}}\mathrm{off}_{>\mathrm{r}}(\sim A) = \mathrm{off}_{\leqslant \mathrm{r}}\mathrm{off}_{\geqslant \mathrm{r}}(\overline{\sim A}).$$

In particular $\mathrm{round}_r(A)$ *is regular closed.*

Proof. Let $\mathbf{x} \in \mathrm{round}_r(A)$, and pick $\mathbf{y} \in A$ with $\mathbf{x} \in \mathrm{B}[\mathbf{y}, r] \subseteq A$. Then $\mathbf{y} \in \mathrm{off}_{>\mathrm{r}}(\sim A)$ and $d(\mathbf{x}, \mathbf{y}) \leqslant r$, and so $\mathbf{x} \in \mathrm{off}_{\leqslant \mathrm{r}}\mathrm{off}_{>\mathrm{r}}(\sim A)$. Conversely, let $\mathbf{x} \in \mathrm{off}_{\leqslant \mathrm{r}}\mathrm{off}_{>\mathrm{r}}(\sim A)$ and pick $\mathbf{y} \in \mathrm{off}_{>\mathrm{r}}(\sim A)$ with $d(\mathbf{x}, \mathbf{y}) \leqslant r$. Then $\mathbf{x} \in \mathrm{B}[\mathbf{y}, r] \subseteq A$, i.e. $\mathbf{x} \in \mathrm{round}_r(A)$. The rest follows by Lemma 5 and Corollary 6. □

Let $A \subseteq \mathbb{R}^n$ be closed. We obviously have $\mathrm{round}_0(A) = A$. For $0 < r \in \mathbb{R}$ it follows immediately from Theorem 7 that $\mathrm{round}_r(A) \subseteq A^*$, and further

$$\bigcup_{0<r\in\mathbb{R}} \mathrm{round}_r(A) \subseteq A^*.$$

This inclusion relation is in general a proper one: Consider the first quadrant $A = \{ (x_1, x_2) \in \mathbb{R}^2 \mid x_1 \geqslant 0 \text{ and } x_2 \geqslant 0 \}$ of real 2-space. It is easily seen to be regular closed, but for all $0 < r \in \mathbb{R}$ the origin $(0,0)$ is not in $\mathrm{round}_r(A)$.

We eventually have to clarify whether r-roundings are actually r-rounded. They are, and moreover they are as close to the original set as possible:

Lemma 8. *Let $A \subseteq \mathbb{R}^n$ be closed, and let $0 \leqslant r \in \mathbb{R}$. Then* $\mathrm{round}_r(A)$ *is the largest subset of A that is r-rounded.*

Proof. We are allowed to form $\mathrm{round}_r(\mathrm{round}_r(A))$, because $\mathrm{round}_r(A)$ is closed by Theorem 7. Since $\mathrm{round}_r(A)$ itself is already defined as union of closed r-balls, it follows that $\mathrm{round}_r(\mathrm{round}_r(A)) = \bigcup\{ \mathrm{B}[\mathbf{b}, r] \mid \mathrm{B}[\mathbf{b}, r] \subseteq \mathrm{round}_r(A) \} = \mathrm{round}_r(A)$, i.e., $\mathrm{round}_r(A)$ is r-rounded from inside. Let $B \subseteq A$ be r-rounded from inside. Then $B = \mathrm{round}_r(B) \subseteq \mathrm{round}_r(A)$. □

Finally, we examine the interplay between roundings from inside and boolean operations.

Lemma 9. *Let A, $B \subseteq \mathbb{R}^n$ be closed, and let $0 < r \in \mathbb{R}$. Then the following assertions hold:*

1. $\mathrm{round}_r(A) \cup \mathrm{round}_r(B) \subseteq \mathrm{round}_r(A \cup B)$
2. $\mathrm{round}_r(A) \cap \mathrm{round}_r(B) \supseteq \mathrm{round}_r(A \cap B)$
3. *If both A and B are r-rounded from inside, then so is $A \cup B$.*

Proof. 1. Let $\mathbf{x} \in \text{round}_r(A) \cup \text{round}_r(B)$, say $\mathbf{x} \in \text{round}_r(A)$. Then there exists $\mathbf{y} \in A$ such that $\mathbf{x} \in \text{B}[\mathbf{y}, r] \subseteq A \subseteq A \cup B$.
2. Let $\mathbf{x} \in \text{round}_r(A \cap B)$. Then there exists $\mathbf{y} \in A \cap B$ with $\mathbf{x} \in \text{B}[\mathbf{y}, r] \subseteq A \cap B \subseteq A, B$.
3. Let $\mathbf{x} \in A \cup B$, say $\mathbf{x} \in A$. Then there exists $\mathbf{y} \in A$ with $\mathbf{x} \in \text{B}[\mathbf{y}, r] \subseteq A \subseteq A \cup B$. □

The inclusions in the Lemma are proper in general: Consider real 2-space and fix $r = 1$. For the first inclusion consider two adjacent squares of side lengths 2. For the second inclusion consider two overlapping closed unit disks centered at $(0, 0)$ and $(1, 0)$. This latter example shows also that the intersection of r-rounded sets need not be r-rounded.

Our first computation example is quite elementary, but gives a good idea of the quality of our results.

Example 10 (First Octant). We compute the r-rounding from inside of the first octant

$$O_1 = S(x_1 \geqslant 0 \wedge x_2 \geqslant 0 \wedge x_3 \geqslant 0).$$

For this, we compute the $\leqslant$-offset of the obvious quantifier-free description $x_1 - r \geqslant 0 \wedge x_2 - r \geqslant 0 \wedge x_3 - r \geqslant 0$ of the $>$-offset of the complement of O_1:

$$\exists u_1 \exists u_2 \exists u_3 \Big(\bigwedge_{i=1}^{3} u_i - r \geqslant 0 \wedge \sum_{i=1}^{3} (x_i - u_i)^2 \leqslant r^2 \Big).$$

We automatically obtain the following equivalent quantifier-free description of $\text{round}_r(O_1)$ with 19 atomic formulas:

$$2r^2 - 2rx_1 - 2rx_2 - 2rx_3 + x_1^2 + x_2^2 + x_3^2 \leqslant 0 \vee (r - x_1 \leqslant 0 \wedge r - x_2 \leqslant 0 \wedge r - x_3 \leqslant 0) \vee (2rx_1 - x_1^2 \geqslant 0 \wedge r - x_2 \leqslant 0 \wedge r - x_3 \leqslant 0) \vee (2rx_2 - x_2^2 \geqslant 0 \wedge r - x_1 \leqslant 0 \wedge r - x_3 \leqslant 0) \vee (r^2 - 2rx_1 - 2rx_2 + x_1^2 + x_2^2 \leqslant 0 \wedge r - x_3 \leqslant 0) \vee (2rx_3 - x_3^2 \geqslant 0 \wedge r - x_1 \leqslant 0 \wedge r - x_2 \leqslant 0) \vee (r^2 - 2rx_1 - 2rx_3 + x_1^2 + x_3^2 \leqslant 0 \wedge r - x_2 \leqslant 0) \vee (r^2 - 2rx_2 - 2rx_3 + x_2^2 + x_3^2 \leqslant 0 \wedge r - x_1 \leqslant 0).$$

This computation takes 30 ms. The first inequality in the above result formula is the ball $\sum_{i=1}^{3} (x_i - r)^2 \leqslant r^2$ at (r, r, r) with radius r including the rounded corner. The conjunction following this inequality is the first octant shifted from the origin to the center (r, r, r) of our ball. The next conjunction can be rewritten as $(x_1 - r)^2 \leqslant r^2 \wedge r \leqslant x_2 \wedge r \leqslant x_3$, which for positive r amounts to

$$0 \leqslant x_1 \leqslant 2r \wedge r \leqslant x_2 \wedge r \leqslant x_3,$$

i.e. a slice of width $2r$ parallel to the Y-Z-plane. The two other conjunctions that look similar describe corresponding slices in the other coordinate directions. The conjunction $r^2 - 2rx_1 - 2rx_2 + x_1^2 + x_2^2 \leqslant 0 \wedge r - x_3 \leqslant 0$ describes the semiinfinite cylinder

$$\sum_{i=1}^{2} (x_i - r)^2 \leqslant r^2 \wedge x_3 \geqslant r$$

continuing our ball to infinity in direction x_3. Again, the two similar conjunctions describe the corresponding cylinders in direction x_1 and x_2, respectively.

6.2 Roundings from Outside

Recall that we have defined roundings from inside of a solid as the union of all closed balls contained in it. This removes certain parts of the solid such as convex edges. Roundings from outside, in contrast, will add certain parts to the solid thus smoothing concave edges. Roughly speaking, the rounding from outside of a solid is obtained by rounding its complement from inside, i.e. as the complement of all balls not intersecting it. Note, however, that both the input solid and the result of the rounding operation should be closed, i.e. the corresponding complements are open. We hence operate with open balls here, and, as a consequence, the notion of rounding from outside is not exactly dual to that of rounding from inside.

Let A be a closed set in $\mathbb{R}^n$, and let $0 \leqslant r \in \mathbb{R}$. Then the *r-rounding from outside* of A is defined as the following set:

$$\begin{aligned}\mathrm{Round}_r(A) &= \bigcap\{\sim \mathrm{B}(\mathbf{x}, r) \mid \mathrm{B}(\mathbf{x}, r) \cap A = \emptyset\} \\ &= \{\mathbf{x} \in \mathbb{R}^n \mid \forall(\mathbf{y} \in \mathbb{R}^n) : \mathrm{B}(\mathbf{y}, r) \cap A = \emptyset \Longrightarrow \mathbf{x} \notin \mathrm{B}(\mathbf{y}, r)\}.\end{aligned}$$

A is called *r-rounded from outside* if $A = \mathrm{Round}_r(A)$.

From the definition it follows that $A \subseteq \mathrm{Round}_r(A)$, and for $A \subseteq B$, we can conclude that $\mathrm{Round}_r(A) \subseteq \mathrm{Round}_r(B)$. As an intersection of closed sets $\mathrm{Round}_r(A)$ is closed.

Similar to roundings from inside, roundings from outside can be characterized as iterated offset computations:

Theorem 11. *Let $A \subseteq \mathbb{R}^n$ be closed, and let $0 \leqslant r \in \mathbb{R}$. Then* $\mathrm{Round}_r(A) = \mathrm{off}_{\geqslant r}\mathrm{off}_{\geqslant r}(A)$.

Proof. Let $\mathbf{x} \in \mathrm{Round}_r(A)$, and let $\mathbf{y} \in \mathrm{off}_{\geqslant r}(A)$. Then $\mathrm{B}(\mathbf{y}, r) \cap A = \emptyset$, thus $\mathbf{x} \notin \mathrm{B}(\mathbf{y}, r)$, and so $d(\mathbf{x}, \mathbf{y}) \geqslant r$. This shows that $\mathrm{Round}_r(A) \subseteq \mathrm{off}_{\geqslant r}\mathrm{off}_{\geqslant r}(A)$. Conversely, let $\mathbf{x} \in \mathrm{off}_{\geqslant r}\mathrm{off}_{\geqslant r}(A)$, and let $\mathbf{y} \in \mathbb{R}^n$ be such that $\mathrm{B}(\mathbf{y}, r) \cap A = \emptyset$. Then $\mathbf{y} \in \mathrm{off}_{\geqslant r}(A)$, thus $d(\mathbf{x}, \mathbf{y}) \geqslant r$, and so $\mathbf{x} \notin \mathrm{B}(\mathbf{y}, r)$. This shows the converse inclusion. □

Also, just as one would expect, rounding from outside yields rounded sets:

Lemma 12. *Let $A \subseteq \mathbb{R}^n$ be closed, and let $0 \leqslant r \in \mathbb{R}$. Then $\mathrm{Round}_r(A)$ is the smallest superset of A that is r-rounded from outside.*

Proof. We know $\mathrm{Round}_r(A) \subseteq \mathrm{Round}_r(\mathrm{Round}_r(A))$. Let now, conversely, $\mathbf{x} \in \mathrm{Round}_r(\mathrm{Round}_r(A))$, and let furthermore $\mathbf{y} \in \mathbb{R}^n$ be such that $\mathrm{B}(\mathbf{y}, r) \cap A = \emptyset$. Then for all $\mathbf{z} \in \mathrm{Round}_r(A)$ we have $\mathbf{z} \notin \mathrm{B}(\mathbf{y}, r)$, i.e., $\mathrm{B}(\mathbf{y}, r) \cap \mathrm{Round}_r(A) = \emptyset$. Hence $\mathbf{x} \notin \mathrm{B}(\mathbf{y}, r)$, and we have shown that $\mathrm{Round}_r(\mathrm{Round}_r(A)) \subseteq \mathrm{Round}_r(A)$. Let $B \supseteq A$ be r-rounded from outside. Then $\mathrm{Round}_r(A) \subseteq \mathrm{Round}_r(B) = B$. □

Compared to roundings from inside, unions and intersections of roundings from outside obey the opposite inclusions. Accordingly, roundings from outside are closed under intersection instead of union.

Lemma 13. *Let A, $B \subseteq \mathbb{R}^n$ be closed, and let $0 < r \in \mathbb{R}$. Then the following assertions hold:*

1. $\mathrm{Round}_r(A) \cup \mathrm{Round}_r(B) \subseteq \mathrm{Round}_r(A \cup B)$
2. $\mathrm{Round}_r(A) \cap \mathrm{Round}_r(B) \supseteq \mathrm{Round}_r(A \cap B)$
3. *If both A and B are r-rounded from outside, then so is $A \cap B$.*

Again, the inclusions are proper in general already in real 2-space: For the first inclusion, consider the two half spaces given by $x_1 \leqslant 0$ and by $x_2 \leqslant 0$. For the second inclusion, consider the complements of the first and of the second quadrant. The first example shows also that roundings from outside are not closed under union.

Example 14 (Closed Complement of First Octant). For the r-rounding from outside of the closed complement of the first octant $S(x_1 \leqslant 0 \vee x_2 \leqslant 0 \vee x_3 \leqslant 0)$ we obtain, similarly to rounding the octant itself from inside in Example 10, 19 atomic formulas in 30 ms.

Example 15 (Cylinder with Orthogonal Box). Our aim is to blend the infinite cylinder $S(x_1^2 + x_2^2 \leqslant 1)$ with the semi-infinite box $S(x_3 \leqslant 0 \wedge -1 \leqslant x_2 \leqslant 1)$. For this, we have to compute the $\geqslant$-r-offset of the $\geqslant$-r-offset

$$S(x_1^2 + x_2^2 \geqslant (1+r)^2 \wedge (x_3 \geqslant r \vee x_2 \leqslant -1-r \vee x_2 \geqslant 1+r))$$

of their union. We obtain 1595 atomic formulas in 36 s. Similarly, we can blend this box with a larger cylinder with radius 4. This yields 1700 atomic formulas in 40 s.

Automatic blending of two orthogonal circular cylinders currently fails due to a violation of the degree restrictions imposed by our quantifier elimination method. The same happens for blendings of half spaces with oblique cylinders.

There appear to be no relations between r-roundings from outside and regular closed or open sets. In fact, there is a regular closed, simply connected set $A \subseteq \mathbb{R}^2$ such that $\mathrm{Round}_r(A)$ is disconnected and not regular closed.

The relation between r-roundings from inside and r-roundings from outside works via the operation of taking closures of complements.

Theorem 16. *Let $A \subseteq \mathbb{R}^n$ be closed, and let $0 < r \in \mathbb{R}$. Then the following assertions hold:*

1. $\mathrm{round}_r(A) = \overline{\sim \mathrm{Round}_r(\overline{\sim A})}$
2. $\mathrm{Round}_r(A^*)^* = \overline{\sim \mathrm{round}_r(\overline{\sim A})}$.

Proof. 1. Using our previous results on offsets and the coding of roundings by offsets, we obtain

$$\begin{aligned}\mathrm{round}_r(A) &= \mathrm{off}_{\leqslant \mathrm{r}}\mathrm{off}_{\geqslant \mathrm{r}}(\overline{\sim A}) = \overline{\mathrm{off}_{<\mathrm{r}}\mathrm{off}_{\geqslant \mathrm{r}}(\overline{\sim A})} = \overline{\sim \mathrm{off}_{\geqslant \mathrm{r}}\mathrm{off}_{\geqslant \mathrm{r}}(\overline{\sim A})}\\ &= \overline{\sim \mathrm{Round}_r(\overline{\sim A})}.\end{aligned}$$

2. We rewrite $\overline{\sim \mathrm{round}_r(\overline{\sim A})}$ according to part (1), and transform it in the following way:

$$\overline{\sim\overline{\sim \mathrm{Round}_r(\overline{\overline{\sim\overline{\overline{\sim A}}}})}} = \overline{\sim\overline{\sim \mathrm{Round}_r(\overline{\sim\sim(A^\circ)})}} = \overline{\sim\overline{\sim \mathrm{Round}_r(A^*)}}.$$

By the same transformation we eventually obtain $\mathrm{Round}_r(A^*)^*$. □

7 Comparison to Other Work

Rossignac and Requicha [12] define solids as regular closed sets $A \subseteq \mathbb{R}^n$. Rounding is defined in terms of the offsetting operations of *expanding* $A \uparrow^* r$ and (regularized) *shrinking* $A \downarrow^* r$. The former is exactly our $\mathrm{off}_{\leqslant \mathrm{r}}$. Shrinking is defined in terms of expanding.

$$A \uparrow^* r = \mathrm{off}_{\leqslant \mathrm{r}}(A), \quad A \downarrow^* r = \overline{\sim(\overline{\sim A} \uparrow^* r)}.$$

Recall from Corollary 6 that $\mathrm{off}_{\leqslant \mathrm{r}}(A)$ is regular closed for closed A. In the definition of shrinking, note that both applications of the closure operator are to open sets. The closures obtained are thus in fact regular closures, i.e., the result of this shrinking applied to closed sets is regular closed. We translate the definition of shrinking to our framework:

$$A \downarrow^* r = \overline{\sim(\overline{\sim A} \uparrow^* r)} = \overline{\sim \mathrm{off}_{\leqslant \mathrm{r}}(\overline{\sim A})} = \overline{\mathrm{off}_{>\mathrm{r}}(\overline{\sim A})} = \mathrm{off}_{>\mathrm{r}}(\overline{\sim A})^* = \mathrm{off}_{\geqslant \mathrm{r}}(\overline{\sim A})^*.$$

Rounding from inside and from outside are defined by Rossignac and Requicha as

$$\mathrm{round}'_r(A) = (A \downarrow^* r) \uparrow^* r \quad \text{and} \quad \mathrm{Round}'_r(A) = (A \uparrow^* r) \downarrow^* r,$$

respectively. As we have seen above both expanding and shrinking yields regular closed sets, and so do hence rounding from inside and from outside.

This means that at least Round'_r must differ in some way from our notion Round_r: Consider, e.g., $\sim]-r, r[\subseteq \mathbb{R}$, which is regular closed. In our sense, it is r-rounded from outside:

$$\mathrm{Round}_r(\sim]-r, r[) = \mathrm{off}_{\geqslant \mathrm{r}}\mathrm{off}_{\geqslant \mathrm{r}}(\sim]-r, r[) = \mathrm{off}_{\geqslant \mathrm{r}}(\{0\}) = \sim]-r, r[\,.$$

In the framework of [12] we obtain in contrast:

$$\mathrm{Round}'_r(\sim]-r, r[) = \mathrm{off}_{\geqslant \mathrm{r}}(\overline{\sim \mathrm{off}_{\leqslant \mathrm{r}}(\sim]-r, r[)})^* = \mathrm{off}_{\geqslant \mathrm{r}}(\overline{\sim \mathbb{R}})^* = \mathrm{off}_{\geqslant \mathrm{r}}(\emptyset)^* = \mathbb{R}.$$

Similar effects occur with roundings from inside. In our framework, closed balls of radius r are r-rounded from inside:

$$\text{round}_r([-r,r]) = \text{off}_{\leqslant r}\text{off}_{>r}(\sim[-r,r]) = \text{off}_{\leqslant r}(\{0\}) = [-r,r].$$

Using round'_r we obtain in contrast the empty set as inner rounding:

$$\text{round}'_r([-r,r]) = \text{off}_{\leqslant r}(\text{off}_{\geqslant r}(\overline{\sim[-r,r]})^*) = \text{off}_{\leqslant r}(\emptyset^*) = \text{off}_{\leqslant r}(\emptyset) = \emptyset.$$

The examples can be lifted to any dimension using corresponding balls instead of intervals. Moreover, for a dimension greater than or equal to one, $\sim \mathrm{B}(\mathbf{0}, r)$, as used in the example for roundings from outside, becomes connected.

The counterintuitive behavior of regularized roundings has already been observed and illustrated by Rossignac [11] by means of more natural examples similar to Figure 1.

Fig. 1. Given the solid on the left hand side, our notions of rounding from inside and from outside lead to the results in the middle. Regularized blending yields the results on the right hand side for these limit cases

Example 17 (Rossignac's Limit Case). We compute the rounding from inside of Rossignac's solid pictured in Figure 1. It is described by

$$(0 \leqslant x_1 \leqslant 8 \wedge 0 \leqslant x_2 \leqslant 3) \vee (0 \leqslant x_1 \leqslant 2 \wedge 3 \leqslant x_2 \leqslant 7) \vee (6 \leqslant x_1 \leqslant 8 \wedge 3 \leqslant x_2 \leqslant 7).$$

Here, we use partial cylindrical decomposition [1] for computing the initial $>$-offset. There is an interface to Hong's implementation QEPCAD of this included in REDLOG. After 78 s we obtain a quantifier-free description containing 2063 atomic formulas. Substituting $r = 1$ and simplifying requires 6 s. The description then obtained for the 1-rounding contains 1298 atomic formulas. If we restrict to the 1-rounding from the beginning, we obtain a description with 1250 atomic formulas in 33 s.

Due to his observation, Rossignac has defined "non-regularized" variants of rounding from inside and from outside in terms of non-regularized shrinking and expanding:

$$A \uparrow r = \text{off}_{\leqslant r}(A), \quad A \downarrow r = \sim \text{off}_{\leqslant r}(\sim A) = \text{off}_{>r}(\sim A).$$

The corresponding rounding from inside amounts to exactly our notion:

$$\text{round}''_r(A) = (A \downarrow r) \uparrow r = \text{off}_{\leqslant r}\text{off}_{>r}(\sim A) = \text{round}_r(A).$$

The new rounding from outside provides, in contrast, a new operator with similar problems as Round′ above:

$$\mathrm{Round}''_r(A) = (A \uparrow r) \downarrow r = \mathrm{off}_{>\mathrm{r}}\mathrm{off}_{>\mathrm{r}}(A).$$

Rossignac thus defines a "modified" rounding from outside in terms of his now working rounding from inside:

$$\begin{aligned}\mathrm{Round}'''_r(A) &= \sim \mathrm{round}''_r((\sim A)^*)^* \\ &= \mathrm{off}_{>\mathrm{r}}\mathrm{off}_{>\mathrm{r}}(\sim(\sim A)^*)^* = \mathrm{off}_{>\mathrm{r}}\mathrm{off}_{>\mathrm{r}}(\mathring{\overline{A}})^* = \mathrm{off}_{\geqslant\mathrm{r}}\mathrm{off}_{\geqslant\mathrm{r}}(\mathring{\overline{A}})^*.\end{aligned}$$

For regular closed input A^*, we have $\mathring{\overline{A^*}} = A^{*\circ}$. It follows that Round‴ is then the regular closure of our rounding from outside:

$$\mathrm{Round}'''_r(A^*) = \mathrm{off}_{\geqslant\mathrm{r}}\mathrm{off}_{\geqslant\mathrm{r}}(A^{*\circ})^* = \mathrm{off}_{\geqslant\mathrm{r}}\mathrm{off}_{\geqslant\mathrm{r}}(\overline{A^{*\circ}})^* = \mathrm{Round}_r(A^*)^*.$$

Both of the references discussed finally turn the regularized versions round′ and Round′ in order to avoid open sets even as intermediate results.

Example 18 (Full Box). We wish to rounding the full box given by the following description:

$$B = (0 \leqslant x_1 \leqslant 2 \land 0 \leqslant x_2 \leqslant 3 \land 0 \leqslant x_3 \leqslant 4).$$

As in Example 17 above, we use QEPCAD for the first offset step. We obtain within 1 s a quantifier-free description of $\mathrm{round}_r(B)$ containing 106 atomic formulas. Substituting $r = 2$ and simplifying yields "false" describing the empty set (10 ms). Substituting $r = 1$ yields a description of the 1-rounded box, which contains only 50 atomic formulas (40 ms). Note that the 1-rounding of this box would be the empty set with the common notion of regularized rounding.

8 Conclusions

We have proposed to use quantifier-free formulas for representing objects in solid modelers. The operations commonly encountered in there and, in addition, offsetting and blending can then be performed by quantifier-elimination. In particular blending has been extensively discussed in the literature during the last years.

Our approach to blending uses in a natural way open sets as intermediate results. This allows us to define very adequate and convenient notions of blending from inside and from outside in terms of some newly developed offsetting operations. We have carefully analyzed the mathematical properties of our various offsets. This has finally enabled us to straightforwardly prove the relevant properties of our notions of blending from inside and from outside. Moreover, we could use our offsets as a tool for analyzing, comparing, and systemizing also the notions of blending discussed elsewhere.

The mathematical tools required for our method are part of a general-purpose software package by the author et al. Some promising sample computations based on first experimental implementations point on the practical applicability of our approach.

References

[1] George E. Collins and Hoon Hong. Partial cylindrical algebraic decomposition for quantifier elimination. *Journal of Symbolic Computation*, 12(3):299–328, September 1991.

[2] Andreas Dolzmann, Oliver Gloor, and Thomas Sturm. Approaches to parallel quantifier elimination. In Oliver Gloor, editor, *Proceedings of the 1998 International Symposium on Symbolic and Algebraic Computation (ISSAC 98)*, pages 88–95, Rostock, Germany, August 1998. ACM, ACM Press, New York, 1998.

[3] Andreas Dolzmann and Thomas Sturm. Redlog: Computer algebra meets computer logic. *ACM SIGSAM Bulletin*, 31(2):2–9, June 1997.

[4] Andreas Dolzmann and Thomas Sturm. Redlog user manual. Technical Report MIP-9905, FMI, Universität Passau, D-94030 Passau, Germany, April 1999. Edition 2.0 for Version 2.0.

[5] Andreas Dolzmann, Thomas Sturm, and Volker Weispfenning. Real quantifier elimination in practice. In B. H. Matzat, G.-M. Greuel, and G. Hiss, editors, *Algorithmic Algebra and Number Theory*, pages 221–247. Springer, Berlin, 1998.

[6] Paul R. Halmos. *Lectures on Boolean Algebras*, volume 1 of *Van Nostrand Mathematical Studies*. D. van Nostrand Company, Inc., Princeton, New Jersey, 1963.

[7] Chistoph Hoffmann and John Hopcroft. Quadratic blending surfaces. *Computer-Aided design*, 18(6):301–306, July and August 1986.

[8] Christoph M. Hoffmann. *Geometric and Solid Modeling*. Computer Graphics and Geometric Modeling. Morgan Kaufmann, San Mateo, California, 1989.

[9] Christoph M. Hoffmann. Algebraic and numerical techniques for offsets and blends. In Wolfgang Dahmen, Mariano Gasca, and Charles A. Micchelli, editors, *Computation of Curves and Surfaces*, pages 499–528. Kluwer Academic Publishers, Dordrecht, Boston, London, 1990.

[10] Rüdiger Loos and Volker Weispfenning. Applying linear quantifier elimination. *The Computer Journal*, 36(5):450–462, 1993. Special issue on computational quantifier elimination.

[11] Jaroslaw R. Rossignac. *Blending and Offsetting Solid Models*. Ph.D. thesis, Department of Electrical Engineering, College of Engineering and Applied Science, University of Rochester, Rochester, New York 14627, July 1985.

[12] Jaroslaw R. Rossignac and Aristides A. G. Requicha. Offsetting operations in solid modelling. *Computer Aided Geometric Design*, 3(2):129–148, August 1986.

[13] Thomas Sturm and Volker Weispfenning. Computational geometry problems in Redlog. In Dongming Wang, editor, *Automated Deduction in Geometry*, volume 1360 of *Lecture Notes in Artificial Intelligence (Subseries of LNCS)*, pages 58–86. Springer-Verlag, Berlin Heidelberg, 1998.

[14] Alfred Tarski. A decision method for elementary algebra and geometry. Technical report, RAND, Santa Monica, CA, 1948.

[15] Janos Vida, Ralph R. Martin, and Tamas Varady. A survey of blending methods that use parametric surfaces. *Computer-Aided Design*, 26(5):341–365, May 1994. Survey.

[16] Volker Weispfenning. Quantifier elimination for real algebra—the quadratic case and beyond. *Applicable Algebra in Engineering Communication and Computing*, 8(2):85–101, February 1997.

Application of Computer Algebra Methods to Some Problems of Theoretical and Applied Celestial Mechanics

Akmal Vakhidov

Lohrmann Observatory, Technical University Dresden,
Mommsenstr.,13, Dresden, 01062, Germany
e-mail: vakhidov@rcs.urz.tu-dresden.de

Abstract. Applications of computer algebra methods to some problems of dynamical astronomy are described. It is shown, how with the help of computer algebra it is possible to solve different important problems of applied celestial mechanics: in particular, to develop efficient analytical and semianalytical theories of motion of celestial bodies, to construct efficient algorithms for computation of some special functions of celestial mechanics, to solve some problems of asteroid dynamics. Possibilities to use computer–algebraic methods for efficient solution of other topical problems of dynamics of Hamiltonian systems are discussed.

1 Introduction

Many important problems of applied celestial mechanics were solved successfully in the recent time with the help of computer algebra methods. For example, we can mention that using computer algebra it was possible to obtain in symbolic form the LeVerrier expansion of planetary disturbing function including the terms of high orders and degrees [1], to construct normal forms of different Hamiltonian systems [2, 3], to solve symbolically the generalized Kepler equation and some other important equations of classical celestial mechanics [4], to develop efficient theories of motion for some types of artificial Earth satellites [5, 6], to construct convenient algorithms for computation of some special functions of celestial mechanics [7].

In the present paper we consider some other applications of computer algebra methods to problems of classical celestial mechanics. In particular, we show the possibility of obtaining (in symbolic form) the averaged Hamiltonian of planetary motion presented via LeVerrier expansion of disturbing function, the possibility of symbolic computation of Hamiltonian of planetary and satellite motion expressed via Kaula expansion of elliptic motion functions, we develop efficient algorithms for computation of such well–known function of celestial mechanics as Hansen coefficients with the help of polynomial approximations and convenient recurrence relations, we describe some applications of computer algebra methods to problems of asteroid dynamics and discuss the possibilities to use computer algebra for solving some other problems of celestial mechanics in the future.

The algorithms and methods described in the present paper can be used also for efficient solution of other problems of Hamiltonian mechanics and can be realized with the help of different systems of computer algebra.

2 Symbolic computation of the averaged Hamiltonian of planetary motion in the LeVerrier form

For constructing semianalytical theories of motion of celestial bodies it is necessary to compute the averaged Hamiltonian describing the behaviour of dynamical system only under the influence of long–periodic, secular and resonant perturbations. The averaging procedure consists in the elimination of short–periodic terms from the initial Hamiltonian system and can be realized, for example, by means of Lie–series method [8, 9]. As a result, we obtain the averaged Hamiltonian, which (because of absence of short–periodic terms) gives a possibility to integrate the equations of motion numerically with a very large step. It allows to study the dynamical behaviour of Hamiltonian system (in particular, evolution of orbits of celestial bodies) on large intervals of time.

If we construct the theory of perturbations of the first order only, then the averaging procedure is very simple and consists in the direct selection of long–periodic, secular and resonant terms from initial Hamiltonian for the future numerical integration. Unfortunately in many cases the accuracy of theory of the first order is not enough for practical purposes, that makes necessary to take into consideration also the perturbations of the second order. If the structure of Hamiltonian is not simple, then the second approximation of the averaging procedure can be complicated and in many practical cases can be realized efficiently only by means of computer algebra.

In our research we develop a computational scheme for averaging procedure of Hamiltonian of planetary motion presented in the LeVerrier form [10, 11]. The LeVerrier expansion of planetary disturbing function has enough complicated structure and consists of more than hundred terms with different trigonometrical arguments. Quite generally, this expansion can be written in the following form

$$R = m' \sum_{i,j} L_i^j C_i \cos(i_1 l + i_2 l' + i_3 g + i_4 g' + i_5 h + i_6 h') \ ,$$

where m' is the mass of disturbing body, L_i^j are the LeVerrier coefficients [10], C_i are some functions of eccentricity and inclination, l, g, h, l', g', h' are angular Delaunay orbit elements of disturbed and disturbing body.

Computation of Poisson brackets required in the second approximation of the averaging procedure [8] generates for LeVerrier expansion many thousands terms in the second order. Algebraic manipulations with such a disturbing function can be realized efficiently only by means of computer algebra.

In our developments we realize the averaging procedure of the second order in symbolic form using computer algebra system MAPLE V [12]. We use here the following procedures and algorithms.

• *Representation of LeVerrier expansion of planetary disturbing function via orbit elements* on the basis of the algorithms by M.Yuasa [11], corrected later in the paper [13]. We take into consideration here also the indirect part of planetary disturbing function [14], which is extremly important in the case of resonances. All computations are realized here in symbolic form.

• *Evaluation of LeVerrier coefficients and their derivatives.* Here we use the representation of LeVerrier coefficients via Laplace coefficients [10], that gives a possibility to obtain symbolically the LeVerrier coefficients as function of semi–major axes of disturbed and disturbing bodies.

• *Selection of long–periodic, secular and resonant terms of the first order* from the initial Hamiltonian and translation of analytical expressions of these terms from MAPLE into Fortran for the forthcoming numerical integration.

• *Analytical solution of Hamilton–Jacobi equation.* The aim of these procedure is to obtain symbolically the generating function of canonical transformation of Lie–series method [8]. Here we use a special algorithm based on the method of variation of arbitrary constants. The description of this algorithm is given in detail in [15]. All computations are realized here in symbolic form.

• *Symbolic computation of Poisson brackets.* Here we compute symbolically the partial derivatives of initial Hamiltonian and generating function with respect to orbit elements and determine (in symbolic form) the Hamiltonian containing the terms, which are responsible for perturbations of the second order.

• *Selection of long–periodic, secular and resonant terms of the second order* and translation of their analytical expressions from MAPLE into Fortran for the forthcoming numerical integration.

After this we integrate differential equations of motion over long–periodic, secular and resonant terms using the well–known Runge–Kutta–Fehlberg method. Our developments can be used for studying the evolution of orbits of minor planets on large intervals of time, for studying the structure of chaotical domains in the main asteroid belt, for investigation of role of different resonant perturbations (also resonant perturbations of the second order) in the origin of chaos in the behaviour of orbits of minor planets. Some procedures (such as, for example, symbolic computation of Poisson brackets or symbolic solution of Hamilton–Jacobi equation using the method of variation of arbitrary constants) can be efficiently used also for solving other problems of celestial mechanics and dynamics of Hamiltonian systems.

3 Symbolic computation of Hamiltonian for Kaula expansion of disturbing function

Kaula expansion of disturbing function [16] can be efficiently used for constructing analytical and semianalytical theories of motion of artificial Earth satellites and also theories of motion of minor planets, especially in the cases, when the

LeVerrier expansion may not be used because of its divergence according to Sundmann–criterion [17]. The Kaula expansion itself converge for all possible cases of elliptic motion, some numerical estimations on the rate of convergence of Kaula expansion are presented in [18].

In the frames of our research we develop a computational scheme in order to obtain in symbolic form the expression for Hamiltonian of planetary motion represented via Kaula expansion of disturbing function. The derivatives of this Hamiltonian with respect to orbit elements give the right parts of canonical equations of planetary motion. These equations of motion can be integrated numerically using traditional numerical techniques. The structure of Hamiltonian in Kaula form is quite simple and can be written quite generally in the following form:

$$R = \frac{m'}{a} \sum_{n,k} \alpha^n F_{n,k} \cos \Phi_{n,k} \ ,$$

where m' is the mass of disturbing body, $\alpha = a/a'$ is the ratio of semi-major axes of disturbed and disturbing bodies, $F_{n,k}$ is a function of inclinations and eccentricities of orbits of disturbed and disturbing bodies (it can be expressed via well–known Kaula inclination functions and Hansen coefficients [16]), $\Phi_{n,k}$ is a linear function of angular orbit elements of disturbed and disturbing bodies.

The averaging procedure for Kaula expansion of disturbing function can be realized without any principial difficulties. The corresponding formulae are presented, for example, in [19]. On the other hand, the orbit elements are included into Kaula expansion of disturbing function in an implicit form via different special functions (inclination functions, Hansen coefficients etc.) and therefore the problem to represent the Hamiltonian analytically in the explicit form via orbit elements (in order to obtain symbolically the derivatives of Hamiltonian and the right parts of equations of motion) is enough complicated and can be solved quite efficiently only by means of computer algebra.

In our developments we prepare a package of computer programs designed in computer algebra system MAPLE V [12] in order to express explicitly the Hamiltonian of planetary motion in Kaula form via orbit elements, to obtain in symbolic form the derivatives of this Hamiltonian, and, as a result, to form symbolically the right parts of usual or averaged equations of motion for the future numerical integration. Our computational scheme consists of the following procedures.

- *Symbolic computation of Hansen coefficients in the form of polynomials of eccentricity* using the different approximating schemes and some special systems of recurrence relations. The detailed description of the corresponding algorithms and techniques is given in the next section of the present paper. As a result, we obtain symbolically the component of Hamiltonian depending on the eccentricity of the orbit.
- *Symbolic computation of Kaula inclination functions in the form of polynomials with respect to inclination.* Here we use the standard algorithms of representation of Kaula inclination functions developed in [20, 21]. As a result, we

obtain symbolically the component of Hamiltonian depending on the inclination of the orbit.

• *Computation of functions of orbit elements of disturbing bodies.* Here there are two possibilities. If we consider the motion of disturbing bodies in the Keplerian approximation (it means, all orbit elements excluding mean anomaly are some constants), then we can evaluate this constant component of the Hamiltonian at once numerically. If we take into consideration the perturbations in motion of disturbing bodies, then we represent the expressions for these perturbations in the form of fragments of trigonometric series with respect to time using the corresponding theories of motion of disturbing bodies (in particular, theory of motion of the pair "Jupiter–Saturn" developed in [22]). Both cases are implemented in our developments.

• *Symbolic computation of trigonometric component of the Hamiltonian* in order to collect the terms with similar trigonometric arguments.

• *Symbolic computation of Hamiltonian of planetary motion in the Kaula form.* This procedure implements the ideas developed in [5], but with the application to the planetary motion. Here we make in symbolic form the summation of harmonical components of Kaula expansion of disturbing function and obtain finally (after some simplifications) the analytical expression of Hamiltonian of planetary motion represented explicitly via orbit elements.

• *Symbolic computation of right parts of equations of planetary motion.* Here we compute symbolically the partial derivatives of Hamiltonian with respect to orbit elements and generate in symbolic form the right parts of equations of motion. The obtained expressions are translated from MAPLE into Fortran for the future integration of equations of motion with the help of numerical techniques.

Our experiments show the high efficiency of described above computational scheme in comparison with the direct numerical evaluation of right parts of equations of motion for solving many problems of celestial mechanics. On the basis of our approach it is possible to develop, in particular, efficient analytical and semianalytical theories of motion of minor planets, artificial Earth satellites, components of double and multiple stars, some other celestial bodies. With the help of described above procedures and techniques we suppose to develop also a package of computer programs in order to realize an explicit semianalytical theory of the second order for asteroid motion presented in [19]. Some remarks concerning the future applications of our developments to problems of asteroid dynamics are given in section 5 of the present paper.

4 Polynomial approximations and recurrence relations for Hansen coefficients

If we construct theories of motion of celestial bodies using the Kaula expansion of disturbing function [16], then we face the problem of computation of Hansen coefficients and their derivatives. The Hansen coefficients appear in expansion of coordinates of elliptic motion into trigonometric series with respect to timo and depend on the eccentricity of elliptic orbit. The problem of computing the Hansen coefficients was studied in many papers (see, for example, [18, 23, 24]), but this problem is rather complicated and up to date we have no standard algorithms for efficient evaluation of these coefficients for all possible practical cases.

One of the problems to use the expansions with Hansen coefficients in theories of motion of celestial bodies is connected with the necessity to recompute these coefficients many times for different (often only a little different) values of eccentricity. In particular, if we use a semianalytical theory of motion and integrate averaged equations for orbit elements, then we should recompute the Hansen coefficients at each step of numerical integration. It requires a lot of computing time and decreases the efficiency of the corresponding theory of motion itself. Therefore in many practical cases it could be much more convenient to construct some simple approximations for Hansen coefficients on the determined interval of eccentricity and to compute these coefficients using the approximating formulae instead of more complicated direct methods.

In the frames of our research we investigate three types of polynomial approximations of Hansen coefficients: approximation by fragments of Taylor series, interpolation by Lagrange polynomials, Chebyshev approximation. For construction of approximating polynomials and for the estimation of accuracy of approximations we use the representation of Hansen coefficients via Hill's formula [25]. In order to analyze the numerical efficiency of different approximating schemes we consider several hundreds of Hansen coefficients from Kaula expansion of disturbing function and investigate the accuracy and some properties of polynomial approximations of these coefficients on different intervals of eccentricity. On the basis of our numerical experiments we can make the following conclusions.

- *Approximation by fragments of Taylor series.* Expanding the analytical expressions for Hansen coefficients into Taylor series in the neighbourhood of the eccentricity e_0, we obtain the result in the following form:

$$X_s^{n,m}(e) = \sum_{k=0}^{N} A_k (e - e_0)^k ,$$

where the coefficients A_k are computed numerically $\forall n, m, s$. Our experiments show, that this type of approximation can be efficiently used only for orbits with small and moderately large ($e < 0.5$) eccentricities. Moreover, the Taylor expansions of Hansen coefficients should be constructed only in the neighbourhood of the point $e_0 = 0$, because in the neighbourhood of other points ($e_0 \neq 0$) the domain of their convergence is very small. For orbits with small eccentricities ($e < 0.1$) approximation by polynomials of the 20th degree is sufficient to provide for the errors about 10^{10} times smaller than the values of coefficients themselves. For orbits with the eccentricity up to $e = 0.4$ the similar accuracy can be obtained in approximations by polynomials of the about 50th degree. The main advantage of this type of approximation is the small relative errors, so that the coefficients with small absolute values can be approximated with sufficiently high accuracy. There are no singular points, where the approximations fail. The main disadvantage of Taylor approximations is the impossibility to use them for large eccentricities. Approximations by fragments of Taylor series can be efficiently constructed by means of computer algebra.

• *Interpolation by Lagrange polynomials.* With the help of Lagrange interpolation we obtain the representation of Hansen coefficients on the interval (a, b) by the following formula:

$$X_s^{n,m}(e) = \sum_{i=0}^{N} X_s^{n,m}(e_i) \prod_{\substack{k=0 \\ k \neq i}}^{N} \frac{e - e_k}{e_i - e_k},$$

where $N + 1$ is the number of basic points for interpolation (corresponds to the degree of interpolating polynomial), $e_i = a + ih$, $e_k = a + kh$ and $h = (b - a)/N$ is the step of interpolation. $X(e_i)$ is the value of the function $X(e)$ in the point $e = e_i$.

This type of approximation can be used also for orbits with large eccentricities, but the interval of interpolation should be not larger than $\Delta e \approx 0.2 - 0.4$. The optimal degree of approximating polynomial depends very substantially on the number of digits in the process of computation. The accuracy of approximation with increasing the degree of approximating polynomial at first increases, but after a certain maximum rapidly decreases. Therefore it is extremly important to evaluate the optimal degree of approximating polynomial beforehand, and, if possible, to construct the interpolating polynomial with large number of digits, although the Hansen coefficients can be computed on the basis of this approximation with a smaller number of digits. The singular point of Lagrange interpolation is $e = 0$. In its vicinity the relative errors of approximations are always large. The main advantage of this type of approximation is the simplicity of its construction. The main disadvantage is the necessity of preliminary estimations on the optimal degree of interpolating polynomial. Construction of Lagrange interpolation can be efficiently realized by means of computer algebra, where we have a possibility to make computations with large number of digits.

• *Chebyshev approximation.* Chebyshev approximations give always small absolute errors, but relative errors can be large, if we approximate the coefficients with small numerical values. The degree of approximating polynomial (for the fixed accuracy of approximation) decreases with decreasing the interval of approximation very slowly. Therefore it is always better to construct one approximation on a large interval of eccentricity instead of many approximations on short intervals. The singularities and, as a result, large relative errors appear in the vicinity of points, where the Hansen coefficients change the sign. Chebyshev approximations can be constructed by means of computer algebra (for example, in the system MAPLE V with the help of built-in function `chebyshev`; the accuracy of approximation is an input parameter of this function).

Our numerical experiments show the high efficiency to use the interpolating polynomials for computation of Hansen coefficients in theories of motion of celestial bodies. Such an approach gives a possibility to economize the computing time about several hundreds times in the process of computing the right parts of equations of motion (in comparison with the use of direct methods of computing the Hansen coefficients). In the frames of our research we have prepared corresponding computer programs for constructing the above indicated polynomial approximations of Hansen coefficients using computer algebra system MAPLE V. Some estimations on the numerical efficiency of considered approximating schemes are given in [26].

On the practice it is, of course, not necessary to compute all Hansen coefficients required in theories of motion of celestial bodies with the help of approximating polynomials. The approximating schemes can be used only for computing a determined set of Hansen coefficients, all other coefficients can be evaluated using recurrence algorithms. The well–known recurrence relations [27] give a connection between Hansen coefficients with the determined (as a rule, neighbouring) indices and allow to construct different recurrence algorithms for evaluation of Hansen coefficients.

If we look now at the structure of disturbing function of planetary or satellite motion, then we can see that this function contains the Hansen coefficients with some selected (and not neighbouring) indices only. Therefore the use of well–known recurrence relations [27] will require evaluation of many "intermediate" coefficients, which are not included into disturbing function and are necessary for continuation of recurrence process only. It leads to decreasing the efficiency of recurrence process and requires a lot of additional computer memory.

In order to avoid this inconvenience we construct on the basis of already known recurrence relations a new system of recurrence formulae for Hansen coefficients with the aim to obtain a connection only between coefficients included into disturbing function. The similar idea to construct efficient recurrence relations by means of computer algebra was already successfully realized earlier in paper [7] for another function of celestial mechanics, and now we apply this approach to classical Hansen coefficients. The idea consists in the consequent substitution of some known recurrence relations into others, the elimination of coefficients not included into disturbing function, and some symbolic manipulations to simplify the obtained result. It can be realized efficiently only by means of computer algebra.

In our research we design a special package of computer programs which allows to generate the recurrence formulae with prior determined properties using the standard package LINALG of the system MAPLE V. With the help of this package we generate new recurrence relations for Hansen coefficients, which do not require the computation of any "intermediate" coefficients and give a possibility to realize the recurrencies quite efficiently for the case of planetary or satellite motion both from high to low and from low to high harmonics of disturbing function. These new recurrence relations and some details of their construction are presented in [28], the formulae are available also from the author upon request.

Finally, we construct the following computational scheme for evaluating the Hansen coefficients.

• Computation of Hansen coefficients for two lowest or two highest harmonics of disturbing function (as initial conditions for recurrencies) with the help of approximating polynomials.

• Computation of Hansen coefficients for the next four or five harmonics using our new recurrence relations.

• Computation of Hansen coefficients for the next two harmonics using polynomial approximations, and computation of Hansen coefficients for the next four or five harmonics using new recurrence relations, etc.

As a result, we have the recurrence algorithm for evaluating the Hansen coefficients with the intermediate correction of coefficients inside the recurrencies with the help of approximating polynomials. The optimal type of approximating polynomial can be choosen in dependence on the values of eccentricity of the orbit and on the interval of possible variations of eccentricity in the process of evolution of orbit of the celestial body.

Our test experiments show the efficiency of such an approach to computing the Hansen coefficients in theories of motion of celestial bodies and the possibility to use it quite efficiently on the practice.

5 Applications to some problems of asteroid dynamics

Let us make now some remarks concerning the application of described above techniques to some concrete problems of asteroid dynamics.

On the basis of Kaula expansion of planetary disturbing function we have a possibility to develop efficient analytical and semianalytical theories of motion of minor planets. One of such theories is presented in [19]. In order to realize this theory on the practice we designed a MAPLE—Fortran package of computer programs for numerical integration of averaged equations of motion taking into consideration secular, long–periodic and resonant perturbations up to the second order inclusively. It is the extension of the package described in the section 3 of the present paper (that was initially intended for theories of the first order only) on the second–order perturbations. For this extension we have prepared some additional procedures for symbolic computation of second derivatives of Hansen coefficients, inclination functions and some other functions of orbit elements, required in theories of the second order. As earlier, the right parts of equations of motion are constructed in symbolic form using computer algebra methods. The motion of disturbing bodies is considered in the Keplerian approximation. The numerical integration of equations of motion is realized with the help of Runge–Kutta–Fehlberg method of the fifth order. Comparison with the numerical integration of equations of motion in Cartesian coordinates shows the acceptable accuracy of our theory of the second order (some estimations are presented in the report at the conference) and possibility to use it for studying the dynamics of minor planets. The next step of our research is the introduction of perturbations in motion of Jupiter and Saturn into our semianalytical theory. Here we suppose to use again the theory of motion of the pair "Jupiter–Saturn" developed in [22].

With the help of this semianalytical theory of the second order we suppose to study the structure of chaotical domains in the main asteroid belt, to investigate the role of different mixed resonances "Asteroid–Jupiter–Saturn" in the evolution of orbits of minor planets, to study the phenomenon of "chaotical diffusion" [29] of asteroid orbits.

Using Kaula expansion we suppose to develop also a pure analytical theory of the first order for asteroid motion, taking into account also the inequalities in motion of the pair "Jupiter–Saturn". For constructing this theory it is necessary to integrate analytically the Lagrange equations of motion in the linear approximation by means of computer algebra (the similar theory of satellite motion was developed in [30, 31]). With the help of this theory we suppose to study the influence of different harmonics of disturbing function on irregular variations of eccentricity of asteroid orbits in the vicinity of domains of strong resonances "Asteroid–Jupiter" [32]. This theory will be designed in computer algebra system MAPLE V.

6 Concluding remarks and acknowledgements

The fast development of computer algebra in the recent time gives now a possibility to solve efficiently many topical problems of the modern science. The successful solution of many such problems even a few years ago was extremly complicated or impossible at all. Also the problems described in the present paper could not be solved efficiently without computer algebra.

The techniques and algorithms presented in this paper can help to solve efficiently also many other problems of theoretical and applied astronomy. They can be used, in particular, for constructing theories of motion of artificial and natural satellites of planets, for studying the motion of components of multiple stars, for investigation of orbits of N–body problem, for computing the normal forms of Hamiltonian systems, for solving some other problems of dynamical astronomy, geodynamics and Hamiltonian mechanics.

The author would like to express his gratitude to Prof. M.Soffel and Dr. I.Tupikova for their help in this research and many stimulating discussions. The author is very grateful also to Dr. N.Vasiliev for his kind attention to this work. The author thanks DFG (Deutsche Forschungsgemeinschaft) for the financial support of these researches.

References

1. Murray, C.: Celest. Mech. **36** (1985) 163.
2. Edneral, V.F.: Proc. ISSAC–93, Kiev (1993) 14.
3. Shevchenko, I.I., Sokolsky, A.G.: Celest. Mech. and Dynam. Astron. **62** (1995) 289.
4. Klioner, S.A.: Proc. 25th Symp. on Celest. Mech., Tokyo (1992) 172.
5. Vakhidov, A.A., Vasiliev, N.N.: Journ. of Symbolic Computation, **24** (1997) 705.
6. Vakhidov, A.A., Vasiliev, N.N.: Comp. Phys. Comm. **118** (1999) 17.
7. Sokolsky, A.G., Vakhidov, A.A., Vasiliev, N.N.: SIGSAM Bulletin (1995) June Special Issue, 16.

8. Hori, G.: Publ. Astron. Soc. Japan, **18** (1966) 287.
9. Deprit, A.: Celest. Mech. **1** (1969) 12.
10. LeVerrier, U.J.: Ann. Obs. Paris, **1** (1855) 258.
11. Yuasa, M.: Publ. Astron. Soc. Japan, **25** (1973) 399.
12. Char, B.W., Geddes, K.O., Gonnet, G.H., Leong, B.L., Monagan, M.B., Watt, S.M.: MAPLE V Library Reference Manual. Springer, New York (1993).
13. Tupikova, I.V., Soffel, M., Klioner, S.A.: Celest. Mech. and Dynam. Astron. **74** (1999) 147.
14. Knezevic, Z.: Celest. Mech. and Dynam. Astron. **55** (1993) 387.
15. Vakhidov, A.A., Tupikova, I.V.: Proc. ISSAC–98, Rostock (1998) 76.
16. Kaula, W.M.: Theory of Satellite Geodesy. Blaisdell Publ. Co., Mass. (1966).
17. Ferraz–Mello, S.: Celest. Mech. and Dynam. Astron. **58** (1994) 37.
18. Klioner, S.A., Vakhidov, A.A., Vasiliev, N.N.: Celest. Mech. and Dynam. Astron. **68** (1998) 257.
19. Tupikova, I.V., Vakhidov, A.A., Soffel, M.: Celest. Mech. and Dynam. Astron. **73** (1999) 87.
20. Jeffreys, B.: Geophys. Journ. **10** (1965) 141.
21. Fominov, A.M.: Bull. ITA, **10** (1980) 621.
22. Bretagnon, P., Simon, J.–L.: Astron. and Astrophys. **239** (1990) 387.
23. Giacaglia, G.E.O.: Celest. Mech. **14** (1976) 515.
24. Hughes, S.: Celest. Mech. **25** (1981) 101.
25. Brumberg, V.A.: Analytical techniques of celestial mechanics. Springer, Berlin – Heidelberg (1995).
26. Vakhidov, A.A.: Comp. Phys. Comm. **124** (2000) 40.
27. Aksenov, E.P.: Special functions in celestial mechanics. Nauka, Moscow (1986).
28. Vakhidov, A.A.: Celest. Mech. and Dynam. Astron. (2000) in press.
29. Morbidelli, A., Nesvorny, D.: Icarus, **139** (1999) 295.
30. Vakhidov, A.A., Vasiliev, N.N.: Astron. Journ. **112** (1996) 2330.
31. Vasiliev, N.N., Vakhidov, A.A., Sokolsky, A.G.: Preprint ITA (1996) No.55.
32. Vakhidov, A.A.: Baltic Astronomy, **8** (1999) 425.

Computing the Frobenius Normal Form of a Sparse Matrix

Gilles Villard

CNRS-LMC, BP53 F38041 Grenoble cedex 9, France

Abstract. We probabilistically determine the Frobenius form and thus the characteristic polynomial of a matrix $A \in \mathbb{F}^{n\times n}$ by $O(\mu n \log(n))$ multiplications of A by vectors and $O\left(\mu n^2 \log^2(n) \log\log(n)\right)$ arithmetic operations in the field $\mathbb{F}$. The parameter μ is the number of distinct invariant factors of A, it is less than $3\sqrt{n}/2$ in the worst case. The method requires $O(n)$ storage space in addition to that needed for the matrix A.

1 Introduction

The known complexity estimates of the computation of the characteristic polynomial and *a fortiori*, of the Frobenius normal form of special – sparse or black box – square matrices A over a field $\mathbb{F}$, seem to not be satisfactory. We refer to Kaltofen [8, Open Problem 3] and to Pan et al. [16,15] for discussions on this subject and survey of current solutions. We denote by $\mathcal{M}(n)$ the number of operations in $\mathbb{F}$ required for $n\times n$ matrix multiplications. The characteristic polynomial of a general matrix A can be computed at cost of $O(n^3)$ or $O(\mathcal{M}(n)\log n)$ operations by the method of Keller-Gehrig [10]. The Frobenius normal form can be computed in $O(n^3)$ as achieved by Storjohann [19,20] and Storjohann and the author [21] while the randomized Las Vegas algorithms of Giesbrecht [7] and of Eberly [3, §4.3] give the best known asymptotic complexity $O(\mathcal{M}(n)\log n)$. The problem we address is to reduce these estimates when A is sparse or more generally, given by a fast (faster than $O(n^2)$) matrix-vector multiplication.

If we rely on the bound $\mathcal{M}(n) \le 2n^3 - n^2$, the worst case estimate for sparse matrices was $O(n^3)$, not better than in the general case. Our paper decreases the bound to $O(\mu n \log(n))$ products of A by vectors and $O\left(\mu n^2 \log^2(n) \log\log(n)\right)$ arithmetic operations in $\mathbb{F}$ where μ is the number of distinct invariant factors of A. Since μ is less than $3\sqrt{n}/2$, we gain a factor almost $\sqrt{n}$ if multiplying A by a vector costs $n\log^{O(1)}$ operations. The algorithm is Monte Carlo randomized, it succeeds with high probability if the field contains a large enough number of distinct elements compared to the dimension of the matrix (see Theorem 3).

Faster solutions exist for particular classes of sparse matrices. A first particular class is given by matrices A defined with their $s(n)$-separator families, the characteristic polynomial can be computed in $O(n^2) + n\mathcal{M}(s(n))$ operations as shown by Reif [17] or Pan et al. [16,15]. Up to a logarithmic factor with the standard matrix multiplication, we reach this estimate even when $s(n) = O(\sqrt{n})$.

In this framework of separable matrices it is not known how to compute the Frobenius normal form.

Another particular class of matrices is formed by those having few invariant factors. Let us call ν the number of non-trivial (non equal to one) invariant factors of A. If $\nu = 1$, *i.e.* if the characteristic polynomial is equal to the minimum polynomial, then Wiedemann [22] has shown that the characteristic polynomial is computed at cost of $O(n)$ multiplications of A by vectors and $O(n^2)$ additional operations in $\mathbb{F}$ (which could be generalized to any value of ν using a block method). The Eberly's algorithm [4] does not require the knowledge of ν, has cost sensitive to ν and applies to the computation of the Frobenius form. Eberly obtains the normal form by $O(n)$ multiplications of A and A^T by vectors and $O(\nu n^2)$ operations over $\mathbb{F}$ with an additional requirement of $O(n^2)$ storage space. His method also provides a corresponding transition matrix. Our result gives a better cost for large values of ν and thus in the worst case where $\nu = O(n)$, but we do not provide a transition matrix. By reducing the problem to computing minimum polynomials we require only $O(n)$ storage space for elements of $\mathbb{F}$ in addition to the storage of A.

The paper is organized as follows. The Frobenius form of A is computed as the list of the μ distinct invariant factors together with the multiplicities m_i, $1 \leq i \leq \mu$, at which they appear in the characteristic polynomial. We first study in Section 2 the effect of a perturbation $A + B$, with B of rank k, on the Frobenius form of A. Section 3 extends the result to random Toeplitz perturbations B on which we rely for the final complexity. We then prove in Section 4 that computing any of the invariant factors reduces to computing the minimum polynomials of A and of $A + B$ for B of rank given by the index of the target invariant factor. Finally, the Frobenius form itself is computed at Section 5. It is found by a binary search of the μ invariant factor degree changes using random Toeplitz perturbations. We refer to Gantmacher for the classical definitions of the *invariant factors* of matrices over $\mathbb{F}^{n\times n}$ or $\mathbb{F}[x]^{n\times n}$ in relation with the *Frobenius normal form* [6, Chap. 7 §5] or with the *Smith normal form* [6, Chap. 6 §2].

2 Rank-k Perturbations

For a given matrix $A \in \mathbb{F}^{n\times n}$, many relationships are known between the invariant factors of A and those of $A + B$, in particular we have:

Lemma 1. *If $s_1, \ldots, s_n$ are the invariant factors of A and if B has rank k then the invariant factors $\sigma_1, \ldots, \sigma_n$ of $A + B$ satisfies:*

$$s_i | \sigma_{i+k}, \; i = 1, \ldots, n-k. \tag{1}$$

Proof. Let U and V be unimodular matrices in $\mathbb{F}[x]^{n\times n}$ such that $U(x - A)V$ is in Smith normal form $S = \text{diag}(s_1, s_2, \ldots, s_n)$ and let R be unimodular in $\mathbb{F}[x]^{n\times n}$ such that the last $n - k$ rows of $R(UBV)$ are zero. We define the matrix

T, equivalent to $x - A - B$, by:

$$T = RS - R(UBV) = \begin{bmatrix} \times & \dots & \dots & \times \\ \vdots & & & \vdots \\ \times & \dots & \dots & \times \\ \alpha_{1,1}s_1 & \alpha_{1,2}s_2 & \dots & \alpha_{1,n}s_n \\ \vdots & & & \vdots \\ \alpha_{n-k,1}s_1 & \dots & \dots & \alpha_{n-k}s_n \end{bmatrix} \in \mathbb{F}[x]^{n\times n}$$

where the "$\times$" stand for any elements in $\mathbb{F}[x]$ and the $\alpha_{i,j}$'s are in $\mathbb{F}[x]$. Let $\overline{T}$ be the matrix formed by the last $n-k$ rows of T and let its first $n-k$ linearly independent columns be those indexed by $j_1, \dots, j_{n-k}$. Considering the left Hermite normal form of $\overline{T}$, we know that for a matrix $\overline{W} \in \mathbb{F}[x]^{(n-k)\times n}$, $\overline{T}$ is left equivalent to $\operatorname{diag}(s_{j_1}, \dots, s_{j_{n-k}})\overline{W}$. Thus for $W \in \mathbb{F}[x]^{n\times n}$, T is equivalent to $\operatorname{diag}(1, \dots, 1, s_{j_1}, \dots, s_{j_{n-k}})W$. Since the invariant factors of two non-singular matrices divide the invariants factors of the product – see Newman [14, Theorem II.14], the invariants factors $\sigma_1, \dots, \sigma_n$ of T and thus of $x - A - B$ are respectively divisible by $1, \dots, 1, s_{j_1}, \dots, s_{j_{n-k}}$. Thus s_{j_i} divides σ_{i+k} and since $j_i \geq i$ implies that s_i divides s_{j_i}, s_i also divides σ_{i+k} for $1 \leq i \leq n-k$. □

In addition to this knowledge on the invariant factors, informations on the characteristic polynomial may be obtained following Lidskii's theory [12] and the construction of Moro et al.[13, §2]. Let $\tilde{A} \in \mathbb{F}^{n\times n}$ be in Jordan normal form $\operatorname{diag}(J, J')$ where

$$J = \operatorname{diag}(J_1^{r_1},\ J_2^{r_1},\ \dots, J_{q_1}^{r_1},\ \dots\dots, J_1^{r_l},\ J_2^{r_l},\ \dots, J_{q_l}^{r_l}) \in \mathbb{F}^{r\times r}$$

gives all the Jordan blocks

$$J_i^{r_j} = \begin{bmatrix} \lambda & 1 & & \\ & \ddots & \ddots & \\ & & \ddots & 1 \\ & & & \lambda \end{bmatrix} \in \mathbb{F}^{r_j\times r_j}$$

arranged in decreasing dimensions $r_1 > r_2 > \dots > r_l$, associated to a given eigenvalue λ. We denote by u_i^j and v_i^j, $1 \leq j \leq l$ and $1 \leq i \leq q_j$, the canonical vectors which are right (column) and left (row) eigenvectors in $\mathbb{F}^n$ associated to the block $J_i^{r_j}$. For an integer $1 \leq s \leq l$ and $\tilde{B} \in \mathbb{F}^{n\times n}$, we define Φ_s by:

$$\Phi_s = \begin{bmatrix} v_1^1 \\ \hline \vdots \\ \hline v_{q_1}^1 \\ \hline \vdots \\ \hline v_1^s \\ \hline \vdots \\ \hline v_{q_s}^s \end{bmatrix} \cdot \tilde{B} \cdot \left[u_1^1|\dots|u_{q_1}^1|\dots|u_1^s|\dots|u_{q_s}^s\right] \in \mathbb{F}^{d_s\times d_s}, d_s = q_1 + \dots + q_s.$$

Let us note that Φ_s is a sub-matrix of Φ_{s+1} and of $\tilde{B}$. Now, for an integer k, $1 \le k \le n$, we define Ψ_k^* and Ψ_k by:

$$\begin{cases} \Psi_k^* = [\,],\ \Psi_k = \Phi_1 \quad \text{if } 1 \le k \le q_1, \\ \Psi_k^* = \Phi_{s-1} \text{ and } \Psi_k = \Phi_s \quad \text{if } q_1 + \ldots + q_{s-1} < k \le q_1 + \ldots + q_s, \\ \Psi_k^* = \Psi_k = \Phi_l \quad \text{if } q_1 + \ldots + q_l < k \le n. \end{cases}$$

As previously, Ψ_k^* is a sub-matrix of Ψ_k, they are both sub-matrices of $\tilde{B}$. Lidskii's conditions on particular minors of Ψ_k and thus of $\tilde{B}$ give relations between the characteristic polynomial of $\tilde{A}$ and the one of $\tilde{A} + \epsilon\tilde{B}$:

Lemma 2. *Let $\tilde{B}$ be of rank k. Let Σ_k be the sum of all determinants of $k \times k$ principal sub-matrices of Ψ_k that contain Ψ_k^*, taking the sum of the $k \times k$ principal minors of Ψ_k if $1 \le k \le q_1$ and $\Sigma_k = \Sigma_{q_1+\ldots+q_l}$ for $k > q_1 + \ldots + q_l$. Then for $k \le q_1 + \ldots + q_l$ we have*

$$\chi_\lambda(x,\epsilon) = \det\Big((x+\lambda) - (\tilde{A} + \epsilon\tilde{B})\Big) = (-1)^k c' \Sigma_k \epsilon^k x^{\rho(k)} + \beta(x,\epsilon) x^{\rho(k)+1} \quad (2)$$

where c' is nonzero in the algebraic closure $\overline{\mathbb{F}}$ of $\mathbb{F}$, $\rho(k)$ is the sum of the dimensions of the $q_1 + \ldots + q_l - k$ smallest (last) Jordan blocks of J and β is in $\overline{\mathbb{F}}[x,\epsilon]$. For $k > q_1 + \ldots + q_l$, taking $\rho(k) = 0$, we have

$$\chi_\lambda(x,\epsilon) = \det\Big((x+\lambda) - (\tilde{A} + \epsilon\tilde{B})\Big) = \Big((-1)^{\tilde{k}} c' \Sigma_{\tilde{k}} + \alpha(\epsilon)\epsilon\Big)\epsilon^{\tilde{k}} + x\beta(x,\epsilon) \quad (3)$$

where $\tilde{k} = q_1 + \ldots + q_l$, c' and β are as above and $\alpha \in \overline{\mathbb{F}}[\epsilon]$ has degree $k - \tilde{k} - 1$.

Proof. Since B has rank k there is no power of ϵ greater than k in the determinant, this gives the bound on the degree of α. We begin with the case $k \le q_1 + \ldots + q_l$. By Lemma 1, with $\tilde{B}$ of rank k, at least $q_1 + \ldots + q_l - k$ invariant factors of $\tilde{A}$ are factors of the characteristic polynomial of $\tilde{A} + \epsilon\tilde{B}$ thus $x^{\rho(k)}$ divides $\chi_\lambda(x,\epsilon)$. We adapt the arguments of Moro et al. [13, Theorem 3.1] to show that the terms in $x^{\rho(k)}\epsilon^i$, $i < k$, are zero and to compute the coefficient of ϵ^k.

For a given i we begin to compute the lowest possible power $x^{\rho(i)}$ involved with ϵ^i. The terms in $\chi_\lambda(x,\epsilon)$ are products of n factors that are elements of $(x+\lambda) - (\tilde{A} + \epsilon\tilde{B})$ in different rows and columns. To produce ϵ^i, $n - i$ factors free of ϵ must be taken in $x + \lambda - J$ or $x + \lambda - J'$. We look at the possible contributions for the term in $x^{\rho(i)}\epsilon^i$. The factors may be "x" or "-1" from the part corresponding to $-J$, or "x", "$\lambda - \lambda'$" or "-1" from the part corresponding to $-J'$ where λ' stands for any eigenvalue of J'. Define $c_{\lambda'} = \lambda - \lambda' \neq 0$. The "$x$" contributing to $x^{\rho(i)}$ must come from the $-J$ part because otherwise the corresponding terms with "$c_{\lambda'}$" would contradict the fact that $\rho(i)$ is minimum. Let j be the number of "ϵ" contributing to ϵ^i and taken in the $-J$ part. If β factors "-1" are taken in the $-J$ part from γ different Jordan blocks then also $\beta + \gamma$ factors "x" are excluded (from the same rows and columns) to choose the remaining factors in the $-J$ part. Since $r - j - \beta$ factors "x" from the $-J$ part will be in the term we have $(r-j-\beta)+(\beta+\gamma)=r-j+\gamma\leq r$ and $\gamma\leq j$. To produce the lowest possible power $\rho(i)$ of x, the term must be formed by taking all the "-1" from j blocks among the largest Jordan blocks of J and j must be as large as possible thus equal to i. The term must be completed by $r-i-\beta$ factors "x" from the $-J$ part and by all the factors "$c_{\lambda'}$" from the $-J'$ part to not introduce additional "ϵ".

We may conclude as done by Moro et al. [13]. The lowest power $\rho(i)$ is the sum of the dimensions of the $q_1+\ldots+q_l-i$ smallest blocks of J. Thus for $i<k$ there is no term in $x^{\rho(k)}\epsilon^i$. If we delete the rows and the columns corresponding to the factor "-1" and "x" from the $-J$ part for the term in $x^{\rho(k)}\epsilon^k$, the remaining elements are in $-\epsilon\Psi_k$. The different possible ways to choose i largest Jordan blocks in J give the different $k\times k$ principal minors to sum and give $(-1)^k\Sigma_k$. The constant c' is the product $\prod_{\lambda'}(\lambda-\lambda')$ taken over all the eigenvalues (with their multiplicities) of J'.

For $k>q_1+\ldots+q_l$, similar arguments show that the lowest power of ϵ in the constant term of χ_λ is $\tilde{k}=q_1+\ldots+q_l$ and give the corresponding coefficient. □

We now apply both Lemma 1 and Lemma 2 for a general $A+\epsilon_0 B$:

Theorem 1. *Let P be such that $P^{-1}AP$ is in Jordan normal form. Let $\lambda_1,\ldots,\lambda_\delta$ be the distinct eigenvalues of A. Let B be of rank k. We denote by $\Gamma^B_{k,\lambda_j}(\epsilon)$ the factor of $\epsilon^k x^{\rho(k)}$ in (2) or of $\epsilon^{\tilde{k}}x^0$ in (3) for $\lambda=\lambda_j$, $1\leq j\leq\delta$, and for $\tilde{B}=P^{-1}BP$. If the invariant factors of A are $s_1,\ldots,s_n$ and if*

$$\Delta_{A,B}(\epsilon_0)=\epsilon_0\prod_{j=1}^{\delta}\Gamma^B_{k,\lambda_j}(\epsilon_0)\neq 0 \tag{4}$$

for ϵ_0 in $\mathbb{F}^*$, then the invariant factors of $A+\epsilon_0 B$ are $t_1,\ldots,t_2,s_1t_{k+1},\ldots,s_{n-k}t_n$ where the t_i's are polynomials in $\mathbb{F}[x]$ relatively prime to the characteristic polynomial of A thus also to $s_1,\ldots,s_{n-k}$.

Proof. The existence of the t_i's is given by Lemma 1. We have to prove their relative primeness to the characteristic polynomial of A. This may be deduced locally at each eigenvalue λ_j, $1\leq j\leq\delta$. Indeed, if (4) holds then by Lemma 2, the valuation of $\chi_{\lambda_j}(x,\epsilon)$ in x is exactly $\rho(k)$. Since $\rho(k)$ is the valuation of $\prod_{i=1}^{n-k}s_i(x+\lambda_j)$, λ_j cannot be a zero of t_i, $1\leq i\leq n$, otherwise the valuation would be strictly greater. □

3 Rank-k Toeplitz Perturbations

For any given matrix A, we first prove that condition (4) is generically satisfied when B is a product of two Toeplitz matrices:

Lemma 3. *Let $\zeta_1, \ldots, \zeta_{n+k-1}$ and $\xi_1, \ldots, \xi_{n+k-1}$ be $2(n+k-1)$ distinct indeterminates over $\mathbb{F}$. Let $\mathcal{B} = \mathcal{U}\mathcal{V}$ be the product of the two Toeplitz matrices*

$$\mathcal{U} = \begin{bmatrix} \zeta_n & \zeta_{n+1} & \cdots & \zeta_{n+k-1} \\ \zeta_{n-1} & \zeta_n & & \zeta_{n+k-2} \\ \vdots & \ddots & \ddots & \vdots \\ \vdots & \ddots & \ddots & \vdots \\ \zeta_2 & \ddots & & \zeta_{k+1} \\ \zeta_1 & \zeta_2 & \cdots & \zeta_k \end{bmatrix}, \quad \mathcal{V} = \begin{bmatrix} \xi_k & \xi_{k+1} & \cdots\cdots & \xi_{n+k-1} \\ \xi_{k-1} & \xi_k & \ddots\ \ddots\ \ddots & \xi_{n+k-2} \\ \vdots & & \ddots\ \ddots & \vdots \\ \xi_1 & \xi_2 & \cdots\cdots & \xi_n \end{bmatrix} \tag{5}$$

of $\mathbb{F}[\zeta_1, \ldots, \zeta_{n+k}]^{n\times k}$ and of $\mathbb{F}[\xi_1, \ldots, \xi_{n+k}]^{k\times n}$. Then for any given $A \in \mathbb{F}^{n\times n}$, $\Delta_{A,\mathcal{B}}$ is a nonzero polynomial in $\overline{\mathbb{F}}[\zeta_1, \ldots, \xi_{n+k-1}, \epsilon]$. Its degree in ϵ is at most $n(k-1)+1$ and its coefficient $\Delta^{(1)}_{A,\mathcal{B}}$ of degree 1 in ϵ is nonzero in $\overline{\mathbb{F}}[\zeta_1, \ldots, \xi_{n+k-1}]$ of total degree at most $2n$.

Proof. We denote by $\Sigma^{\mathcal{B}}_{k,\lambda_j}$ the quantity Σ_k of Lemma 2 for $\lambda = \lambda_j$, $1 \le j \le \delta$, and $\tilde{B} = P^{-1}BP$. We prove that $\Delta^{(1)}_{A,\mathcal{B}}$ and thus $\Delta_{A,\mathcal{B}}$ is nonzero. It is equivalent to show that any $\Sigma^{\mathcal{B}}_{k,\lambda_j}$ is nonzero since – up to a nonzero constant – their product gives $\Delta^{(1)}_{A,\mathcal{B}}$. By definition, $\Sigma^{\mathcal{B}}_{k,\lambda_j}$ is a sum of $k \times k$ (or $\tilde{k} \times \tilde{k}$) minors built on different rows and columns of $P^{-1}\mathcal{B}P$. For a matrix A and sets I, J of indexes, we denote by $A_{I,J}$ the determinant of the sub-matrix of A built on the rows whose indexes are in I and columns whose of indexes in J. By the Binet-Cauchy formula – see Gantmacher [6, p9], the $k \times k$ minor of $P^{-1}\mathcal{B}P$ built on sets I and J is:

$$\mathcal{D} = \sum_{\substack{L = \{l_1, \ldots, l_k\} \\ 1 \le l_1 < \ldots < l_k \le n}} \sum_{\substack{M = \{m_1, \ldots, m_k\} \\ 1 \le m_1 < \ldots < m_k \le n}} P^{-1}_{I,L}\mathcal{B}_{L,M}P_{M,J}$$

Now, if the terms of the sum $\Sigma^{\mathcal{B}}_{k,\lambda_j}$ are built on $(I_1, J_1), \ldots (I_\iota, J_\iota)$ then

$$\Sigma^{\mathcal{B}}_{k,\lambda_j} = \sum_{i=1}^{\iota} \mathcal{D}_i = \sum_{i=1}^{\iota}\sum_{L}\sum_{M} P^{-1}_{I_\iota,L}\mathcal{B}_{L,M}P_{M,J_\iota} = \sum_{L}\sum_{M}\left(\sum_{i=1}^{\iota} P^{-1}_{I_\iota,L}P_{M,J_\iota}\right)\mathcal{B}_{L,M}.$$

Using ideas similar to those Kaltofen and Pan [9, Theorem 2], to prove that the latter sum is non zero we observe that there exist sets L_0 and M_0 with $\sum_i P^{-1}_{I_\iota,L_0}P_{M_0,J_\iota} \ne 0$ and that the $\mathcal{B}_{L,M}$'s are linearly independent over $\mathbb{F}$. Indeed, if $\sum_i P^{-1}_{I_\iota,L}P_{M,J_\iota} = 0$ for all L and M then $\sum_i (P^{(-1)}TP)_{I_\iota,J_\iota} = 0$ for any matrix T. But taking $T = PEP^{-1}$ with E a matrix whose only $k \times k$ nonzero minor is $E_{I_{\iota_0},J_{\iota_0}}$ for some i_0, $1 \le i_0 \le \iota$, $\sum_i (P^{(-1)}TP)_{I_\iota,J_\iota} = E_{I_{\iota_0},J_{\iota_0}} \ne 0$. Therefore, L_0 and M_0 exist as announced. For the independence of the $\mathcal{B}_{L,M}$'s, let us first notice that:

$$\mathcal{B}_{L,M} = (\mathcal{U}\mathcal{V})_{L,M} = \mathcal{U}_{L,\{1,\ldots,k\}}\mathcal{V}_{\{1,\ldots,k\},M}.$$

We may view these minors of $\mathcal{U}$ and $\mathcal{V}$ as polynomials in $\mathbb{F}[\zeta_1,\ldots,\zeta_{n+k-1}]$ and $\mathbb{F}[\xi_1,\ldots,\xi_{n+k-1}]$ with lexicographically ordered terms using the variable orders

$$\zeta_{n-k+1} > \ldots > \zeta_2 > \zeta_1, \quad \xi_{n-k+1} > \ldots > \xi_2 > \xi_1.$$

For given $L = \{l_1,\ldots,l_k\}$ and $M = \{m_1,\ldots,m_k\}$, the diagonal terms

$$\zeta_{n-l_1+1}\cdot\zeta_{n-l_2+2}\cdot\ldots\cdot\zeta_{n-l_k+k} \text{ and } \xi_{k+m_1-1}\cdot\xi_{k+m_2-2}\cdot\ldots\cdot\xi_{m_k}$$

are the lexicographically smallest terms in the minor expansions of $\mathcal{U}_{L,\{1,\ldots,k\}}$ and of $\mathcal{V}_{\{1,\ldots,k\},M}$ and uniquely correspond to L and M thus the polynomials $\mathcal{B}_{L,M}$ are linearly independent over $\mathbb{F}$. This establishes that $\Delta^{(1)}_{A,\mathcal{B}}$ and $\Delta_{A,\mathcal{B}}$ are nonzero polynomials. Moreover, each $\Sigma^{\mathcal{B}}_{k,\lambda_j}$ has degree at most two times the number of Jordan blocks with eigenvalue λ_j thus their product has total degree at most $2n$. By the definition of the $\Gamma^{\mathcal{B}}_{k,\lambda_j}$ in Theorem 1 and by the degree bound for α in (3), $\Gamma^{\mathcal{B}}_{k,\lambda_j}$ has degree in ϵ at most k minus the number d_{λ_j} of Jordan blocks with eigenvalue λ_j thus the product $\Delta_{A,\mathcal{B}}$ has degree in ϵ at most $1+\sum_j(k-d_{\lambda_j}) \le nk-n+1$. □

Lemma 3 indicates that the condition of application of Theorem 1 is generically satisfied for a special class of matrix B, for random such perturbations it follows that:

Theorem 2. *Let A be a matrix in $\mathbb{F}^{n\times n}$ with invariant factors $s_1,\ldots,s_n \in \mathbb{F}[x]$ and let S be a finite subset of $\mathbb{F}$. Let $U \in \mathbb{F}^{n\times k}$ and $V \in \mathbb{F}^{k\times n}$ be Toeplitz matrices – as in (5) – with entries chosen uniformly and independently from S. The minimum polynomial of $A+B = A+UV$ is $s_{n-k}t$, where $t \in \mathbb{F}[x]$ has degree less than $\sum_{n-k+1}^{n} \deg s_i$ and is relatively prime to the minimum polynomial of A, with probability at least $1-(nk+n+1)/|\mathsf{S}|$.*

Proof. By Lemma 3, $\Delta^{(1)}_{A,\mathcal{B}}$ has degree bounded by $2n$. Therefore, using the Schwartz-Zippel Lemma [18, Lemma 1], if the indeterminates $\zeta_1,\ldots,\zeta_{n-k+1}$, $\xi_1,\ldots,\xi_{n-k+1}$ are replaced by uniformly and independently chosen elements in S to give $B = UV \in \mathbb{F}^{n\times n}$ then $\Delta^{(1)}_{A,B}$ is nonzero with probability at least $1-2n/|\mathsf{S}|$. The same argument then gives that for a uniformly and independently chosen ϵ_0 in S, $\Delta_{A,B}(\epsilon_0)$ is nonzero with probability at least $1-(2n+nk-n+1)/|\mathsf{S}| \ge 1-(nk+n+1)/|\mathsf{S}|$. Theorem 1 implies that the minimum polynomial of $A+\epsilon_0 B$ is $s_{n-k}t$, where t is relatively prime to the characteristic polynomial of A. In case of successful random choices, ϵ_0 must be nonzero thus the probability bound is valid for $A+B$. Theorem 1 also implies that the remaining invariant factors are $t_1, t_2, \ldots,$ $s_1t_{k+1}, \ldots, s_{n-k-1}t_{n-1}$, their product has degree at least $\sum_1^{n-k-1} \deg s_i$ which, taking the degree of s_{n-k} into account, proves the degree bound for t. □

4 Computing One of the Invariant Factors

For convenience we denote by f_i the $(n-i+1)$-th invariant factor of A. If ν invariant factors are non-trivial (not equal to one) then the polynomials f_i are the characteristic polynomials of the ν blocks of the Frobenius form of A, $f_{i+1}|f_i$ and if $\deg f_i = d_i$ then $\sum_{i=1}^{\nu} d_i = n$. Theorem 2 reduces the computation of f_i to the computation of the minimum polynomial of A and of $A+B$ for B of rank $i-1$:

Function InvFact

> *Input*: an index $1 \le i \le n$.
> *Output*: the invariant factor f_i.
>
> Compute the minimum polynomial f_1 of A and if $i = 1$ return f_1
> Choose random Toeplitz matrices U and V of rank $k = i - 1$ as in (5)
> Compute g the minimum polynomial of $A + UV$
> Return $\gcd(f_1, g)$. □

Since the function InvFact consists in computing two minimum polynomials, many results are available for its cost analysis, especially:

Lemma 4. *Let* S *be a subset of* $\mathbb{F}$. *The degree* d *and the coefficients of the minimum polynomial of* $A \in \mathbb{F}^{n\times n}$ *may be probabilistically computed by* $O(d)$ *multiplications of* A *by vectors and* $O(dn)$ *arithmetic operations in* $\mathbb{F}$. *The algorithm returns correct answers with probability at least* $1 - 2d/|\mathsf{S}|$ *and requires* $O(n)$ *space in addition to that required for the matrix coefficient.*

Proof. The version of Lanczos's method given by Lambert [11, Algorithm 3.5.1], for two input vectors u_{curr} and v_{curr}, provides a factor of the minimum polynomial within the announced cost and space bounds. As proved by Kaltofen and Pan [9, Lemma 2], if u_{curr} and v_{curr} are randomly chosen with entries in S then the computed factor coincides with the minimum polynomial with probability more than $1 - 2d/|\mathsf{S}|$. □

This lemma together with the function InvFact gives the following.

Lemma 5. *Let* $A \in \mathbb{F}^{n\times n}$ *and* $\mathsf{S} \subset \mathbb{F}$. *We may probabilistically compute the* $(n-i+1)$*-th invariant factor* f_i *of* A *by computing* $O(d_1+d_2+\ldots+d_i)$ *multiplications of* A *by vectors and* $O((d_1+d_2+\ldots+d_i)n\log(n)\log\log(n))$ *arithmetic operations in* $\mathbb{F}$. *The algorithm returns* f_i *with probability at least* $1-(n^2+5n+1)/|\mathsf{S}|$.

Proof. By Theorem 2, with probability at least $1-(n^2+n+1)/|\mathsf{S}|$, the minimum polynomial g of $A+UV$ is f_it with $d = \deg g \le d_1 + \ldots + d_i$. In addition, t is relatively prime to f_1 and $f_i|f_1$ thus $\gcd(f_1, g) = \gcd(f_1, f_it) = f_i$. The cost of the matrix-vector multiplications in the computation of g dominates the overall cost. By Lemma 4, $O(d)$ multiplications of $A+UV$ by vectors are needed. For a vector u, $(A+UV)u = Au + (U(Vu))$, the cost of one multiplication is thus the cost of one multiplication by A plus two times the cost of multiplying a Toeplitz matrix by a vector. This latter operation reduces to polynomial multiplication – see Bini and Pan [1, p133] – and costs $O(n\log(n)\log\log(n))$ using the algorithm of Cantor and Kaltofen [2]. This proves the first assertion. By Theorem 2 and Lemma 4, the probability that the algorithm fails can be bounded by $((n^2+n+1)+2\times 2d)/|\mathsf{S}| \le (n^2+5n+1)/|\mathsf{S}|$. □

5 Computing the Frobenius Normal Form

The Frobenius normal form of A is computed as the set of the invariant factors of A. Each invariant factor may be computed by the function InvFact. To reduce the overall cost of the computation we remark that in case of repeated invariant factors it is sufficient to explicitly compute only one of them and their number. For A having exactly μ distinct invariant factors ϕ_i, our algorithm Frobenius will precisely compute few copies of each ϕ_i (a logarithmic number) together with the powers m_i at which they appear in the characteristic polynomial, so that:

$$\chi_A = \phi_1^{m_1}\phi_2^{m_2}\dots\phi_\mu^{m_\mu}, \quad \phi_{i+1}|\phi_i, \; 1 \le i < \mu. \tag{6}$$

We denote by $\phi_i^{(1)},\dots,\phi_i^{(m_i)}$ the m_i copies of each ϕ_i, $1 \le i \le \mu$. We develop a binary search using rank-k perturbations to detect that – for every ϕ_i – a rank-$k_i^{(1)}$ perturbation provides $\phi_i^{(1)}$ and that a rank-$k_i^{(m_i)}$ provides $\phi_i^{(m_i)} = \phi_i^{(1)}$. From where everything will be known since then the multiplicities m_i must be $k_i^{(m_i)} - k_i^{(1)} + 1$. To know that f_j is $\phi_i^{(1)}$ or $\phi_i^{(m_i)}$ for some i is equivalent to know that $f_{j-1} \neq f_j$ or $f_j \neq f_{j+1}$ ($f_0 = f_{n+1} = 0$). This leads to the function SearchThresholds which recursively computes, for two different invariant factors f_l and f_m, $l < m$, a decomposition of $f_l f_{l+1}\cdots f_m$

$$\begin{cases} f_l f_{l+1}\cdots f_m = \psi_1^{(1)}\dots\psi_1^{(l_1)}\psi_2^{(1)}\dots\psi_2^{(l_2)}\dots\psi_\kappa^{(1)}\dots\psi_\kappa^{(l_\kappa)}, \\ \psi_1^{(1)} = f_l, \;\; \psi_\kappa^{(l_\kappa)} = f_m, \\ \psi_i^{(1)} = \psi_i^{(2)} = \dots = \psi_i^{(l_i)} = \psi_i, \;\; 1 \le i \le \kappa, \\ \psi_{i+1}|\psi_i, \;\; 1 \le i < \kappa, \end{cases} \tag{7}$$

as a partial decomposition (6).

Function SearchThresholds

Input: $[(l, f_l), (m, f_m)]$, $l < m$.
Output: $[l_1, \psi_1, \dots, l_\kappa, \psi_\kappa]$ ** *The decomposition (7)* **

If $l = m - 1$
    if $f_l = f_m$ then Return $[2, f_l]$
(a)   if $f_l \neq f_m$ then Return $[1, f_l, 1, f_m]$
$k := \lceil (l+m)/2 \rceil$
(b) $f_k :=$InvFact(k)
(c) If $f_l \neq f_k$ then $[k_1, \psi_1, \dots, k_r, \psi_r] :=$ SearchThresholds$((l, f_l), (k, f_k))$
(d)   else $r := 1$, $[k_1, \psi_1] := [k - l + 1, f_l]$
(c') If $f_k \neq f_m$ then $[m_1, \tilde{\psi}_1, \dots, m_s, \tilde{\psi}_s] :=$ SearchThresholds$((k, f_k), (m, f_m))$
(d')   else $s := 1$, $[m_1, \tilde{\psi}_1] := [m - k + 1, f_m]$
** *Here,* $\psi_r = \tilde{\psi}_1$ **
Return $[k_1, \psi_1, \dots, k_r + m_1 - 1, \psi_r, m_2, \tilde{\psi}_2, \dots, m_s, \tilde{\psi}_s]$. □

The function basically finds all the threshold indexes j such that $f_j \neq f_{j+1}$, $l \le j < m$. They are found at step (a) – with $j = l$, $j + 1 = m$ – which is the deepest level of the recursion. The multiplicities l_i of the invariants factors at accumulated at each level when it is discovered – at steps (d) and (d') – that for two indexes – (l,k) or (k,m) – the corresponding invariant factors are identical. Let us bound the number of recursive calls to the function and thus the number of invariant factor computations by $\mathsf{InvFact}$ at step (b). Since k is an index bound of the created search intervals, the function may go through (b) with $f_l \neq f_k$ and $f_k \neq f_m$ at most $\kappa - 2$ times. In this latter situation, a new invariant factor f_k is found. Then, at its deeper levels, the binary search will refine at most two intervals with index bounds corresponding to invariant factors identical to f_k. The function will thus go $O(\sum_{i=1}^{\kappa} \log l_i)$ times through (b) and either through (c) or (c'), this may be bounded by $O(\kappa \log(m-l))$.

Since decompositions (6) and (7) coincide for $l = 1$ and $m = n$, the Frobenius normal form can be computed by:

Algorithm Frobenius

> *Input*: $A \in \mathbb{F}^{n \times n}$.
> *Output*: the Frobenius normal form of A (decomposition (6)).
> $f_1 := \mathsf{InvFact}(1)$
> $f_n := \mathsf{InvFact}(n)$
> $\mathsf{SearchThresholds}((1, f_1), (n, f_n))$. □

Before giving the final theorem we may run the algorithm on an example.

Example 1. Let us compute the Frobenius normal form of

$$A = \begin{bmatrix} 2 & 0 & 3 & 0 & 3 & 0 \\ 0 & 2 & -1 & 0 & -1 & 0 \\ 0 & 0 & 3 & 0 & 1 & 0 \\ 0 & 0 & 1 & 2 & 1 & 0 \\ 0 & 0 & -2 & 0 & 0 & 0 \\ 0 & 0 & 0 & 0 & 0 & 2 \end{bmatrix} \in \mathbb{Q}^{6 \times 6}.$$

The first call to $\mathsf{InvFact}$ gives the minimum polynomial $f_1 = x^2 - 3x + 2$ of A. For Toeplitz matrices

$$U = \begin{bmatrix} 1 & 0 & 3 & 1 & -1 \\ 2 & 1 & 0 & 3 & 1 \\ 4 & 2 & 1 & 0 & 3 \\ -2 & 4 & 2 & 1 & 0 \\ 3 & -2 & 4 & 2 & 1 \\ 5 & 3 & -2 & 4 & 2 \end{bmatrix}, V = \begin{bmatrix} 3 & 1 & -4 & 2 & 1 & 2 \\ 0 & 3 & 1 & -4 & 2 & 1 \\ -1 & 0 & 3 & 1 & -4 & 2 \\ 3 & -1 & 0 & 3 & 1 & -4 \\ 1 & 3 & -1 & 0 & 3 & 1 \end{bmatrix}$$

of rank $n-1=5$, the second call to InvFact computes the minimum polynomial g of $A+UV$:

$$g(x)=x^6+28\,x^5+612\,x^4-16993\,x^3-499329\,x^2+1938360\,x-434012.$$

Since g is relatively prime to f_1, $f_n=f_6=\gcd(f_1,g)=1$. The call to the recursive function SearchThresholds then leads to the execution:

· SearchThresholds($(1,x^2-3x+2),(6,1)$)
  $k=4,\ f_4=x-2$
  ·· SearchThresholds($(1,x^2-3x+2),(4,x-2)$)
    (*) $k=3,\ f_3=x-2$
    ··· SearchThresholds($(1,x^2-3x+2),(3,x-2)$)
      $k=2,\ f_2=x-2$
      ···· SearchThresholds($(1,x^2-3x+2),(2,x-2)$)
        $f_1\neq f_2$
        (a) Return $[1,x^2-3x+2,1,x-2]$
      ···· (d') $f_2=f_3$, found $[2,x-2]$
      Return $[1,x^2-3x+2,2,x-2]$
    ··· (d') $f_3=f_4$, found $[2,x-2]$
    Return $[1,x^2-3x+2,3,x-2]$
  ·· SearchThresholds($(4,x-2),(6,1)$)
    $k=5,\ f_5=x-2$
    ··· (d) $f_4=f_5$, found $[2,x-2]$
    ··· SearchThresholds($(5,x-2),(6,1)$)
      $f_5\neq f_6$
      (a) Return $[1,x-2,1,1]$
    Return $[2,x-2,1,1]$
  Return $[1,x^2-3x+2,4,x-2,1,1]$.

The numbers of dots indicate the levels of the recursion and the labels (a), (d) or (d') – corresponding to the definition of SearchThresholds – are given for the deepest levels. The (*) indicates one of the steps (b) and thus one of the internal calls to InvFact. Here, for $k=3$, InvFact(3) generates two Toeplitz matrices of rank $k-1=2$, for instance:

$$U=\begin{bmatrix}2&3\\1&2\\0&1\\3&0\\-1&3\\-1&-1\end{bmatrix},V=\begin{bmatrix}1&6&-1&2&1&3\\4&1&6&-1&2&1\end{bmatrix}.$$

The minimum polynomial of $A+UV$ is

$$g=x^4-42\,x^3+349\,x^2+80\,x-1236,$$

this implies that $f_3 = \gcd(f_1, g) = x - 2$.

The resulting list $[1, x^2 - 3x + 2, 4, x - 2, 1, 1]$ provides the invariant factors of A:

$$f_1 = x^2 - 3x + 2, f_2 = f_3 = f_4 = f_5 = x - 2, f_6 = 1$$

according to identities (7). □

Theorem 3. *Let $A \in \mathbb{F}^{n \times n}$ and $\mathsf{S} \subset \mathbb{F}$. If A has μ distinct invariant factors then we may probabilistically compute the Frobenius normal form of A by computing $O(\mu n \log(n))$ multiplications of A by vectors and $O\left(\mu n^2 \log^2(n) \log\log(n)\right)$ arithmetic operations in $\mathbb{F}$. Since μ is always less than $3\sqrt{n}/2$, this gives $O(n^{3/2}\log(n))$ multiplications of A by vectors plus $O\left(n^{5/2}\log^2(n)\log\log(n)\right)$ operations in the worst case. The algorithm returns the correct answer with probability at least $1 - O(n^{5/2}\log n)/|\mathsf{S}|$ and requires $O(n)$ storage space in addition to that necessary to store the input matrix.*

Proof. Taking into account the invariant factors equal to one, the number μ of distinct invariant factors must satisfy:

$$(\mu - 1) + (\mu - 2) + \ldots + 1 \leq \deg(\phi_1) + \deg(\phi_2) + \ldots + \deg(\phi_{\mu-1}) \leq n,$$

thus μ is less than $3\sqrt{n}/2$. The execution of Frobenius will generate $O(\sqrt{n}\log(n))$ calls to InvFact. Lemma 5 then imply the cost and storage assertion. In the same way, the failure probability is in $O(n^2 \times \sqrt{n}\log n)$. □

Our strategy, based on rank-k perturbations combined to binary searches, may be applied in other situations. A paper of Eberly, Giesbrecht and the author [5] demonstrates the technique over the integers.

Acknowledgements. Grateful thanks to Erich Kaltofen for his questions.

References

1. D. Bini and V. Pan. *Polynomial and matrix computations.* Birkhäuser, 1994.
2. D.G. Cantor and E. Kaltofen. On fast multiplication of polynomials over arbitrary algebras. *Acta Informatica*, 28(7):693–701, 1991.
3. W. Eberly. Asymptotically Efficient Algorithms for the Frobenius Form. Technical report, Department of Computer Science, University of Calgary, Canada, TR2000/649/01, may 2000.
4. W. Eberly. Black Box Frobenius Decompositions over Small Fields. In *Proc. International Symposium on Symbolic and Algebraic Computation, St Andrews, Scotland*, ACM Press, august 2000.
5. W. Eberly, M. Giesbrecht, and G. Villard. Computing the determinant and Smith form of an integer matrix. In *Proc. 41st Annual IEEE Symposium on Foundations of Computer Science, Redondo Beach, CA*, 2000.
6. F.R. Gantmacher. *Théorie des matrices.* Dunod, Paris, France, 1966.
7. M. Giesbrecht. Nearly optimal algorithms for canonical matrix forms. *SIAM Journal on Computing*, 24(5):948–969, 1995.

8. E. Kaltofen. Challenges of symbolic computation my favorite open problems, 1998. Manuscript submitted for publication, North Carolina State Univ., Dept. Mathematics.
9. E. Kaltofen and V. Pan. Processor efficient parallel solution of linear systems over an abstract field. In *Proc. 3rd Annual ACM Symposium on Parallel Algorithms and Architecture*. ACM-Press, 1991.
10. W. Keller-Gehrig. Fast algorithms for the characteristic polynomial. *Theoretical Computer Science*, 36:309–317, 1985.
11. R. Lambert. *Computational aspects of discrete logarithms*. PhD thesis, University of Waterloo, Ontario, Canada, 1996.
12. V.D. Lidskii. Perturbation theory of non-conjugate operators. *U.S.S.R. Comput. Maths. Math. Phys.*, 1:73–85, 1965.
13. J. Moro, J. Burke, and M. Overton. On the Lidskii-Vishik-Lyusternik perturbation theory for eigenvalues of matrices with arbitrary Jordan structure. *SIAM J. Matrix Anal. Appl.*, 18:793–817, 1997.
14. M. Newman. *Integral Matrices*. Academic Press, 1972.
15. V.Y. Pan and Z.Q. Chen. The complexity of the matrix eigenproblem. In *The 21rst Annual ACM Symposium on Theory of Computing, Atlanta, Georgia*, pages 507–516. ACM Press, May 1999.
16. V.Y. Pan, Z.Q. Chen, and A. Zheng. The complexity of the algebraic eigenproblem. MSRI Preprint 1998-071, Mathematical Sciences Research Institute, California, USA, dec. 1998.
17. J.H. Reif. Efficient parallel computation of the characteristic polynomial of a sparse, separable matrix. In *The 36th Annual IEEE Conf. Found. Comp. Sc., Milwaukee, WI*, pages 123–132, October 1995.
18. J.T. Schwartz. Fast probabilistic algorithms for verification of polynomial identities. *J. ACM*, 27:701–717, 1980.
19. A. Storjohann. An $O(n^3)$ algorithm for Frobenius normal form. In *International Symposium on Symbolic and Algebraic Computation, Rostock, Germany*, pages 101–104. ACM Press, August 1998.
20. A. Storjohann. PhD Thesis, Institut für Wissenschaftliches Rechnen, ETH-Zentrum Zürich, Switzerland, to appear, 2000.
21. A. Storjohann and G. Villard. Algorithms for similarity transforms. In *The Seventh Rhine Workshop on Computer Algebra, Bregenz, Austria*, March 2000.
22. D. Wiedemann. Solving sparse linear equations over finite fields. *IEEE Transf. Inform. Theory*, IT-32:54–62, 1986.

Lessons Learned from Using CORBA for Components in Scientific Computing

Andreas Weber[1*], Gabor Simon[2], Wolfgang Küchlin[2], and Jörg Hoss[2]

[1] Fraunhofer-Institut für Graphische Datenverarbeitung
Rundeturmstr. 6, Darmstadt, Germany
E-mail: aweber@igd.fhg.de
[2] Wilhelm-Schickard-Institut für Informatik
Universität Tübingen, Tübingen, Germany
E-mail: {simon,kuechlin}@informatik.uni-tuebingen.de
WWW: http://www-sr.informatik.uni-tuebingen.de

Abstract. The use of component architectures to solve the problem of reuse and interoperability in scientific computing has been investigated by various research groups during the last years. Moreover, architectures for Internet accessible mathematical services have been proposed. In this paper we give a brief abstract requirements analysis with respect to these problems and show that there is an existing technology that solves most of the requirements. The *Common Object Request Broker Architecture* (CORBA) provides a component framework that can be used for Internet accessible mathematical services and also for the efficient reuse of medium grained size functionality. We give some examples on the use of CORBA with respect to both applications. We provide measurement data which show that components of a granularity of less than 100 ms can be reused with an acceptibly small overhead.

1 Introduction

In the development of a computer algebra algorithm one would frequently like to use (perhaps only temporarily) existing algorithms implemented in different, possibly incompatible systems. This means, that one needs possibilities which allow the cooperation and/or communication between programs, applications, objects and environments, despite differences in the implementation language, execution environment and model abstraction. All these facilities are meant by the notion of interoperability. The basic aspects of interoperability are:

(i) the control aspect, caring about the coordination of the inter-operating programs,
(ii) the data aspect, establishing type compatibility for the shared data objects, and

* Partially supported by *Deutsche Forschungsgemeinschaft* under grants Ku 966/4-1 and Ku 966/6-1. Most of the presented research was done as member of the Symbolic Computation Group at the University of Tübingen.

(iii) choosing transport services and low-level communication protocols achieving an efficient transmission of the data.

There are several groups working on such problems in computer algebra. Most of the research activities in this field concentrate on the data aspect (or type model), looking for a single universal representation, either on a character or a byte stream basis; many of these activities have influenced the OpenMath standard [17], which e. g. offers both representations.

Recently, also the control aspect of the communication has received some attention, for example Internet accessible mathematical services [26,27] and other general aspects such as parallization, callback mechanisms, exception handling, load balancing, etc.

Although the transport aspect is not really computer algebra specific, there has been some work in this direction [1].

The state of the art in general software development suggests the use of *software components*. A vision of computer algebra using components is shown in [4]. The benefits of having existing symbolic computation systems in some form of components have also been described in [10,27].

In those papers only some of the possibilities have been treated: The use of CORBA for generating components that can be linked together [10], the use of Java for algebraic components [4], or using Java RMI to generate Internet accessible mathematical services [27].

In this paper we concentrate on the aspect of interoperability between computer algebra components. We try to identify the main requirements for them and we will propose the use of CORBA as a solution, cf. Sec. 3. Some preliminaries will be given in Sec. 2: CORBA and the most widely used exchange formats for mathematical information, MathML and OpenMath. Some examples, in which we used various features of CORBA to cope with the requirements of components of symbolic computation are described in Sec. 4. In this section also some time measurements are given to discuss the problem of infrastructure overhead for medium grained components. We conclude by an overall discussion in Sec. 5.

2 Preliminaries

2.1 CORBA

The *Common Object Request Broker Architecture* (CORBA) [15] is a standard for a system infrastructure for software components. It has the support of almost all vendors of workstations and major software developers (such as the vendors of the database systems). Some commercial object brokers that adhere to the CORBA standard (in its version 2.3) are Iona's ORBIX, ObjectBroker of Digital Equipment Corporation, Visibroker by Inprise, Component Broker by IBM, etc. Several object brokers that can be obtained free of charge are also available, e. g. ILU[1] [11], MICO [21], omniORB [3], or ORBacus [16].

[1] ILU is an acronym for "inter language unification system" and is developed at XEROX PARC.

We will only sketch some major features of CORBA. For more information we refer to [19,18,22,14].

The interfaces between the different software components are described in a standardized *Interface Definition Language (IDL)*. The object request brokers contain tools that allow a mapping of interfaces in IDL format to various languages, such as C, C++, Java, or Smalltalk. If a language mapping is supported by an object-broker system, it will contain an IDL compiler that generates client stubs and server skeletons in that language for the client side or server side out of the IDL definition of the interface. Thus a CORBA object-broker provides also an *inter language unification* for those languages for which it has a language mapping, because the server code and the client code can be written in different languages.

The communication methods and the method of invocation of a client function is provided by the object-broker and hidden in it. The *referential transparency* given by this object-broker architecture allows to use more efficient communication methods between client side and server side with very little programming effort in special situations, such as where client and server run on the same machine or where server code and client code can be linked together into a common address space.

In several of the examples described in this paper we used the CORBA-like object broker ILU that adheres in many respects to the CORBA standard, but has a few omissions and several extensions, such as network garbage collection. ILU has C, C++, Java, Python and Common Lisp language mappings. Among others, it supports the IIOP protocol too. It can be used free of charge.

IIOP (Internet Inter-ORB Protocol) is an *object-oriented* protocol that makes it possible for distributed programs written in different programming languages to communicate over the Internet. Any CORBA 2.0 ORB must support the IIOP protocol. IIOP passes requests or receives replies through the Internet's transport layer using the Transmission Control Protocol TCP. A Common Data Representation (CDR) provides a way to encode and decode data so that it can be exchanged in a standard way [18].

Java RMI (Remote Method Invocation) [25] provides a communication infrastructure allowing applications or Java applets to establish a dialogue. Originally, RMI was designed on the top of RRL (Remote Reference Layer) to interface with transport layers. Recently, this layer has been more and more replaced by RMI over IIOP (JDK 1.2), so as to provide interoperability with CORBA.

2.2 OpenMath and MathML

MathML is the W3C proposal for a markup language for mathematical content in Web pages [28]. As other extensions to HTML, it is based on XML. MathML comes in two fashions: *presentation markup* and *content markup*. The presentation markup is primarily concerned with describing the rendering ("presentation") of mathematical information on Web pages. As it is not suitable for the exchange of mathematical information (other than via a human being, who "understands" the displayed information) we will not consider this variant of MathML in the following.

The data coded in content markup is suitable both for displaying mathematical information and for the exchange of it. Mathematical information is stored in form of an abstract syntax tree (using XML); this abstract syntax tree can be rendered according to the specification given in a style sheet. Moreover, the representation of the content in form of an abstract syntax tree is suitable for its interpretation by algebra programs, as they store the mathematical expressions in similar syntax trees. In general these could be generated from the MathML content markup automatically. The newest versions of various algebra programs such as Mathematica 4.0 support MathML.

Some aspects of OpenMath [17] are now closely related to the content markup of MathML. Although the origins of OpenMath are older than those of MathML it can now be seen as an extension of the content markup mechanism of MathML. OpenMath contains the content markup of MathML as a subset. It uses so called *content dictionaries* to give the abstract syntax trees a well-understood meaning, a mechanism that makes the intuitive meaning that is implicitly used in the understanding of MathML content markup expression trees more precise. In addition, these content dictionaries allow the extension of the markup to other mathematical areas: new syntactic symbols can be introduced and specified by the definition of new content dictionaries.

In addition to the representation of mathematical information as XML syntax trees, i. e. in ASCII form, OpenMath defines a binary encoding of the syntax trees, too. In this binary representation the encoding of arbitrary precision integers is binary in contrast to the decimal ASCII representation used in the XML representation. As will be shown in Sec. 4.2 such a binary encoding can be crucial for the performance of components of medium grained that interoperate.

Although the various exchange formats have often been developed independently, some basic ideas occur in one form or the other in several of these. For an excellent discussion of the relationship of the concepts of the MathBus to the ones of MathML and OpenMath we refer to [29].

3 Symbolic Computation Components

It is now widely accepted that the reuse of existing symbolic computation systems in some form of components is highly desirable [10,4,27]. An object oriented design approach has some distinct advantages when dealing with complexity as usually experienced in symbolic computations. Such a design has to provide for explicit separation of the concepts of interface, communication and implementation. Even more if the new infrastructure must support existing systems, providing transparency of location, implementation language, memory management etc.

Let us summarize the most important requirements for software components performing operations on distributed collections of resources. There has to be a strict distinction between

- the *resource interface*, the specification of operations that a component supports,
- the *communication protocol* used to carry these operations from point to point,
- and the *implementation* that carries out the computation associated with the operations.

When existing symbolic computation systems have to work together (in the form of components or in some other form) the following issues have to be considered:

- *Location transparency.* It allows a designer to perform operations on components without specific regard to whether the component is local to the program, in a different process, or on a distant machine. The infrastructure deals with all the low-level details of establishing network connection, marshalling arguments, un-marshalling return values, etc. This allows the programmer to concentrate on the real problem rather than the low-level details. Location transparency allows to build systems using different subsystems for different parts. In this setup each part can run its own arithmetic or memory management.
- *Language transparency.* It allows a system to be constructed whose parts may be implemented in different programming languages, choosing the most appropriate language for various parts of the overall task.
- *Efficiency.* Even though location transparency makes it easy to ignore details of distribution and networking, this does not mean that one can pretend that distribution and networking do not exist. Calls to remote systems take roughly an order of magnitude longer time than calls that occur between processes on a single system, and these in turn take roughly an order of magnitude longer than calls that occur within a process. The cost of sending remote messages is mostly determined by two factors: call latency and marshalling time. Because call latency depends on a large number of parameters, such as the underlining network technology, CPU speed, operating system, TCP/IP implementation, CORBA implementation etc., the only way to make reliable comparisons between different setups is the use of benchmarks. Marshalling performance depends on the type of transmitted data. Simple types marshal the fastest. On the other hand, marshalling highly structured data, such as nested user-defined types or object references is usually much slower. Similar to call latency, the marshalling rate can be highly different using different networks, hardware-architectures, programming languages, compilers, and CORBA implementations.

Unfortunately, the goals of scalability, performance and maintainability are in conflict. Better performance often implies coding or design technique that results in source code that compromises maintainability. Increased scalability—the ability to grow with the number of users or clients, the amount of data, and the required functionality—often implies reduction in performance. Of course, the optimal compromises between these conflicting goals will be different for different applications. In addition, rather small changes in an application might cause considerable shifts in these optimal settings.

In the following section we will discuss some examples that show some of the possibilities CORBA offers to master this situation.

4 Examples

4.1 Connecting LiDIA and Saclib

Our first experiments with CORBA go back to [24,23]. These papers summarize experiences with a prototype implementation for a communication interface between SAC-2 (on ALDES) [13,5] and LiDIA [12], using ILU. The two systems were running on different architectures (Intel PC and SUN Sparc workstations) under different operating systems (WinNT and UNIX) using their own arbitrary precision arithmetic and memory management. SAC-2 is written in ALDES (implementated on top of C) and LiDIA is implemented in C++.

These experiments proved that the CORBA infrastructure simplifies programming distributed applications, providing extensibility and flexibility. The performance evaluation of the implementations showed that using CORBA has a potential for applicability even over network connections, presuming a reasonable modularisation.

4.2 Medium Sized Granularity in a Parallel Environment

The following example involves the reuse of medium grained algebraic software components, which run in a parallel environment involving different memory management mechanisms. In this example we will characterize some of the typical pitfalls in these context, and we will describe the way how we used CORBA to solve the conflicts.

The CORBA implementation of ILU was used to connect an extension to parameterized Gröbner bases of a high-performance Gröbner solver [2] and the SACLIB [5]. The high-performance Gröbner solver was built over a computer algebra nucleus called CANNES, which gives a very efficient implementation of arbitrary precision integers on the basis of GNU/MP and of lists, including an automatic garbage collection. It uses the parallel version PARCAN of CANNES and shows good speed-ups on shared-memory multiprocessors. Even super-linear speed-ups due to strategy effects have been observed [2] in an older version. It currently runs on multi-processor SUN SPARC servers.

The extension of the Gröbner solver to the parameterized case required several functions that were not available in CANNES. Instead of reimplementing these functions in CANNES it was decided to reuse their implementation in SACLIB and to use ILU to objectify their SACLIB implementation.

The following SACLIB functions had to be called remotely, because their counterparts are not implemented in CANNES.

- IPQ: Integral Polynomial Quotient.

- IPGCD: Integral Polynomial Greatest Common Divisor.
- IPGCDC: Integral Polynomial Greatest Common Divisor and Cofactors.

The IDL definition of our CORBA interface to the remote functions is given in Fig. 1.

The data types and interface definitions in the IDL file could be kept quite close to the used data types of SACLIB and also of CANNES. Thus the conversion functions from the internal representations in CANNES or SACLIB could be implemented quite efficiently.

On the basis of ILU a good prototype for the system for parameterized Gröbner basis computation was obtained relatively quickly, which uses the parallel PARCAN system for most computations and the SACLIB functions only for the few additional function calls that are not available in PARCAN. A description of the entire system can be found in [9].

Since the results of the prototype were promising and many possible optimizations were at hand, it was decided to investigate these optimizations. Since the setting that was found will apply to many other computer algebra applications, we will not only give the results of the final optimizations, but also of other settings. Each of the settings might arise in the specific context of another application. Thus we think that providing a broader range of empirical results will give others the opportunity to estimate whether a reuse of library code via CORBA is feasible for their own application.

Setup of the Experiments Experiments with different conversion functions, e. g. with one that used the already available decimal conversions from the integer coefficients of the polynomials in CANNES and SACLIB were performed. Since SACLIB uses 2^{29} as basis for its long integers and CANNES uses the 2^{32} base of the GNU MP library, a more efficient conversion than the decimal one is possible and also these conversion functions were implemented.

The referential transparency given by a CORBA like object-broker such as ILU allows to switch with very little programming effort between different modes of connecting the software components:

- Separate processes running on different machines.
- Separate processes running on the same machine.
- The components are linked into a common address space.

Although the CANNES based Gröbner solver and SACLIB have separate garbage collectors, it was nevertheless possible to use all of these possibilities, as long as the multi-threaded version of the Gröbner system was not used.

Two examples of the PoSSo test suite [20] have been used, which have quite different characteristics for the timing tests. For both examples the parameterized Gröbner basis with respect to the lexicographic ordering was computed.

- Example 1: The so called "Bini example": two equations in two variables and six parameters.

```
module CA {
  typedef sequence<unsigned long> Number;
  struct BigInteger {
    boolean sign;
    Number  digits;
  };

  typedef sequence<long> PowerProduct;

  struct Monomial {
    BigInteger   coefficient;
    PowerProduct pp;
  };

  typedef sequence<Monomial> Polynomial;

  interface SAClib {
    void Init();
    void End();

    // integral polynomial gcd and cofactors
    void IPolyGCDandCofac(in  Polynomial poly1,
                          in  Polynomial poly2,
                          out Polynomial gcd,
                          out Polynomial cofactor1,
                          out Polynomial cofactor2);

    // integral polynomial gcd
    Polynomial IPolyGCD(in  Polynomial poly1,
                        in  Polynomial poly2);

    // integral polynomial quotient
    Polynomial IPolyQuotient(in  Polynomial poly1,
                             in  Polynomial poly2);

  };

  interface Factory {
    SAClib CreateSAClib();
    void   StopCAServer();

  };
};
```

Fig. 1. Interface definition in IDL format

– Example 2: The so called "Liu example": four equations in four variables and one parameter.

The following settings with respect to ILU and conversion functions have been used.

#1: separate address spaces (same machine), conversion via decimal format
#2: common address space, conversion via decimal format
#3: common address space, binary integer conversions

In these examples a single threaded-version of the libraries was used. In recent work, we have extended the experiments to the multi-threaded versions of the Gröbner package. Setting #1 also worked in this case, as did the setting in which binary integer conversions were used in separate address spaces. Settings #2 and #3 did not work, as had to be expected, because of the effects of the garbage collector of the parallel CANNES system.

Because of the referential transparency given by a CORBA like object-broker such as ILU only minimal changes in the code were necessary in order to switch between the different settings.

These experiences show one of the major advantages of the referential transparency given by the CORBA approach: switching back to a setting in which the components run in separate address spaces can be done with almost no effort, if an attempt to use them in a common address space fails. Such an option is not present in the traditional library approach to software reuse!

Table 1. Statistics over all remote methods

overall averages	Example 1 (Liu) (times in ms)			Example 2 (Bini) (times in ms)		
	#1	#2	#3	#1	#2	#3
number of remote calls	3769	3769	3769	122	122	122
av. total calling time	723	465	398	259413	260638	255015
av. conv. time per call (client)	34	34	4	19	19	7
av. conv. time per call (server)	38	39	7	13	13	11
av. comm. time per call	262	0.03	0.03	73	0.03	0.03
av. exec. time of SACLIB function	389	392	387	259308	260606	254998
overhead due to comm. and conv. % of total calling time	46	16	3	0.04	0.01	0.01

#1: separate address spaces (same machine), conversion via decimal format
#2: common address space, conversion via decimal format
#3: common address space, binary integer conversions

Empirical Results Statistics over all remote functions are given in Table 1. They are dominated by the measurements for IPGCDC and IPGCD. The granularity of these functions is coarse enough so that also the use of two address

Table 2. Integral polynomial quotient statistics

IPQ statistics	Example 1 (Liu) (times in ms)			Example 2 (Bini) (times in ms)		
	#1	#2	#3	#1	#2	#3
number of remote calls	712	712	712	22	22	22
av. total calling time	253	91	54	154	77	52
av. conv. time per call (client)	26	26	3	28	28	11
av. conv. time per call (server)	18	17	3	14	23	13
av. comm. time per call	161	0.03	0.03	85	0.04	0.04
av. execution time of IPQ	48	48	48	27	27	27
overhead due to comm. and conv. % of total calling time	81	47	11	82	66	46

#1: separate address spaces (same machine), conversion via decimal format
#2: common address space, conversion via decimal format
#3: common address space, binary integer conversions

spaces for client and server and the use of an intermediate decimal format to convert the long integers gives small communication and conversion times with respect to the costs of the function themselves.

The remote function with the smallest granularity is IPQ. The results in Table 2 show that in its case the use of a common address space and of a direct conversion of the long integers from a base 2^{32} format in CANNES to a base 2^{29} in SACLIB—without an intermediate conversion into a decimal format—reduces the communication and conversion costs compared to the costs of the function itself significantly.

The example of IPQ shows that CORBA can be used in principle to reuse code down to a granularity of less than 100 ms!

4.3 Internet Accessible Services—Java RMI and CORBA

In [7,8,27] we have developed an infrastructure for Internet accessible mathematical services using a standard exchange format (the MathBus) and the Java RMI mechanism. We wrapped existing coarse grained mathematical services—Gröbner basis programs, programs for quantifier elimination on real closed fields—as Java software components that can be accessed over the Internet from other programs via a small Java proxy.

These examples show one of the benefits of using standards: They were developed using the original RMI mechanism that uses a transport protocol of its own. However, these services can now be accessed from other CORBA based components via IIOP without any programming effort: instead of generating the client stubs and server skeletons for the RMI protocol using the `rmic` compiler one simply has to generate the corresponding stubs and skeletons for RMI over IIOP using the tools in JDK 1.2!

5 Discussion

We have shown that using ideas present in CORBA complex systems can be generated with moderate programming effort. These complex systems are not only useful for fast prototyping, but can also be used for the final system, because the communication overhead induced by them is often negligible.

In the case of Internet accessible services using a standard interface and not using a protocol of your own gives a major advantage. New protocols can be used if these become available without causing new programming effort for the provider of the scientific components, cf. Sec. 4.3.

As the measurements given in [24] indicate the use of a high-level protocol such as IIOP instead of a direct socket communication might reduce the communication performance up to an order of magnitude. However, as a communication over the Internet is always relatively slow, the absolute performance is not of central importance in this case. Having access to an otherwise unavailable service is the most important aspect by an access over the Internet. There are many important scientific services which are very coarse grained—Gröbner basis computations or quantifier-elimination procedures to name just two of them—which can be accessed over the Internet with comparable little communication costs, cf. Sec. 4.3 and [27,7]. Moreover, these services themselves can run on a parallel architecture resulting in a lower overall computation time than having all of them installed on the local desktop workstation or PC [27].

The object oriented encapsulation of the software systems and their strict access via interfaces guarantee a good software maintenance and also easy expandability: An existing service can be substituted by a new one without affecting the overall system, as long as the contract given by the interface is fulfilled. Extending the interface by new functionality is also possible without breaking the contract with existing clients.

The data types specified in the interfaces can be one of the existing exchange formats, e. g. OpenMath, cf. [10]. The OpenMath standard [17] defines an XML encoding of its objects and also a binary encoding.

However, in addition to these standard object encodings special IDL definitions can be used if required by efficiency, such as the ones used in Sec. 4.2.

To exchange data between cooperating computer applications, one should use an efficient way of defining and storing data. Traditionally these definitions of data formats are realized with an interface definition language. In contrast to markup languages, which are used for long-term storage of mostly human readable data, an interface definition language defines smaller standard units of transient data in machine representation, which are exchanged between the components of a distributed application, triggered by some events.

Using IDL and CORBA allows you to set up a persistent connection, send requests back and forth and maintains the transaction state, supporting long-term association of clients with state at the server end. CORBA also provides user authentication, resource control, and error handling. Thus many services that are important for large scale applications come for free using CORBA, whereas it would be a major effort to provide these in a specialized environment for Internet accessible mathematical services.

For fine grained to medium grained applications that do not have harmful side-effects, a connection in the same address space is possible and is an alternative to reimplementing the desired functionality in the client application. We have shown that even two systems that have their separate garbage collectors can be connected in many circumstances by the ILU system in this way and that the communication overhead is negligible even for functions of such a fine granularity like integral polynomial quotients.

On the other hand harmful interactions of the components in a common address space result only in a loss of performance and not in a failure of the entire attempt of software reuse. The programmer just has to switch back to use separate address spaces for the components, a task that can be accomplished almost without any effort. Since problems because of interacting garbage collectors or multi-threaded software are sometimes hard to predict, we assume that this option was not only valuable for us but could also be useful for many other potential applications in the context of high-performance scientific computing.

Thus for any setting CORBA can compete as well with a traditional library approach on the one hand and a traditional client-server architecture on the other hand. Having one single architecture for all settings in CORBA, a programmer is given additional degrees of freedom to the tasks of reusing software in new systems or substituting parts of existing systems by new and "better" components.

There is a good interoperability between Java based services and CORBA. Thus from a user's point of view the Java wrapped services described in Sec. 4.3 look like parts of an algebra system written in Java. So an integration into a Java based algebra system [4] is possible, if conversion between the data types is supported. So these developments for symbolic components point into the same direction as ours and even their interoperability can be achieved.

References

1. ABBOT, J. PossoXDR, 1995. Available at `www.rrz.uni-koeln.de/themen/Computeralgebra/OpenMath/related.html`.
2. AMRHEIN, B., GLOOR, O., AND KÜCHLIN, W. A case study of multi-threaded Gröbner basis completion. In *Proceedings of the 1996 International Symposium on Symbolic and Algebraic Computation (ISSAC '96)* (Zürich, July 1996), Y. N. Lakshman, Ed., Association for Computing Machinery.
3. AT&T LABORATORIES CAMBRIDGE. omniORB, 1999. `http://www.uk.research.att.com/omniORB/`.
4. BERNARDIN, L., CHAR, B., AND KALTOFEN, E. Symbolic computation in Java: An appraisement. In Dooley [6], pp. 237–244.
5. BUCHBERGER, B., COLLINS, G. E., ENCARNACIÓN, M. J., HONG, H., JOHNSON, J. R., KRANDICK, W., LOOS, R., MANDACHE, A., NEUBACHER, A., AND VIELHABER, H. *SACLIB User's Guide.* Johannes Kepler Universität, 4020 Linz, Austria, Mar. 1993. Available via anonymous ftp at `melmac.risc.uni-linz.ac.at` in `pub/saclib`.

6. Dooley, S., Ed. *Proceedings of the 1999 International Symposium on Symbolic and Algebraic Computation (ISSAC '99)* (Vancouver, BC, Canada, 1999), The Association for Computing Machinery, ACM.
7. El Kahoui, M., and Weber, A. Deciding Hopf bifurcations by quantifier elimination in a software-component architecture. *Journal of Symbolic Computation* (2000). To appear.
8. Göbel, M., Küchlin, W., Müller, S., and Weber, A. Extending a Java based framework for scientific software-components. In *Computer Algebra in Scientific Computing (CASC '99)* (München, June 1999), V. G. Ganzha, E. W. Mayr, and E. V. Vorozhtsov, Eds., Springer-Verlag, pp. 207–222.
9. Hoss, J. Parallele Berechnung von Gröbner Basen in parametrisierten Ringen. Diplomarbeit, Universität Tübingen, Fakultät für Informatik, Jan. 1998.
10. Iglio, P., and Attardi, G. Software components for computer algebra. In *Proceedings of the 1998 International Symposium on Symbolic and Algebraic Computation (ISSAC '98)* (Rostock, Germany, 1998), O. Gloor, Ed., The Association for Computing Machinery, ACM, pp. 62–69.
11. Janssen, B., and Spreitzer, M. *ILU Reference Manual.* XEROX, 1997. For Version 2.08. Available at `ftp:/ftp.parc.xerox.com/pub/ilu/2.0/20a8-manual-html/`.
12. LiDIA Group. *LiDIA Manual Version 1.3—A library for computational number theory.* TH Darmstadt, Alexanderstr. 10, 64283 Darmstadt, Germany, 1997. `http://www.informatik.th-darmstadt.de/TI/LiDIA/Welcome.html`.
13. Loos, R. G. K., and Collins, G. E. Revised report on the algorithm description language ALDES. Tech. Rep. WSI–92–14, Wilhelm-Schickard-Institut für Informatik, Universität Tübingen, 72076 Tübingen, Germany, 1992.
14. Mowbray, T., and Malveau, R. *CORBA Design Patterns.* Wiley, 1996.
15. Object Management Group (OMG). *The Common Object Request Broker: Architecture and specification*, Dec. 1998. Revision 2.3.
16. Object Oriented Concepts, Inc. ORBacus, 1999. `http://www.ooc.com/ob/`.
17. OpenMath Consortium. OpenMath. `http://www.openmath.org/`, 1999.
18. Orfali, R., and Harkey, D. *Client/Server Programming with JAVA and CORBA*, second ed. Wiley, 1998.
19. Orfali, R., Harkey, D., and Edwards, J. *The Essential Distributed Objects Survival Guide.* John Wiley & Sons, 1995.
20. PoSSo Group. Polynomial systems library. Available at `ftp://posso.dm.unipi.it`, 1996.
21. Römer, K., and Puder, A. Mico is Corba, 1998. `http://www.vsb.cs.uni-frankfurt.de/~mico`.
22. Siegel, J. *CORBA Fundamentals and Programming.* John Wiley & Sons, New York, 1996.
23. Simon, G. A. Communication Interface between SAC-2 and LiDIA. In *Proc. of OpenMath Workshop 6* (Zürich, 27-28th July 1996), J. Abbott, Ed. `http://www.openmath.org/V1/History/workshops/proceedings-6.html`.
24. Simon, G. A. Interoperability between computer algebra systems. Tech. Rep. WSI-96-8, Wilhelm-Schickard-Institut für Informatik, Universität Tübingen, 72076 Tübingen, Germany, 1996.
25. Sun Microsystems. *Java Remote Method Invocation — Distributed Computing for Java*, 1997. `http://java.sun.com:80/marketing/collateral/javarmi.html`.
26. Wang, P. Design and protocol for internet accessible mathematical computation. In Dooley [6].

27. WEBER, A., KÜCHLIN, W., AND EGGERS, B. Parallel computer algebra software as a Web component. *Concurrency: Practice and Experience 10*, 11–13 (1998), 1179–1188.

28. WORLD WIDE WEB CONSORTIUM. Math Home Page. `http://www.w3.org/Math/`, 1999.

29. ZIPPEL, R. The MathBus. In *Internet Accessible Mathematical Computation* (Vancouver, Canada, July 1999). `http://SymbolicNet.mcs.kent.edu/icm/research/iamc99paper/Zippel.pdf`.

Deciding Linear-Transcendental Problems

Volker Weispfenning

Fakultät für Mathematik und Informatik,
Universität Passau
D-94030 Passau, Germany

e-mail: `weispfen@uni-passau.de`

Abstract. We present a decision procedure for linear-transcendental problems formalized in a suitable first-order language. The problems are formalized by formulas with arbitrary quantified linear variables and a block of quantifiers with respect to mixed linear-transcendental variables. Variables may range both over the reals and over the integers. The transcendental functions admitted are characterized axiomatically; they include the exponential function applied to a polynomial, hyperbolic functions and their inverses, and the arcustangent. The decision procedure is explicit and implementable; it is based on mixed real-integer linear elimination, the symbolic test point method, elementary analysis, and Lindemann's theorem. As a byproduct we obtain sample solutions for existential formulas and a qualitative description of the connected components of the satisfaction set wrt. a mixed linear-transcendental variable. Potential applications include reachability problems for linear differential systems.

1 Introduction

The decision problem for the first-order theory of the real numbers as ordered field was solved by Tarski. It asks for an algorithm to decide the validity in the reals of an arbitrary first-order formula in the language of ordered rings. In his famous monograph (see [13]) Tarski posed the extended problem, whether one can algorithmically decide the validity in the domain of real numbers of an arbitrary first-order formula involving the exponential function. This problem remained open for a long time (compare [14,15]). It was conditionally solved in the positive sense in the monumental paper [7], compare also [24,25]. The positive solution relies on Schanuel's conjecture in transcendental number theory. The conjecture asserts that for arbitrary real numbers $r_1, \ldots, r_n$ that are linearly independent over the field $\mathbb{Q}$ of rationals at least n of the real numbers $r_1, \ldots, r_n, e^{r_1}, \ldots, e^{r_n}$ are algebraically independent over $\mathbb{Q}$. The case $n = 1$ of the conjecture is a consequence of Lindemann's theorem; it asserts that for an arbitrary non-zero complex number z not both z and e^z can be algebraic over $\mathbb{Q}$ (see [12], p. 51).

For algorithmic or relative algorithmic solutions of decision problems of a more specialized form see [8, 18, 9–11]; for related problems also [16, 17, 24]. Here too, Schanuel's conjecture often plays a decisive role.

In this paper we consider the decision problem for a fragment of the full first-order theory of the reals as ordered field with one specific transcendental function. Let F be a fixed subfield of the field $\mathbb{R}_a$ of real algebraic numbers. We consider two disjoint sets of variables $x_1, \ldots, x_n$ and $y_1, \ldots, y_m$. Terms are restricted to the form $p(x_1, \ldots, x_n, y_1, \ldots, y_m) + c \cdot \mathrm{trans}(q(y_1, \ldots, y_m))$, where p and q are a linear polynomials with coefficients in F and $c \in F$. Atomic formulas are equations or inequalities between terms. Quantifier-free formulas are obtained from atomic formulas by means of the boolean operations $\wedge, \vee, \neg$. Normal formulas are closed formulas of the form

$$\exists y_1 \ldots \exists y_m Q_1 x_1 \ldots Q_n x_n(\psi)$$

with arbitrary quantifiers Q_i and a quantifier-free formula ψ.

We specify some requirements for the real transcendental function $\mathrm{trans}(x)$ axiomatically. These axioms are satisfied e.g. for exponential functions of type $\exp(p(x))$, where $p(x)$ is a polynomial with real algebraic coefficients, hyperbolic functions, and the arcustangent. Moreover the decision procedure can be extended to the case of inverse functions of the above, when quantifiers are restricted to the natural domains of these functions.

We present an unconditional decision procedure for normal formulas. In contrast to the general decision procedure for real exponential formulas in [7] our decision algorithm does not require Schanuel's conjecture. Its role is taken over by the classical theorem of Lindemann (see [12], page 70) asserting that the values of the exponential function for algebraic arguments are transcendental. We also consider the decision problem for normal formulas, where quantifiers range over some intermediate field $F \subseteq G \subseteq \mathbb{R}_a$ of real algebraic numbers.

Finally we extend the decision procedure for normal formulas to extended normal formulas, where terms may involve the integer-part operation. By means of this additional operation we can code the integers inside the reals and thus also quantifiers ranging over the integers. Hence we refer to this case as the *mixed real-integer case.* In this case we require the argument $(q(y_1, \ldots, y_m)$ of $\mathrm{trans}(q(y_1, \ldots, y_m))$ to be bounded by a rational constant s.

The proofs for the decision methods involve

- real and mixed real-integer linear quantifier elimination [19, 20, 6, 21, 23] for a reduction to the case of a single mixed linear-transcendental variable,
- convexity arguments and elementary analysis for the univariate case,
- the evaluation of the sign of mixed linear-transcendental expressions at symbolic test points,
- Lindemann's theorem from transcendence theory for the decision of relations between transcendental and real algebraic constants.

For the purely real case and a transcendental function of type $\exp(p(x))$, where $p(x)$ is a polynomial with real algebraic coefficients, our decision problem may be considered as a special case of the problem solved in [18] in combination with algorithmic quantifier elimination for Tarski algebra. In contrast to the highly sophisticated machinery of real algebraic geometry and non-standard analysis required there and in [7], our method is quite elementary in character and well-suited for implementation (e.g. on top of the REDLOG-package of REDUCE [4]). The case of other admissible transcendental functions such as the arcustangent is not covered by [18]. The mixed real-integer case cannot be subsumed under any previous work. For the special case of the exponential function as trans(x) our results were announced in [22].

Potential applications include reachability checking for solutions of linear differential systems with constant coefficients.

2 The formal framework

Let F be an arbitrary subfield F of the field $\mathbb{R}_a$ of real algebraic numbers. We consider the following fragment L of a first-order language for the real numbers: *Linear terms* $s(y_1, \ldots, y_m, x_1, \ldots, x_n)$ are expressions of the form $s := \sum_{i=1}^m b_i y_i + \sum_{i=1}^n c_i x_i + c_0$ with $b_i, c_i \in F$. Fix a real function trans(x) and a linear term $q(y_1, \ldots, y_m)$ involving only the variables $y_1, \ldots, y_m$. Then a *transcendental term* is of the form $c \cdot \operatorname{trans}(q(y_1, \ldots, y_m))$ with $c \in F$. Arbitrary *terms* are sums of linear and transcendental terms. *Atomic formulas* are equations or inequalities between terms. *Quantifier-free formulas* are obtained from atomic formulas by means of the boolean operations $\wedge, \vee, \neg$. *Normal formulas* are formulas without free variables of the form

$$\exists y_1 \ldots \exists y_m Q_1 x_1 \ldots Q_n x_n (\psi)$$

with arbitrary quantifiers Q_i and a quantifier-free formula ψ.

Our goal is a *decision procedure* for normal formulas in L i. e. an algorithm that takes normal formulas in L as inputs and outputs the truth value of the input formula over the reals. This goal is analogous to that in [20] for almost linear integer problems and to [1] for linear-trigonometric problems. We will also consider analogous decision procedures for normal formulas, where the quantifiers are regarded as ranging over some fixed intermediate field $F \subseteq G \subseteq \mathbb{R}_a$. In contrast to the purely linear case the decision of a normal formula in $\mathbb{R}$ and in G may differ.

In order for our decision procedures to work we need some requirements on the real function trans(x) that we specify axiomatically. In the following we call a real number *manageable* if it is either algebraic, given as a root of a squarefree rational polynomial together with an isolating interval with rational endpoints, or transcendental with a recursive binary expansion. Notice that by interval bisection and Sturm's theorem, equalities and inequalities between two algebraic manageable numbers are decidable (see [2]). Moreover inequalities between an algebraic and a transcendental manageable number are decidable by iterated bisection of the isolating interval for the algebraic real and computation of a large enough initial segment of the transcendental manageable number. In the following we will frequently need these decisions. By way of contrast we will never run into the problem of deciding equalities or inequalities between two transcendental manageable numbers, a problem that is known to be undecidable.

The *axioms on trans*(x) are as follows:

1. trans(x) is a real analytic function defined on all of $\mathbb{R}$ (see [5]). There is a recursive function $\delta(n)$ such for all integers n, $\delta(n)$ is a modulus of continuity (see [3]) for trans(x) and its derivative on the interval $[n, n+1]$. Moreover for every rational number q, trans(x) and its derivative are given by recursive power series centered at q with an explicit rate of convergence. Finally $\lim_{x \longrightarrow \infty}$ trans(x) and $\lim_{x \longrightarrow -\infty}$ trans(x) are given as a manageable real number or $\pm\infty$. We collect all these properties by saying that trans(x) is a *well-behaved* real analytic function.
2. There exist finitely many manageable reals $r_1 < \ldots < r_k$, such that all zeros of the second derivative trans$''(x)$ of trans are among the r_i. The values trans(r_i) are manageable real numbers and if r_i is transcendental then trans(r_i) and trans$'(r_i)$ are algebraic. We call these points the *critical points* of trans(x).
3. For any real $r \neq r_1, \ldots, r_k$ it is never the case that both r and trans(r) are algebraic.
4. The graph of trans(x) has no tangent lines with real algebraic coefficients except possibly at the critical points r_i.

The last axiom is a consequence of axiom (3) and the following axiom:

5. For all reals $r \neq r_1, \ldots, r_k$ the value of trans(r) is algebraically dependent on the value of the derivative trans$'(r)$ over the field $\mathbb{R}_a$.

Indeed if the line $cx + d$ is a tangent to trans(x) at point r and trans$'(r) = c$ is algebraic, then by (5) trans(r) is algebraic, and so by (3) r is transcendental. Since $d =$ trans(r)$-cr$, it follows that d must be transcendental.

We call a function trans(x) satisfying the axioms an *admissible function.* Using the remark above, [3], and Lindemann's theorem (see [12], p. 51) it is not difficult to verify that the following transcendental functions trans(x) satisfy the axioms and hence are admissible:

1. $e^{p(x)}$ for any polynomial $p(x)$ with real algebraic coefficients.
2. $\sinh(x)$, $\cosh(x)$ and $\tanh(x)$. More generally, every real function of the form $ae^{mx} + be^{nx}$ for integers m, n and real algebraic numbers a, b.
3. $\arctan(x)$

By way of contrast trigonometric functions fail to satisfy axiom (2). The analogous problems for these functions have been solved in [1].

Our goal is a *decision procedure* for normal formulas in L i. e. an algorithm that takes normal formulas in L as inputs and output the truth value of the input formula over the reals.

In fact we will consider also formulas in a more general sense, where we admit besides the "ordinary" quantifiers $\exists, \forall$ also the "counting quantifiers"

$\exists^{\leq n}, \exists^{n}, \exists^{\geq n}, \exists^{<\infty}, \exists^{\infty}$ interpreted as "there exist at most n, exactly n, at least n, only finitely many, infinitely many", respectively for $n \in \mathbb{N}$. Formulas, closed formulas, normal formulas in this extended sense will be refered to as formulas, closed formulas, normal formulas with counting quantifiers. Notice that the first three types of counting quantifiers are definable using ordinary quantifiers, while the last two are not definable in this way.

In the following we describe two *reduction steps* that will reduce the decision problem for normal formulas to normal formulas with a single mixed linear-transcendental variable y.

Reduction step 1:
Consider a normal formula

$$\varphi := \exists y_1 \ldots \exists y_m Q_1 x_1 \ldots Q_n x_n(\psi)$$

with arbitrary quantifiers Q_i and a quantifier-free formula ψ.

Then φ is equivalent over the reals to the following normal formula

$$\varphi' := \exists y \exists y_1 \ldots \exists y_m Q_1 x_1 \ldots Q_n x_n (y = q(y_1, \ldots, y_m) \ \wedge \ \psi'),$$

where ψ' results from ψ by replacing everywhere in ψ the transcendental term $\mathrm{trans}(q(y_1, \ldots, y_m))$ by $\mathrm{trans}(y)$.

Notice that φ' is again a normal formula, but contains only y as mixed linear-transcendental variable; all other variables are linear in φ', moreover the transcendental term in φ' is of the simple form $\mathrm{trans}(y)$.

Reduction step 2:
Consider a normal formula with a single mixed linear-transcendental variable y and $\mathrm{trans}(y)$ as transcendental term:

$$\varphi' := \exists y Q_1 x_1 \ldots Q_n x_n(\psi).$$

Let ψ_1 result from ψ by replacing the transcendental term $\mathrm{trans}(y)$ by a new variable z. Then ψ_1 is a linear formula in the sense of [19, 6]. So by linear real quantifier elimination, $Q_1 x_1 \ldots Q_n x_n(\psi_1)$ is equivalent to a quantifier-free linear formula ψ_2 with y and z as the only linear variables. Let ψ_3 result from ψ_2 by backsubstitution of the transcendental term $\mathrm{trans}(y)$ for z. Then $\varphi'' := \exists y(\psi_3)$ is a formula with only one quantifier that is equivalent to φ'. Notice that linear combinations of terms with coefficients in F can be rewritten as a single term. Hence the formula φ'' can be rewritten as a normal formula $\exists y(\psi_4)$.

Both reduction steps work similarly if the given formula involves counting quantifiers by virtue of linear elimination by test points. So for example a counting quantifier $\exists^{\infty} x_i(\sigma(x_i))$ can be eliminated by restricting the disjunction over test terms used for the elimination of the quantifier $\exists x_i(\sigma(x_i))$ to test terms of the form $\pm\infty$ and $s \pm \epsilon$, where s is a linear term (compare [6]). Moreover if all quantifiers $Q_i x_i$ are existential, then extended linear elimination produces sample points for solutions x_i in dependence of $\mathrm{trans}(y)$.

So the problem of deciding the truth value of normal formulas is reduced to normal input formulas of the form $\exists y(\psi)$, where ψ is quantifier-free and has trans(y) as transcendental term. So it suffices to decide univariate normal formulas of this type over the reals. In fact we will do more: In case the truth value of $\exists y(\psi)$ is 'true', then we will provide a sample point $r \in \mathbb{R}$ for y. In fact if M is the set of reals satisfying $\psi(y)$, then we will also determine the number of connected components of M and their types as a non-empty intervals $[b,c]$, (b,c), $[b,c)$ or $(b,c]$.

As a consequence we can also in an obvious way decide a normal formula $Qy(\psi)$ with quantifier-free ψ, where Q is a counting quantifier.

3 Geometry of admissible functions and straight lines

In this section we will solve the decision problem for univariate normal formulas φ of type $\exists x(\psi(x))$ with quantifier-free ψ and trans(x) as transcendental term. It will turn out that this is essentially a study of the geometry of the graph of trans(x) in relation to a finite system of straight lines. In order to carry out this study algorithmically, we require the axioms on the admissibility of the function trans(x).

The atomic subformulas of ψ can be written in the form trans$(x)+cx+d\ \rho\ 0$, or $cx+d\ \rho\ 0$, where c,d are real algebraic numbers in F and ρ is one of the relations $=,<,>,\leq,\geq$. Let $r_1 < \ldots < r_k$, be the critical points of the function trans(x) specified in axiom (2). We call the intervals

$$I_{-\infty} = (-\infty, r_1),\ I_{\infty} = (r_k, \infty),\ I_1 = (r_1, r_2), \ldots, I_{k-1} = (r_{k-1}, r_k)$$

the *basic intervals* of ψ. Let ϵ be a symbol for a positive infinitesimal in an elementary extension field $\mathbb{R}^*$ of $\mathbb{R}$. Equivalently ϵ may be regarded as a positive real small enough for the given situation.

In order to decide φ it suffices

- to evaluate the truth value of $\psi(r_i)$ for $1 \leq i \leq k$,
- to decide the satisfiability of $\psi(x)$ on each basic interval of ψ.

The *first task* amounts to deciding all atomic subformulas of ψ at the critical points r_i i. e. the order relation between manageable reals of the form $r :=$ trans(a) and b. Since by axiom (2) not both r and b are transcendental, this is algorithmically possible.

For the *second task* we may assume that all atomic formulas in $\psi(x)$ are among $c_i x + d_i\ \rho_i\ 0$ and trans$(x) + c_i x + d_i\ \rho_i\ 0$ for $1 \leq i \leq m$. We refer to the straight lines $y = c_i x + d_i,\ 1 \leq i \leq m$ as the *lines* of ψ.

Suppose now we can solve the following subtasks on each basic interval $I := (r, r')$:

1. Decide whether two lines $y = c_i x + d_i$ and $y = c_{i'} x + d_{i'}$ intersect on I, and if yes evaluate the sign of another line $y = c_j x + d_j$ of ψ and decide the order relation between the graph of $c_j x + d_j$ and of trans(x) at the intersection point ξ and infinitesimally to the right of this point.

2. Compute the number of intersection points between a line $y = c_i x + d_i$ of ψ and the graph of the function trans(x) on the given basic interval I.
3. For each intersection point ξ of a line $y = c_i x + d_i$ and the graph of the function trans(x) on I and for each other line $y = c_j x + d_j$ of ψ evaluate the sign of $c_j x + d_j$ and decide the order relation between the graph of $c_j x + d_j$ and of trans(x) at the point ξ and infinitesimally to the right of this point.
4. Determine the order relation between test points ξ arising in (1), (2), (3) for different atomic subformulas of ψ.
5. Evaluate the truth value an atomic subformula of ψ infinitesimally to the right of a critical point r_i.
6. For each line $y = c_j x + d_j$ of ψ decide the order relation between the graph of $c_j x + d_j$ and of trans(x) at $\pm\infty$.

We refer to the points r_i in the first task and the points ξ mentioned in subtasks (1) and (3) as the *proper test points* for $\psi(x)$. Point of the form $\xi + \epsilon$ and $\pm\infty$ are referred to as *improper test points* for $\psi(x)$.

Obviously $\psi(x)$ is satisfiable on a basic interval I iff $\psi(x)$ holds for at least one proper or improper test point in I. Hence $\psi(x)$ is satisfiable on $\mathbb{R}$ iff it is satisfied at some point r_i or it is satisfiable on some basic interval of $\psi(x)$.

From the evaluation of the atomic formulas of $\psi(x)$ at all test points we can moreover in a finite combinatorial manner determine the number of connected components of the satisfaction set of ψ on I and the type of interval of each such component. This involves in addition the subtask to determine the order relation between test points on the same basic interval of ψ arising from different atomic subformulas of ψ. By piecing together this information on each basic interval together with the evaluation of ψ at all points r_i we obtain the number of connected components of the global satisfaction set of ψ and the type of interval of each such component. As a consequence we can now evaluate the existential or counting quantifier over the variable x, and thus decide the given normal formula φ. Moreover, any proper or improper test point at which $\psi(x)$ holds in $\mathbb{R}$ is a sample solution of $\psi(x)$. Together with the sample solutions provided by extended linear quantifier elimination, this provides sample solutions for arbitrary multivariate existential normal formulas. Notice that these sample solutions are produced in an exact symbolic description; upon request they can be evaluated numerically to arbitrary precision.

For the rest of this section we will be concerned with *solutions of the subtasks* listed above.

The *first subtask* involves only easy computations with algebraic real numbers and the approximate evaluation of trans(a) at a real algebraic point ξ. We call the intersection points ξ of two lines arising in this subtask a point of *type 0*.

The *last subtask* requires the knowledge of an approximation of $\lim_{x \longrightarrow -\infty}$ trans(x) provided by axiom (1).

For the *second subtask* let $I = (r, r')$ and recall that by axiom (2) trans(x) is either strictly convex or strictly concave on I. Moreover the decision between convexity and concavity of trans(x) on I can be made by the order relation between trans$'(r)$ and trans$'(r')$.

Suppose to begin with that trans(x) is strictly convex. Then trans$'(x)$ is strictly increasing on I. Hence $y = c_i x + d_i$ has at exactly one intersection point with the graph of trans(x) on I, iff

$$(\text{trans}(r) < c_i r + d_i \text{ and } c_i r' + d_i < \text{trans}(r')) \text{ or}$$

$$(\text{trans}(r) > c_i r + d_i \text{ and } c_i r' + d_i > \text{trans}(r')).$$

In the first case we call the intersection point of type 1, in the second case of type 2. For an intersection point ξ of type 1 we have trans$(\xi + \epsilon) > c_i\xi + d_i$, and for an intersection point ξ of type 2 we have trans$(\xi + \epsilon) < c_i\xi + d_i$.

Notice that by axiom (4) a tangent situation between the line $y = c_i x + d_i$ and the graph of trans(x) on I is excluded.

Two intersection points between the line $y = c_i x + d_i$ and the graph of trans(x) on I occur iff

$$\text{trans}'(r) < c_i < \text{trans}'(r') \text{ and}$$

$$\text{trans}(s) < c_i s + d_i \text{ for the unique point } s \in I \text{ such that trans}'(s) = c_i.$$

The point s can be computed to arbitrary precision by iterated bisection of intervals and evaluation of the order relation between trans$'(x)$ and c_i. This is algorithmically possible by axiom (1); moreover the first condition can be decided by axiom (2). Hence all of the condition above can be tested effectively. If it is satisfied then we call the smaller of the two intersection points a point of type 3 and the larger one a point of type 4. Notice that for an intersection point ξ of type 3 we have trans$(\xi + \epsilon) < c_i\xi + d_i$, and for an intersection point ξ of type 4 we have trans$(\xi + \epsilon) > c_i\xi + d_i$.

In the remaining case there is no intersection point between the line $y = c_i x + d_i$ and the graph of trans(x) on I.

The case, where trans(x) is strictly concave on I is treated analogously.

The *third subtask*:

Suppose we have identified the intersection point ξ of a line $y = c_i x + d_i$ and the graph of the function trans(x) by its basic interval I, the information on its type 1-4, and the information whether trans(x) is strictly convex or strictly concave on I. Then we know whether trans$(x) - c_i x + d_i$ is increasing or decreasing at the intersection point ξ.

Let $y = c_j x + d_j$ be another straight line. Again we restrict to the case where trans(x) is strictly convex on I; The concave case is handled analogously.

In order to decide the sign of $c_j\xi + d_j$, it suffices to determine the order relation between ξ and the intersection point $-d_j/c_j$ of the line $y = c_j x + d_j$ with the x-axis, in case this point is in I. (Notice that the exceptional cases $c_j = 0$ or $-d_j/c_j \notin I$ are trivial.) This in turn can be done once one knows the type of the point ξ and the order relation between $c_i(-d_j/c_j) + d_i$ and trans$(-d_j/c_j)$ and the order relation between $c_i(-d_j/c_j) + \epsilon + d_i$ and trans$(-d_j/c_j + \epsilon)$

Finally we need to decide the order relation between $c_j\xi + d_j$ and trans(ξ). If the line $y = c_jx + d_j$ has no intersection point with the graph of trans(x) on I, then this order relation is the same as that between $c_j(r+\epsilon) + d_j$ and trans($r + \epsilon$), and hence can be handled by the fifth subtask. Otherwise the line has one intersection point η or two intersection points $\eta_1 < \eta_2$ with the graph of trans(x) on I. Then our problem is reduced to knowing the types of ξ and of η, or η_1 and η_2 and the order relation between these points. Notice that we know for sure by axiom (3) that ξ can not coincide with one of the other points unless the lines $y = c_ix + d_i$ and $y = c_jx + d_j$ are identical. If these two lines have no intersection point on I, then the order relation between ξ and η, or η_1, or η_2 is obvious. If they have an intersection point p on I, then the order relation between $c_ip + d_i$ and trans(p) will determine the order relation between ξ and η, or η_1, or η_2.

Notice that all tests required in this subtask are again algorithmic by our axioms. The decisions are easily visualized by drawing pictures of the possible geometric situations; they are formally verified by elementary analysis.

For the *fourth subtask* let ξ_1 and ξ_2 be two test points of given type 0 - 4 as in subtasks (1) - (3) arising from different atomic subformulas of ψ. Then equality or an order relation between ξ_1 and ξ_2 is easily decided by using subtasks (1) - (3) together with the type information on the points.

The *fifth subtask* is handled similar to the first task taking into account also the evaluation of the derivative trans$'(r_i)$.

This completes the evaluation of all atomic formulas of $\psi(x)$, and hence of $\psi(x)$ at all test points. As a corollary we obtain an overview over the number and types of connected components of the satisfaction set of the total formula ψ. Thus we have proved all the claims at the end of section 2.

Finally we remark that our decision procedure can also be construed as a quantifier elimination procedure in a liberal sense: If we admit parameters in the coefficients of terms in normal formulas φ, then the result of our decision procedure would turn into a boolean combination φ' of equations and inequalities about these parameters. The terms of φ' are, however, more complicated expressions than those allowed in normal formulas.

4 Inverses of admissible functions

In this section we consider the decision problem for normal formulas, where the role of an admisible function trans(x) is replaced by the inverse function trans$^{-1}(x)$. In general such an inverse may not exist or it may be defined only on some real interval. So we have to make a further restriction in the fomation of normal formulas involving such fucntions, in order to guarantee that the truth-value of it is well-defined.

Let trans(x) be an admissible function, let I be an interval such that trans(x) is strictly monotonic on I, and let the interval J be the image of I under trans(x). Then the inverse function trans$^{-1} : J \longrightarrow I$ exists and is also strictly monotonic.

For fixed inverse function trans^{-1} and linear term $q(y_1, \ldots, y_m)$ we define now a *transcendental term* as an expression of the form $c \cdot \text{trans}^{-1}(q(y_1, \ldots, y_m))$ with $c \in F$.

Arbitrary terms are again sums of linear and transcendental terms. *Atomic formulas* are equations or inequalities between terms. *Quantifier-free formulas* are obtained from atomic formulas by means of the boolean operations $\wedge, \vee, \neg$ as before. *Normal formulas* are now of the form

$$\exists y_1 \ldots \exists y_m Q_1 x_1 \ldots Q_n x_n (q(y_1, \ldots, y_m) \in J \ \wedge \ \psi)$$

with arbitrary quantifiers Q_i and a quantifier-free formula ψ. Here the expression $z \in J$ is of course only a short hand for the inequality or conjunction of two inequalities defining J.

We are going to show that normal formulas in this sense can be equivalently reduced to normal formulas in the previous sense. Hence all the results proved for normal formulas in the previous sense apply also to normal formulas in the current sense.

The *first reduction step* is as in section 1:

Consider a normal formula

$$\varphi := \exists y_1 \ldots \exists y_m Q_1 x_1 \ldots Q_n x_n (q(y_1, \ldots, y_m) \in J \ \wedge \ \psi)$$

with arbitrary quantifiers Q_i and a quantifier-free formula ψ.

Then φ is equivalent over the reals to the following normal formula

$$\varphi' := \ \exists y \exists y_1 \ldots \exists y_m Q_1 x_1 \ldots Q_n x_n (y \in J \ \wedge \ y = q(y_1, \ldots, y_m) \ \wedge \ \psi'),$$

where ψ' results from ψ by replacing everywhere in ψ the transcendental term $\text{trans}^{-1}(q(y_1, \ldots, y_m))$ by $\text{trans}^{-1}(y)$.

In a next reduction step we see that this formula is equivalent to

$$\varphi'' := \ \exists x \exists y_1 \ldots \exists y_m Q_1 x_1 \ldots Q_n x_n (x \in I \ \wedge \ \text{trans}(x) = q(y_1, \ldots, y_m) \ \wedge \ \psi''),$$

where ψ'' results from ψ' by replacing everywhere in ψ' the transcendental term $\text{trans}^{-1}(y)$ by x. Here again the expression $z \in I$ is only a short hand for the inequality or conjunction of two inequalities defining I. Hence φ'' is indeed a normal formula in the previous sense equivalent to φ. Hence it can be decided algorithmically.

So this decision procedure applies e. g. to linear-transcendental problems involving the real functions $\ln : (0, \infty) \longrightarrow \mathbb{R}$, or $\text{Arsinh} : \mathbb{R} \longrightarrow \mathbb{R}$, or $\text{Arcosh} : [1, \infty) \longrightarrow \mathbb{R}$, or $\tan : (-\frac{\pi}{2}, \frac{\pi}{2}) \longrightarrow \mathbb{R}$. The more general case of linear-transcendental problems involving the tangent function on $\mathbb{R} \setminus \{\frac{\pi}{2} + \pi\mathbb{Z}\}$ has been discussed in [1].

5 The mixed real-integer case

In this section we turn to a considerable extension of linear-transcendental problems treated in sections 2 and 3 given by an extension of the concept of term used so far. We admit as new operation symbol in our extension language the integer-part operation (compare [23]). Similar as in the language L'' considered in [23], extended terms are now expressions obtained from terms by means of addition, scalar multiplication by rational constants and the integer-part operation in arbitrary nesting. Atomic formulas in the extended sense are then of the form $t = 0,\ t > 0,\ t \geq 0$, for extended terms t. Formulas, closed formulas and normal formulas in the extended sense are obtained from atomic formulas as before with the following important *restriction*:

The argument $q(y_1, \ldots, y_m)$ of the transcendental term

$$c \cdot \mathrm{trans}(q(y_1, \ldots, y_m))$$

with $c \in F$ has to be bounded in absolute value by a rational constant s.

By means of the integer-part operation we are now able to code also quantifiers ranging over the integers. So the problems that can be expressed by normal formulas in the extended sense are now *mixed real-integer problems.*

We are now going to describe a decision procedure for normal formulas in the extended sense:

By an argument similar to the one in section 1 we can use reduction step 1 to reduce the input formula to a normal formula with at most one mixed linear-transcendental variable. Next the mixed real-integer linear quantifier elimination of [23] can be applied to reduce the decision problem for normal formulas with one mixed linear-transcendental variable to the case of normal formulas with just one bounded existential quantifier $\exists x(-s < x < s\ \wedge\ \psi(x))$, where $\psi(x)$ is quantifier-free. The bound s provides an obvious bound for all integer parts of terms in $\psi(x)$. So every occurence of an integer part operation can be eliminated by a finite disjunction over all integer constants within the corresponding bound. So the evaluation of the truth value of $\exists x(\psi(x)$ boils down to the evaluation of a finite disjunction of existentially quantified conjunctions of the form

$$\exists x(\bigwedge_{i=1}^{k} [t_i(x)] = m_i\ \wedge\ \psi'(x))$$

with integer constants m_i, and a non-extended quantifier-free formula $\psi'(x)$. Replacing each equation $[t_i(x)] = m_i$ equivalently by $m_i \leq t_i(x) < m_i + 1$, we reduce this extended normal formula to a normal formula in the original sense, and can now decide the result by the decision method of the previous sections.

6 Decision procedure over a subfield

Let as before F be a subfield of the field $\mathbb{R}_a$ of real algebraic numbers. We suppose as before that all linear terms have coefficients in F. In addition we assume that all the points $r_1, \ldots, r_k$ occuring in the axioms for the function trans(x) are in F. Normal formulas meeting these conditions are called *normal formulas defined over* F.

We consider the modified decision problem for normal formulas defined over F, where we interpret all quantifiers as ranging not over $\mathbb{R}$, but over some intermediate field G with $F \subseteq G \subseteq \mathbb{R}_a$. Notice that both the real linear elimination in [6] and the mixed real-integer linear elimination in [23] are valid in arbitrary subfields of the reals. This situation changes, however, radically for univariate normal formulas $\exists x(\psi(x))$: Consider e. g. the normal formula $\varphi := \exists x(e^x - 2 = 0)$. This formula is true over $\mathbb{R}$, but false over G due to Lindemann's theorem.

In contrast to this situation, we will show now that the truth-value of a normal formula defined over F is the same in all intermediate fields $F \subseteq G \subseteq \mathbb{R}_a$. Moreover there is again a decision procedure for normal formulas over such a field G.

It suffices to modify the original decision procedure in such a way that we drop all proper test terms arising from points that are not real algebraic. Thus we keep proper test terms arising from intersections of lines, but drop the test terms arising from intersections of lines with trans(x) on some basic interval. Notice that all improper test point can be regarded as representing points in F, since F is dense and cofinal in $\mathbb{R}$.

7 Examples

Example 1. Put trans$(x) := e^x$ and consider the normal formula $\exists x(cx + \frac{1}{2} > e^x \ \wedge \ -x < e^x)$ for a rational constant c. Then our decision procedure will say that this formula holds in the reals iff

$$c \leq -1 \ \vee \ (c > -1 \ \wedge \frac{1}{2(c+1)} > e^{\frac{-1}{2(c+1)}}) \ \vee \ (c > 1 \ \wedge c \leq e^{1-\frac{1}{2c}})$$

By Lindemann's theorem the equations

$$\frac{1}{2(c+1)} = e^{\frac{-1}{2(c+1)}}, \quad c = e^{(1-\frac{1}{2c})}$$

are impossible for rational c. Hence the inequalities above can be decided by computing enough of the expansion of $e^{\frac{-1}{2(c+1)}}$ and of $e^{(1-\frac{1}{2c})}$. So the formula holds e.g. for $c = -1$, $c = -1/4$, $c = 2$, and fails for $c = -0.1$ and for $c = 1$.

Example 2. The next example involves two variables and illustrates the role of linear elimination in the decision procedure. We put trans$(x) := \arctan(x)$, and observe that $\arctan(x)$ has unique inflection point 0 and basic intervals $I_1 = (-\infty, 0)$, $I_2 := (0, \infty)$; furthermore $\arctan(x)$ is strictly convex on I_1 and strictly concave on I_2.

Let φ be the normal formula $\exists y \exists x(x \geq \arctan(y) \ \wedge \ 10x \leq -10y - 1 \ \wedge cy > \arctan(y) \ \wedge \ -y - 1 < \arctan(y))$. Linear elimination applied to the formula $\exists x(x \geq z \ \wedge \ 10x \leq -10y - 1)$ yields the equivalent formula $10y \leq 10z - 1$. Hence

φ is equivalent to $\exists y(10y \leq 10\arctan(y) - 1 \;\wedge\; cy > \arctan(y) \;\wedge\; -y-1 < \arctan(y))$. Our decision procedure applied to this formula will yield "true" iff $c > -1 \;\wedge\; \arctan(-1/(c+1)) < -c/(c+1)$. So φ holds e. g. for $c \geq -0.5$ and fails for $c = -0.9$.

Example 3. The final example illustrates the decision of extended normal formulas involving the integer-part operation. The transcendental function here is the Gaussian error function $\mathrm{erf}(x) := e^{-x^2}$. It satisfies the axioms with critical points $-1/\sqrt{2}, 0, 1/\sqrt{2}$. It is strictly concave on the interval $(-1/\sqrt{2}, 1/\sqrt{2})$, and strictly convex on the intervals $(-\infty, -1/\sqrt{2})$ and $(1/\sqrt{2}, \infty)$. Consider the extended normal formula $\varphi := \exists x(-10 < x < 10 \;\wedge\; [10\,\mathrm{erf}(x)] = x + c)$. Then elimination of the integer-part operation yields the equivalent normal formula $\bigvee_{i=0}^{9} \exists x(-10 < x < 10 \;\wedge\; i1 \leq 10\,\mathrm{erf}(x) < (i+1)1 \;\wedge\; x + c = i1)$, which is in turn is equivalent to $\bigvee_{0,c-10 \leq i \leq 9} i1 \leq 10\,\mathrm{erf}(i1 - c) < (i+1)1$. So φ fails for e. g. $c > 18$, and holds for $c = 9.5$.

8 Conclusions and Applications

We have presented an explicit, unconditional and implementable decision procedures for large classes of mixed linear-transcendental real and mixed real-integer problems formulated by first-order formulas. As a by-product we have obtained the structure of the satisfaction set of formulas with a single non-linear variable, and sample points for existential formulas. The main restriction on the transcendental part is that only a single, "well behaved" transcendental function is allowed in formulas. This may e.g. be the exponential function applied to a polynomial with real algebraic coefficients, hyperbolic functions, and the arcustangent.

Our formulas can be used e.g. to express semilinear reachability conditions on solutions of homogeneous linear differential systems with constant coefficients. Consider e.g. the linear system $\dot{y} = Ay$ with constant 3×3 matrix

$$A = \frac{1}{7}\begin{pmatrix} -2 & 12 & -4 \\ 0 & 7 & 0 \\ 1 & 15 & 2 \end{pmatrix}$$

and the solution $y(t) = (y_1(t), y_2(t), y_3(t))^t$ with initial value $(1, 1, -1)^t$. It has the components $y_1(t) = 2t + 1$, $y_2(t) = e^t$, $y_3(t) = 3e^t - t - 4$. Let S be a (not necessarily convex) semilinear subset of $\mathbb{R}^3$ described by a boolean combination $\varphi(x_1, x_2, x_3)$ of linear inequalities. Then for given time bounds $a < b$, we can check, whether the solution enters the set S in the time interval $[a, b]$ by deciding the validity of the normal formula

$$\exists t(a \leq t \leq b \;\wedge\; \varphi(y_1(t), y_2(t), y_3(t)).$$

Using the analogous results in [1] one can decide similar reachability problems for solutions involving trigonometric functions.

An implementation of the decision procedures is planned in REDUCE on top of the REDLOG-package [4] that provides the necessary tools for formula handling and for linear quantifier elimination.

The general case of real formulas involving several transcendental functions obtained by superposition of exponential functions and arithmetic operations has been solved under the condition of Schanuel's conjecture in [7]. The proof is, however, very indirect using a powerful model-theoretic machinery; so it is far away from a possible implementation in a computer algebra system.

References

1. ANAI, H., AND WEISPFENNING, V. Deciding linear-trigonometric problems. Technical Report MIP-0001, FMI, Universität Passau, D-94030 Passau, Germany, Feb. 2000. to appear in Proc. ISSAC'2000.
2. BECKER, T., WEISPFENNING, V., AND KREDEL, H. *Gröbner Bases, a Computational Approach to Commutative Algebra*, corrected second printing ed., vol. 141 of *Graduate Texts in Mathematics*. Springer, New York, 1998.
3. BISHOP, E., AND BRIDGES, D. *Constructive Analysis.* Grundlehren der math. Wissenschaften. Springer-Verlag, Berlin, 1985.
4. DOLZMANN, A., AND STURM, T. Redlog: Computer algebra meets computer logic. *ACM SIGSAM Bulletin 31*, 2 (June 1997), 2–9.
5. KRATZ, S., AND PARKS, H. *A Primer of Real Analytic Fuctions.* Basler Lehrbücher. Birkhauser, Basel, 1992.
6. LOOS, R., AND WEISPFENNING, V. Applying linear quantifier elimination. *The Computer Journal 36*, 5 (1993), 450–462. Special issue on computational quantifier elimination.
7. MACINTYRE, A., AND WILKIE, A. On the decidability of the real exponential field. In *Kreiseliana: About and around Georg Kreisel.* A.K. Peters, 1996, pp. 441–467.
8. RICHARDSON, D. The elementary constant problem. In *Proceedings of the 1992 International Symposium on Symbolic and Algebraic Computation* (Berkeley, California, July 1992), P. S. Wang, Ed., pp. 108–116.
9. RICHARDSON, D. A zero structure theorem for exponential polynomials. In *Proceedings of the 1993 International Symposium on Symbolic and Algebraic Computation* (Kiev, Ukraine, July 1993), M. Bronstein, Ed., pp. 144–151.
10. RICHARDSON, D. An ISSAC'95 tutorial: Algorithmic methods for finding real solution of systems involving exponential and other elementary functions. Preprint, July 1995.
11. RICHARDSON, D. Local theories and cylindrical decompositon. In *Quantifier Elimination and Cylindrical Algebraic Decomposition*, B. Caviness and J. Johnson, Eds., Texts and Monographs in Symbolic Computation. Springer, Wien, New York, 1998, pp. 351–364.
12. SHIDLOVSKII, A. B. *Transcendental Numbers.* Walter de Gruyter, Berlin, New York, 1989.
13. TARSKI, A. A decision method for elementary algebra and geometry. In *Quantifier Elimination and Cylindrical Algebraic Decomposition*, B. Caviness and J. Johnson, Eds., Texts and Monographs in Symbolic Computation. Springer, Wien, New York, 1998, pp. 24–84.

14. VAN DEN DRIES, L. Remarks on Tarski's problem concerning (r,+,.,exp). In *Logic Colloquium* (Amsterdam, u.a., 1982), G. Longi, G. Longo, and A. Marcja, Eds., North-Holland, pp. 97–121.
15. VAN DEN DRIES, L. A generalization of the Tarski-Seidenberg theorem and some nondefinability results. *Bulletin of AMS 15* (1986), 189–193.
16. VAN DEN DRIES, L., MACINTYRE, A., AND MARKER, D. The elementary theory of restricted analytic fields with exponentiation. *Annals of Mathematics 140* (1994), 183–205.
17. VAN DEN DRIES, L., AND MILLER, C. On the real exponential field with restricted analytic functions. *Israel Journal of Mathematics 85* (1994), 19–56.
18. VOROBJOV, N. The complexity of deciding consistency of systems of polynomials in exponent inequalities. *J. Symbolic Computation 13*, 2 (1992), 139–173.
19. WEISPFENNING, V. The complexity of linear problems in fields. *Journal of Symbolic Computation 5*, 1–2 (Feb.–Apr. 1988), 3–27.
20. WEISPFENNING, V. The complexity of almost linear Diophantine problems. *Journal of Symbolic Computation 10*, 5 (Nov. 1990), 395–403.
21. WEISPFENNING, V. Complexity and uniformity of elimination in Presburger arithmetic. In *International Symposium on Symbolic and Algebraic Computation, Maui, Hawaii* (New York, July 1997), W. W. Küchlin, Ed., ACM Press, pp. 48–53.
22. WEISPFENNING, V. Deciding linear-exponential problems. Poster presentation at ISSAC'99, Vancouver, July 1999. Poster presentation at ISSAC'99, Vancouver.
23. WEISPFENNING, V. Mixed real-integer linear quantifier elimination. In *ISSAC'99* (1999), S. Dooley, Ed., ACM-Press, pp. 129–136.
24. WILKIE, A. Model completeness results for expansions of the ordered field of real numbers by restricted. *Journal of American Mathematical Society 9* (1996), 1051–1094.
25. WILKIE, A. Schanuel's conjecture and the decidability of the real exponential field. In *Algebraic Model Theory*, B. Hart, A. Lachlan, and M. Valeriote, Eds., Advanced Science Institutes Series. Kluwer Academic Publishers, 1997, pp. 223–230.

Author Index